Pollution Control Technology

POLLUTION CONTROL TECHNOLOGY

- AIR POLLUTION CONTROL
- WATER POLLUTION CONTROL
- SOLID WASTE DISPOSAL

Staff of Research and Education Association

REA
Research and Education Association
342 Madison Avenue
New York, N. Y. 10017

POLLUTION CONTROL TECHNOLOGY

Printed in the United States of America

Library of Congress Catalog Card Number 72-97640

International Standard Book Number 0-87891-504-4

2nd Printing, 1974

PREFACE

Historically, the public had not been particularly aware of the problems of controlling pollution of our environment, until the last several years. Pollution of our natural environment by industries and the consumer has been ocurring for decades. It was not, however, considered a crucial problem until recently when it came to the fore after a vast increase in the number of consumer products and services that pollute directly or indirectly. Thus, the increase in the number of automobiles per family, and the sharp rise in food products and household items that are packaged in disposable containers, are examples of the numerous sources of pollution that have become important recently.

Increase in population is only partially responsible for the increased usage of consumer goods that pollute directly upon being consumed or indirectly while being manufactured. Among the main factors which are responsible, are the advances in technology that have been made in the fabrication of these products to enable the consumer to purchase them at relatively low cost. Advances in technology, however, have not progressed to the same degree in combating the pollution problems associated with the products. Until recently, manufacturers were not motivated to find solutions to the pollution problems created by their manufacturing processes and the disposal of fabricated products. Local and Federal legislation, however, now require that the manufacturer meet minimum anti-pollution requirements or cease manufacture.

At first glance, one might advocate suspension of the production and use of products if their associated pollution problems cannot be readily solved. This negative approach, however, would retard progress in our standard of living by depriving us of products that make for easier and improved living conditions. Furthermore, the use of the pollution-prone automobile is an essential part of our present living habits.

The solution to the pollution problem appears to be not in deprivation, but rather in development of technologies that will enable us to combat the problem. The development of such technologies will probably require a vast number of engineers, scientists, and technical personnel comparable to the number that has been involved in defense and space activities.

A pollution control effort of this magnitude will not only result in a higher standard of living and improved environment, but also in an expanding economy with increase in available job opportunities.

Satisfactory solutions to the environmental pollution problems remain far off. It will take large numbers of engineers, scientists, and technical support personnel to produce solutions that are both technically and economically feasible. Considerable ingenuity, time, as well as research and development will be necessary to arrive at solutions that meet these requirements of technical and economic feasibility. For this reason, the pollution control field has grown by a factor of 20 within five years, and is expected to reach $75 billion in the next several years.

For those who are interested in working on the pollution problems, this book furnishes guidance and direction in this rapidly growing field. It reviews the state of the art—what is known, what is being used, and what is being worked on. The technical and economic feasibility of processes, equipment, and plants are analyzed. In turn, this should enable the reader to arrive at an assessment of requirements and opportunities in this field.

Leaders in government and industry who have attended seminars given by REA, believe that the pollution control field now offers vast opportunities for industrial organizations and their personnel in almost all technical areas. They have found that opportunities exist in the chemical, electronics, mechanical, and construction/civil areas. The rewards are there for those willing and able to meet the challenge.

****

"POLLUTION CONTROL TECHNOLOGY" is organized to provide detailed working knowledge to those seeking successful solutions to the environmental pollution problems. In pollution control, a problem is usually not solved successfully if the solution works technically, but is too costly to be acceptable. For this reason, economic considerations are included throughout the book where they are applicable.

In preparing the contents of this book, a large amount of available information was screened, analyzed, and reduced into form believed to be most useful to the reader. Contributions are gratefully acknowledged from the many persons and organizations referenced throughout the book. Particularly significant contributions are thankfully included from the staffs of the organizations below.

Max Fogiel, Ph. D., Editor

*CONTRIBUTORS TO "POLLUTION CONTROL TECHNOLOGY"

Air Resources Board-Calif... Alabama U... Albany Medical College... Allied Chemical Corp... American Cyanamid... American Electric Power... American Geological Institute.. American Institute of Chemical Engineers... American Meteorology Ass'n.. Alabama Polytechnic Inst...Atlantic Richfield... Atomic Energy Commission... Bell & Howell Research...Bethlehem Steel...Boyce Thompson Inst... Bureau of Mines... Bureau of Solid Waste Management... Battelle Memorial Inst... California Institute of Technology... Case Western Reserve U... Chicago Dept. of Air Pollution Control... Chrysler Corp... Cincinnati Water Research Lab... Cold Regions & Engineering Lab.. Colorado State U... Combustion Engineering...Continental Oil... Cornell U... Crown Zellerbach.. Detrex Chemical.. Detroit Edison.. Dow Chem- E. I. duPont... Eastman Kodak... Albert Einstein College of Medicine... Environmental Control Admin... Environmental Protection Agency... Esso Research & Engineering... Federal Water Pollution Control Admin... Fischer and Porter.. Ford Motor... General Motors... Halliburton... Harvard U.. Hooker Labs... Johns Hopkins U... Ideal Cement... Illinois Power... Incinerator Institute of America... Ionics... Iowa State U... Karolinska Inst... Kennecott Copper...Koppers... Lehigh Portland Cement... Arthur D. Little... Litton Systems... Lone Star Cement... Los Angeles County Air Pollution Control Ass'n... MIT... Mayo Grad. School of Medicine McGill U.. Floyd H. Meller Research... Merrimack College Midwest Research Inst... Mississippi Geological Society... Missouri Botanical Gardens... Mobile Oil...Monsanto... Mt. Sinai School of Medicine.. National Air Pollution Control Admin... National Center for Atmospheric Research... New York U... Oak Ridge National Lab... Ohio State U... Owens-Illinois... Oxford U...Peabody Coal.. Portland State College Public Health Service... Purdue U... Research - Cottrell... Resource Engineering Assoc... Reynolds Metals... David B. Smith Engineers... Southwest Research Inst... Stanford Research Institute... Systems Development Corp... Robert A. Taft Water Research Center... Texas A and M University... Travelers Research... Union Carbide... University of Calif. University of Florida...University of Michigan...University of North Carolina... University of Pittsburgh...University of Washington.. University of Wisconsin... U. S. Dept. of Agriculture... U. S. Steel... Water Resources Inst... Westinghouse Electric... York University... Zurn Industries

CONTENTS

PART I

AIR POLLUTION CONTROL

CHAPTER 1

METEOROLOGY AND TOPOGRAPHY

Air pollution is the gross effect of the contribution of pollutants emitted by all sources in a given area. The concentration of a pollutant in the atmosphere at any given location will vary from one point of time to another even if the pollutant is discharged from its source at a uniform rate. This is due to local topography and changing weather conditions. Also, the interpretation of results from air sampling would be impossible without the knowledge of what meteorological conditions existed at the time of sampling and how the local topography affected the dispersion of pollutants at the sampling site. It is evident, therefore, that some knowledge of meterology and topography are required to assess pollution effects from sources in a given area.

The air pollution cycle can be considered to consist of three phases: the release of air pollutants at the source, the transport and diffusion in the atmosphere, and the reception of air pollutants in reduced concentrations by people, plants, animals, or inanimate objects. The influence of meteorology is to the greatest extent during the diffusion and transport phase. The motions of the atmosphere which may be highly variable in all dimensions are responsible for the transport and diffusion of air pollutants.

Although the distribution with time of a cloud of pollutant material will depend on the summation of all motions of all sizes and periods acting upon the cloud, it is convenient to first consider some mean atmospheric motions over periods on the order of an hour.

Wind Direction. What effect will the mean wind direction have on an air pollutant? If the wind direction is measured at the height at which the pollutant is released, the mean direction will be indicative of the direction of travel of the pollutants. In meteorology it is conventional to consider the wind direction as the direction from which the wind blows, therefore a northwest wind will move pollutants to the southeast of the source.

Wind Speed. The effect of wind speed is twofold. The wind speed will determine the travel time from a source to a given receptor, e.g., if a receptor is located 1000 meters downwind from a source and the wind speed is 5 meters/second, it will take 200 seconds for the pollutants to travel from the source to the receptor. The other effect of wind speed is a dilution in the downwind direction. If a continuous source is emitting a certain pollutant at the rate of 10 grams/second and the wind speed is 1 meter/second, then, downwind, a 1 meter length of the plume will contain 10 grams of pollutant since 1 meter of air moves past the source each second. Next, consider that the conditions of emission are the same, but the wind speed is 5 meters/second. In this case, since 5 meters of air moves past the source each second, each meter of plume length contains 2 grams of pollutant. Therefore it can be seen that the dilution of air pollutants released from a source is proportional to the wind speed. This may be restated in another form: The concentration of air pollutants is inversely proportional to wind speed.

Variability of the Wind. In the preceding paragraphs consideration of only the mean speed and direction of wind has been made. Of course, there are deviations from this mean velocity. There are velocity components in all directions so that there are vertical motions as well as horizontal ones. These random motions of widely different scales and periods are essentially responsible for the movement and diffusion of pollutants about the mean downwind path. These motions can be considered atmospheric turbulence. If the scale of a turbulent motion, i.e., the size of an eddy, is larger than the size of the pollutant plume in its vicinity, the eddy will move that portion of the plume. If an eddy is smaller than the plume, its effect will be to diffuse or spread out the plume. This diffusion caused by the eddy motion is widely variable in the atmosphere, but even when the effect of this diffusion is least, it is on the order of three orders of magnitude greater than the diffusion by molecular action alone.

Mechanical Turbulence. Mechanical turbulence is the induced eddy structure of the atmosphere due to the roughness of the surface over which the air is passing. Therefore, the existence of trees, shrubs, buildings, and terrain features will cause mechanical turbulence. The height and spacing of the elements causing the roughness will affect the turbulence. In general, the higher the roughness elements, the greater the mechanical turbulence. In addition the mechanical turbulence increases as wind speed increases.

Thermal Turbulence. Thermal turbulence is that induced by the stability of the atmosphere. When the earth's surface is heated by the sun's radiation, the lower layer of the atmosphere tends to rise and thermal turbulence becomes greater, especially under conditions of light wind. On clear nights with winds, heat is radiated from the earth's surface resulting in cooling of the ground and air adjacent to it. This results in extreme stability of the atmosphere near the earth's surface. Under these conditions turbulence is at a minimum.

Relation of Turbulence to Wind Records. Attempts to relate different measures of turbulence of the wind to atmospheric diffusion have been made for quite some time. The distance of the maximum concentration has been related to the standard deviation of wind direction over 10 to 15 minute periods. Diffusion experiment results have also been compared with statistics of wind direction fluctuations in both the horizontal and vertical. Direct methods of relating wind statistics to estimates of dispersion show promise and attempts at developing suitable instrumentation to yield the necessary wind statistics directly, have been made.

Relation of Turbulence to Atmospheric Stability. Relations of a more qualitative type have been noted between atmospheric diffusion and the stability of the atmosphere. Measurement of atmospheric stability by temperature difference measurements on a tower are frequently utilized as an indirect measure of turbulence, particularly where climatological estimates of turbulence are desired. Under strong lapse or super-adiabatic conditions of temperature change with height, strong vertical and horizontal mixing takes place in the atmosphere contrasted to inversion conditions with slight horizontal mixing but extremely limited vertical mixing.

Variations of Wind Speed and Direction With Height. Wind speed is generally found to increase with height above the ground and wind direction to veer (turn clockwise) with height (in the northern hemisphere at latitudes above the tropical zone) due to the effects of friction with the earth's surface. The amount of these increases in speed and veering in direction are widely variable and to a great degree related to the roughness of the surface and the stability of the atmosphere.

Effect of Surface Roughness. Consider the surface wind as measured at 10 meters compared to the wind above the influence of the earth's friction, for example about 1000 meters. Over smooth terrain such as the great plains or over the ocean the speed at the surface is on the order of 0.9 the upper wind and the degree of veering with height on the order of 10°. (See figure 1-1.) Over average terrain with small changes in elevation and with some trees and shrubs, the surface speed is more like 4/5 of the upper wind and the amount of veering with height about 15° to 20°. Over rough terrain, quite hilly or mountainous or with numerous buildings and vegetation, the surface speed may be only half the speed of the upper wind and the amount of veering with height as much as 40° to 45°.

Diurnal Variation. During the daytime, solar heating causes turbulences to be at a maximum and vertical motions to be strongest. This causes the maximum amount of momentum exchange between various levels in the atmosphere. Because of this, the variation of wind speed with height is least during the daytime. Also the amount of veering with height is least (on the order of 15° to 20° over average terrain). The thickness of the friction layer will also be greatest during the day due to the vertical exchange. At night the vertical motions are least and effect of friction is not felt through as deep a layer as during the day. The surface speed over average terrain is much less than the free atmosphere wind (on the order of ¼ to ⅓ that of the 1000 meter wind), and the amount of veering with height may be on the order of 40° to 45°. Figure 1-2 shows the diurnal variation of wind speed at two different levels on a meteorological tower.

Stability and Instability. Whether the atmosphere has a tendency for large vertical motions or to damp out vertical motions is important to atmospheric processes which produce weather as well as to the effects upon air pollutants. The stability of the atmosphere is highly dependent upon the vertical distribution of temperature with height.

a. **Adiabatic Lapse Rate.** Due to the decrease of pressure with height, a parcel of air lifted to a higher altitude will encounter decreased pressure and will expand and, in undergoing this expansion, will cool. If this expansion takes place without loss or gain of heat to the parcel, the change is adiabatic. Similarly a parcel of air forced downward in the atmosphere will encounter higher pressures, will contract, and will become warmer. This rate of cooling with lifting or heating with descent is the dry adiabatic process lapse rate and is 5.4°F per 1000 feet or approximately 1° per 100 meters. This process lapse rate is the rate of heating or cooling of any descending or rising parcel of air in the atmosphere and should not be confused with the existing temperature variation with height at any one time, the environmental lapse rate.

b. **Environmental or Prevailing Lapse Rate:**

(1) The manner in which temperature changes

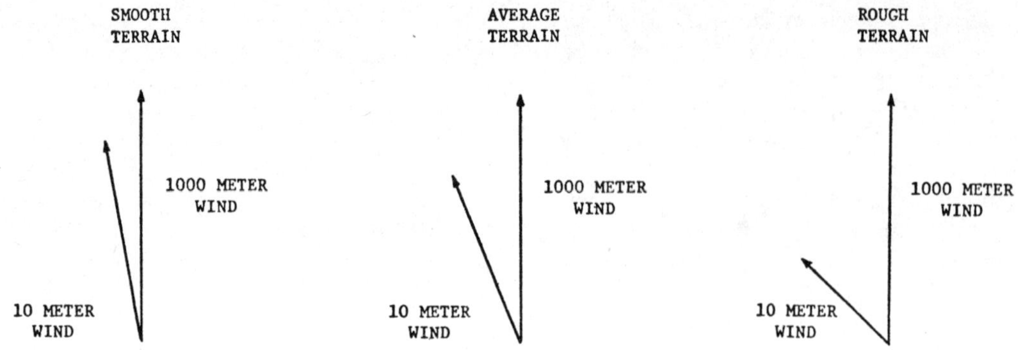

Fig. 1-1 Effect of roughness on variation of wind speed with height

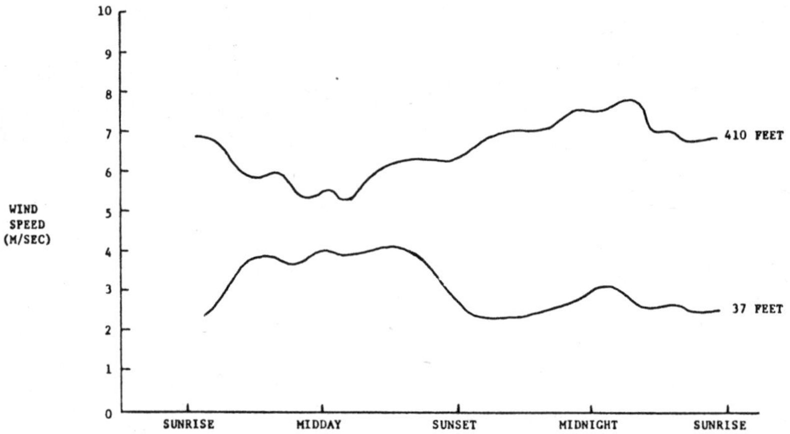

Fig. 1-2 Diurnal Variation of wind speed

with height at any one time is the prevailing lapse rate. This is principally a function of the temperature of the air and of the surface over which it is moving and the rate of exchange of heat between the two. For example, during clear days in midsummer the ground will be rapidly heated by solar radiation resulting in rapid heating of the layers of the atmosphere nearest the surface, but further aloft the atmosphere will remain relatively unchanged. At night radiation from the earth's surface cools the ground and the air adjacent to it, resulting in only slight decrease of temperature with height, or if surface cooling is great enough, temperature will increase with height.

(2) If the temperature decreases more rapidly with height than the dry adiabatic lapse rate, the air has a super-adiabatic or strong lapse rate and the air is unstable. If a parcel of air is forced upwards, it will cool at the adiabatic lapse rate, but will still be warmer than the environmental air. Thus it will continue to rise. Similarly, a parcel which is forced downward will heat dry adiabatically, but will remain cooler than the environment and will continue to sink.

(3) For environmental lapse rates that decrease with height at a rate less than the dry adiabatic lapse rate (sub-adiabatic or weak lapse), a lifted parcel will be cooler than the environment and will sink; a descending parcel will be warmer than the environment and will rise. Figure 1-3 shows the relative relation between the environmental lapse rates of superadiabatic (strong lapse), sub-adiabatic (weak lapse), isothermal, and inversion with the dry adiabatic process lapse rate.

(4) Lifting motions which will cause cooling at dry adiabatic lapse rates may be due to up-slope motion over mountains or rising over a cooler air mass. Descending motion (subsidence) may occur to compensate for the lateral spreading of air in high pressure areas.

The Influence of Vertical Temperature Structure Upon Stack Effluents. The manner in which stack effluents diffuse is primarily a function of the stability of the atmosphere. The behavior of smoke plumes has been typed into six classes, taking into account inversions aloft.

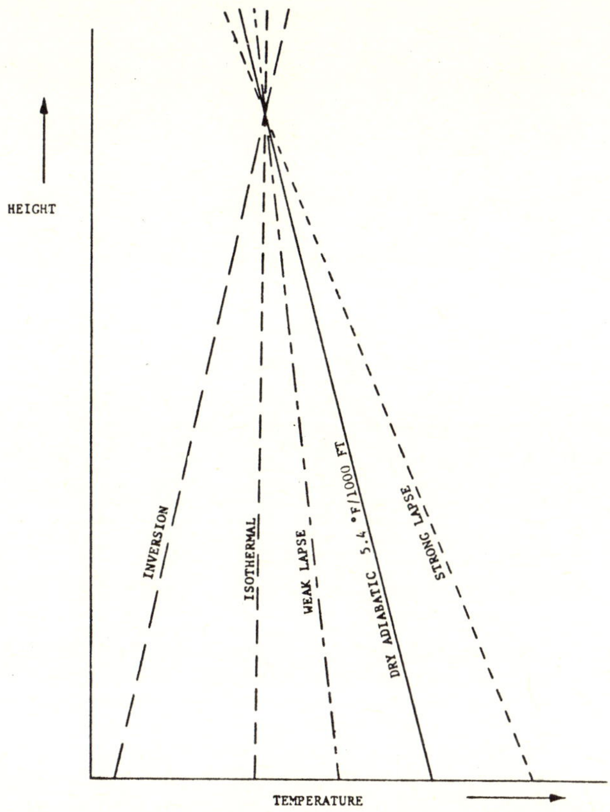

Fig. 1-3 Types of temperature structure with height related to the dry adiabatic process lapse rate

a. **Looping.** Looping occurs with a superadiabatic lapse rate. Large thermal eddies are developed in the unstable air, and high concentrations may be brought to the ground for short time intervals. Diffusion is good, however, when considering longer time periods. The superadiabatic conditions causing looping occur only with light winds and strong solar heating. Cloudiness or high winds will prevent such unstable conditions from forming.

b. **Coning.** With a vertical temperature gradient between dry adiabatic and isothermal, slight instability occurs with both horizontal and vertical mixing but not as intense as in the looping situation. The plume tends to be cone shaped, hence the name. The plume reaches the ground at greater distances than with looping. Coning is prevalent on cloudy or windy days or nights.

c. **Fanning.** If the temperature increases upward, the air is stable and vertical turbulence is suppressed. Horizontal mixing is not as great as in coning, but still occurs. The plume will therefore spread horizontally, but little if any vertically. Since the winds are usually light, the plume will also meander in the horizontal. Plume concentrations are high, but little effluent from elevated sources reaches the ground with this situation except when the inversion is broken due to surface heating, or terrain at the elevation of the plume is encountered. Clear skies with light winds during the night are favorable conditions for fanning.

d. **Lofting.** Lofting occurs when there is a superdiabatic layer above a surface inversion. Under this condition diffusion is rapid upward, but downward diffusion does not penetrate the inversion and so is damped out. With these conditions gases will not reach the surface, but particles with appreciable settling velocities will drop through the inversion. Sunset on a clear evening in open country is most favorable for lofting. Lofting is generally a transition situation and as the inversion deepens is replaced by fanning.

e. **Fumigation:**

(1) As solar heating increases, the lower layers are heated, and a super-adiabatic lapse rate occurs through a deeper and deeper layer. When the layer is deep enough to reach the fanning plume, thermal turbulence will bring high concentrations to the ground along the full length of the plume. This is favored by clear skies and light winds, and is apt to occur more frequently in summer due to increased heating.

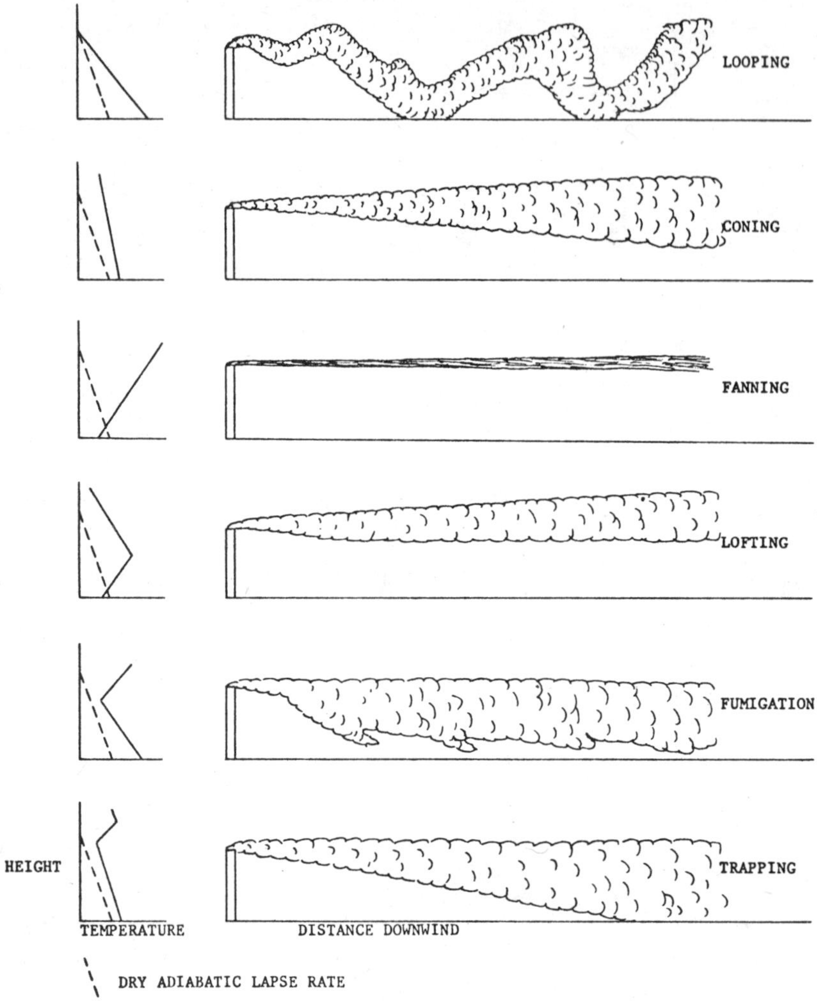

Fig. 1-4 The six plume classes

(2) Another type of fumigation may occur in the early evening over cities. Heat sources and mechanical turbulence due to surface roughness cause a lapse condition in the lower layers of the stable air moving into the city from non-urban areas where radiation inversions are already forming. This causes a fumigation until the city loses enough heat so that the lapse condition can no longer be maintained.

f. **Trapping.** When an inversion occurs aloft, such as a frontal or subsidence inversion, a plume released beneath the inversion will be trapped. Even if the diffusion is good beneath the inversion, such as a coning plume, the limit to upward diffusion will increase concentration in the plume and at ground level.
The six plume classes are diagrammed in figure 1-4.

Variation of Pollutant Concentrations Due to Meteorologic Variations.

a. An example of the diurnal variation of pollutant concentrations is given in figure 1-5. These are the concentrations some distance downwind from a continuous elevated source on a day when stability reaches extremes, i.e., on a clear day with light winds. This shows only the variations on the order of an hour's duration, rather than the rapid variations which may occur over a few minutes duration.

b. The primary maximum around 10 AM is due to fumigation. The rapid decrease in concentration following this is due to the heating of a progressively deeper layer and mixing of pollutants through this layer. The increase of concentrations during the late afternoon are due to the slight increase in stability after the period of maximum heating. During this period the lapse rate is generally changing from strong lapse to weak lapse.

c. The secondary maximum that occurs in the evening is a phenomena observed only in the urban area. During the late afternoon and early evening, a radiation inversion begins to form at the earth's surface in the non-urban areas, i.e.,

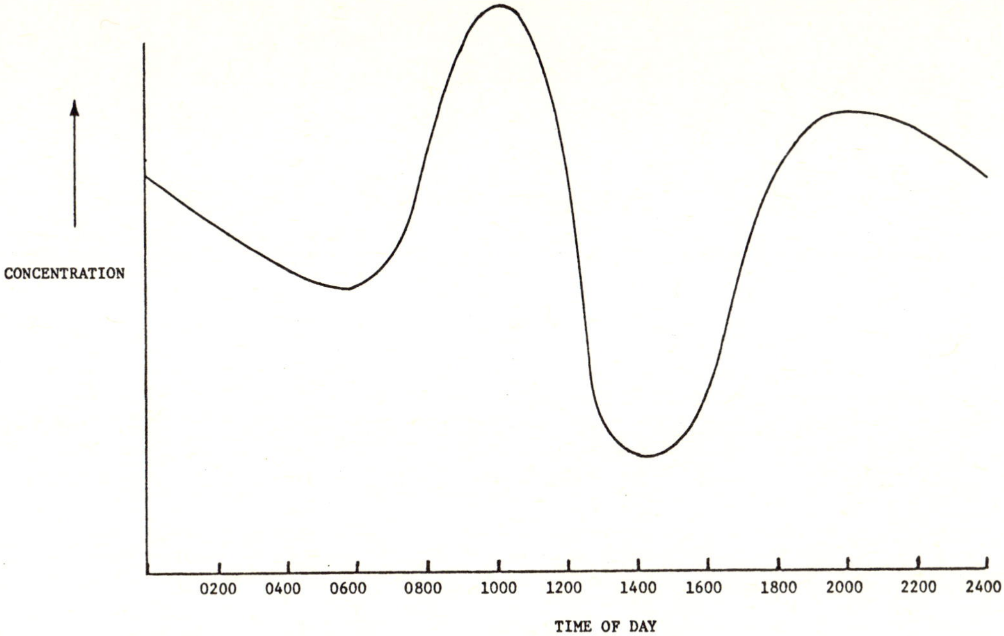

Fig. 1-5 Diurnal variation of ground-level concentrations from elevated sources

the surrounding countryside. The air over the city, however, does not have a radiation inversion in the lower layers due to release of heat from the buildings and pavings of the city. However, later in the evening, an inversion above the weak lapse layer forms above the city, and a mixing of the pollutants in this layer produces the higher concentrations.

d. When the base of this inversion aloft lowers enough so that the elevated emissions are into the inversion layer, the concentrations decrease and continue to decrease until after sunrise. Then surface heating is sufficient to produce mixing beneath the inversion, and pollution previously in the inversion layer is mixed through the layer of superadiabatic lapse rate beneath the remaining upper portion of the inversion, resulting in high ground level concentrations.

Variation in Concentration as Affected by Sampling Time.

a. Since the motions of the atmosphere are not constant with time, the concentrations of pollutants are also variable with time. Mean concentrations of pollutants measured over different time periods are not directly comparable. This is due principally to the fluctuation of the horizontal wind direction. If sampling is conducted downwind from a source, a short sample of 3 to 5 minutes duration will yield a certain concentration, but as the sampling interval is increased to say an hour, the mean concentration will decrease due to short periods of time when the wind is not blowing directly from the source to the receptor, even though the mean wind throughout the sampling interval is from source to receptor.

b. Relations have been experimentally determined to correlate the peak short term concentration with larger sampling intervals. Figure 1-6 shows some of these findings. This shows that under neutral conditions, the peak 3 minute concentration over an hour is about 2½ times the hourly mean. Under unstable conditions when the vertical direction of the wind is also rapidly changing, the peak 3 minute concentration can be expected to be about 5½ times the hourly concentration.

c. Where there are multiple sources of the measured pollutant, the variation of the wind does not have as large an effect on the changes of concentration with sampling time. A recent study on carbon monoxide concentrations in Nashville, Tennessee, has shown, as an example, 3-minute peak concentrations on the order of 1.5 times hourly concentrations.

Topography. In many cases the transport and diffusion of air pollutants is complicated by terrain features. Most large urban areas are located either in river valleys or on the shores of lakes or oceans. These features alter meteorologic conditions.

a. **Valley Effects:**

(1) *Channeling.* Although the more extreme effects of a valley location occur when the general flow is light, valleys tend to channel the general flow along the valley axis resulting in a bidirectional wind frequency distribution.

(2) *Slope and Valley Winds:*

(a) When the general wind flow is light and

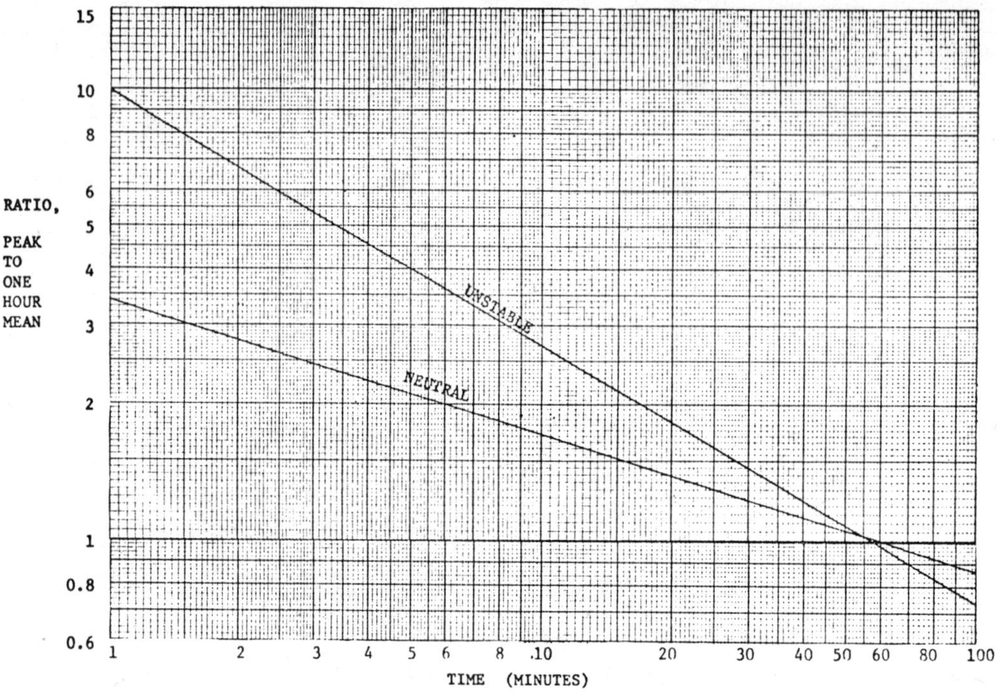

Fig. 1-6 Relation of peak short term concentrations to one hour mean concentrations

skies are clear, the differences in rates of heating and cooling of various portions of the valley floor and sides cause slight density and pressure differences resulting in small circulations. During the evening hours radiation of heat from the earth's surface and consequent cooling of the ground and air adjacent to the ground causes density changes. The air at point A (figure 1-7) is more than at point B, since point A is nearer the radiating surface. Therefore, the more dense air at point A tends to flow in the general direction of B and similarly at other points along the slope. This is the slope wind.

(b) If the slope in figure 1-7 is a side of a valley as in figure 1-8, the cold air moving down the slopes will tend to drain into the valley floor and deepen with time, intensifying the radiation inversion that would form even without the addition of cold air. Any pollutants that are emitted into the air, because of the inversion structure, will have very limited vertical motion.

(c) If, in addition, the valley floor has some slope, the cold air will have a tendency to move downhill along the valley axis. This is usually referred to as the valley wind as shown in figure 1-9. Because of the necessity of some accumulation of cold air from slope winds, the onset of the valley wind usually lags several hours behind the onset of the slope wind.

(d) The steeper the slopes of the valley, the stronger the slope winds can become. Vegetation will tend to reduce the flow both due to impeding the flow and also restricting the amount of radiation that can take place.

(e) On a clear day with light winds, the heating of the valley may cause upslope and upvalley winds. However, the occurence of upslope and upvalley winds is not as frequent nor as strong as the downslope and downvalley winds, principally due to the fact that downslope and downvalley winds, due to their density, hug the surfaces over which they travel. Complex valley systems, where several valleys merge at angles or slopes vary, usually require special observations to determine flow under various meteorologic conditions.

(3) *Inversions Aloft.* The trapping of air pollutants beneath inversions aloft is also a problem encountered in valleys. Two types of inversions, warm frontal and subsidence inversions, are of particular concern since they are usually slow moving. High concentrations may occur particularly if the layer of air beneath the inversion becomes unstable enough to mix pollutants from elevated sources to ground level.

b. *Shoreline Winds.* The differences in heating and cooling of land and water surfaces and the air above them result in the setting up of circulations if the general flow is light.

(1) *Sea or Lake Breeze:*

(a) On summer days with clear skies and light winds the heating of the land surface adjacent to a large lake or the ocean is much more rapid than the heating of the body of water. This results in a temperature difference and consequently a density and pressure difference between

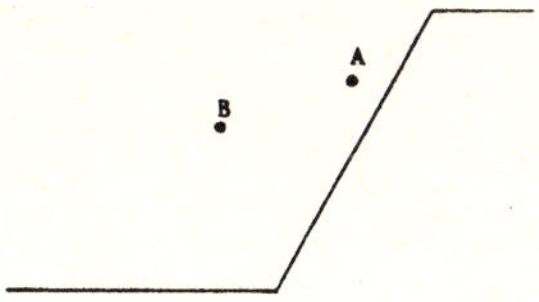

Fig. 1-7 Slope wind

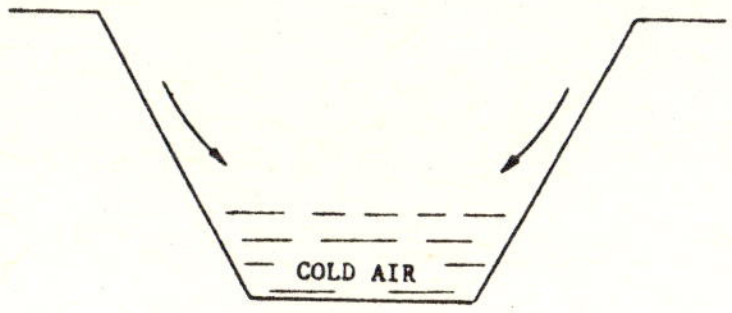

Fig. 1-8 Slope wind into a valley

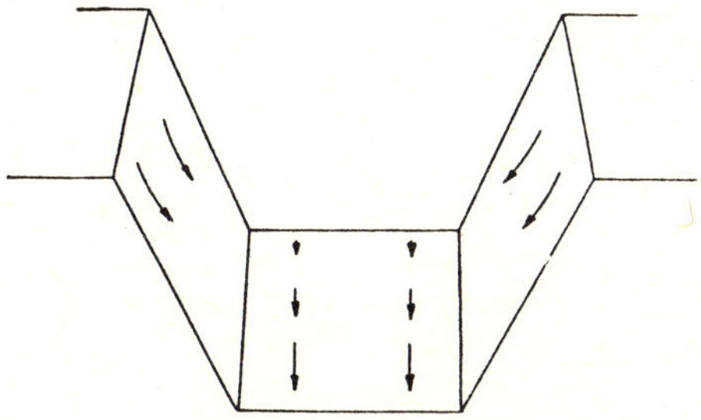

Fig. 1-9 Valley wind

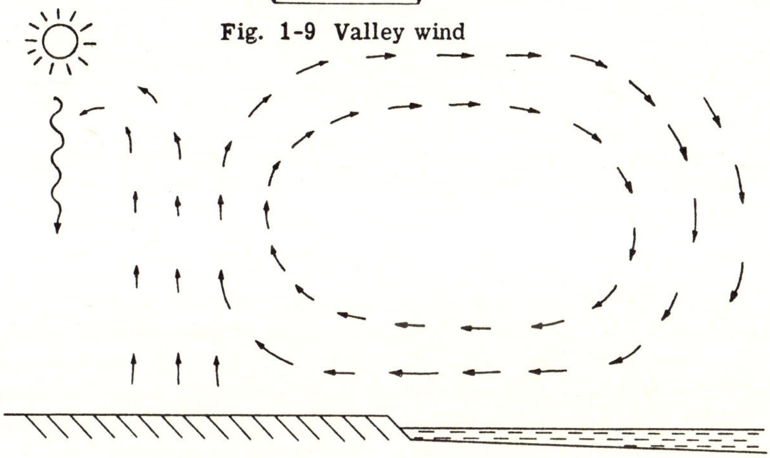

Fig. 1-10 Sea or lake breeze

the air just above the land surface and the air over the water. Because of the pressure gradient forces, a local circulation is set up with wind from the water toward the land. There is usually some upward motion over the land and subsidence over the water accompanying the sea breeze. There may result a weak transport from land to water aloft completing a cellular structure to the sea breeze as shown in figure 1-10.

(b) In cases where a strong lake breeze occurs, air from quite some distance out over the water may be brought toward the land, and due to Coriolus forces acting over the long trajectory, the resulting flow will become nearly parallel to the shoreline. This occurs just after the sea breeze is the strongest and results in decreasing the flow normal to the coastline and subsequent breaking down of the sea breeze.

(2) *Land Breeze:*

(a) At night the rapid radiational cooling of the land causes lower temperatures above the land surface than over the water. Thus a reverse flow, the land breeze, may result. The land breeze does not usually achieve as high a velocity as the lake breeze and is usually shallower than the sea or lake breeze. Of course, any wind flow due to the large scale pressure pattern will alter the local circulation and flow will be resultant of the two effects. Usually a light general flow is enough to overshadow the effects of land and sea breezes.

c. **Modification of Thermal Structure by Bodies of Water:**

(1) At different seasons of the year and also different times of the day, the temperature of bodies of water and adjacent land surfaces may be quite different. For example, during the late spring, large bodies of water are still cold relative to adjacent land surfaces and during mid-afternoon this difference is greatest due to the more rapid heating of the land surface. If the general flow in the area is such that the wind has a lengthy trajectory over the water and is blowing toward the shore, an interesting modification of the temperature structure takes place. Because of the passage over the cold water surface, the air will have an inversion in the lower layer as it

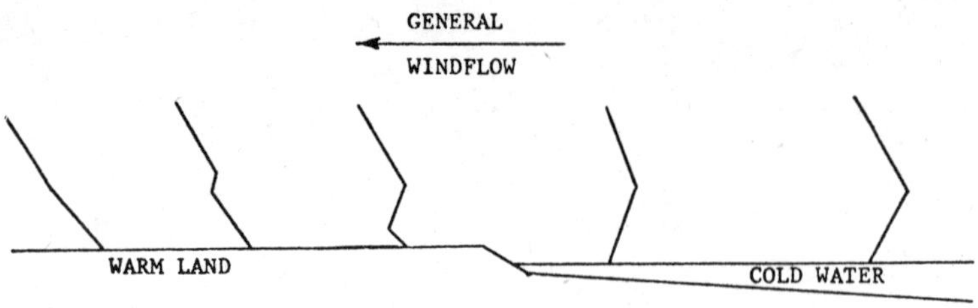

Fig. 1-11 Modification of Vertical Temperature Structure Due to Flow Over Differently Heated Surfaces (Late Spring, Afternoon).

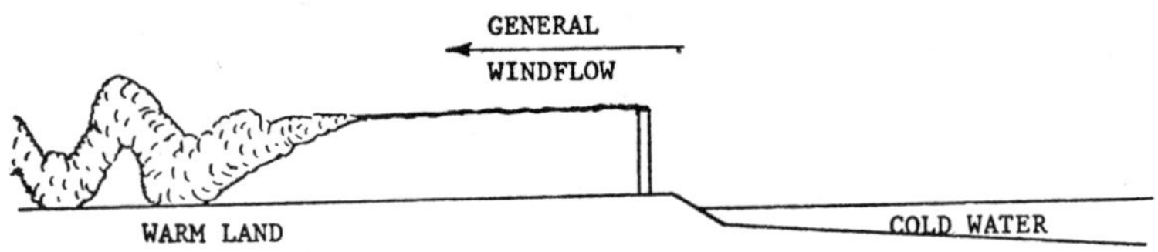

Fig. 1-12 Effect Upon Plume Characteristics of Flow Over Differently Heated Surfaces (Late Spring, Aternoon).

reaches the shoreline. Any air pollutants released into this inversion will essentially have the characteristics of a fanning plume. As the air passes over the warm land, a strong lapse replaces the inversion near the surface. The depth of this lapse layer becomes deeper as the air moves over more heated land surface. At the point where the lapse layer is deep enough to reach the fanning effluent from an elevated source, fumigation will occur. Fumigation of this type may last considerably longer than the usual diurnal breakup of nocturnal inversions; this fumigation will occur as long as the temperature difference between land and water is maintained and flow from water to land occurs. At greater distances from the shoreline the inversion will be eliminated, and looping-type of plume behavior will occur. On the other hand, if the source is high enough to be above the lake induced inversion, lofting of the plume would occur until enough distance and consequently enough heating takes place to eliminate the inversion.

(2) Figure 1-11 indicates the difference in vertical temperature structure that occurs in the above example, and figure 1-12 indicates the effect this has on the plume characteristics of an elevated shoreline source.

(3) At other times such as when the water is warmer than the land surface, offshore flow will result in fumigation over the water.

d. **Influence of Hills:**

(1) The influence of hills upon transport and diffusion depends upon a number of factors. Whether the source is on the windward or lee side of the hill or ridge is important. A smooth hill will alter the flow least; one with sharp ridges will cause turbulent eddies to form. The stability of the atmosphere will affect the influence of hills. During stable conditions, the flow will tend to flow around obstructions. Under unstable conditions the tendency is for air to move over obstructions.

(2) When a source is located upwind of a hill or ridge, the pollutants may come in contact with the facing slope, particularly under stable conditions. If the ridge is quite rough, induced turbulence may cause mixing down to the slope even when the general flow is over the ridge. Wind tunnel studies or field trials with constant level balloons may be desirable to determine the flow under given circumstances.

(3) For a source downwind from a hill or ridge, lee eddies will generally cause considerable downwash of the effluent near the source. If turbulent flow is induced by the hillside, diffusion will be increased but high concentrations very near the stack will result periodically due to the downwash.

CHAPTER 2

ESTIMATING ATMOSPHERIC DISPERSION

During recent years methods of estimating atmospheric dispersion have undergone considerable revision, primarily due to results of experimental measurements. In most dispersion problems the relevant atmospheric layer is that nearest the ground, varying in thickness from several hundred to a few thousand meters. Variations in both thermal and mechanical turbulence and in wind velocity are greatest in the layer in contact with the surface. Turbulence induced by buoyancy forces in the atmosphere is closely related to the vertical temperature structure. When temperature decreases with height at a rate higher than 5.4°F per 1000 ft (1°C per 100 meters), the atmosphere is in unstable equilibrium and vertical motions are enhanced. When temperature decreases at a lower rate or increases with height (inversion), vertical motions are damped or reduced. Examples of typical variations in temperature and wind speed with height for daytime and nighttime conditions are illustrated in Figure 2-1.

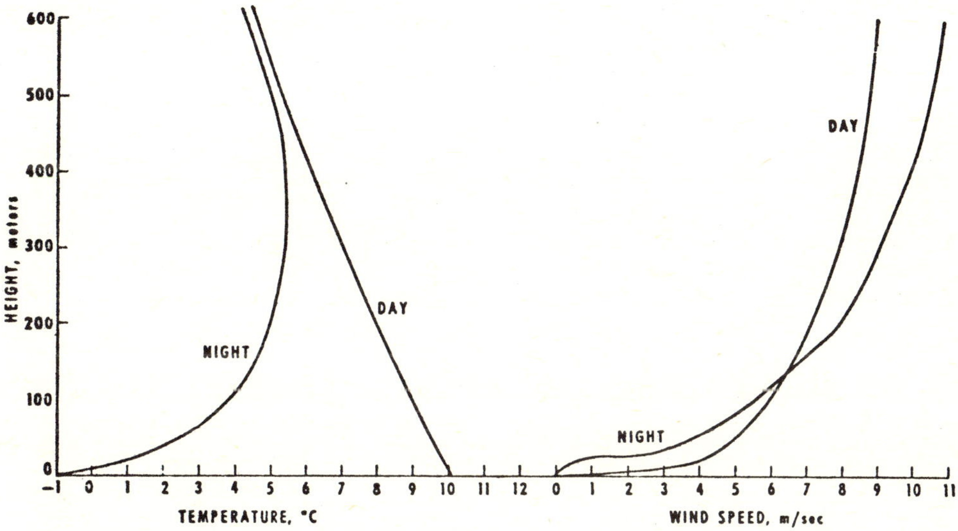

Figure 2-1 Examples of variation of temperature and wind speed with height (after Smith, 1963).

The transfer of momentum upward or downward in the atmosphere is also related to stability; when the atmosphere is unstable, usually in the daytime, upward motions transfer the momentum "deficiency" due to eddy friction losses near the earth's surface through a relatively deep layer, causing the wind speed to increase more slowly with height than at night (except in the lowest few meters). In addition to thermal turbulence, roughness elements on the ground engender mechanical turbulence, which affects both the dispersion of material in the atmosphere and the wind profile (variation of wind with height). Examples of these effects on the resulting wind profile are shown in Figure 2-2.

As wind speed increases, the effluent from a continuous source is introduced into a greater volume of air per unit time interval. In addition to this dilution by wind speed, the spreading of the material (normal to the mean direction of transport) by turbulence is a major factor in the dispersion process.

The procedures presented here to estimate atmospheric dispersion are applicable when mean wind speed and direction can be determined, but measurements of turbulence, such as the standard deviation of wind direction fluctuations, are not available. If such measurements are at hand, techniques such as those outlined by Pasquill (1961) are likely to give more accurate results. The diffusion param-

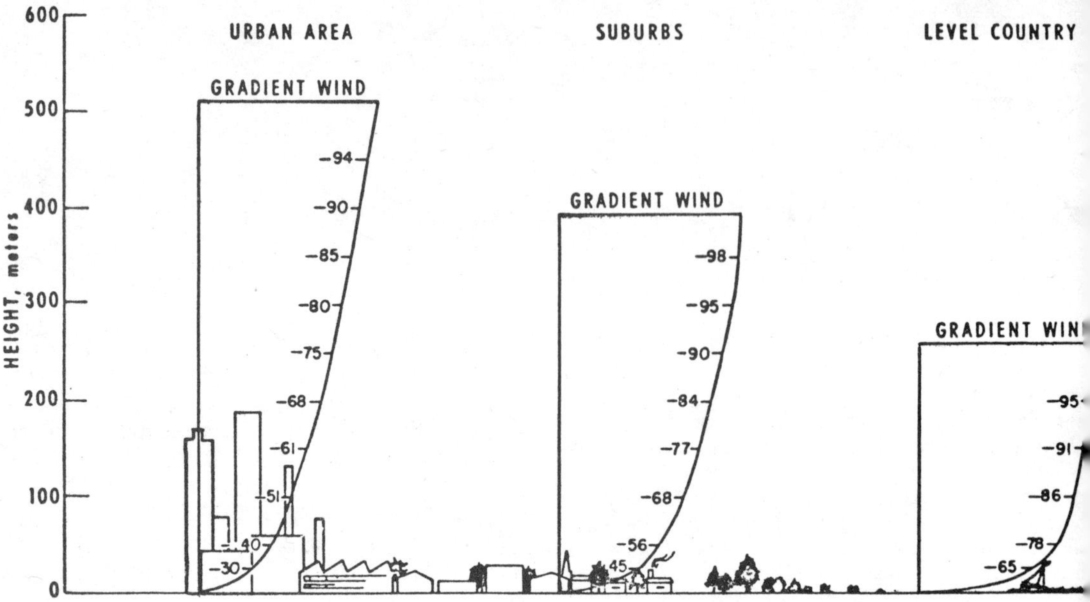

Figure 2-2 Examples of variation of wind with height over different size roughness elements (figures are percentag of gradient wind); (from Davenport, 1963).

eters presented here are most applicable to ground-level or low-level releases (from the surface to about 20 meters), although they are commonly applied at higher elevations without full experimental validation. It is assumed that stability is the same throughout the diffusing layer, and no turbulent transfer occurs *through* layers of dissimilar stability characteristics. Because mean values for wind directions and speeds are required, neither the variation of wind speed nor the variation of wind direction with height in the mixing layer are taken into account. This usually is not a problem in neutral or unstable (e.g., daytime) situations, but can cause over-estimations of downwind concentrations in stable conditions.

ANALYTICAL METHOD

Hay and Pasquill (1957) have presented experimental evidence that the vertical distribution of spreading particles from an elevated point is related to the standard deviation of the wind elevation angle, σ_E, at the point of release. Cramer (1957) derived a diffusion equation incorporating standard deviations of Gaussian distributions: σ_y for the distribution of material in the plume across wind in the horizontal, and σ_z for the vertical distribution of material in the plume. These statistics were related to the standard deviations of azimuth angle, σ_A, and elevation angle, σ_E, calculated from wind measurements made with a bi-directional wind vane (bivane). Values for diffusion parameters based on field diffusion tests were suggested by Cramer, et al. (1958) (and also in Cramer 1959a and 1959b). Hay and Pasquill (1959) also presented a method for deriving the spread of pollutants from records of wind fluctuation. Pasqui (1961) has further proposed a method for est mating diffusion when such detailed wind data ar not available. This method expresses the heigh and angular spread of a diffusing plume in terms c more commonly observed weather parameters. Sug gested curves of height and angular spread as function of distance downwind were given for sev eral "stability" classes. Gifford (1961) converte Pasquill's values of angular spread and height int standard deviations of plume concentration distri bution, σ_y and σ_z. Pasquill's method, with Gifford' conversion incorporated, is used here for diffusio estimates.

Advantages of this system are that (1) only tw dispersion parameters are required and (2) result of most diffusion experiments are now being re ported in terms of the standard deviations of plum spread. More field dispersion experiments are bein conducted and will be conducted under condition of varying surface roughness and atmospheric sta bility.

COORDINATE SYSTEM

In the system considered here the origin is a ground level at or beneath the point of emission with the x-axis extending horizontally in the direc tion of the mean wind. The y-axis is in the hori zontal plane perpendicular to the x-axis, and th z-axis extends vertically. The plume travels alon or parallel to the x-axis. Figure 2-3 illustrates th coordinate system.

DIFFUSION EQUATIONS

The concentration, χ, of gas or aerosols (parti

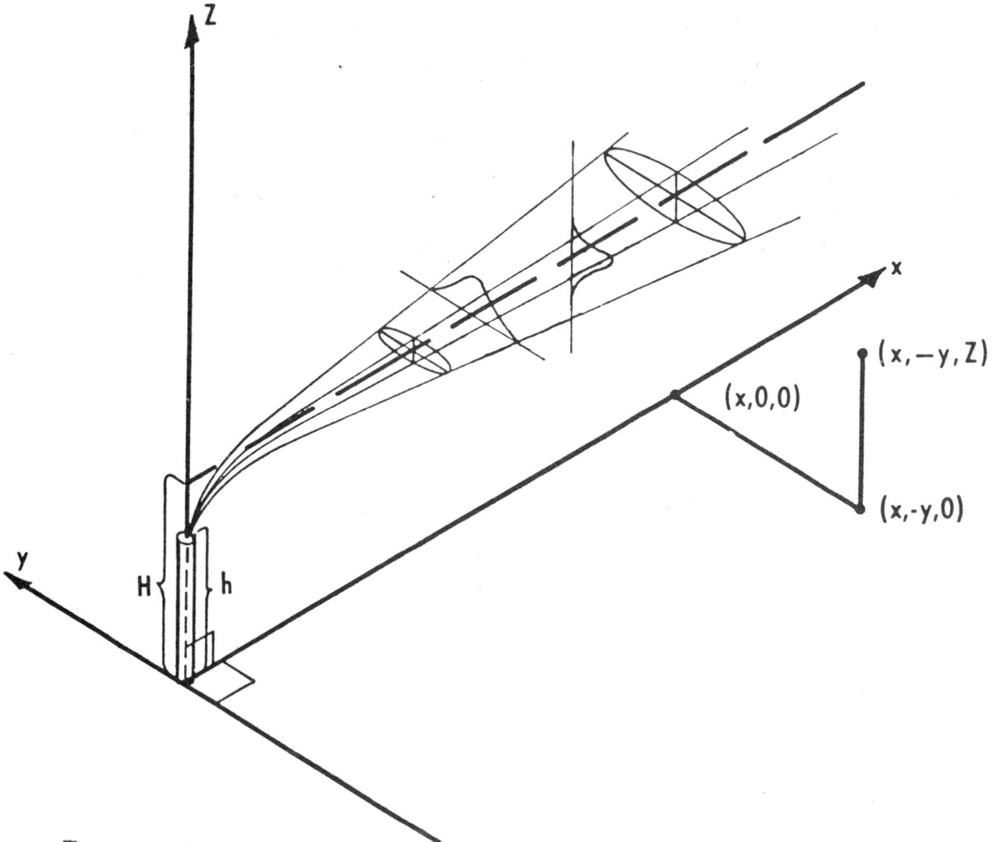

Figure 2-3 Coordinate system showing Gaussian distributions in the horizontal and vertical.

cles less than about 20 microns diameter) at x,y,z from a continuous source with an effective emission height, H, is given by equation **2. 1.** The notation used to depict this concentration is χ (x,y,z;H). H is the height of the plume centerline when it becomes essentially level, and is the sum of the physical stack height, h, and the plume rise, ΔH. The following assumptions are made: the plume spread has a Gaussian distribution in both the horizontal and vertical planes, with standard deviations of plume concentration distribution in the horizontal and vertical of σ_y and σ_z, respectively; the mean wind speed affecting the plume is u; the uniform emission rate of pollutants is Q; and total reflection of the plume takes place at the earth's surface, i.e., there is no deposition or reaction at the surface.

$$\chi\,(x,y,z;H) = \frac{Q}{2\pi\,\sigma_y\,\sigma_z\,u}\,\exp^*\left[-\frac{1}{2}\left(\frac{y}{\sigma_y}\right)^2\right]\left\{\exp\left[-\frac{1}{2}\left(\frac{z-H}{\sigma_z}\right)^2\right]+\exp\left[-\frac{1}{2}\left(\frac{z+H}{\sigma_z}\right)^2\right]\right\} \qquad (2.\ 1)$$

Note: exp —a/b = $e^{-a/b}$ where e is the base of natural logarithms and is approximately equal to 2.7183.

Any consistent set of units may be used. The most common is:

χ (g m^{-3}) or, for radioactivity (curies m^{-3})
Q (g sec^{-1}) or (curies sec^{-1})
u (m sec^{-1})
σ_y, σ_z, H,x,y, and z (m)

χ is a mean over the same time interval as the time interval for which the σ's and u are representative. The values of both σ_y and σ_z are evaluated in terms of the downwind distance, x.

Eq. (3.1) is valid where diffusion in the direction of the plume travel can be neglected, that is, no diffusion in the x direction.

This may be assumed if the release is continuous or if the duration of release is equal to or greater than the travel time (x/u) from the source to the location of interest.

For concentrations calculated at ground level, i.e., z = 0, the equation simplifies to:

$$\chi\,(x,y,0;H) = \frac{Q}{\pi\,\sigma_y\,\sigma_z\,u}\,\exp\left[-\frac{1}{2}\left(\frac{y}{\sigma_y}\right)^2\right]\exp\left[-\frac{1}{2}\left(\frac{H}{\sigma_z}\right)^2\right] \qquad (2.\ 2)$$

Where the concentration is to be calculated along the centerline of the plume (y = 0), further simplification results:

$$\chi\,(x,0,0;H) = \frac{Q}{\pi\,\sigma_y\,\sigma_z\,u}\exp\left[-\frac{1}{2}\left(\frac{H}{\sigma_z}\right)^2\right] \quad (2.3)$$

For a ground-level source with no effective plume rise (H = 0),

$$\chi\,(x,0,0;0) = \frac{Q}{\pi\,\sigma_y\,\sigma_z\,u} \quad (2.4)$$

EFFECTS OF STABILITY

The values of σ_y and σ_z vary with the turbulent structure of the atmosphere, height above the surface, surface roughness, sampling time over which the concentration is to be estimated, wind speed, and distance from the source. For the parameter values given here, the sampling time is assumed to be about 10 minutes, the height to be the lowest several hundred meters of the atmosphere, and the surface to be relatively open country. The turbulent structure of the atmosphere and wind speed are considered in the stability classes presented, and the effect of distance from the source is considered in the graphs determining the parameter values. Values for σ_y and σ_z are estimated from the stability of the atmosphere, which is in turn estimated from the wind speed at a height of about 10 meters and, during the day, the incoming solar radiation or, during the night, the cloud cover (Pasquill, 1961). Stability categories (in six classes) are given in Table 2-1. Class A is the most unstable, class F the most stable class considered here. Night refers to the period from 1 hour before sunset to 1 hour after sunrise. Note that the neutral class, D, can be assumed for overcast conditions during day or night, regardless of wind speed.

Table 2-1 KEY TO STABILITY CATEGORIES

Surface Wind Speed (at 10 m), m sec^{-1}	Day: Incoming Solar Radiation — Strong	Moderate	Slight	Night: Thinly Overcast or ≥4/8 Low Cloud	≤3/8 Cloud
< 2	A	A-B	B		
2-3	A-B	B	C	E	F
3-5	B	B-C	C	D	E
5-6	C	C-D	D	D	D
> 6	C	D	D	D	D

The neutral class, D, should be assumed for overcast conditions during day or night.

"Strong" incoming solar radiation corresponds to a solar altitude greater than 60° with clear skies; "slight" insolation corresponds to a solar altitude from 15° to 35° with clear skies. Table 170, Solar Altitude and Azimuth, in the Smithsonian Meteorological Tables (List, 1951) can be used in determining the solar altitude. Cloudiness will decrease incoming solar radiation and should be considered along with solar altitude in determining solar radiation. Incoming radiation that would be strong with clear skies can be expected to be reduced to moderate with broken (5/8 to 7/8 cloud cover) middle clouds and to slight with broken low clouds. An objective system of classifying stability from hourly meteorological observations based on the above method has been suggested (Turner, 1961).

These methods will give representative indications of stability over open country or rural areas, but are less reliable for urban areas. This difference is due primarily to the influence of the city's larger surface roughness and heat island effects upon the stability regime over urban areas. The greatest difference occurs on calm clear nights; on such nights conditions over rural areas are very stable, but over urban areas they are slightly unstable or near neutral to a height several times the average building height, with a stable layer above (Duckworth and Sandberg, 1954; DeMarrais, 1961).

Some preliminary results of a dispersion experiment in St. Louis (Pooler, 1965) showed that the dispersion over the city during the daytime behaved somewhat like types B and C; for one night experiment σ_y varied with distance between types D and E.

ESTIMATION OF VERTICAL AND HORIZONTAL DISPERSION

Having determined the stability class from Table 2-1, one can evaluate the estimates of σ_y and σ_z as a function of downwind distance from the source, x, using Figures 2-4 and 2-5. These values of σ_y and σ_z are representative for a sampling time of about 10 minutes. Figures 2-4 and 2-5 apply strictly only to open level country and probably underestimate the plume dispersion potential from low-level sources in built-up areas. Although the vertical spread may be less than the values for class F with very light winds on a clear night, quantitative estimates of concentrations are nearly impossible for this condition. With very light winds on a clear night for ground-level sources free of topographic influences, frequent shifts in wind direction usually occur which serve to spread the plume horizontally. For elevated sources under these extremely stable situations, significant concentrations usually do not reach ground level until the stability changes.

A stable layer existing above an unstable layer will have the effect of restricting the vertical diffusion. The dispersion computation can be modified for this situation by considering the height of the base of the stable layer, L. At a height 2.15 σ_z above the plume centerline the concentration is one-tenth the plume centerline concentration at the same distance. When one-tenth the plume centerline concentration extends to the stable layer, at height L, it is reasonable to assume that the distribution starts being affected by the "lid." The following method is suggested to take care of this situation. Allow σ_z to increase with distance to a value of L/2.15 or 0.47 L. At this distance x_L, the plume is assumed to have a Gaussian distribution in the vertical. Assume that by the time the plume travels twice this far, 2 x_L, the plume has become uniformly distributed between the earth's surface and the height L, i.e., concentration does not vary with height (see Figure 2-6). For the distances greater han 2 x_L, the concentration for any height between he ground and L can be calculated from:

$$\chi\,(x,y,z;H) = \frac{Q}{\sqrt{2\pi}\,\sigma_y\,L\,u}\,\exp\left[-\frac{1}{2}\left(\frac{y}{\sigma_y}\right)^2\right] \quad (2.5)$$

for any z from 0 to L
for $x > 2\,x_L$; x_L is where $\sigma_z = 0.47\,L$

Note that Eq. (2. 5) assumes nor- nal or Gaussian distribution of the plume *only* in he horizontal plane. The same result can be ob- ained from the following equation where σ_{zL} is an ffective dispersion parameter because $\sqrt{2\pi}\,L = .5066\,L$ and $0.8\,\pi L = 2.51\,L$.

$$\chi\,(x,y,z;H) = \frac{Q}{\pi\,\sigma_y\,\sigma_{zL}\,u}\left[\exp-\frac{1}{2}\left(\frac{y}{\sigma_y}\right)^2\right] \quad (2.6)$$

for any z from 0 to L
for $x > 2x_L$; x_L is where $\sigma_z = 0.47\,L$
The value of $\sigma_{zL} = 0.8\,L$

VALUATION OF WIND SPEED

For the wind speed, u, a mean through the ver- ical extent of the plume should be used. This vould be from the height $H - 2\,\sigma_z$ through $H + \sigma_z$. Of course, if $2\,\sigma_z$ is greater than H then the vind can be averaged from the ground to $H + 2\,\sigma_z$. lowever, the "surface wind" value may be all that s available. The surface wind is most applicable o surface or low-level emissions, especially under table conditions.

LOTS OF CONCENTRATIONS GAINST DISTANCE

To gain maximum insight into a diffusion prob- m it is often desirable to plot centerline concen- rations against distance downwind. A convenient rocedure is to determine the ground-level center- ne concentrations for a number of downwind dis- ances and plot these values on log-log graph paper. y connecting the points, one may estimate con- entrations for intermediate downwind distances.

CCURACY OF ESTIMATES

Because of a multitude of scientific and techni- al limitations the diffusion computation method resented here provides *best estimates* but ot infallible predictions In the unstable and table cases, severalfold errors in estimate of σ_z an occur for the longer travel distances. In some ases the σ_z may be expected to be correct within a actor of 2, however. These are: (1) all stabilities or distance of travel out to a few hundred meters; 2) neutral to moderately unstable conditions for istances out to a few kilometers; and (3) unstable conditions in the lower 1000 meters of the atmos- phere with a marked inversion above for distances out to 10 km or more. Uncertainties in the esti- mates of σ_y are in general less than those of σ_z. The ground-level centerline concentrations for these three cases (where σ_z can be expected to be within a factor of 2) should be correct within a factor of 3, including errors in σ_y and u. The relative confidence in the σ's (in decreasing order) is indicated by the heavy lines and dashed lines in Figures **2-4** and **2-5**.

Estimates of H, the effective height of the plume, may be in error because of uncertainties in the esti- mation of ΔH, the plume rise. Also, for problems that require estimates of concentration at a specific point, the difficulty of determining the mean wind over a given time interval and consequently the location of the x-axis can cause considerable un- certainty.

PLOTTING GROUND-LEVEL CONCENTRATION ISOPLETHS

Often one wishes to determine the locations where concentrations equal or exceed a given mag- nitude. First, the axial position of the plume must be determined by the mean wind direction. For plotting isopleths of ground-level concentrations, the relationship between ground-level centerline concentrations and ground-level off-axis concentra- tions can be used:

$$\frac{\chi\,(x,y,0;H)}{\chi\,(x,0,0;H)} = \exp\left[-\frac{1}{2}\left(\frac{y}{\sigma_y}\right)^2\right] \quad (2.7)$$

The y coordinate of a particular isopleth from the x-axis can be determined at each downwind dis- tance, x. Suppose that one wishes to know the off-axis distance to the 10^{-3} g m^{-3} isopleth at an x of 600 m, under stability type B, where the ground- level centerline concentration at this distance is 2.9×10^{-3} g m^{-3}.

$$\exp\left[-\frac{1}{2}\left(\frac{y}{\sigma_y}\right)^2\right] = \frac{\chi\,(x,y,0;H)}{\chi\,(x,0,0;H)} = \frac{10^{-3}}{2.9 \times 10^{-3}} = 0.345$$

when $\exp\left[-\frac{1}{2}\left(\frac{y}{\sigma_y}\right)^2\right] = 0.345$, $y/\sigma_y = 1.46$

From Figure **2-4**, for stability B and x = 600 m, $\sigma_y = 92$. Therefore y = (1.46) (92) = 134 meters. This is the distance of the 10^{-3} isopleth from the x-axis at a downwind distance of 600 meters.

This can also be determined from:

$$y = \left\{2\,\ln\left[\frac{\chi\,(x,0,0;H)}{\chi\,(x,y,0;H)}\right]\right\}^{1/2} \sigma_y \quad (2.8)$$

The position corresponding to the downwind dis- tance and off-axis distance can then be plotted. After a number of points have been plotted, the concentration isopleth may be drawn.

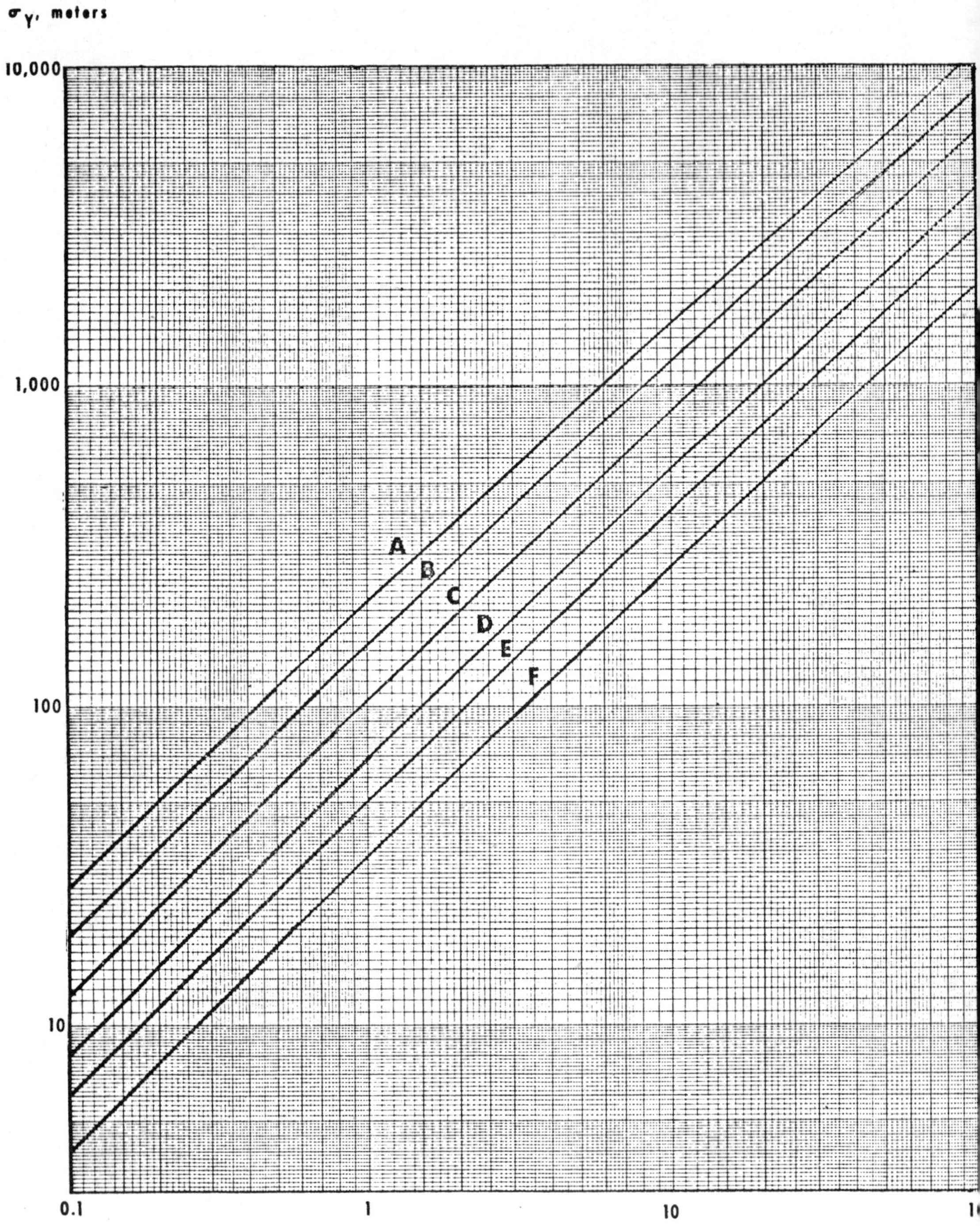

Fig. 2-4 Horizontal dispersion coefficient as a function of downwind distance from the source.

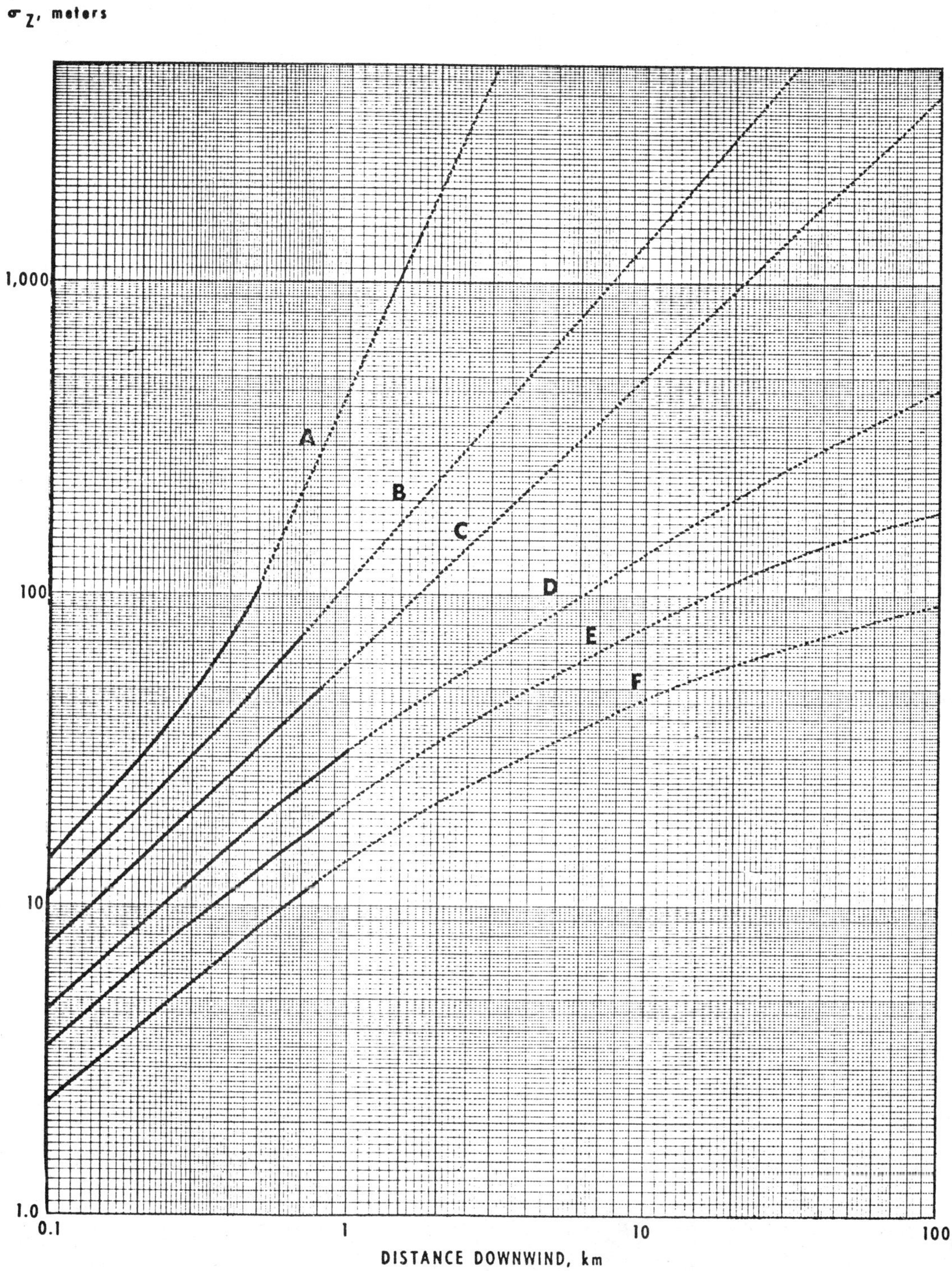

Fig. 2-5 Vertical dispersion coefficient as a function of downwind distance from the source.

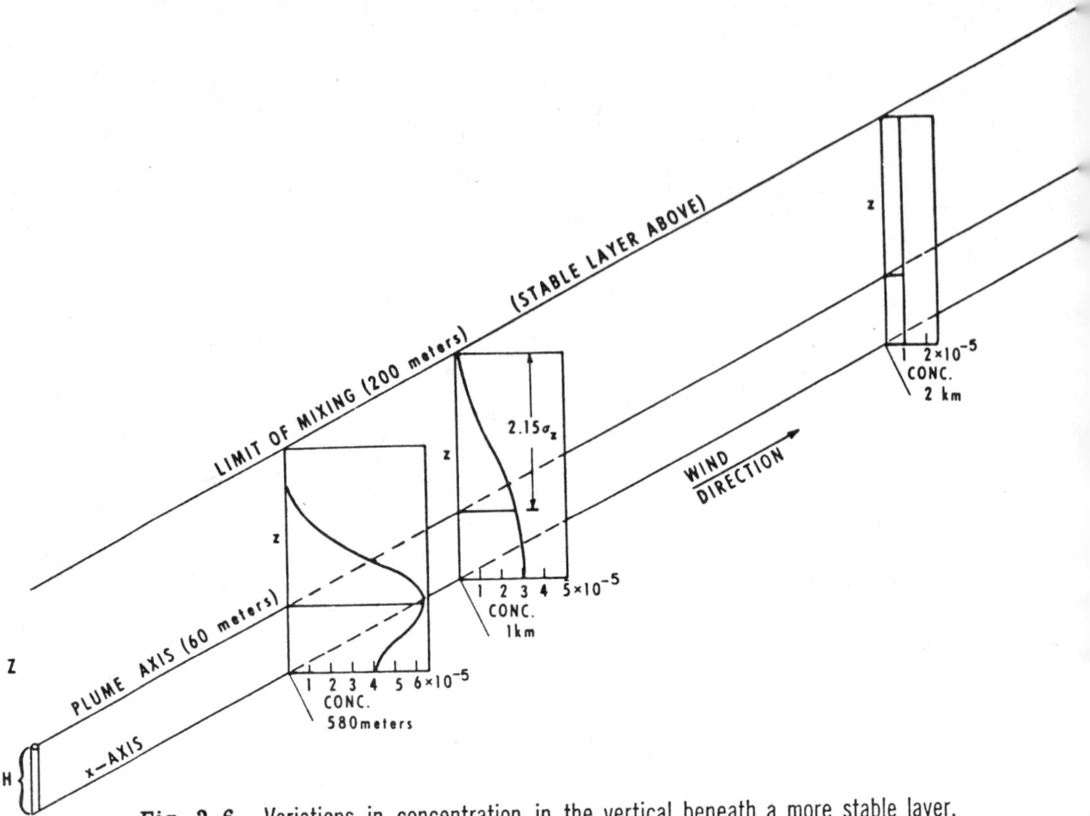

Fig. 2-6 Variations in concentration in the vertical beneath a more stable layer.

EFFECTIVE HEIGHT OF EMISSION

In most problems one must estimate the effective stack height, H, at which the plume becomes essentially level. Rarely will this height correspond to the physical height of the stack, h. If the plume is caught in the turbulent wake of the stack or of buildings in the vicinity of the stack, the effluent will be mixed rapidly downward toward the ground (aerodynamic downwash). If the plume is emitted free of these turbulent zones, a number of emission factors and meteorological factors influence the rise of the plume. The emission factors are: velocity of the effluent at the top of the stack, v_s; temperature of the effluent at the top of the stack, T_s; and diameter of the stack opening, d. The meteorological factors influencing plume rise are wind speed, u; temperature of the air, T_a; shear of the wind speed with height, du/dz; and atmospheric stability. No theory on plume rise takes into account all of these variables; even if such a theory were available, measurements of all of the parameters would seldom be available. Most of the equations that have been formulated for computing the effective height of emission are semi-empirical. For a recent review of equations for effective height of emission see Moses, Strom, and Carson (1964).

Moses and Strom (1961), having compared actual and calculated plume heights by means of six plume rise equations, report "There is no one formula which is outstanding in all respects." Th formulas of Davidson-Bryant (1949), Hollan (1953), Bosanquet-Carey-Halton (1950), and Bo sanquet (1957) all give generally satisfactory re sults in the test situations. The experiments con ducted by Moses and Strom involved plume ris from a stack of less than 0.5 meter diameter, stac gas exit velocities less than 15 m sec^{-1}, and effluen temperature not more than 35°C higher than tha of the ambient air.

The equation of Holland was developed wit experimental data from larger sources than thos of Moses and Strom (stack diameters from 1.7 t 4.3 meters and stack temperatures from 82 t 204°C); Holland's equation frequently under estimates the effective height of emission therefore its use often provides a sligh "safety" factor.

Holland's equation is:

$$\Delta H = \frac{v_s\, d}{u} \left(1.5 + 2.68 \times 10^{-3}\, p\, \frac{T_s - T_a}{T_s}\, d\right) \quad (2.9$$

where:

ΔH = the rise of the plume above the stack, r

v_s = stack gas exit velocity, m sec^{-1}

d = the inside stack diameter, m
u = wind speed, m sec^{-1}
p = atmospheric pressure, mb
T_s = stack gas temperature, °K
T_a = air temperature, °K

and 2.68 x 10^{-3} is a constant having units of mb^{-1} m^{-1}.

Holland (1953) suggests that a value between .1 and 1.2 times the ΔH from the equation should be used for unstable conditions; a value between 0.8 and 0.9 times the ΔH from the equation should be used for stable conditions.

Since the plume rise from a stack occurs over some distance downwind, Eq. (2. 9) should not be applied within the first few hundred meters of the stack.

EFFECTIVE HEIGHT OF EMISSION AND MAXIMUM CONCENTRATION

If the effective heights of emission were the same under all atmospheric conditions, the highest ground-level concentrations from a given source would occur with the lightest winds. Generally, however, emission conditions are such that the effective stack height is an inverse function of wind speed as indicated in Eq. (2. 9). The maximum ground-level concentration occurs at some intermediate wind speed, at which a balance is reached between the dilution due to wind speed and the effect of height of emission. This critical wind speed will vary with stability. In order to determine the critical wind speed, the effective stack height as a function of wind speed should first be determined. The maximum concentration for each wind speed and stability can then be calculated as a function of effective height of emission and stability. When the maximum concentration is a function of wind speed is plotted on log-log graph paper, curves can be drawn for each stability class; the critical wind speed corresponds to the point of highest maximum concentration on the curve.

ESTIMATES OF REQUIRED STACK HEIGHTS

In some situations high concentrations upon the property of the emitter are of little concern, but maximum concentrations beyond the property line are of the utmost importance. For first approximations it can be assumed that the maximum concentration occurs where $\sqrt{2}\,\sigma_z$ = H and that at this distance the σ's are related to the maximum concentration by:

$$\sigma_y\,\sigma_z \simeq \frac{Q}{\pi\, u\, e\, \chi_{max}} \simeq \frac{0.117\,Q}{u\,\chi_{max}} \qquad (2.\,10)$$

Knowing the source strength, Q, and the concentration not to be exceeded χ_{max}, one can determine the necessary $\sigma_y\,\sigma_z$ for a given wind speed.

EFFECT OF EVAPORATIVE COOLING

When effluent gases are washed to absorb certain constituents prior to emission, the gases are cooled and become saturated with water vapor. Upon release of the gases from the absorption tower, further cooling due to contact with cold surfaces of ductwork or stack is likely. This cooling causes condensation of water droplets in the gas stream. Upon release of the gases from the stack, the water droplets evaporate, withdrawing the latent heat of vaporization from the air and cooling the plume. The resulting negative buoyancy reduces the effective stack height.

EFFECT OF AERODYNAMIC DOWNWASH

The influence of mechanical turbulence around a building or stack can significantly alter the effective stack height. This is especially true with high winds, when the beneficial effect of high stack-gas velocity is at a minimum and the plume is emitted nearly horizontally. The region of disturbed flow surrounds an isolated building, generally to at least twice its height and extends downwind 5 to 10 times its height. Building the stack 2.5 times the height of the highest building adjacent to the stack usually overcomes the effects of building turbulence (Hawkins and Nonhebel, 1955). Ensuring that the exit velocity of the stack gas is more than 1.5 times the wind speed will usually prevent downwash in the wake of the stack. Most of the knowledge about the turbulent wakes around stacks and buildings has been gained through wind tunnel studies (Sherlock and Lesher, 1954; Strom, 1955-1956; Strom, et al, 1957; and Halitsky, 1962). By use of models of building shapes and stacks, one may determine the wind speeds required to cause downwash for various wind directions. With a wind tunnel the meteorological variables most easily accounted for are wind speed and wind direction (by rotation of the model within the tunnel). The emission factors that may be considered are the size and shape of the plant building; the shape, height, and diameter of the stack; the amount of emission; and the stack-gas velocity.

Through wind tunnel studies, the critical wind speeds that will cause downwash from various directions can be determined for a given set of plant factors. The average number of hours of downwash per year can then be calculated by determining the frequency of wind speeds greater than the critical speeds for each direction (Sherlock and Lesher, 1954) if climatological data representative of the site are available.

Maximum downwash about a rectangular structure occurs when the direction of the wind is at an angle of 45 degrees from the major axis of the structure; minimum downwash occurs with wind flow parallel to the major axis of the structure.

Halitsky (1961, 1963) has shown that the effluent from flush openings on flat roofs frequently flows in a direction opposite to that of the free atmospheric wind, owing to counter-flow along the roof in the turbulent wake above the building. In addition to the effect of aerodynamic downwash upon the release of air pollutants from stacks and buildings, one must also consider the effects of aero-

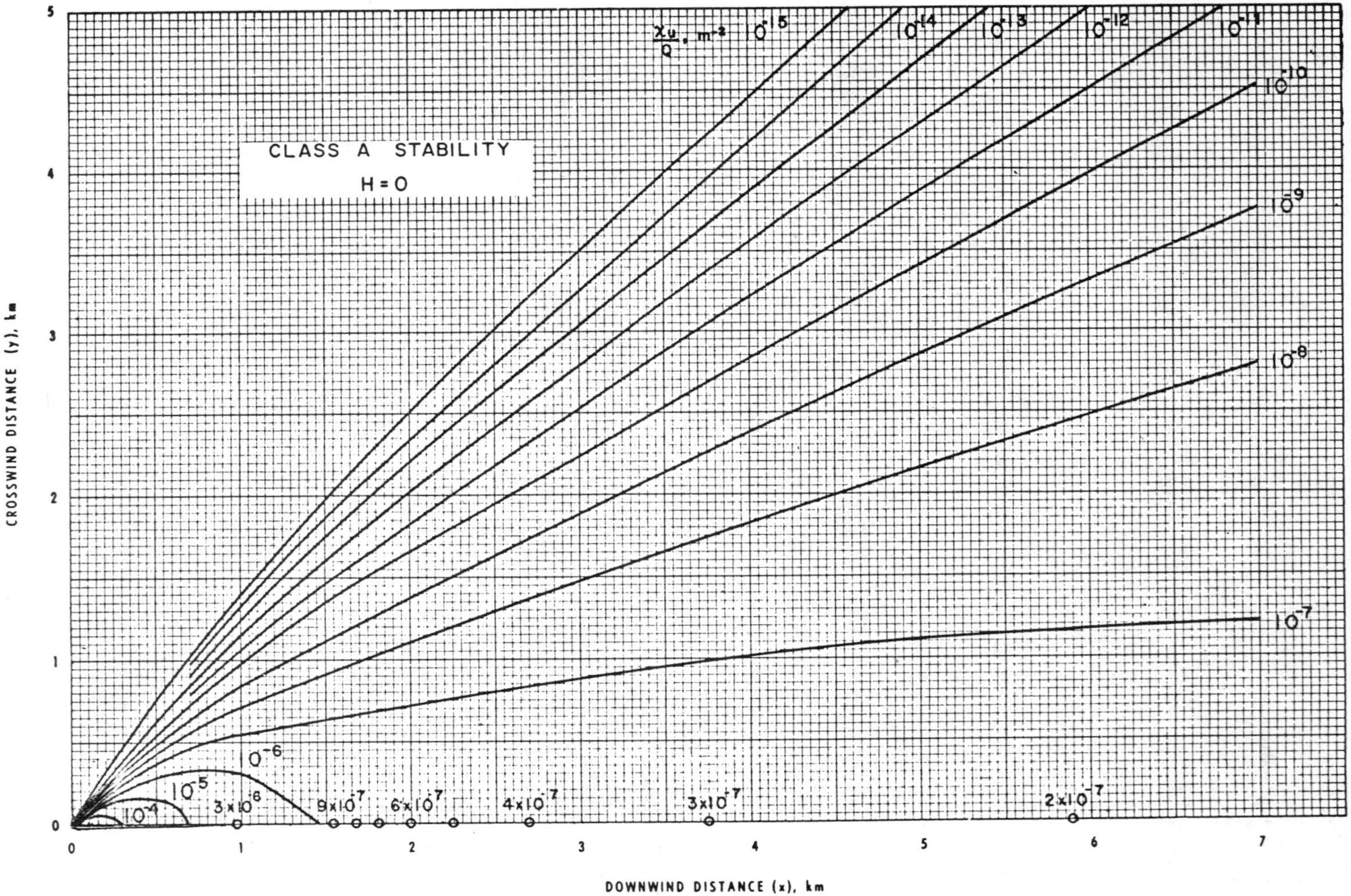

Fig. 2-7 Isopleths of $\chi u/Q$ for a ground-level source, A stability.

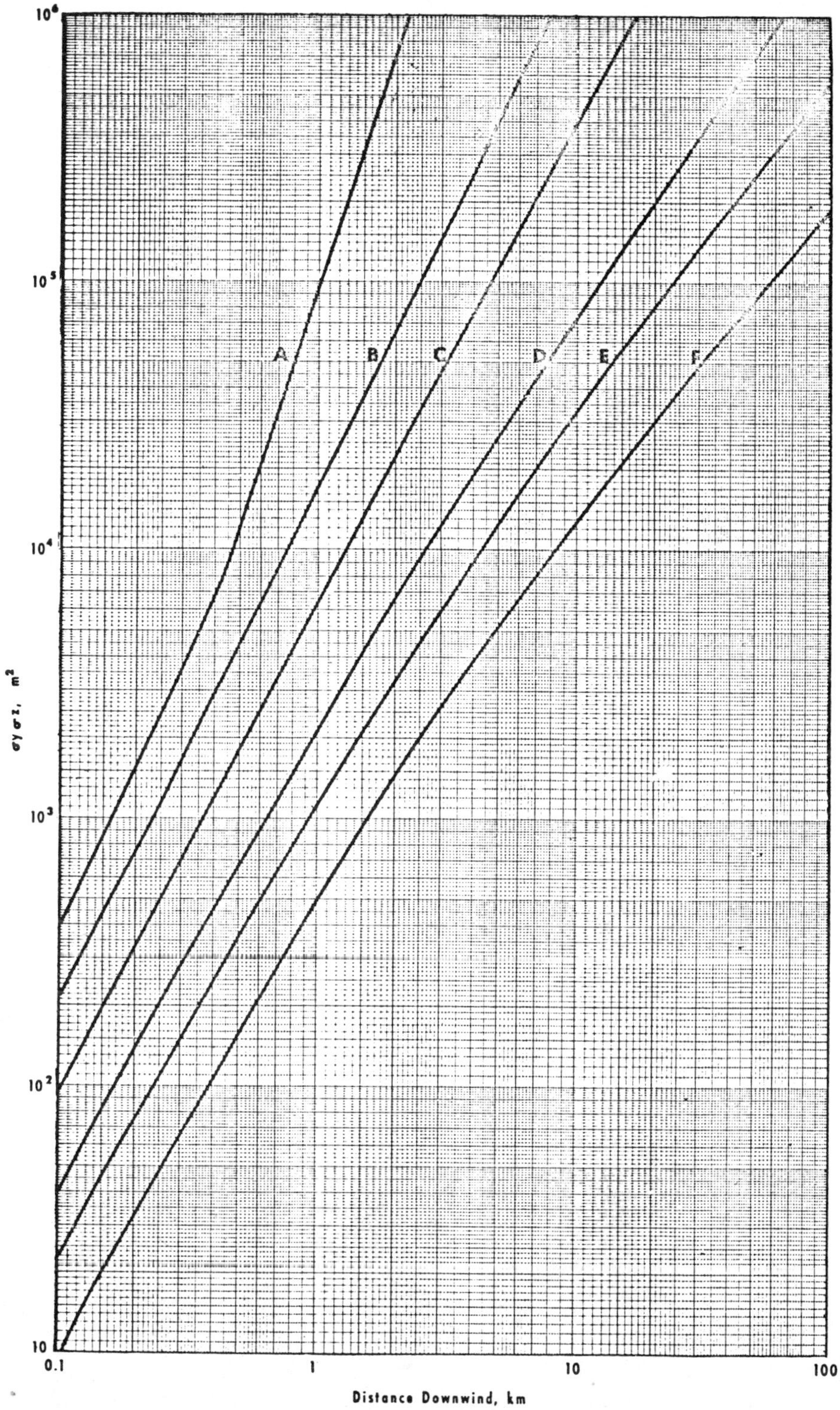

Fig. 2-8 The product of $\sigma_y\sigma_z$ as a function of downwind distance from the source.

dynamic downwash when exposing meteorological instruments near or upon buildings.

Where the pollution is emitted from a vent or opening on a building and is immediately influenced by the turbulent wake of the building, the pollution is rapidly distributed within this turbulent wake. To account for mixing in the turbulent wake, one may assume binormal distributions of concentrations at the source, with horizontal and vertical standard deviations of σ_{yo} and σ_{zo}. The standard deviations are related to the width and height of the building, for example, letting 4.3 σ_{yo} equal the width of the building and 2.15 σ_{zo} equal the height. Values other than 4.3 and 2.15 can be used. When these values are used 97% of the distribution is included within these limits. Virtual distances x_y and x_z can be found such that at x_y, $\sigma_y = \sigma_{yo}$ and at x_z, $\sigma_z = \sigma_{zo}$. These x's will differ with stability. Equations applicable to point sources can then be used, determining σ_y as a function of $x + x_y$ and σ_z as a function of $x + x_z$.

REFERENCES

Bosanquet, C. H., W. F. Carey, and E. M. Halton, 1950: Dust from chimney stacks. Proc. Inst. Mech. Eng., *162*, 355-367.

Bosanquet, C. H., 1957: The rise of a hot waste gas plume. J. Inst. Fuel, *30*, 197, 322-328.

Davidson, W. F., 1949: The dispersion and spreading of gases and dust from chimneys. Trans. Conf. on Ind. Wastes, 14th Ann. Meeting, Ind. Hygiene Found. Amer., 38-55.

Halitsky, J., 1961: Wind tunnel model test of exhaust gas recirculation at the NIH Clinical Center. Tech. Rep. No. 785.1, New York Univ.

Halitsky, J., 1962: Diffusion of vented gas around buildings. J. Air Poll. Cont. Assoc., *12*, 2, 74-80.

Halitsky, J., 1963: Gas diffusion near buildings, theoretical concepts and wind tunnel model experiments with prismatic building shapes. Geophysical Sciences Lab. Rep. No. 63-3. New York Univ.

Hawkins, J. E., and G. Nonhebel, 1955: Chimneys and the dispersal of smoke. J. Inst. Fuel, *28*, 530-546.

Holland, J. Z., 1953: A meteorological survey of the Oak Ridge area. 554-559 Atomic Energy Comm., Report ORO-99, Washington, D.C., 584 pp.

Moses, H., and G. H. Strom, 1961: A comparison of observed plume rises with values obtained from well-known formulas. J. Air Poll. Cont. Assoc., *11*, 10, 455-466.

Moses, H., G. H. Strom, and J. E. Carson, 1964: Effects of meteorological and engineering factors on stack plume rise. Nuclear Safety, *6*, 1, 1-19.

Scorer, R. S., 1959: The behavior of plumes. Int. J. Air Poll., *1*, 198-220.

Sherlock, R. H., and E. J. Lesher, 1954: Role of chimney design in dispersion of waste gases. Air Repair, *4*, 2, 1-10.

Strom, G. H., 1955-1956: Wind tunnel scale model studies of air pollution from industrial plants. Ind. Wastes, Sept. - Oct. 1955, Nov. - Dec. 1955, and Jan. - Feb. 1956.

Strom, G. H., M. Hackman, and E. J. Kaplin, 1957: Atmospheric dispersal of industrial stack gases determined by concentration measurements in scale model wind tunnel experiments. J. Air Poll. Cont. Assoc., *7*, 3, 198-203.

CHAPTER 3

ATMOSPHERIC TRANSPORT PROCESSES

In studies of the general circulation, it has long been recognized that the transport processes that maintain the climatic balance of the atmosphere occur not only in mean meridional motions but also in horizontal "austausch" or eddy-exchange processes, which involve the big whirls of cyclones and anticyclones (Defant, 1921; Jeffreys, 1933). Which of the two processes was the more important long remained a matter of argument. With the improvement of the aerological network after World War II, computations of transport quantities from actual observations tipped the scales in favor of the eddy transport, at least in the middle and high latitudes.

Riehl and Fultz (1957, 1958) used geophysical-model experiments to demonstrate that the relative efficiency of mean and eddy transports is of secondary importance. A meandering jet stream in a "dishpan" experiment with a symmetric three-wave configuration showed, for instance, that the circulation in middle latitudes underneath the mean jet-stream position is indirect if flux terms are averaged along

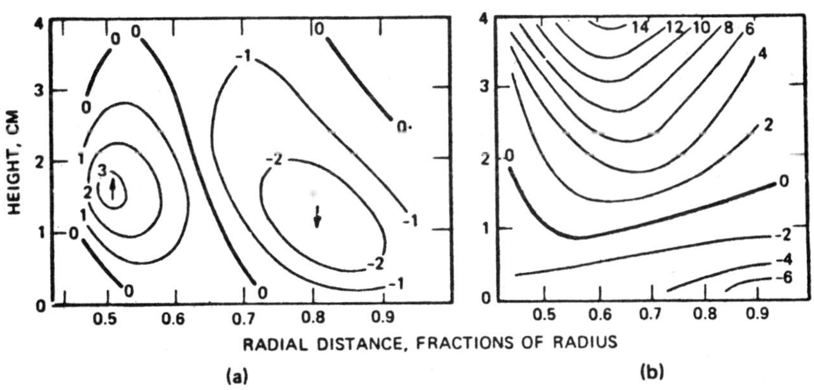

Fig. 3-1 Zonal average of (a) vertical motions (per mille of linear speed of equator) and (b) horizontal motions (percent of linear speed of equator) for the case of three symmetric waves in a dishpan. [From H. Riehl and D. Fultz, *Quarterly Journal of the Royal Meteorological Society (England)*, 83(356): 228 (1957).]

latitude circles (Fig. 3-1). Mean meridional cross sections through such a flow configuration show three circulation cells, such as those postulated by Palmén (1954) and shown in Fig. 3-2, for the earth's atmosphere. (In Fig. 3-2, rising motion along the outer rim of the dishpan and sinking motion along the inner rim are assumed because of the heating at the "equator" and the cooling at the "pole" of the pan.)

However, if the hydrodynamic variables are averaged with respect to a coordinate system whose point of origin is located in the jet axis, following its meanders a totally different circulation pattern evolves (Fig. 3-3). In this curvilinear coordinate system, one single direct circulation cell now occupies the space between the equator and the pole of the dishpan.

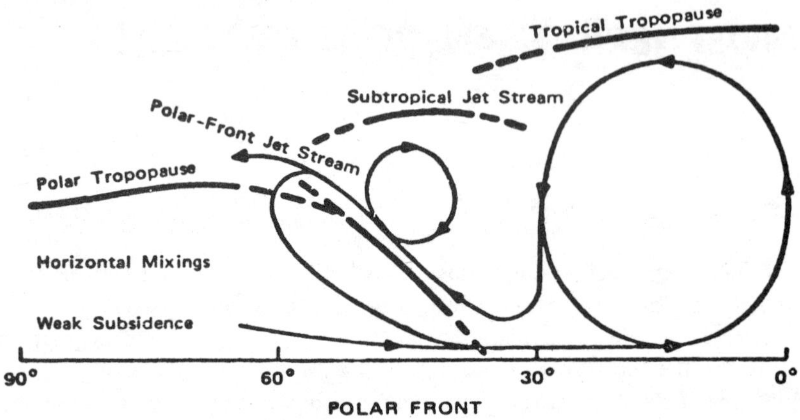

Fig. 3-2 Schematic diagram of the mean meridional circulation in the northern hemisphere during winter. Heavy lines indicate tropopauses and the polar front. (After E. Palmén, 1954.)

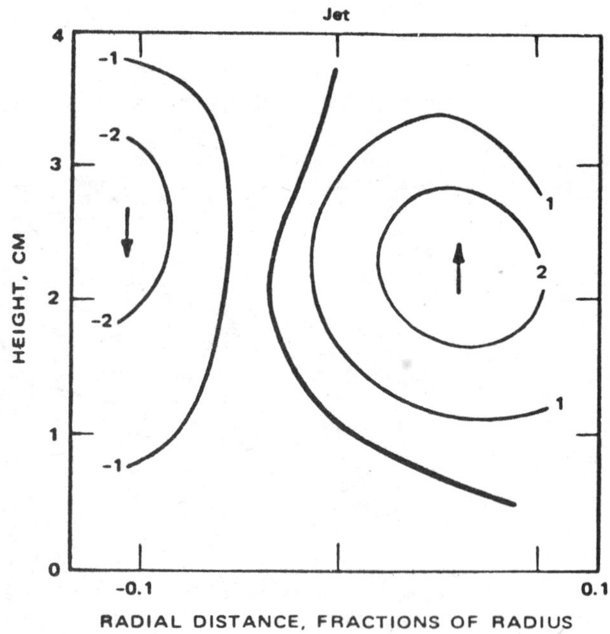

Fig. 3-3 Vertical motions (per mille of linear speed of equator) in a curvilinear coordinate system for three symmetric waves in a dishpan. [From H. Riehl and D. Fultz, *Quarterly Journal of the Royal Meteorological Society (England)*, 83(356): 228 (1957).]

In the geographic coordinate system, the indirect-circulation cell underneath the jet stream implies that the jet stream continuously loses kinetic energy and generates potential energy. Similar considerations could be made for the polar-front jet stream in Fig. 3-2. For this jet stream to be sustained against frictional dissipation as well as the loss of kinetic energy in the indirect circulation, eddy-transport processes must resupply this energy.

The curvilinear system, on the other hand, follows the large-scale eddy motions prescribed by the jet axis, thus filtering the influence of these eddies from the transport computations. The task of maintaining the jet stream against frictional dissipation falls now to the mean cross-axis circulation. Only at some distance from the jet axis will eddy motions gain some importance because of departures of the wind vectors there from the wind direction in the jet axis. Examples for heat transports are given in Figs. 3-4 and 3-5. These two figures refer to a steady three-wave pattern in a dishpan experiment. The pattern migrates slowly eastward, showing only rather diffuse jet maxima. Circles and crosses in these two diagrams refer to actual measurements in the dishpan, obtained from thermistors at various depths and from "winds" at the surface of the pan made visible by floating aluminum powder. "Winds" at depth were computed from the thermal wind equation. Some smoothing of these measured data, indicated by the heavy curves, proved to be necessary in order to approximate the theoretical magnitude of the transports given by the straight, slightly slanting lines. In the real atmosphere, where well-developed jet maxima lead to appreciable magnitudes of the departures of velocities from the mean velocity measured along reference lines parallel to the jet axis, one should expect eddy-

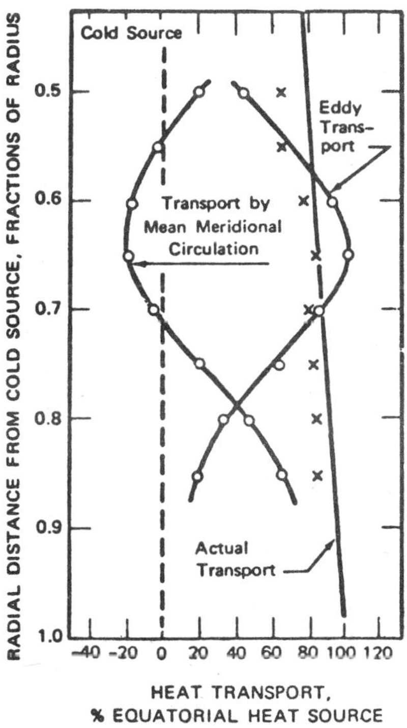

Fig. 3-4 Mean meridional heat transport in cylindrical (geographical) coordinates (percent of the equatorial heat source) for three symmetric waves in a dishpan. The ordinate marks the radial distance from the cold source (total distance cold source–equator = 1). O, transports by mean meridional circulation and by the wave motion (eddy transport). X, observed total transport. The straight line is the computed total transport. This line is not vertical because it incorporates heat losses at the free-water surface of the dishpan. [From H. Riehl and D. Fultz, *Quarterly Journal of the Royal Meteorological Society (England)*, 84(362): 390 (1958).]

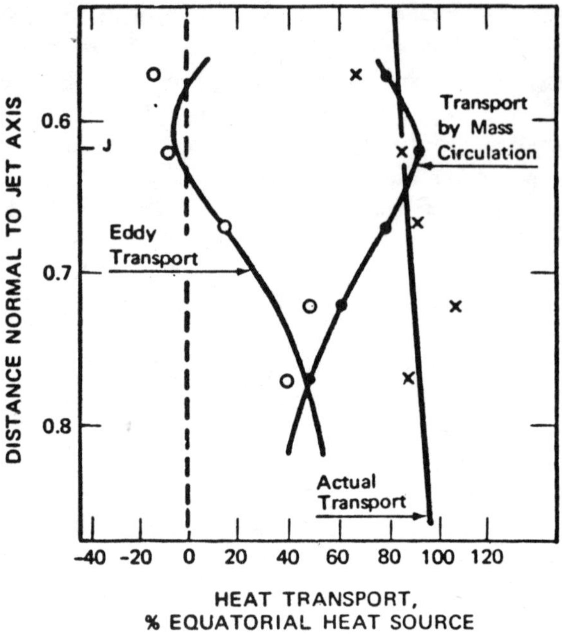

Fig.3-5 Same as Fig. 1.4 except curvilinear coordinate system. Ordinate shows relative distance from mean jet axis (J). [From H. Riehl and D. Fultz, *Quarterly Journal of the Royal Meteorological Society (England)*, 84(362): 397 (1958).]

transport terms of some size, even in the jet axis, more so than indicated in Fig. **3-5**. This will not distract, however, from the fact that the mean mass circulation across the jet axis will carry the bulk of the flux of quasi-conservative air-mass characteristics across the jet stream.

From the studies by Riehl and Fultz (1957, 1958), it appears that the relative importance of mean- and eddy-transport terms depends decisively on the choice of the coordinate system. The total transport, as the sum of mean and eddy transports, remains unaffected by the choice of reference frame, however.

Even though only the magnitude of the combined transport processes is important in maintaining the general circulation, there are some far-reaching implications on the transport of nuclear debris associated with the various modes of these processes. For instance, if the total transport of heat, momentum, and trace substances were carried out solely by a mean meridional (Hadley) circulation—as it would be in a slowly rotating dishpan—the trajectories of contaminated air parcels would follow a spiral motion: ascent near the equator, northeastward displacement aloft, sinking near the pole, and southwestward flow near the surface of the earth. The gradual spread of an injection of nuclear or other debris along these spiral paths would be controlled mainly by small-scale diffusion processes.

This rather unrealistic model of an atmospheric circulation will now be modified by introducing a number of symmetric planetary waves, such as in the experiments by Riehl and Fultz (1957, 1958). The effect is a certain distortion of the original spiral trajectories by the regular wave pattern. The gradual spread of a plume of debris still is controlled largely by the small-scale diffusion processes. The large-scale eddy motions in the symmetric wave pattern, as well as local horizontal and vertical wind shears, produce distortions in the plume only when the diffusion processes reach dimensions commensurate with the scale of these local shears and of the large waves. The absence of eddies or waves in the spectrum range between the long planetary waves and the small-scale (molecular) diffusion processes would make the spreading of a plume of contaminants an extremely slow process.

Conditions become more realistic if we allow a continuous spectrum of eddies in our model atmosphere, ranging from planetary waves to cyclone waves to meso-scale disturbances and to small-scale turbulent eddies and molecular diffusion (Kao, 1962; Vinnichenko, 1969). Such a continuous spectrum of perturbation motions will cause a rather rapid diffusion of any trace substance. The diffuse depletion of radioactivity concentrations and of the concentrations of any other tracer will strongly depend on the intensity, or kinetic energy, of the eddy motions in the wavelength band commensurate with the scale of the diffusing plume. Eddies smaller than those in this wavelength band will contribute toward homogeneity of the tracer within the contaminated volume of air that vacillates in its motion around the hemisphere. Thus, with the increase in size of a debris cloud, its further spread can be controlled by a changing range of eddy scales.

The kinetic energies contained in the various "wave bands" of large-scale eddies cannot be assumed to be constant. They depend strongly on the state of the general circulation and on the processes that generate, dissipate, and redistribute these kinetic energies. A better understanding of the energetics of the general circulation, therefore, will aid in the comprehension of the intricate mechanisms that control the global and small-scale spread of atmospheric contaminants.

From this rather general discussion, we may arrive at the following statements:

1. The relative magnitudes of eddy and mean meridional transport processes are important in the spread of aerosols. These relative magnitudes affect the large-scale diffusion and the latitudinal and vertical displacement of the debris.

2. The spectrum distribution of the eddy motions in the atmosphere affects the time and space scales of diffusion processes.

3. Standing and transient eddies will have different effects on diffusion. (In transient eddy motions trajectories will be different from streamlines.) The two types of eddies, therefore, may well be considered separate from each other.

CHAPTER 4

DATA AQUISITION SYSTEMS IN METEOROLOGY

A typical data acquisition system comprises four major subsystems: *sensors*, which provide electrical or mechanical analogs of meteorological variables; *telemetry*, which provides for the transfer of sensor information to a central collection point; *central control*, which provides for the interrogation of sensors, the recording and processing of sensor outputs, and the routing of processed data to displays or storage; and *displays*, which present processed data in a convenient form. Fig. 4-1 illustrates these relationships.

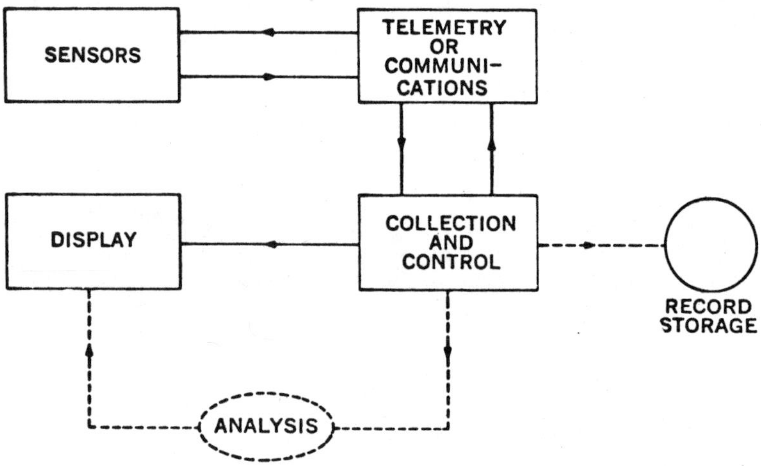

Fig. 4-1— Schematic Diagram of Basic Components for Data Acquisition System

BASIC CONSIDERATIONS IN THE DESIGN AND OPERATION OF METEOROLOGICAL DATA ACQUISITION SYSTEMS

GEOPHYSICAL CONSIDERATIONS

The basic purpose to be served by a meteorological data acquisition system is generally to provide a satisfactory description of atmospheric structure within a specified reference volume. In air pollution problems the horizontal dimensions of the reference volume are fixed by the areal extent of an urban complex, for example, and the vertical dimension is set by the maximum height attained by any pollutants that are transported across the complex by the wind. Relevant properties of atmospheric structure within the reference volume include the three-dimensional distribution of mean air temperature, moisture, wind speed, wind direction, and the turbulent fluctuations of the latter two variables. In mathematical terms, the mean value, $\overline{M}$ of a meteorological variable obtained from a time series at a fixed point is expressed as

$$\overline{M} = \frac{1}{T} \int_{t_0 - T/2}^{t_0 + T/2} M(t)\,dt \qquad (1)$$

where T is the length of record and t_0 is an arbitrary reference. It also follows from the argument presented above that the value of M at any arbitrary time t is given by

$$M = \overline{M} - M' \qquad (2)$$

and

$$\overline{M'} = 0 \qquad (3)$$

where M' is the departure of M from the mean. Generally, M is a function of both space and time variables:

$$M = f(x,y,z,t) \qquad (4)$$

The choice of appropriate integration limits is dictated both by the scale of the problem to be investigated and the form of the energy spectrum of the meteorological variables. Most characteristic air properties exhibit a spectrum of variability that is at least quasi-continuous (consists of disconnected, continuous segments) over a very wide range of time or space frequencies. Since the spectrum is quasi-continuous over a broad range of frequencies, measurements of its properties is limited at high frequencies by the response characteristics of the instrumentation and at low frequencies by the length of record. Pasquill (1962) has shown that the omitted portions of the spectrum can be approximated by weighting functions of the form

$$\frac{\sin^2 \pi n \bar{t}}{(\pi n \bar{t})^2} \quad \text{and} \quad \frac{1 - \sin^2 \pi n \overline{T}}{(\pi n \overline{T})^2}$$

where $\bar{t}$ is the response time of the measurement system, $\overline{T}$ is the length of record, and n is an arbitrary frequency.

To derive meaningful relationships between meteorological variables and diffusion patterns, for example, it is usually necessary to choose an averaging time $\overline{T}$ such that $\overline{M}$ is reasonably stable.

The space and time variability of meteorological variables is only partially understood at present, and such choices as we have been discussing are usually not routine. Determination of upper and lower frequency limits that will include the "significant" portion of the spectrum for a given variable is still, in the last analysis, a matter for experienced judgment.

Because of the importance of these scaling considerations, it is usually necessary in meteorological applications to relate the functional requirements to three scales of observations: macroscale, mesoscale, and microscale. Macroscale data are those normally used in describing the general weather conditions prevailing over a large area, perhaps 100 miles on a side. Mesoscale data pertain to the general environmental conditions within a few miles distance and in particular to the deviations of the local parameters from the general macroscale weather. Microscale data deal primarily with the fine structure of the local atmosphere for distances of 1 mile or less. With each of these scales of observation are associated different sensor and system input requirements, different data rates, and different processing and display requirements. A properly conceived system will provide for adequate integration of all three data streams.

SYSTEMS ENGINEERING CONSIDERATIONS

The proper development of an automatic data acquisition system for meteorological use involves both meteorology and data systems engineering techniques. Failure to recognize this fact at the outset results in a system based on many practial compromises that may fail to meet the application requirements optimally or even at all. The engineering design of such a system should reflect both the realities of current data systems technology and the ultimate application for which the measurement system is intended. It is well worth the effort to develop a systematic plan for the implementation of the system in advance of choosing specific pieces of hardware. This effort requires technical skills in the areas of operations research, communications systems, computer systems engineering, and programming as well as meteorology to develop a definition of the system concept adequate for determination of subsystem specifications and requirements.

The first step in such a study involves ascertaining and specifying precisely what is to be measured, when, where, and why. In particular, this includes descriptions of the significant range of the spectrum of each variable, the appropriate averaging times, and the portion of the complete time cycle during which information on each variable is wanted. On this basis one can determine data rates in the various parts of the information-gathering network — the fundamental consideration upon which engineering design considerations must be based. It is, of course, also the basis for specification of the meteorological sensors that comprise the sensor subsystem of the data system. The next step involves the specification of the engineering philosophy and logical procedures to be followed in data transmission, as dictated by the inherent data rates; by requirements for mobility, expandability, and change in the system; and by cost. For example, it makes considerable difference whether each meteorological measurement is to be recorded on a punched paper tape for later computer analysis and study, or whether these measurements are connected on-line to a computer for automatic real-time sorting and processing. The choice should be based upon a broad, scientific examination of the total information-handling problem and total cost. Frequently it is less expensive to replace a variety of special devices with a central general-purpose control element, even though the speed and flexibility requirements of the system do not require this.

The third step in this activity entails analysis of data recording and data processing requirements, including necessary mathematical modeling and calculations. A fourth step involves a thorough analysis of requirements for display of the processed information to a potential user.

Such information, properly organized to reveal the interdependence of the various functions, in proper detail, is the necessary foundation for specifying system performance criteria; it is also essential for an organized approach to the technical problems of subsystem hardware and for the integration of subsystem interface requirements into the design approach. It is of the utmost importance for the success of a given meteorological system that this be done in advance of procuring pieces of equipment for the system and in the light of the broader uses of the meteorological measurement program that this system is intended to implement.

The foregoing factors must be investigated against the background of various constraints on the design and implementation of the system. These constraints reflect known limitations in physical, engineering, economic, and human factors, which will significantly affect the feasibility of utilizing specific techniques or equipment in the system. These factors may be grouped in the following categories:

1. Economics — The approximate amount of money available for the development of the data system is usually the overriding constraint on system design. A thorough analysis of the problem, such as we have described, is primarily aimed at obtaining maximum performance from the proposed system within the available funds.

2. Physical Environment — The geographical and climatological regime in which the system must operate is a factor that automatically rules out certain approaches. Included in this category are considerations of morbidity, or the ease with which the measuring system is available for relocation.

3. Personnel — The training, capabilities, and number of personnel required to operate and install the system can have a significant effect on system design philosophy and procedures.

4. Interfaces of the Meteorological System with Other Functions — A meteorological data collection system designed for furnishing information to an air traffic controller entails significantly different problems from one designed for study of atmospheric pollution around a city.

With the information developed from analysis of these factors we can establish system functional objectives consistent with the broad application requirements, technological state-of-the-art, and so-called "practical" limitations on the system design and development. These functional objectives properly described with their relationship to one another, form the basis for more detailed determination of the specific requirements for each of the subsystems: sensors, communications, data processing, and display.

The final phase of the design study involves specification of requirements for each subsystem and its important components. For the sensor subsystem characteristics such as the following must be specified:

1. Form of the sensor outputs (digital or analog).
2. Accuracy, resolution, and range of the sensor readings.
3. Required ruggedness and reliability, including protection from the environment, and electrical and mechanical functioning.

For the communication subsystem, major considerations will be:

1. Function of the communication system over fixed-wire channels or by radio telemetry.
2. Necessary bandwidth as determined both by the data rates generated from the meteorological measuring instruments and by reliability considerations.
3. Coding techniques, particularly whether signals are to be transmitted in analog or digital form.
4. Power requirements.
5. Physical maintenance requirements.

For data processing subsystem we must specify:

1. Which parts of the processing are to be automatic and which are to be handled manually.
2. Appropriate processing speeds and memory requirements for the automatic data handling.

3. Necessary automatic (real-time) computer inputs, if any.
4. Appropriate forms of recording data not directly entered into the computer, such as punch cards, punch paper tapes, or strip-chart graphs.
5. Signal conversion operations, such as analog-to-digital conversion.
6. Computational requirements, such as objective forecast models, turbulent diffusion models, and atmospheric statistics.
7. Data recording requirements and appropriate forms, such as printed copy, magnetic tape, punch cards, etc.

For the information display subsystem of our data collection system, appropriate forms and numbers for display devices and materials must be indicated. These include:

1. Necessity for automatic moving displays, such as cathode ray tubes and various types of projection systems.
2. Printed record requirements.
3. Automatic alarm signal requirements.
4. Physical location and number of displays required.

CHAPTER 5

AIR POLLUTION SOURCES

Classification of Sources:

a. In assessing the status of air pollution at any installation or in any region it is necessary to evaluate the sources which are emitting the air contaminants. This would mean determining where all of the sources were located, how much contaminating material was being released from each source per unit of time under various conditions, e.g., time of day, time of year, and what future emissions could be expected from present and future sources. To do all of this, the staff seeking such information should have a general idea as to what facilities are potential sources so that none would be overlooked.

b. In approaching air pollution problems it is convenient to divide sources into eight categories. These are:

(1) Industry.

(2) The combustion of fuel for heat and power.

(3) The combustion of refuse and other wastes.

(4) The storage, handling, and dispensing of fuels.

(5) The combustion of fuel for transportation.

(6) Community-type activities.

(7) Natural sources.

(8) Miscellaneous.

Total US Air Pollution By Source in 1966 as reported by the Department of Health, Education and Welfare were: Motor Vehicles—86 million tons per year (60.6%); Industry—23 million tons per year (16.8%); Power Plants—20 million tons per year (14.1%); Space Heating—8 million tons per year (5.6%) and Refuse Disposal—5 million tons per year (3.5%). At Department of Defense installations the air pollution sources of prime importance are space heating, refuse disposal and to some lesser degree industry and power plants.

Significance of Various Sources:

a. **Industry.** Industrial activities constitute one of the main sources of air pollutants. The quantities of contaminants and their potential to produce various effects depend upon the size of the industry, the type of industry, the controls which the industry maintains on its processes and emissions, local topography, local meteorological conditions, and how the surrounding property is being utilized. A plant located in a desert may cause little or no damage if it is surrounded by unused land, but the same plant in the center of a large city may well be totally unacceptable with respect to air pollution. Depending on the factors just mentioned, industrial emissions may or may not be the principal source of air contaminants in a given area. For example, it was felt at one time that the smog problem in Los Angeles, California, was a direct result of industrial activities. A program was initiated to control industrial emissions but this did little to reduce the problem. It was later discovered that the exhaust products from automobiles was the significant contributor to the production of smog. It is seen, then, that one must be careful in placing the blame of air pollution on industry alone.

b. **Combustion of Fuel for Heat and Power.** Steam generating plants using coal or oil as the main source of energy have become serious problems in some localities due to the large quantities of fuel burned per unit of time. Other sources are large buildings using coal or oil for heating purposes. Little or no difficulty is encountered from any of these sources if natural gas is used for fuel. One large source such as a power or heating plant may be obvious, but in some areas a problem may arise due to the many contributions of a large number of small sources such as home furnaces. This is no problem if natural gas is the predominant fuel.

c. **Combustion of Refuse and Other Wastes.** Many installations dispose of their refuse and other wastes by burning them in incinerators. Others burn their trash in open dumps. Both of these activities may produce great quantities of smoke and blowing materials which can give rise to serious problems in the immediate area of activity. Home incinerators also contribute to the overall air pollution problem of any community or region. They are seldom the major source of contaminants in problem areas, but they can cause significant increases in the overall level of contamination.

d. **Storage, Handling, and Dispensing of Fuels.** Most liquid hydrocarbon fuels, such as gasoline, are volatile and are continually being released into the atmosphere. Fuel vapors escape from vented storage tanks, some automobile and truck fuel tanks, and operations where fuels are being transferred from one tank to another. These vapors not only contribute to the overall contaminant levels in the atmosphere but represent economic losses to the people whose fuel is evaporating. Evaporation of fuels is seldom a major source of pollution, but it does add to the overall contaminant level in any atmosphere.

e. **Combustion of Fuel for Transportation.** Automobiles, trucks, buses, motorcycles, trains, boats, and aircraft using hydrocarbon fuels all release exhaust products into the air. Each vehicle in itself does not produce a very large quantity of contaminants; but when they are all considered as a group this source can become quite significant in some localities, especially in large cities and adjacent areas.

f. **Community-Type Activities.** Community activities such as street sweeping, demolition, vegetation burning, insect control, weed control, and construction often contribute to air pollution problems. Usually, these activities only cause difficulties in the immediate area in which they are located.

g. **Natural Sources.** Decaying plant and animal life such as found on beaches of lakes, oceans, river banks, or other areas can give rise to offensive odors. Various plants produce pollens which irritate people affected by allergies and asthma. Dust can be blown up from barren areas. All of these contribute to the overall problem. They may become critical in some instances such as in the dust storms encountered in the Great Plains areas of the United States, but in most cases they are of minor importance.

h. **Miscellaneous.** Other sources may not seem to fit in any one of the above categories or they may seem to fit in a number of them. These can be grouped together under miscellaneous sources. These could include dirt roads which are sources of dust during periods of use, chemical laboratories, animal shelters, sewage and industrial waste disposal, etc. These contribute to the overall pollution load of any given area, but they are seldom major sources of contaminants.

Primary and Secondary Pollutants. Primary pollutants are emitted directly to the atmosphere from specific sources. Secondary pollutants are those produced by the reaction of two or more primary pollutants or material components of the atmosphere through either photochemical or non-photochemical reactions.

a. **Photochemical Reactions.** Solar radiation influences the composition of the atmosphere. A very good example of this is the formation of smog. Emissions from many sources, inorganic and organic in nature, are modified after they are released to the atmosphere. For this reason the list of primary pollutants cannot be relied upon to provide a list of all of the pollutants which may appear in any one area. Chemical reactions involving pollutants and other substances in the atmosphere are affected by the intensity of the existing sunlight, the reactivity of the molecules in question, the concentration of those molecules, the type of light which affects the molecules (wave length), and the residence time of the molecules in the atmosphere.

(1) *Reaciants:*

(a) The four reactants usually considered in the formation of secondary pollutants are atomic oxygen, nitrogen oxides, ozone, and hydrocarbons.

(b) An unbound atom of oxygen is very reactive in photochemical reactions. It is known to be produced by three atmospheric reactions; the photochemical dissociation of atmospheric oxygen, ozone, and nitrogen dioxide.

$$O_2 + hV \longrightarrow O + O$$

$$O_3 + hV \longrightarrow O_2 + O$$

$$NO_2 + hV \longrightarrow NO + O$$

[hV: photochemical energy; h (Planck's Constant) times V (Radiation Frequency)] tomic oxygen will react with all but most inert rganic compounds to form a new product.

(c) The nitrogen oxides considered here onsist of nitric oxide (NO) and nitrogen dioxide NO_2). Nitric oxide is emitted into the atmosphere rimarily as a by-product of high temperature mbustion. Nitric oxide in the air is slowly idized to nitrogen dioxide. The nitrogen dioxide photochemically active and is involved in veral important atmospheric reactions, two of hich are listed below.

$$NO_2 + h\nu \longrightarrow NO + O$$

$$RCO + NO_2 \longrightarrow \underset{\underset{O}{\|}}{RCONO}$$

(free radical)

(d) Ozone is another reactant which be- mes involved in photochemical processes. It is rmed when atmospheric oxygen reacts with olecular oxygen.

$$O_2 + O \rightleftharpoons O_3$$

zone itself corrodes metals, cracks certain rubber mpounds, and reacts with nitric oxide to form itrogen dioxide. It also reacts with hydrocarbons form free radicals.

$$O_3 + NO \rightleftharpoons NO_2 + O_2$$

$$\text{ydrocarbons} + O_3 \longrightarrow \text{free radicals}$$

(e) Hydrocarbons are the last reactants hich will be considered. They are emitted into the atmosphere primarily in the exhaust streams of vehicles using fuels made from petroleum. Hydrocarbons do not react with each other but react in the presence of other compounds such as ozone, nitric oxide, etc., to form noxious pollutants.

(2) *Factors Affecting Reactions:*

(a) A photochemical reaction begins with the act of sunlight absorption. The sunlight must have a certain degree of intensity to initiate the reaction. There is an intensity below which photochemical reactions are negligible. The speed of the reaction is directly proportional to the intensity, i.e., the reaction speeds up as the intensity increases. The wave length of the light supplied is also important. Light which produces electronic activation is the only light which will initiate photochemical reactions. These changes take place only in the presence of radiation in the visible and ultraviolet regions (3000—7000 Angstroms) of the solar spectrum.

(b) As with any chemical process, the speed of the reaction is directly proportional to temperature, i.e., the higher the temperature, the greater the rate of the reaction.

(c) Concentration and time are also relevant factors. When an inversion occurs, the escape of pollutants into the upper atmosphere is impeded. This allows the emitted pollutants to become concentrated and therefore, provides more of an opportunity for photochemical reactions to occur and, it also provides the sun sufficient time to initiate the photochemical reactions. This explains the formation of smog during inversion conditions.

(d) The more important organic compounds entering into photochemical reactions are classified as reactive hydrocarbons. These include olefins, substituted aromatics, and aldehydes.

Example:

Olefin-Ozone Reactions

$$H_2C{=}\overset{H}{C}{-}\overset{H}{C}{=}CH_2 + 2O_3 \xrightarrow{h\nu} CH_2{=}CHCHO + HCHO + 2O_2$$

(1,3 butadiene) + (ozone) = (acrolein) + (formaldehyde) + (oxygen)

lefins are the most important beginning class f organic compounds which are capable of pro- ucing irritants and phytotoxicants.

b. Non-Photochemical Reactions. Aside from rimary photochemical reactions and associated secondary reactions, there are additional atmospheric reactions which contribute to the formation of secondary pollutants. It is rather difficult to clearly categorize atmospheric reactions into photochemical, thermal or other organic and inorganic reactions because once the pollutant is in the atmosphere, it may be involved in numerous reactions. For the sake of convenience, then, the division of "Photochemical" and "Non-Photochemical Reactions" was chosen. These other reactions, include absorption, catalysis, polymerization, condensation, oxidation and reduction. A few examples are:

(1) Oxidation of SO_2 to SO_3 under catalytic influence.

$$2SO_2 + O_2 \rightarrow 2SO_3$$

(2) Hydrolysis of nitrogen oxides to nitrous and nitric acid.

$$2NO_2 + H_2O \rightarrow HNO_2 + HNO_3$$

(3) Formation of sulfuric acid aerosol through the oxidation of SO_2 to SO_3 and hydrolysis to aqueous sulfuric acid.

$$SO_3 + H_2O \rightarrow H_2SO_4$$

(4) Formation of NH_4Cl by the reaction of NH_3 and HCl

$$NH_3 + HCl \rightarrow NH_4Cl$$

(5) Particle growth by adsorption of polar molecules such as water followed by coagulation.

CHAPTER 6

MEASURING EMISSIONS FROM AIR POLLUTION SOURCES

Stack Sampling:

a. Introduction

(1) The purpose of stack sampling is to determine the quantity and types of pollutants that are contained in the gases that are emitted from a source. The data from stack sampling surveys can be used to determine compliance or non-compliance with emission standards, to design air pollution control equipment, to determine emission factors for use in emission inventories, to determine efficiencies of air pollution control equipment, and to determine efficiencies of manufacturing processes.

(2) It can be seen from the uses of stack sampling data that the principles outlined in this section are not restricted to vertical stack sampling per se, but are valid when sampling ductwork prior to the actual stack. This includes sampling in horizontal ducts.

(3) Many problems arise in stack sampling as a result of the characteristics of the flow that can exist in stacks. For example, a rate of stack effluent of 300,000 cubic feet per minute is not an unusual flowrate to encounter. It is impractical if not impossible to collect the entire effluent, even with a short sampling time.

(4) From a practical standpoint only a very small portion of the total stack effluent can be sampled. For accurate results the samples removed from the stack must represent the actual conditions in the stack. While it is theoretically possible to have one sample represent the entire stack effluent, flow variations necessitate many samples with each sample representing a known portion of the effluent.

(5) Characterization of stack flow to establish flow patterns existing in the stack is required before actual sampling is undertaken.

b. Sampling Principles

(1) A factor that must be considered in stack sampling is this relationship between a sample collected and the existing effluent in the stack. Although theoretically possible, it is highly improbable because of flow variations, that a single sample can ever be representative of the total stack effluent. Figure **6-1** illustrates a profile plot of one of many possible plots of stack conditions. Because of this variation it is wise to take several samples at different locations in the cross-section of the stack so that any sample error is reduced to a minimum.

(2) The amount of particulate pollutant collected can be altered if the velocity of the sample as it enters the sampling nozzle is different from the gas velocity in the stack. Figure **6-2** illustrates the effect upon the streamlines of a parent gas stream of varying the velocity with which a sample is drawn into a nozzle.

(a) When the gas is drawn into the nozzle with a velocity less than the approach-stream velocity figure **6-2A**, a number of streamlines diverge so that they do not enter the nozzle. The fine particulates are transported along the deflected streamlines, but the coarse ones, because of their greater inertia, leave the streamlines during the curvature and enter the nozzle. Hence, the volume of gas extracted (i.e., the gas sample) will contain more coarse particles than the same volume contains in the approach stream. As a result, the particulate pollutant concentration in the gas sample will be higher than that of the parent stream. Moreover, the particulate matter collected from the gas sample would present a false picture regarding particle-size distribution.

(b) When gas is drawn into the nozzle with a velocity greater than the approach stream velocity, figure **6-2B**, a number of streamlines converge and crowd into the nozzle. The fine particulates follow the deflected streamlines, but the coarse ones continue along their original straight paths and miss the nozzle completely. Therefore, the gas sample will contain fewer coarse particles than actually exist in the same volume in the approach stream. As a result, the particulate pollutant concentration in the gas sample will be lower than that in the parent stream, and the results of particle size analysis will be false.

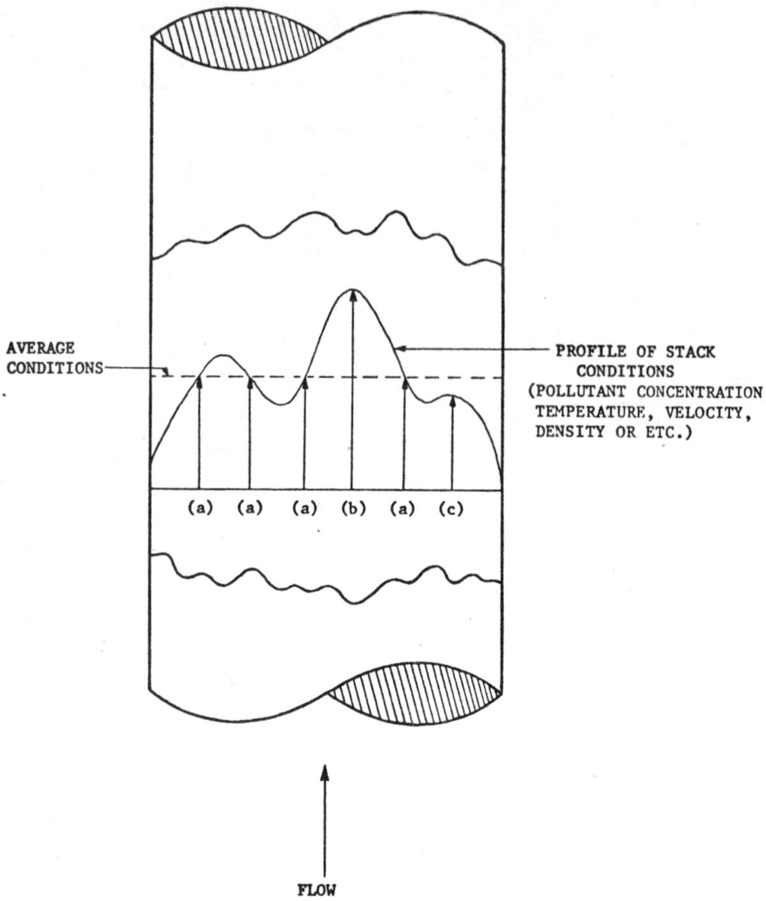

Fig. 6-1 Profile of Stack Conditions

(c) When gas is drawn into the nozzle with a velocity equal to the approach stream velocity, (figure **6-2C**), the streamlines will not be disturbed. As a result, the gas sample will contain the same number of coarse and fine particulates that actually exist in the same volume in the approach stream. Therefore, the gas sample will have the same particulate pollutant concentration as the parent stream and the results of particle size analysis will be true. This condition, when the velocity of the gas across the nozzle tip is equal to that of the approach stream, is essential for the accurate determination of the concentration of particulate matter and of particle size distribution in the parent stream. Such a procedure is called isokinetic sampling. Normally, gaseous effluents contain particles of a wide size range. Therefore, neither concentration nor particle size distribution of the sample withdrawn is likely to represent those actually existing at the sampling point unless true isokinetic sampling is achieved.

(3) *Sampling Without Altering Flow Patterns.* The mere presence of the nozzle and probe in the gas stream, together with disturbances that arise when gas is drawn into the sampling nozzle, tend to distort the natural flow pattern of the parent gas stream. Because the flow patterns are distorted, a situation similar to non-isokinetic sampling is set up, which results in a sample whose particle size distribution is distorted. It is not possible to eliminate this distortion completely. To minimize this distortion it is important to reduce aerodynamic disturbances by using a streamlined nozzle and probe, by keeping the area of the nozzle as small as practically possible, and by sampling so that the velocity in the nozzle is equal to the velocity in the approach stream (isokinetic sampling).

(4) *Sampling Without Physically or Chemically Changing the Sample.* When the parent gas or pollutants are of such nature or composition that they are still reacting physically or chemically

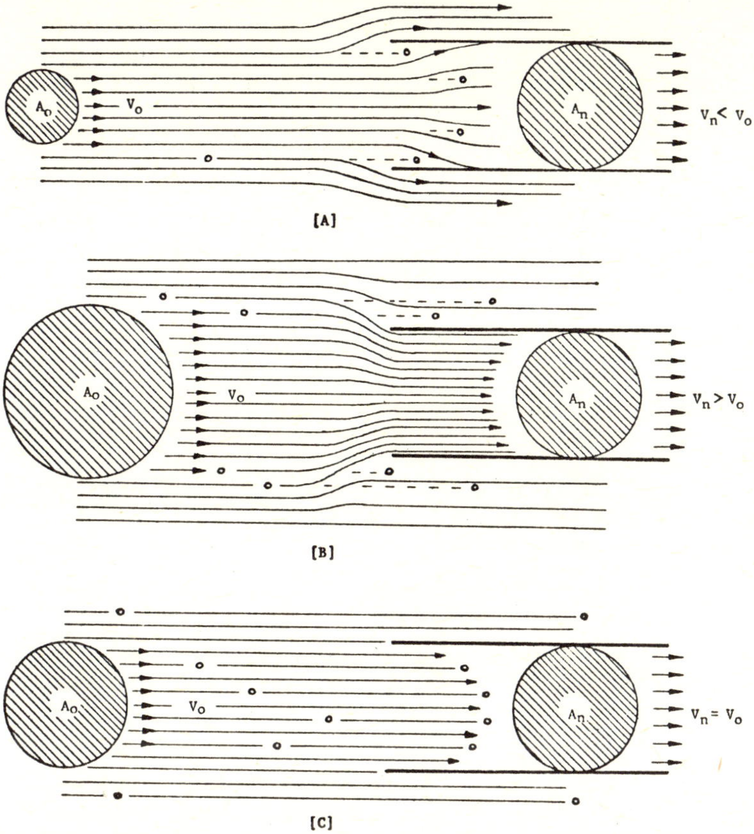

Fig. 6-2 Influence of Sampling Rate on Flow Patterns

with one another at the sampling point, the reactions may continue inside the probe, tubing or pollutant collector. If such occurs, the composition or nature of the pollutants collected by the sampling train or the composition of the sample gas will not be that which exists at the sampling point. Also, if condensation of sample gas is permitted in the probe tubing or pollutant collector, conditions not representative of the sampling point are set up in the sampling train. Still another factor to consider is the possibility of the material of which the nozzle, tubing, probe and pollutant collector are constructed acting as a catalyst. Ideally, the arresting of change in parent gas and pollutants should be instantaneous upon extraction from the parent.

c. **Selection of Sampling Site:**

(1) *Uniform Cross Sectional Patterns*:

(a) Usually a sampling site with uniformity in pressure, temperature, velocity, and pollutant concentration will enable sampling to be done with the least labor and time and with the greatest accuracy. Therefore, in general, an attempt should be made to find a cross section of duct most likely to provide uniform cross sectional conditions.

(b) Sites very near exits of equipment tend to have the greatest variations in pressure, temperature, velocity, and pollutant concentration over the cross section of the duct. Thus, the sampling site should be downstream from such equipment where these patterns are expected to be uniform. Non-uniform patterns tend to exist near duct configurations, such as bends, enlargements, contractions, and protrusions within certain distances upstream and down-stream from their locations. Therefore, sampling sites should not be chosen at such duct configurations nor within the zone of their influence. In most cases in the field, a straight section of pipe about 10 diameters in length, free flow disturbing obstructions, provides a satisfactory location for flow measurement. Vortex flow may occur when there are two elbows in series but in different planes, in the exit from some cyclones and fans, and in many other places. The measurement of flow when vortices exist, although not necessarily insurmountable, is a

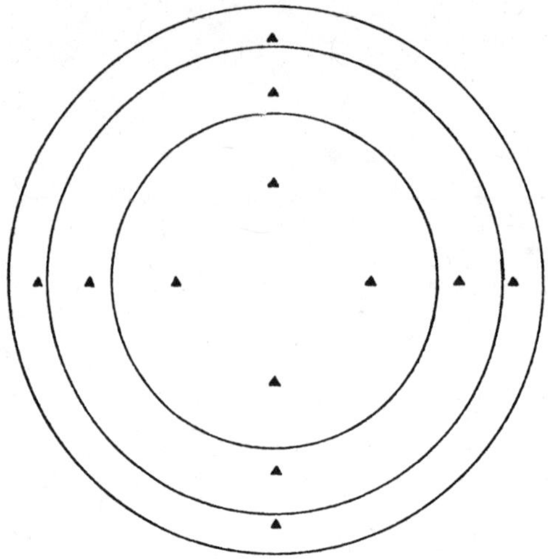

Fig. 6-3 Location of Traverse Points When Dividing Circular Stack into Three Concentric Equal Areas.

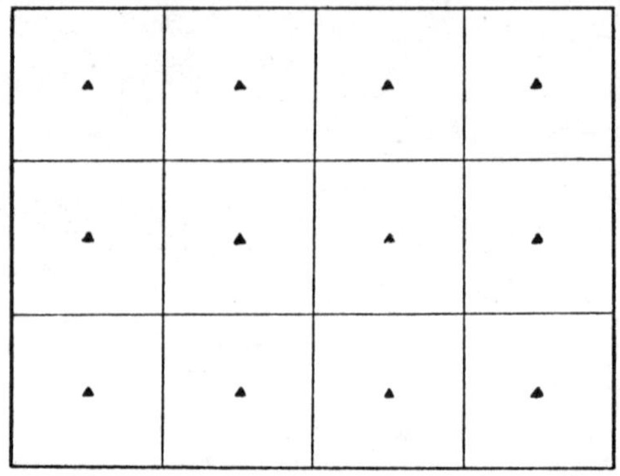

Fig. 6-4 Location of Traverse Points When Divid Rectangular Stack into 12 Equal Areas.

difficult task. Non-uniform pollutant concentration patterns may exist near the outlet of a fan. A fanwheel tends to concentrate particulates at certain points within the casing so that stratification is produced at the outlet. Non-uniform patterns also tend to exist in horizontal or inclined sections of duct when gas velocity is not sufficient to keep the particulates in uniform suspension. If possible, the sampling site should be located in a long, straight, circular, vertical, insulated section of duct. Advantages include:

1. The sampling probes may be oriented in a horizontal position at all times and will be easier to handle.

2. A uniform velocity pattern is more likely to exist in a circular cross section than in any other type geometric configuration.

3. Severe stratification in particulate pollutant concentration and particle size is least likely to occur because of the vertical orientation and existence of a uniform velocity pattern.

4. Temperature will not be readily influenced by changes in ambient air because of the insulation.

(2) *Physical Considerations.* A site should be selected where it will be convenient to set up the sampling train and provide sufficient room for freedom of movement.

(a) *Size of Sampling Port.* The sampling port must be large enough to accommodate the right angle bend of the sampling probes without damage. The size may be dictated by the thickness of the wall of the duct. For example, a "thin" wall may require a port of 2-inches diameter, but a "thick" wall may require a port of 5-inches diameter to accommodate the probes. After a site has been selected as probably the best for sampling based upon information provided by a thorough visual inspection, it is necessary to check the conditions existing within the cross section to determine if it really is a satisfactory one. Preliminary traverses will provide the data necessary to determine whether to accept or reject the site.

d. **Stack Velocity and Flow Determinations**

(1) Velocity measurements within a stack are normally made using a pitot tube and manometer. The pitot tube measures velocity head only at the point in the gas stream where it is placed; therefore, readings must be made at a number of points in the stack cross section so that the average gas velocity may be calculated. The number of points to be used and their location can be readily determined.

(2) The cross-sectional area of the stack at the test station is divided into a number of concentric equal area zones for circular stacks, figure **6-3,** or rectangular equal areas for rectangular stacks, figure **6-4.** The number of areas used depends on the size of the stack. For circular stacks, with fairly uniform flow, the usual practice is to employ the number of areas shown in table **6-1.**

(3) The traverse points for velocity head measurements are located in each area as suggested in figure **6-3** and **6-4.** With rectangular ducts or stacks, these points are at the center of each rectangular area, figure **6-4.** For circular

Table 6-1

Per Cent of Circular Stack Diameter From Inside Wall to Traverse Point

Point number	Number of areas selected				
	2	*3*	*4*	*5*	*6*
1	6.7	4.4	3.3	2.5	2.1
2	25.0	14.7	10.5	8.2	6.7
3	75.0	29.5	19.4	14.6	11.8
4	93.3	70.5	32.3	22.6	17.7
5		85.3	67.7	34.2	25.0
6		95.6	80.6	65.8	35.5
7			89.5	77.4	64.5
8			96.7	85.4	75.0
9				91.8	82.3
10				97.5	88.2
11					93.3
12					97.9

stacks, the location of each point along a stack diameter is shown in table **6-1**.

c. **Particulate Sampling:**

(1) *Equipment & Procedure:*

(a) The procedure for particulate source sampling is essentially that described in the ASME Power Test Code. The specific equipment and procedures employed in particulate sampling as discussed below were adapted from the US Public Health Service, National Center for Air Pollution Control. The particulate sampling equipment is shown schematically in figure **6-5**. The equipment consists of a probe assembly, a sample train unit, an umbilical cord, a pump and metering unit, and a rail support system.

(b) The probe assembly consists of an interchangeable button-hood type probe tip and a heated tyrex glass probe. The glass probe is housed in a stainless steel jacket. Integral with the probe assembly is a type-S pitot tube and dial thermometer. With this assembly, stack velocity pressure temperature measurements are made concurrently with sample collection. The probe is heated to prevent condensation and minimize particle deposition on the probe wall.

(c) The sample train assembly consists of a glass cyclone and a filter holder located in an electrically heated box. The box is maintained at a temperature greater than the dew point of the gas, again to prevent condensation and to minimize particle deposition on the walls of the connecting glassware between the probe, cyclone, and filter. Glass fiber filters are used to collect particles passing through the cyclone. Downstream from the filter holder is an impinger train consisting of four large all-glass impingers. The first two are liquid filled, the third impinger is dry to trap entrained moisture, and the fourth impinger contains silica-gel. The four impingers are located in an ice bath. In this way moisture content determinations can be made from the measurement of condensed water in the impinger train.

(d) The pump and metering assembly consists of a rotary vane-type pump, a dry gas meter, and an orifice. Rate control is achieved by by-passing the suction and discharge of the pump through a throttling valve. The orifice pressure drop is measured along with the pitot tube velocity pressure on a dual inclined-vertical manometer.

(e) A 50-foot "umbilical cord" connects the sampling train unit with the pump and meter unit. In this way only the probe assembly and sampling train must be located at the sampling port and the heavier pump-meter unit can be stationed and operated from a lower level, in many cases on ground level.

(f) The above description represents one of several different variations of sampling train assemblies that could be used to accomplish the same purpose.

(2) *Particulate Sampling Calculations:*

(a) *Introduction.* Stack emissions may be expressed in a number of ways. From the weight of material collected and the volume of gas sampled, the average concentration may be determined. Since stack flow can be determined from the velocity traverse data, the emission in weight per unit time can be determined. When the fuel combustion rate is known, the emission in *weight per unit weight of fuel* may be determined. For heating plants, emission standards in Title

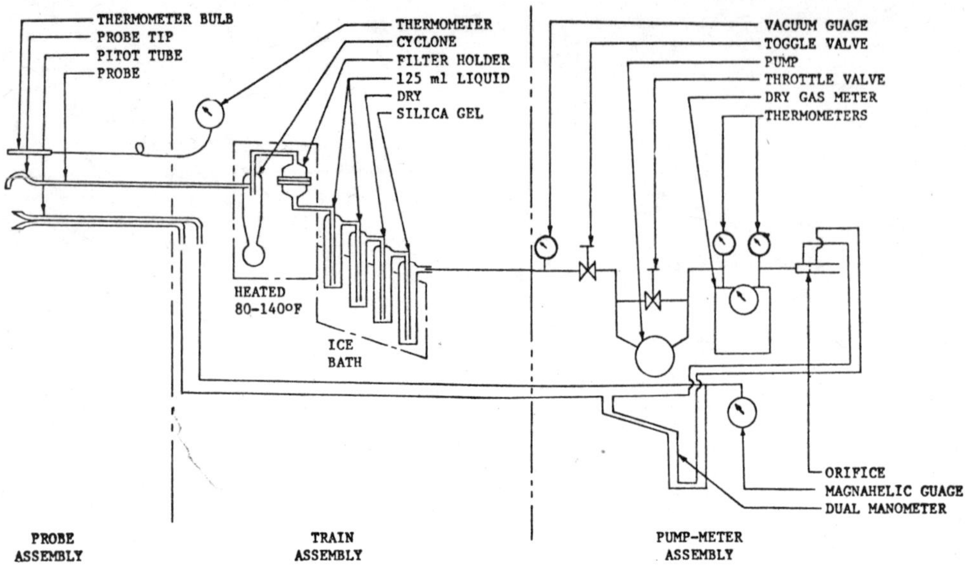

Fig. 6-5 Equipment for Particulate Source Sampling

42, Code of Federal Regulations, Part 76 (42CFR76) are expressed in terms of *pounds per million Btu heat input.* Using the emission in weight per unit weight of fuel and knowing the heat content of the fuel, the emission in pounds per million Btu heat input may be calculated. For incinerators, the standard in 42CFR76 is given in terms of grains per standard cubic foot, corrected to 12 percent CO_2. Hence the CO_2 concentration of the incinerator stack gas must be determined.

A major calculation problem in isokinetic particulate sampling is relating gas flow at stack conditions with the equivalent gas flow at metering conditions. Stack gases may be of the order of 400–700° F with around 5 percent (by volume) moisture, while the sample gas that is pumped and metered may be 70–100° F and very dry. Therefore, a certain volume, as measured by the dry gas meter, may represent nearly twice that volume at stack conditions.

(b) *Determining Stack Flow Rate:*

Determination of stack flow rate requires first determining gas velocities at sampling points across the velocity traverse and then obtaining the average velocity. Data required are stack temperature and pressure, stack gas composition or average molecular weight, pitot tube velocity pressures, and the pitot tube coefficient. Velocities may be calculated from:

$$V_s' = 5130 C_p \left[\frac{T_s \Delta hp}{P_s M_s}\right]^{1/2} \qquad \text{Eq 6.1}$$

Where V_s' = Stack gas velocity – ft/min
C_p = Pitot Coefficient
T_s = Stack Temperature – °R
Δhp = Velocity Pressure – in. H_2O
P_s = Stack absolute pressure – in. Hg.
M_s = Stack gas average molecular weight

Since, using the equal area traverse of paragraph 5–2d, the average stack gas velocity becomes the arithmetic average of the individual velocities. The stack gas flow is then given by:

$Q_s' = A_s V_s'$

Where Q_s' = Stack gas flow – cfm
A_s = Cross section area of stack – ft²
V_s' = Stack gas velocity – ft/min.

(c) Relating volume sampled at metering conditions to volume sampled at stack conditions. For a sampling train as shown in figure **6-5**, the volume of gas sampled is determined from the dry gas meter readings. If silica gel is used in the sampling train, as described above, the metered gas can be considered to be dry. However, the metered volume at temperature T_m may be significantly different from that at the stack temperature T_s. By material balance and perfect gas laws then:

$$\text{Vol (stack conditions)} = \frac{\text{Vol (metered)} \times \frac{T_s}{T_m} \times \frac{P_m}{P_s}}{F_d} \qquad \text{Eq 6.2}$$

F_d in this equation is the fraction of dry gas in the sample. This fraction may be determined from the amount of moisture condensed from the sample as follows:

$$F_d = \frac{\text{Vol (metered)}}{\text{Vol (metered)} + 0.00268 w \frac{T_m}{P_m}} \qquad \text{Eq 6.3}$$

Where: w = Weight of water condensed from sample—gm.
$T_m + P_m$ = Temperature (°R) and pressure (in. Hg) of metering conditions

(d) *Example Problem.* To illustrate the calculations outlined above, let us assume a stack 36 inches in diameter is sampled isokinetically along a 6-point traverse. The following data is obtained:

Barometric pressure: 29.36 in Hg

Pitot coefficient, C_p = 0.88 (this value may or may not be stated on equipment, if not, type-S pitot coefficients normally range from 0.83 to 0.93; standard pitot coefficients are normally considered to be 1.0).

Sampling Point	*Time Interval (Min)*	*Volume (Dry Meter) (cu. ft)*	T_s(°F)	Δhp (in H_2O)	T_m(°F)
1	7	4.28	380	0.21	85
2	7	5.16	390	0.29	86
3	7	5.44	385	0.33	88
4	7	5.35	385	0.31	90
5	7	5.06	390	0.28	93
6	7	4.10	380	0.19	95
Total (or avg)	42	29.39	Avg = 385		Avg = 90°F

Total weight of particulate collected: 0.862 gm

Total weight of water condensed: 27 gm

Calculate particulate concentration per cu. ft. of stack gas:

Volume of sampled gas at stack conditions = total dry gas meter volume = 29.39 cu. ft. @ 550°R, 29.36″ Hg

$$F_d(\text{Eq 5-3}) = \frac{\text{Vol}}{\text{Vol} + \dfrac{0.00268 w T_m}{P_m}}$$

$$= \frac{29.39}{29.39 + \dfrac{0.00268(27)(550)}{29.36}} = 0.955$$

$$\text{Vol (stack conditions) (Eq 5-2)} = \frac{\text{Vol meter} \times \dfrac{T_s}{T_m} \times \dfrac{P_m}{P_s}}{F_d}$$

$$= \frac{29.39 \times \dfrac{845}{550} \times \dfrac{29.36}{29.36}}{.955} = 47.2 \text{ cu. ft.}$$

$$\text{Concentration} = \frac{0.862 \text{ gm}}{453 \text{ gm/lb} \times 47.2} = 0.403 \times 10^{-4} \text{ lb/cu ft}$$

Determine Stack Flow Rate:

Utilizing Equation 6–1, the velocity for each sampling point is determined as shown below for sample point No. 1

Note: For M_s the value of 29 is assumed—for most combustion processes, gas molecular weight does not differ greatly from that for air.

$$\text{Point No. 1} \quad V_s' = 5130 C_p \left[\frac{T_s \Delta hp}{P_s M_s}\right]^{1/2}$$

$$= 5130 \times 0.88 \left[\frac{840 \times 0.21}{29.36 \times 29}\right]^{1/2}$$

$$V_s' = 2050 \text{ ft/min}$$

Velocities for other sampling points are determined in the same manner. The average velocity for these six points then is 2320 ft/min.

$$\text{Stack Flow} = Q_s' = A_s V_s' = \left(\frac{36}{12}\right)^2 \times \frac{\pi}{4} \times 2320$$

Stack Flow = 16, 400 cfm

$$\text{Emission} = 0.403 \times 10^{-4} \text{ lb/cu ft} \times 1.64 \times 10^4 \text{ cu ft/min} \times 60 \frac{\text{min}}{\text{hr}} = 39.7 \text{ lb particulate/hr}$$

Case I. If this stack discharges gases from a coal fired heating plant, stack emission would be calculated in lb/10^6 Btu. The following data is obtained:

coal firing rate: 3500 lb/hr

coal heating value: 13,500 Btu/lb (as Fired)

Heat input = 3500 lb/hr × 13,500 Btu/lb
= 47.3 × 10^6 Btu/hr

$$\text{Emission} = \frac{39.7 \text{ lb particulate/hr}}{47.3 \times 10^6 \text{ Btu/hr}}$$

$$= 0.84 \text{ lb}/10^6 \text{ Btu}$$

Case II. If this stack discharges gases from an incinerator, stack emission would be calculated as a concentration in grains per dry standard cubic foot, corrected to 12% CO_2. The following data is obtained:

CO_2 measured in stack gas = 3%

Then:

$$\text{Concentration} = \frac{\text{weight of particulate in grains}}{\text{Dry sample volume (cu ft) corrected to std conditions} + 12\%\ CO_2}$$

Note: For standard conditions, use 70°F, 29.29 in Hg.

$$\text{Concentration} = \frac{0.862 \times 15.43}{29.39 \times \dfrac{3\%}{12\%} \times \dfrac{530}{550} \times \dfrac{29.36}{29.92}}$$

$$= 1.92 \text{ gr/SCF, corrected to } 12\%\ CO_2$$

f. Gas Sampling:

(1) Combustion Effluents—Orsat Analysis

(a) An effluent containing only gases resulting from combustion of hydrocarbon fuels has CO, CO_2, O_2, N_2 and H_2O as its main gaseous constituents. *Although gases considered as air pollutants may be present also, it may be confidently assumed that their contribution to the mass of total effluent is negligible.*

(b) The Orsat analyzer may be used for the measurement of CO, CO_2, and O when they are significant constituents of a gaseous effluent. A sketch of an Orsat Apparatus is shown in figure **6-6.** A graduated measuring burette (A) is used to measure differences in sample volume. Pipettes contain chemical solutions which absorb CO_2, O_2, and CO. Pipette (B) contains a solution of KOH and absorbs CO_2; pipette (C) contains an alkaline solution of pyrogallic acid and removes the O_2; and pipette (D) contains metallic copper an ammoniacal solution of cuprous chloride for the removal of CO_2, O_2, and CO since each of the succeeding solutions will remove any portion of the CO and O_2 which is allowed to remain in the sample. Each pipette is filled with metallic wool which distributes the gas sample and provides a large chemically wet sutface for absorption to take place.

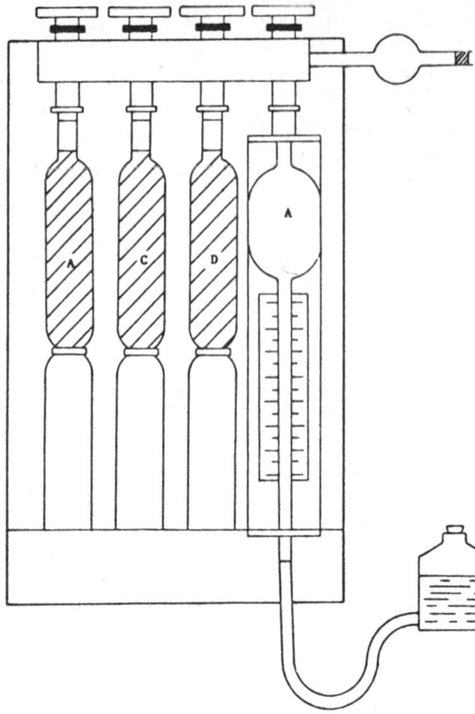

Fig. 6-6 Schematic sketch of an orsat analyser

(c) The gas sample extracted from the duct may not be saturated with water vapor. However, within the Orsat apparatus, the gas sample is necessarily saturated because of the method of analysis employed. Therefore, each measurement of volume by the burette is made, not only at the same pressure and temperature, but also of a gas mixture that is always saturated with water vapor. As a result, the proportion (by volume) of water vapor in the gas sample measured by the burette is constant.

(d) The gas sample is led consecutively through pipettes (B), (C), and (D), so that CO_2, O_2, and CO are removed in that order. For each pass, the decrease in volume of gas sample may be calculated from measurements provided by the burette. The decrease in gas volume for a given pass, divided by the volume of gas sample extracted from the duct (as measured by the burette) is the proportion (by volume) of the gas sample occupied by the gas component referred to a dry basis (gas mixture less water vapor).

(e) Combustion gas analyses may also be performed using other absorption equipment or gas chromatographic procedures.

(2) *Other Effluents—Grab Sampling and Chemical Analysis Techniques*

(a) By definition a grab sample is a gas sample extracted from a gas stream over a period of 10–60 seconds. The four basic techniques for collecting a grab sample are:

1. Use of evacuated containers.
2. Use of displacement techniques.
3. Use of inflatable containers.
4. Use of purging techniques.

(b) *Evacuation Techniques.* The collection vessel used in the evacuated container technique for collecting a grab sample is commonly a round bottomed flask constructed of plastic, metal, or glass. A typical evacuated flask sampling train is illustrated in figure **6-7.** To collect a sample, the flask is evacuated, connected to the sampling train, then the probe is conditioned to prevent condensation, the flask is then opened thus drawing the sample out of the stack into the flask.

(c) *Displacement Techniques.* A type of displacement bottle is shown in figure **6-8.** The gas sample is collected in bottle A; bottle B is used to collect the displaced liquid from bottle A. The liquid must be saturated by the gases before the sample is taken. To collect a sample, bottle B is lowered thus causing the liquid in bottle A to siphon into bottle B. As the liquid is withdrawn from bottle A, it is replaced by the gas sample which is pulled through the probe assembly.

(d) *Inflation Techniques.* Flexible polyethylene bags are usually used as the collecting container in inflation techniques. A typical flex-

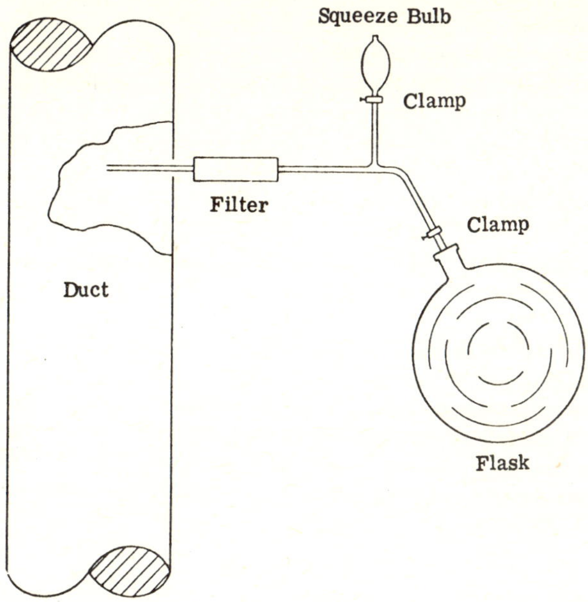

Fig. 6-7 Evacuated flask sampling train

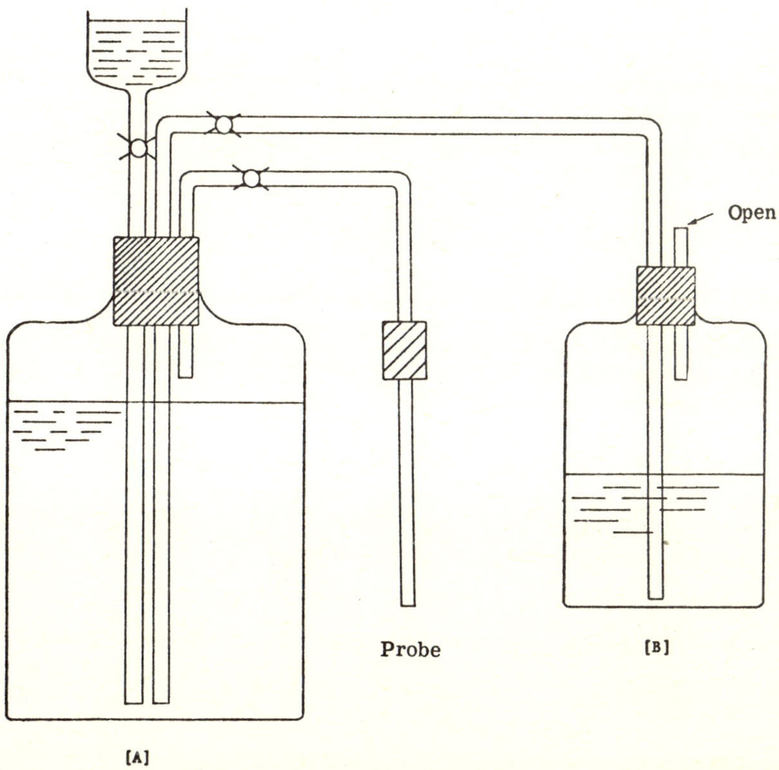

Fig. 6-8 Displacement Bottle

ible container sampling train is illustrated in figure 6-9. A sample is collected by evacuating the air from the grab sample box, thus causing the polyethylene bag to inflate drawing the sample out of the duct. By reversing the flow of air from the pump which forces air into the box, the sample is expelled from the bag for analysis.

(e) *Purging Techniques.* A purging sampling train is shown in figure 6-10. Gas is pulled through the sampling train until enough gas has been pulled through to completely flush out the old gas that was present in the collector. The valves at each end of the collector are then closed, thus confining the sample in the sample collector.

(f) *Chemical Analysis.* Analysis of grab samples can be performed in one of two ways. Appropriate absorbing solutions can be added to the rigid containers prior to collecting the sample. After the sample is collected, the container is agitated and allowed to stand until complete absorption takes place. The absorbing solution can than be analyzed or pressure measurements can be made to determine the volume of gas removed by the absorbing liquid. The samples collected in the flexible containers can be expelled and run through another collecting device such as an absorber or an adsorber with subsequent analysis being made on the collected material. Specific sampling procedures should be coordinated with the laboratory providing analytical service.

g. Continuous Sampling:

(1) *Introduction.* A "continuous sample" is defined as a sample taken from the effluent stream over a relatively long period of time. The objective

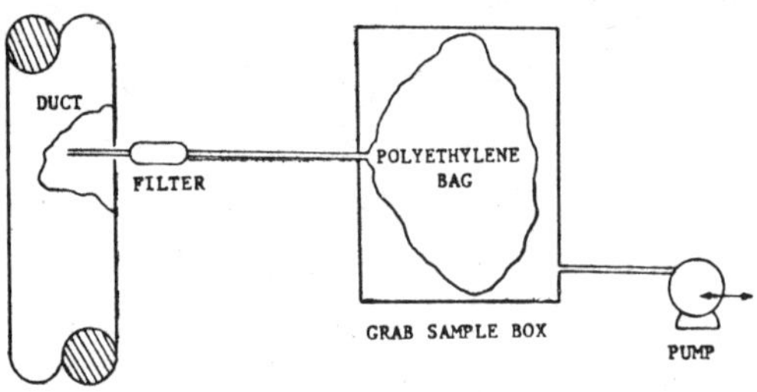

Fig. 6-9 Inflation sampling train

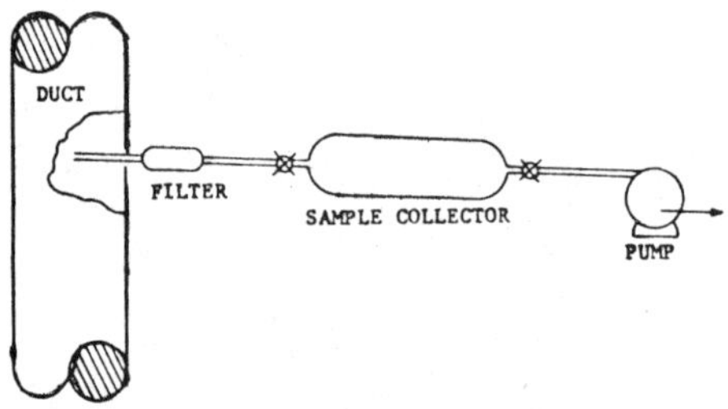

Fig. 6-10 Purging sampling train

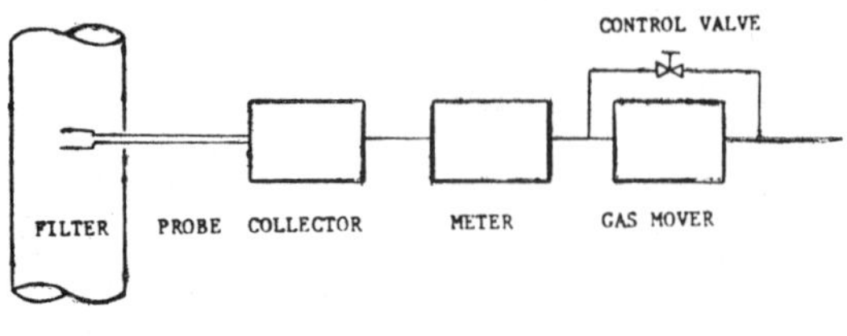

Fig. 6-11 Gas sampling train

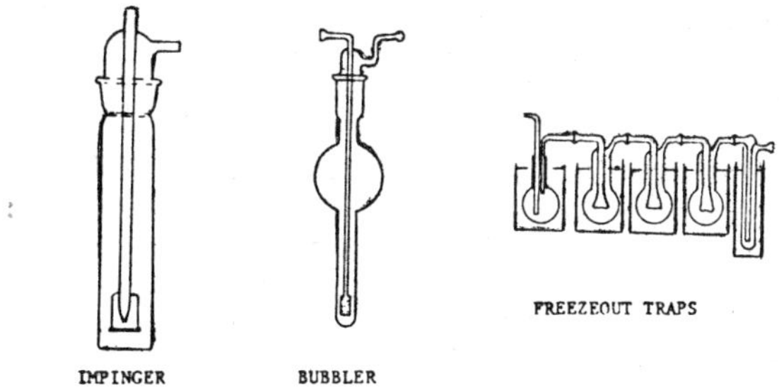

Fig. 6-12 Common Pollutant Collectors

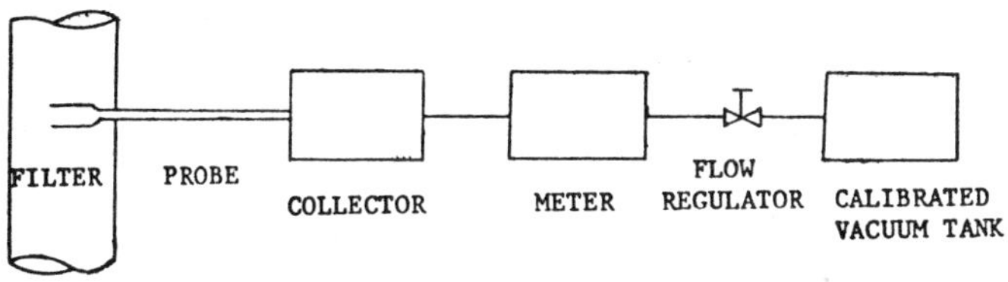

Fig. 6-13 Variation of Sampling Train

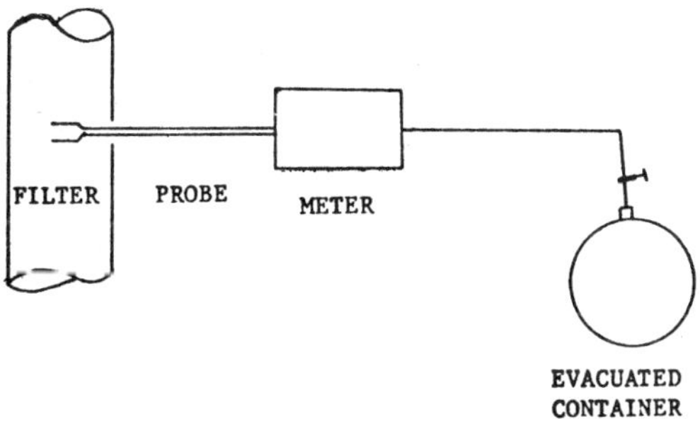

Fig. 6-14 Grab Sampling Train

of taking the sample is to obtain the average gaseous composition during the time interval for which the sample was taken, Continuous samples are collected by the use of a sampling train at a regulated flow rate. Grab sampling techniques can be modified to take a continuous sample by the utilization of devices to indicate and regulate the sampling flow rate.

(2) *Discussion of Continuous Sampling.* Continuous sampling is useful when large sample volumes are required or when an average concentration over a period of time is desired. It is applicable when stack conditions are uniform or vary relatively slowly. If the stack flow fluctuates rapidly and widely without any consistency grab-sampling should be considered or the continuous sampling time should be long enough to allow the fluctuations to average out.

(3) *Conventional Sampling Trains:*

(a) Most gaseous sampling trains are basically composed of a (1) filter, (2) probe, (3) gaseous pollutant collector, (4) meter, (5) flow regulator and (6) gas mover (see figure **6-11).** The filter serves to remove any particulate matter that may interfere in the anlaysis. However, if the possibility that the filter will remove any of the gaseous pollutants of concern exists, the filter should not be used. The probe transports the stack gas to the collector and the meter measures the gas volume sampled and/or the flow rate (separate meters, one to indicate flow and the other to indicate volume, can be used). The flow regulator maintains the flow at the desired rate. The gas mover provides the driving force for the sampling train. Common pollutant collectors are absorbers (such as impingers and fritted glass scrubbers, adsorbers, and freezeout traps as shown in figure **6-12.**

(b) A variation of a sampling train that is ideal for situations where the total gas sample volume required is 1 to 2 cubic feet is shown in figure **6-13.** The flow meter is used only to indicate the flow rate. The calibrated vacuum tank measures the volume of gas sampled and provides the driving force for the sampling train. Still another variation in sampling trains is the grab-sampling technique using an evacuated container which is suitable for small gas volume samples as shown in figure **6-14.** Again the flow meter is used only to indicate flow rate. The evacuated flask measures the sample volume, provides the driving force, and serves as the pollutant collector.

(c) The final selection of the components and type of sampling train depends on the pollutant to be collected and the source conditions. Again, it should be stressed that the specific sampling procedure should be coordinated with the laboratory providing analytical service.

CHAPTER 7

ESTIMATING EMISSION FROM SOURCES OF AIR POLLUTION

Purpose of an Emission Inventory

a. Knowledge of the types and rates of emissions is fundamental to any study of community or installation air pollution. An emission inventory specifies the daily mass rate at which each primary pollutant is emitted to the overall area atmosphere from each class of sources. An example of an emission inventory is shown in table **7-1.**

b. Information provided by an emission inventory is useful:

(1) For estimation of atmospheric concentration under specific meteorological conditions.

(2) For interpretation of the results of atmospheric sampling in terms of both air composition and fluctuations in such composition.

(3) For the establishment of a baseline against which to relate future trends in both pollutant emissions and atmospheric concentrations.

(4) To establish the relative contribution of each major source.

(5) For determination of the type of air-sampling and air-analysis program to be utilized in an area-wide survey.

(6) As one of the basic tools for the establishment of emission standards.

Factors Involved in an Emission Inventory. An emission inventory involves:

a. A classification of the pollutants emitted from the activities of the community.

b. A classification of the sources of emission of pollutants to the community atmosphere.

c. A determination of the average mass of pollutants emitted from each source per unit quantity of material handled, processed, or burned (i.e., a determination of "emission factors").

d. A determination of the quality and daily quantity of materials handled, processed, or burned by each activity identified as a source of emission of pollutants.

e. A computation of the rate (weight units per day) at which each pollutant is emitted to the community atmosphere.

The Use of Emission Factors:

a. **Emission Factor Explained.** The emission factor is a statistical average of the rate at which pollutants are emitted from the burning or processing of a given quantity of material or on the basis of some other meaningful parameter.

b. **Determination of Emission Factors.** Emission factors are determined:

(1) By sampling the effluents at the sources of pollution.

(2) By analyzing data provided by supervisory personnel at the sources of pollution.

(3) By analyzing data reported in technical publications.

c. **Limitations of Use of Emission Factors.** Emission factors are affected by the techniques employed in the processing, handling, or burning operations; by the quality of material used; and by the efficiency of air pollution control. Since the combination of these factors tends to be unique to a locality, emission factors for one community may not be satisfactory for another community. Also, in a dynamic community, the accuracy of emission factors is subject to continuously changing conditions.

Determination of Quality and Quantity of Materials Used. Estimates of the quality and quantity of materials handled, processed, or burned on an installation may be made by analysis of data acquired from direct queries of supervisory personnel and review of records.

Computation and Tabulation of Pollutant Emission:

a. **Computation:**

(1) By application of emission factors and data regarding quantities of materials handled, processed, or burned, pollutant emissions for each source may be computed.

(2) The emissions are commonly expressed as tons or pounds per day.

b. **Tabulation.** The emissions are tabulated according to source and pollutant classification. An example is shown in table **7-11.**

Table 7-1

SUMMARY OF AIR CONTAMINANT EMISSION IN LOS ANGELES COUNTY IN TONS PER DAY

Source	*Organic gases*			*Aerosols*	*Inorganic gases*			
	Hydro-carbons	*Aldehydes & ketones*	*Other organics*		*Oxides of nitrogen*	*Oxides of sulfur*	*Carbon monoxide*	*Other inorganics*
Transportation—Automobile—Exhaust	680	10	12	25	250	19	6850	n
—Blowby	140	u	u	u	u	n	n	n
—Evaporation (Gasoline)	145	0	0	0	0	0	0	0
Trucks and Buses—Exhaust	130	3	4	8	80	6	2100	n
—Blowby	40	u	u	u	u	n	n	n
—Evaporation	45	0	0	0	0	0	0	0
Trucks and Buses (Diesel)	8	n	n	2	8	2	2	n
Ships and Railroads	n	n	n	n	n	n	n	n
Aircraft—Jet	n	r	n	1	1	0	1	0
—Piston	15	n	n	n	5	n	110	n
Petroleum—Refining	90	2	n	5	1	40	700	4
—Marketing	100	0	0	0	0	0	0	0
—Production	60	n	0	0	u	n	u	0
Organic Solvent Uses—								
—Surface Coating	190	25	35	7	0	0	0	0
—Degreasing	20	n	40	0	0	0	0	0
—Dry Cleaning	20	n	8	n	0	0	0	0
—Plastic and Rubber	17	4	3	n	0	0	0	0
—Other	50	14	13	n	0	0	0	0
Combustion of Fuel—								
—Liquid —Type 1	0	0	0	n	3	2	0	0
—Gaseous—Type 1	5	1	3	11	184	8	1	0
—Liquid —Type 2	2	1	6	38	187	454	n	0
—Gaseous—Type 2	4	1	2	14	150	13	1	0
Chemical —Sulfur Plants	0	0	0	0	n	30	0	0
—Sulfuric Acid Plants	0	0	0	n	n	20	0	0
—Other	16	8	16	8	n	1	0	1
Metals	0	0	0	6	3	n	180	0

Table 7-1 (continued)

Incineration—								
—Municipal	n	n	3	3	3	1	3	n
—Industrial	n	0	1	n	1	n	1	n
—Other	0	0	n	1	n	n	n	0
Minerals	2	0	0	4	6	0	0	0
Miscellaneous	u	0	u	2	0	u	0	u
TOTALS—Rounded	1775	70	145	125	695	585	9950	5

n—negligible (0.05–0.5 ton per day). u—unknown.

Interpretation of Pollutant Emission:

a. Maximum and Minimum Emission:

(1) *Maximum emission.* Maximum emission of pollutant results from the total engagement of normal installation activities without the application of control measures.

(2) *Minimum Emission*:

(a) Minimum emission of pollutant results from the total engagement of normal installation activities with the application of the most advanced control techniques and devices available.

(b) The attainment of minimum pollutant emission merely means that emission rates are cut to an irreducible minimum for the present. Such attainment does not necessarily mean, therefore, that air pollution has been eliminated, nor that it will remain at the same level.

b. Relative Magnitudes of Emissions. Because an emission inventory provides no indication of severity or seriousness of effect upon the community for any of the pollutants inventoried, the relative magnitude of emission among the various pollutant classifications has no meaning. However, the inventory does provide an indication of *relative source contribution* of a *given* pollutant to the community atmosphere and, therefore, provides guidance and direction in the design of a control program.

Source Emission Factors. The source emission factors presented in this section were compiled primarily for use in conducting an air pollution inventory. In general, factors are weighted averages of values found in the literature. Thus emission factors should be used as guidelines for estimating pollution and not as a substitute for a source testing program.

a. **Automotive and Diesel Exhaust Emissions.** The composition of automotive and diesel exhausts is characterized by greater amounts of carbon monoxide and hydrocarbons than that of emissions from other fuel burning processes. Automotive and diesel exhaust emissions, in pounds per 1,000 gallons of fuel consumed, are given in Table **7-2**. Another way of presenting automobile emission factors is in pounds per vehicle mile.

b. **Combustion:**

(1) *Coal*:

(a) The burning of coal produces several different kinds of gaseous pollutants including carbon monoxide, nitrogen oxides, sulfur oxides, aldehydes, and hydrocarbons. Table **7-3** gives average emission factors for the gaseous pollutants in the three major categories of coal usage. As a rule of thumb, for these three categories, boiler capacities for power plants are generally above 100×10^6 Btu per hour; industrial boilers are in the range of 10 to 100×10^6 Btu per hour, and domestic and commercial boilers are generally below 10×10^6 Btu per hour capacity. It is generally advisable to deal with large coal burning plants ($<100 \times 10^6$ Btu) on an individual basis.

(b) The burning of coal also produces a particulate emission. The quantity of the emission is mostly dependent upon the ash content of the coal and the type of unit and control equipment used. Table **7-4** provides average total particulate emission factors for various types of coal firing units without control equipment. The factors are expressed as pounds of emission per ton of coal with a heat content of 13,000 Btu per pound. The factors in tables **7-3** and **7-4** should not be used if the heat value of the coal differs significantly from 13,000 Btu per pound. The nomographs presented in figures **7-1** and **7-2** have been constructed to convert the emission values to those appropriate for the coal used in a particular area. If only the type or rank of coal is known, table **7-5** can be used to approximate the composition of the coal.

Table 7-2

GASOLINE AND DIESEL ENGINE EMISSION FACTORS (lb/1000 gal)

Pollutant	*Gasoline Engines*	*Diesel Engines*
Aldehydes	4	1[illegible]
Carbon Monoxide	2300	6[illegible]
Hydrocarbons	200	13[illegible]
Oxides of Nitrogen	113	22[illegible]
Oxides of Sulfur	9	4[illegible]
Organic Acids	4	3
Particulates	11	11[illegible]

Table 7-3

GASEOUS EMISSION FACTORS (lb/ton of coal burned)

Pollutant	*Power Plants*	*Industrial*	*Domestic and commercial*
Aldehydes (HCHO)	0.005	0.005	0.00[illegible]
Carbon monoxide	.5	3	50
Hydrocarbons (CH_4)	0.2	1	10
Oxides of nitrogen (NO_2)	20	20	8
Oxides of sulfur (SO_2)	38 S[a]	38 S[a]	38 S[a]

[a] S equals percent sulfur in coal.
Example: If the sulfur content is 2 percent, the oxides of sulfur emission would be 2×38, or 76 lb. of sulfur oxides/tons of coal burned.

Table 7-4

PARTICULATE EMISSION FACTORS (WITHOUT CONTROL EQUIPMENT)

Type of unit	*Particulate emission, lb/ton of coal burned*
Pulverized—general	16A[a]
Dry bottom	17A[a]
Wet bottom	
Without reinjection	15A[a]
With reinjection	24A[a]
Cyclone	2A[a]
Spreader stoker	
Without reinjection	13A[a]
With reinjection	20A[a]
All other stokers	5A[a]
Hand-fired equipment	20

[a] A equals percent ash in coal.
Example: If the percent ash in the coal is 10 percent, the ash emission for a cyclone unit would be 2×10 20 lb/ton of coal burned.

Table 7-5

HEATING VALUES AND SULFUR CONTENT OF SELECTED COALS

Rank	*State*	*BTU/lb*	*Ash %*	*Sulfur %*	*Carbon %*
Anthracite					
meta-anthracite	Rhode Island	9,620	28.3	----	67
anthracite	Pennsylvania	13,500	8.5	0.5	84
Bituminous					
low-volatile	West Virginia	14,550	3.9	0.6	75
high-volatile A	Pennsylvania	13,610	9.1	1.3	58
high-volatile B	Kentucky	12,950	4.2	2.6	49
high-volatile C	Illinois	11,480	8.6	4.3	39
Lignitic					
lignitic	North Dakota	7,210	6.2	1.7	31
	(Minot AFB)	6,700	7.6	0.3	50

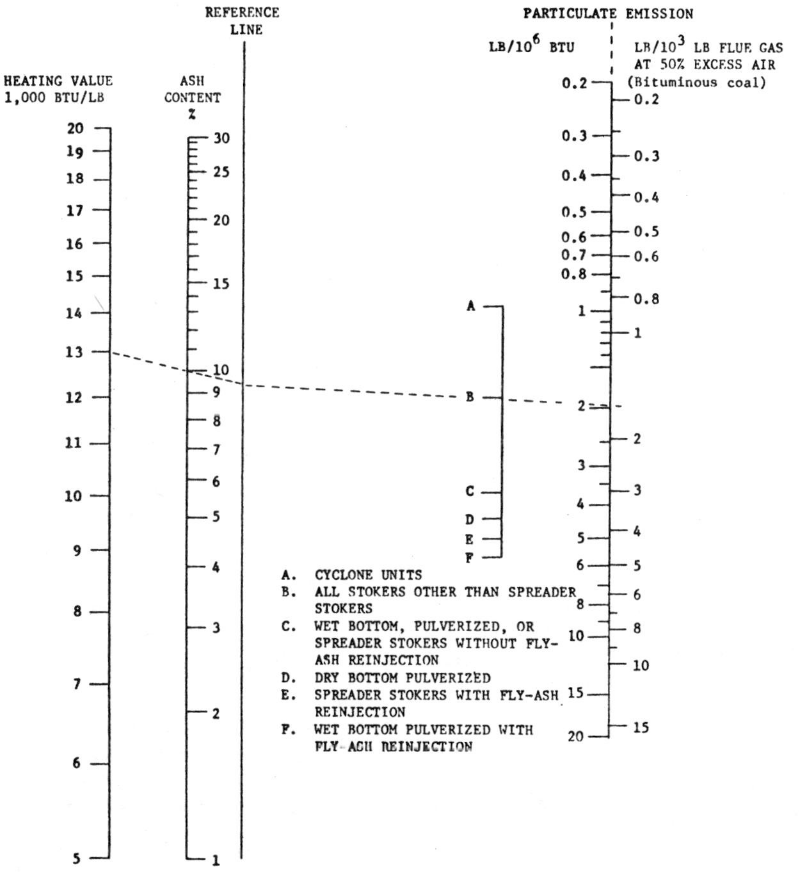

Fig. 7-1 Nomograph for estimating particulate emissions from coal combustion (without air pollution control equipment)

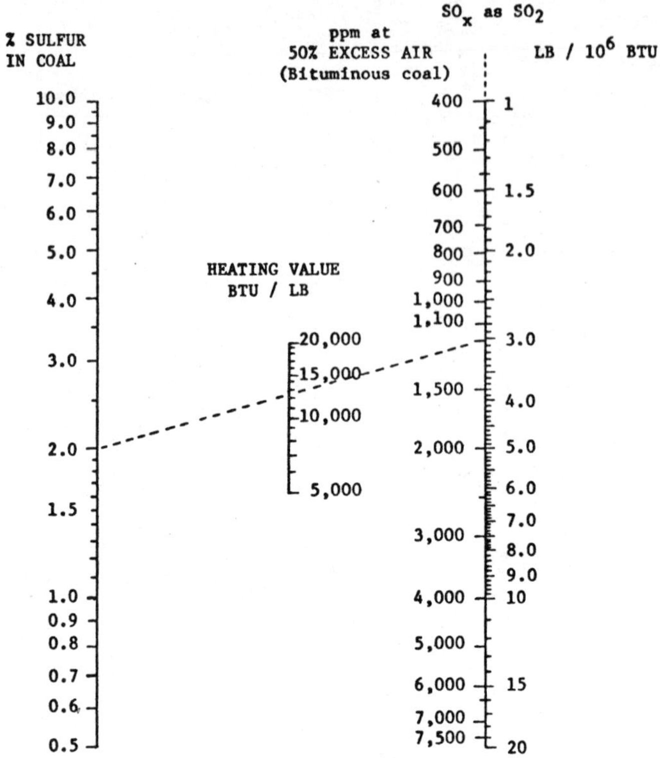

Fig. 7-2 Nomograph for calculating SO_X Emissions

(c) Table **7-6** gives the ranges of fly ash collection efficiencies for various types of control equipment when used on several types of furnaces.

(2) *Natural Gas.* Table **7-7** shows the emission factors for the combustion of natural gas in power plants, industrial boilers, and domestic and commercial heating units. The calculations are based upon a density for natural gas of 0.052 pound per standard cubic foot and a heating value of 1000 Btu per standard cubic foot, 70° F and 14.7 pounds per square inch absolute. The average constituents of three types of natural gas are presented in table **7-8.**

(3) *Oil.* The pollutants emitted in oil combustion are much the same as in coal combustion. Particulate emissions are considerably lower, however. As with coal, the sulfur oxide emissions vary with the sulfur content of the fuel. Table **7-9** gives the emission factors for large and for small sources. The characteristics of the six grades of fuel oil are presented in table **7-10.**

c. **Gasoline Evaporation:**

(1) A study of the typical pattern of motor gasoline storage and handling reveals five major points of gasoline emission. These are:

(a) Breathing and filling losses from storage tanks at refineries and bulk terminals.

(b) Filling losses from loading tank conveyances at refineries and bulk terminals.

(c) Filling losses from loading underground storage tanks at service stations.

(d) Spillage and filling losses in filling automobile gas tanks at service stations.

(e) Evaporative losses from the carburetor and gas tank of motor vehicles.

(2) Breathing loss has been defined as that loss associated with the thermal expansion and contraction of the vapor space resulting from the daily temperature cycle. Filling loss has been defined as the vapors expelled from a tank (by displacement) as a result of filling.

(3) In splash fill, the gasoline enters the top of the fill pipe and then has a free fall to the liquid surface in the tank. The free falling tends to break up the liquid stream into droplets. As these droplets strike the liquid surface, they carry entrained air into the liquid, and a "boiling" action results as this air escapes up through the liquid surface. The net effect of these actions is the creation of additional vapors in the tank.

(4) In submerged filling, the gasoline flows

Table 7-6

AVERAGE OF COLLECTION EFFICIENCIES FOR COMMON TYPES OF FLY ASH CONTROL EQUIPMENT

Type of furnace	Average of collection efficiencies [a]			
	High efficiency cyclone and electrostatic precipitator	High efficiency cyclone	Low resistance cyclone	Settling chamber expanded chimney bases
Cyclone furnace	65–99[b]	30–40	20–30	NA
Pulverized unit	80–99.9[b]	65–75	40–60	NA
Spreader stoker	NA	85–90	70–80	20–30
Other stokers	NA	90–95	75–85	25–50

[a] Estimated operating efficiencies based on normal day to day operation.
[b] High values attained with high efficiency cyclones in series with electrostatic precipitators.

Table 7-7

EMISSION FACTORS
($lb/10^6\ ft^3$ of natural gas burned)

Pollutants	Power plants	Industrial boilers	Domestic and commercial heating units
Aldehydes	1	2	negligible
Carbon monoxide	negligible	0.4	0.4
Hydrocarbons	negligible	negligible	negligible
Oxides of nitrogen	390	214	116
Oxides of sulfur	0.4	0.4	0.4
Other organics	3	5	negligible
Particulate	15	18	19

Table 7-8

NATURAL GAS—ANALYSIS OF COMPONENTS

Gas	Constituents of gas % by volume			
	CO_2	N_2	CH_4	C_2H_6
Natural gas, Texarkana	0.80	3.20	96.00	-----
Natural gas, Cleveland		1.30	80.50	18.20
Natural gas, Oil City, Pa		1.10	67.60	31.30

Typical sulfur content of natural gas used in the Pennsylvania area = 0.034 to 0.035%.

Table 7-9

EMISSION FACTORS
(lb/1000 gal of oil burned) [a]

Pollutants	*Power plant*	*Small sources (1000 h.p. or less)*
Aldehydes	0.6	2
Carbon monoxide	0.04	2
Hydrocarbons	3.2	2
Oxides of nitrogen (NO_2)	104	72
Sulfur dioxide	157 S[b]	157 S[b]
Sulfur trioxide	2.4 S[b]	2 S[b]
Particulate	10	23[c] & 15[d]

[a] Density of fuel oil equals 8 lb/gallon and there are 42 gallons per barrel.
[b] S equals percent sulfur in oil.
Example: If the sulfur content is 2 percent, the sulfur dioxide emission would be 2×157, or 314 lb. SO_2/1000 gal oil burned.
[c] Emission factor for residual unit.
[d] Emission factor for distillate unit.

Table 7-10

FUEL OIL—AVERAGE PROPERTIES

Grades	*Gravity*	*Weight per volume (at 60°F)*		*Sulfur content (Decimal)*
	*API**	*lbs/gallons*	*Kilograms/gallon*	
1	42	6.79	3.08	.0013
2	35	7.08	3.21	.0035
4	21	7.73	3.50	.009
5	15	8.04	3.65	.013
6	10	8.33	3.78	.016

to the bottom of the tank through the fill pipe and enters below the surface of the liquid. This method of filling creates very little disturbance in the liquid bath and, consequently, less vapor formation than splash filling.

(5) Emission factors are given in table **7-11** for splash and submerged fill in tank vehicles and service station tanks storing gasoline.

(6) Vapor emission from tanks storing volatile fuels such as gasoline will vary depending on quantity of fuel stored, storage temperature of fuel, annual throughput of product, and type of vapor emission control. Vapor emission from floating roof tanks are usually less than cone roof tanks. Vapor loss from cone roof tanks equipped with a pressure/vacuum conservation valve are approximately 50 percent of open vented tanks. Vapor loss emission of storage tanks at a specific locality can be calculated using methods in the American Petroleum Institute Bulletin 2517, "Evaporation Loss from Floating Roof Tanks," and Bulletin 2518, "Evaporation Loss from Fixed Roof Tanks." These are available from the American Petroleum Institute, Division of Technical Services, 1271 Avenue of the Americas, New York NY 10020.

Table 7-11 summarizes the emission factors for gasoline evaporation at the other four major points of emission.

d. Solvent Evaporation from Dry Cleaning Plants:

(1) Almost all dry cleaning is performed with three solvents: tetrachloroethylene, Stoddard solvent, and safety 140° F solvent. Stoddard solvent has a minimum flash point of 100° F and a distillation range within 100° to 410° F. Recently a new petroleum solvent, called safety 140° F solvent, has been introduced that has a minimum

Table 7-11

GASOLINE EVAPORATION EMISSION

Point of emission	*Lb/1000 gal of throughput*	*Percent loss by volume* [a]
Filling tank vehicles		
Splash fill	8.2	0.14
Submerged fill	4.9	0.08
50% splash fill and 50% submerged fill	6.4	0.11
Filling service station tanks		
Splash fill	11.5	0.19
Submerged fill	7.3	0.12
50% splash fill and 50% submerged fill	9.4	0.15
Filling automobile tanks	11.6	0.19
Automobile evaporation losses (gas tank and carburetor)	92	1.50

[a] An average gasoline specific gravity of 0.73 is assumed.

Table 7-12

EMISSION FACTORS FOR DRY CLEANING PLANTS

	LA data [a] *January 1963*	*BAAPCD* [b] *data for 1963*
Population	*6,492,000*	*3,691,000*
Tons chlor-hydrocarbons emitted/day	15	7.9
Tons petroleum solvents emitted/day	20	11.5
Total	35	19.4
Tons of clothes cleaned/calendar day	158	92
Pounds of clothes cleaned/capita/year	18	18.3
Pounds of chlor-hydrocarbons emitted/capita/year	1.7	1.5
Pounds of hydrocarbon vapors emitted/capita/year	2.2	2.3
Pounds of total organic solvents emitted/capita/year	3.9	3.8

[a] Los Angeles County Air Pollution Control District data.
[b] San Francisco Bay Area Air Pollution Control District.

flash point of 140° F, thus lessening the explosion hazard.

(2) Chlorinated hydrocarbons are widely used as cleaning solvents. They are nonflammable and dissolve greases and oils rapidly, including substances not soluble in petroleum solvents. Originally, carbon tetrachloride was used, but had to be discarded because of its toxicity and corrosiveness. Trichloroethylene is less toxic and not as corrosive as carbon tetrachloride. Tetrachloroethylene (perchloroethylene) is the most widely used chlorinated dry cleaning agent. It has a lower toxicity, is less corrosive than carbon tetrachloride, and does not bleed acetate dyes as trichloroethylene does. Because it is expensive, and a health hazard, tetrachloroethylene is often recovered by use of carbon adsorption beds.

(2) Table 7-12 gives emission factors for chlorinated and non-chlorinated hydrocarbon dry cleaning solvents expressed in pounds per captia per day.

e. **Refuse Disposal and Open Burning:**

(1) *Incineration of Waste:*

(a) Multiple chamber incinerators are made up of three refractory lined chambers in series. The first chamber, into which the combustible refuse is charged, is called the charging or ignition chamber. In the middle or mixing chamber, additional air is added to help consume any organic matter not completely burned in the ignition chamber. In the last chamber, called the combustion or expansion chamber, the combustion of gaseous organic materials is completed and the greater part of the flyash is settled out. This chamber sometimes leads to a collector equipped with

Table 7-13

INCINERATORS
EMISSION FACTORS (lb/ton of refuse burned)

Pollutant	Municipal	Industrial and commercial	
	multiple chamber	*single chamber*	*multiple chamber*
Aldehydes	0.3	1.0	0.2
Carbon monoxide	1.0	44	10
Hydrocarbons (hexane)	.3	0.8	0.5
Oxides of nitrogen	2.0	3	2.
Oxides of sulfur	2.0	2	1.0
Ammonia	0.3	0.4	0.1
Organic acids	0.6	3	3
Particulate	17	10	3

Pollutant	Flue-fed		Domestic	
	No control	*After burner*	*No control*	*After burner*
Aldehydes	3	2	5.5	2.5
Carbon monoxide	27	NA	200	30
Hydrocarbons	2	NA	2	1.0
Oxides of nitrogen	0.3	10	1.0	2
Oxides of sulfur	0.2	0.2	0.4	0.4
Ammonia	0.4	0.4	0.3	0.3
Organic acids	25	6	7	2
Particulate	28	NA	15	NA

(NA: not available)

Table 7-14

OPEN BURNING
EMISSION FACTORS (lb/ton of refuse burned)

Pollutants	*Municipal Refuse*	*Landscape & Agricultural Refuse* [a]
Aldehydes	0.1	.01
Carbon monoxide	85	60
Hydrocarbons (hexane)	5	2
Oxides of nitrogen	11	2
Organic acids	15	13
Particulate	16	17

[a] Factor can be used for leaves, grass, and various agriculture wastes.

sprays and wetted baffles than reduce flyash emissions further, but have little effect on gaseous emissions.

(b) Single chamber incinerators are generally simple contrivances, consisting of a firebox, door, grate, and flue. Emissions are generally high, but can be significantly reduced with auxiliary gas afterburners. Table **7-13** gives emission factors for the common types of municipal, industrial, commercial, apartment, and domestic incinerators.

(2) *Open Burning.* Open burning, whether in dumps or in very simple backyard contrivances, gives rise to emissions that are extremely variable and difficult to measure. Dump fires generally smolder and burn less efficiently than backyard fires and thus have somewhat higher emissions. The data for backyard burning was derived from tests on burning a mixture of 50 percent newspapers and 50 percent grass clippings. Table **7-14** gives factors for burning dumps and for backyard burning.

Emission Guide. Tables 7-15 thru 7-19 indicate expected pollutants and their concentrations from various industrial processes.

Table 7-15

ESTIMATION OF POLLUTION LEVELS FROM VARIOUS OPERATIONS

Process	*Contaminant*	*Pollution load associated with process*	*Sources of information*
Automobile burning (Dump).	Carbonaceous matter.	250 lbs per automobile	Assumption.
Concrete batching	Cement	0.25–1% cement used, depending on handling.	Survey of 12 plants.
Core baking	Acrolein [1]	6 lb core gas per gal of core oil used.	Core Oil and Core Gas Evaluation, Hiram Brown, American Foundryman, Vol. 18, page 52—Aug., 1950.
Degreasing	Trichloroethylene, Perchloroethylene	95–100% of amount used lost by evaporation.	Correspondence: E. I. duPont de Nemours & Co., Detroit, Michigan.
Dry cleaning	Perchloroethylene	86.5 gal per plant per month average (1 washer, 4–8 hrs per day operation).	Survey of 36 plants.
Dry cleaning	Stoddard solvent (petroleum distillate).	384 gal per plant per month average (1 washer, 8 hrs per day operation).	Survey of 23 plants.
Electric arc	Iron oxide [1]	1–2% of weight of electrode.	Correspondence: The Lincoln Electric Company, Cleveland, Ohio.
Electroplating and metal cleaning.	Various	Based on rate of gassing and make-up solution required.	Gassing Rate Data listed in: Electroplating Engineering Handbook, edited by A. K. Graham, Reinhold Publishing Corp., New York (Tables 6, 7 and 8, pages 519–523).

Gassing rate	*Loss % of make-up*
High	5%
Medium	3%
Low	2%
Nil	0

Table 7-15—*Continued*

ESTIMATION OF POLLUTION LEVELS FROM VARIOUS OPERATIONS

Process	*Contaminant*	*Pollution load associated with process*	*Sources of information*
Iron melting (Cupola)	Iron oxide [1]	20 lbs per ton of grey iron produced (average value).	A Study of Cupola Design and Operating Factors that Influence the Emission Rates From Foundry Cupolas, R. C. Origies, American Air Filter Co., Louisville, Kentucky. Control of Metallurgical and Mineral Dusts and Fumes in Los Angeles County, California by G. L. Allen, F. H. Viets and L. C. McCabe, Bureau of Mines Circular 7627.
Lead melting	Lead oxide	1.5% of charge	Dust and Fume Standards, I. C. McCabe, A. H. Rose, W. J. Hamming and F. H. Vista, Industrial and Engineering Chemistry, Vol. 41, page 2388, Nov., 1949.
Brass melting	Copper oxide, zinc oxide.	1.5% charge	
Magnesium melting	Magnesium oxide	5% of charge	Survey data (1 plant).
Oil quench	Oil fumes	10% of make-up oil	Assumption.
Painting	Thinners	60% of paint volume used	Value calculated on basis of composition of paints used in study area.
Soldering	Varies	0.5% of weight of solder used.	Assumption.
Steel-Bessemer converter.	Carbon iron oxide	0.5% of steel produced	Assumption.
Steel-electric arc	Iron oxide [1]	33 lb per ton of steel produced.	Test data at plant in survey area.
Steel-open hearth	Iron oxide [1]	0.7 grains per S.C.F.M. and out.	Control of Metallurgical and Mineral Dusts and Fumes in Los Angeles County, California, by G. L. Allen, F. H. Viets and L. C. McCabe, Bureau of Mines Circular 7627.
Steel-scarfing	Iron oxide [1]	29 lbs per ton of steel (oxyacetylene scarfing).	Test data at plant in survey area.

[1] Main constituent.

Table 7-16

ESTIMATED EMISSIONS FROM INDUSTRIAL OPERATIONS AND PROCESSES CLASSIFIED BY OPERATIONS AND/OR PROCESSES

(1) Specific Operation or Process	(2) Air Contaminant	(3) Basis Used for Expressing Process Exhaust Loading	(4) Process Exhaust Loading Factors **	(5) Type of Air Cleaning Equipment, if any	(6) Collector Efficiency Wt. %
CRUSHING, PULVERIZING, MIXING, SCREENING (SOLIDS)					
Cement mill	Cement dust	Gr/CF exhaust	5.25 Gr/CF	Electrostatic precipitator.	95–99
Coal screening	Coal dust	Gr/CF exhaust	0.88 Gr/CF	Venturi scrubber	99+
Crushing and grinding	Beryllium dust	Gr/CF exhaust	0.029–0.14 Gr/CF	Cloth collector	98–99+
Granite chipping	Granite dust	Gr/CF exhaust	0.032 Gr/CF	Cloth collector	99+
Kraft pulp mill	Soda ash fume	Gr/CF exhaust	0.5–2.0 Gr/CF	Electrostatic precipitator.	90–95
DISTILLATION					
Black liquor recovery furnace	Chemical fume	Gr/CF exhaust	1–2.5 Gr/CF	Venturi scrubber	90
Coke oven	Tars	Gr/CF exhaust	0.1–1.0 Gr/CF	Electrostatic precipitator.	95–99
Kraft pulp recovery boiler	Sulfur dioxide gas	% by volume of CF exhaust.	0.029% by volume	Venturi scrubber	90
Wood distillation	Tar and acetic acid	Gr/CF exhaust	16.7 Gr/CF	Venturi scrubber	95
			1.45 Gr/CF	Venturi and cyclone scrubbers.	99
DRYING AND BAKING					
Activated carbon rotary drier	Carbon dust	Gr/CF exhaust	1.9 Gr/CF	Venturi scrubber	98
Alumina steam drying	Aluminum dust	Gr/CF exhaust	0.53 Gr/CF	Air tumbler gas washer.	76
Catalyst regenerator (petroleum)	Catalyst dust	Ton of production	0.4 lb/ton	Cyclone and electrostatic precipitator.	(95)
	Catalyst dust	Gr/CF exhaust	0.1–25 Gr/CF	Electrostatic precipitator.	90–99+
Cement drier	Cement dust	Gr/CF exhaust	1–15 Gr/CF	Electrostatic precipitator.	95–99
Cement kiln	Cement dust	Gr/CF exhaust	0.5–15 Gr/CF	Electrostatic precipitator.	85–99
Copperas (ore) roasting kiln	Sulfuric acid mist	Gr/CF exhaust	3.1 Gr/CF	Venturi scrubber	99
Core baking	Acrolein	Gallons core oil consumed.	6.0 lb/gal oil consumed.	----------	----------
Kraft mud lime kiln	Lime dust	Gr/CF exhaust	7.7 Gr/CF	Cyclone scrubber	97

Table 7-16 (continued)

Lime raw stone kiln	Lime dust	Gr/CF exhaust	7.0 Gr/CF	Venturi scrubber	99+
Orange pulp feed drier	Pulp dust	Ton of production	75.3 lb/ton	Cyclone	(85)
Sand drying kiln	Silica dust	Gr/CF exhaust	18.7 Gr/CF	Cyclone	78
Sand and gravel drying	Silica dust	Gr/CF exhaust	5.8 Gr/CF	Inertial collector	50
Sand and stone drying	Silica dust	Gr/CF exhaust	3.6–6.4 Gr/CF	Cylindrical tower washer.	73–92
	Silica dust	Gr/CF exhaust	5.8 Gr/CF	Cyclone	74
Stone drying kiln	Silica dust	Gr/CF exhaust	16.8 Gr/CF	Cyclone	86
HEAT TREAT AND ANNEALING					
Oil quench	Oil mist	Gallons of make-up oil.	Oil gal/gal	----------	----------
MATERIAL STORAGE, CONVEYING, HANDLING					
Gypsum kettles and conveyors	Gypsum dust	Gr/CF exhaust	1.5–5.0 Gr/CF	Electrostatic precipitator.	90–98
MIXING FLUIDS					
Asphalt mixing	Sand and gravel dust	Gr/CF exhaust	5.8–38 Gr/CF	Cyclone	50–86
		Ton of production	10.7 lb/ton	Cyclone	(85)
Bituminous concrete mixing	Sand and stone dust	Gr/CF exhaust	4.0 Gr/CF	Cyclone and scrubber.	(95)
Concrete batching	Cement dust	Ton of process weight.	5.20 lb/ton	----------	----------
MOLTEN MATERIALS HANDLING					
Aluminum sweating furnace	Aluminum oxide fume	Ton of production	71 lb/ton	Baghouse	(99)
Brass melting	Zinc and copper oxide fume.	Ton of process weight.	25.2 lb/ton	----------	----------
Brass (red) melting	Copper and zinc oxide fume.	Ton of process weight.	22.3 lb/ton	----------	----------
Brass (red) indirect fired furnace	Copper and zinc oxide fume.	Ton of production	4.6 lb/ton	----------	----------
Brass (yellow)	Copper and zinc oxide fume.	Ton of process weight.	27.1 lb/ton	----------	----------
Brass (yellow) direct fired rotary furnace	Copper and zinc oxide fume.	Ton of production	54 lb/ton	----------	----------
Ferro-silicon furnace	Iron oxide fume	Gr/CF exhaust	0.97 Gr/CF	Spray tower	75
Frit rotary melter	Fume	Ton of production	17 lb/ton	----------	----------
Glass (bottle) reverberatory furnace	Fume	Ton of production	3 lb/ton	----------	----------
Iron (gray) melting cupola	Iron oxide fume and coke dust.	Ton of production	14.6 lb/ton	Baghouse	99+
Iron (gray) furnace	Iron oxide fume	Ton of process weight.	19.3 lb/ton	----------	----------
Iron blast furnace	Iron ore and coke dust	Gr/CF exhaust	3–24 Gr/CF	Cyclone scrubber	99
				Venturi scrubber	97–99+

Table 7-16—*Continued*

ESTIMATED EMISSIONS FROM INDUSTRIAL OPERATIONS AND PROCESSES CLASSIFIED BY OPERATIONS AND/OR PROCESSES

(1) *Specific Operation or Process*	(2) *Air Contaminant*	(3) *Basis Used for Expressing Process Exhaust Loading*	(4) *Process Exhaust Loading Factors ***	(5) *Type of Air Cleaning Equipment, if any*	(6) *Collector Efficiency Wt. %*
Iron melting cupola	Iron oxide dust and fume.	Ton of melt	10.4 lb/ton	Collector	---------
	Iron oxide dust and fume.	Ton of production	20 lb/ton	---------------	---------
Lead (secondary) blast furnace	Lead oxide fume	Gr/CF exhaust	2–6 Gr/CF	Venturi scrubber	95
Lead reverberatory furnace	Lead oxide fume	Gr/CF exhaust	0.5–2 Gr/CF	Cyclone scrubber	95–98
	Lead and tin oxide	Gr/CF exhaust	1–3 Gr/CF	Venturi scrubber	91
Lead melting	Lead oxide fume	Ton of process	36 lb/ton	---------------	---------
Magnesium melting	Magnesium oxide fume	Ton of charge	100 lb/ton	---------------	---------
Steel Bessemer converter	Carbon and iron oxide dust.	Ton of production	10 lb/ton	---------------	---------
Steel electric arc furnace	Iron oxide fume	Ton of production	33 lb/ton	---------------	---------
	Iron oxide fume	Gr/CF exhaust	0.05–3 Gr/CF	Electrostatic precipitator.	90–99
Steel electric furnace	Iron oxide fume	Ton of production	9.2 lb/ton	Deflector washer	60
	Iron oxide fume	Ton of production	4.8 lb/ton	Disintegrator	98+
Steel open hearth furnace	Iron oxide fume	Ton of production	17.5 lb/ton	Electrostatic precipitator.	98+
		Ton of process weight.	12.9 lb/ton	---------------	---------
		Ton of process weight.	11.2 lb/ton	---------------	---------
		Gr/CF exhaust	0.05 Gr/CF	Electrostatic precipitator.	90–99
Steel open hearth furnace (oxygen lanced)	Iron oxide fume	Gr/CF exhaust	1–6 Gr/CF	Venturi scrubber	92–99
Steel open hearth scrap furnace	Iron and zinc oxide fume.	Gr/CF exhaust	0.5–1.5 Gr/CF	Venturi scrubber	95
Steel oxygen blasting process	Iron oxide and mineral dust.	Ton of production	46.1 lb/ton	Disintegrator	99+
Zinc oxide reducing furnace	Zinc oxide fume	Ton of production	87 lb/ton	---------------	---------
POLISHING, BUFFING, GRINDING AND CHIPPING					
Buffing and polishing	Iron oxide dust and lint.	Gr/CF exhaust	0.0048–0.0094 Gr/CF	Cloth filter	98

Table 7-16(continued)

Grinding (aluminum)	Aluminum dust	Gr/CF exhaust	0.7 Gr/CF	Cyclone	89
Grinding (iron)	Iron scale and sand	Gr/CF exhaust	0.15 Gr/CF	Cyclone	56
Grinding (machine shop)	Dust	Gr/CF exhaust	0.011 Gr/CF	Dynamic precipitator.	91
REACTING—CHEMICAL					
Fertilizer reactor and mixer	Fertilizer dust	Ton of production	35 lb/ton	Scrubber	(80)
Fertilizer ammoniator (TVA type)	Ammonia gas	Ton of production	4 lb/ton	----------	----------
Sulfuric acid concentrator	Sulfuric acid mist	Gr/CF exhaust	2.1 Gr/CF	Venturi scrubber	99+
Superphosphate den and mixer	Fluorine compounds	Gr/CF exhaust	4.8 Gr/CF	Cyclone scrubber	98
				Venturi scrubber	98
SURFACE COATING					
Spray, dip and brush coating	Organic solvents	Gallons of coating material consumed.	0.6 gal/gal coating material.	----------	----------
Rubber dusting	Fluffy zinc stearate	Gr/CF exhaust	0.1–1 Gr/CF	----------	----------
	Fluffy zinc stearate	Gr/CF exhaust	0.6–1.7 Gr/CF	Dynamic precipitator.	78–88
	Talc dust	Gr/CF exhaust	4.3 Gr/CF	Cloth filter	99+
SURFACE TREATMENT—PHYSICAL					
Abrasive cleaning	Talc dust	Gr/CF exhaust	2.2 Gr/CF	Cyclone	93
Abrasive stick trimming and shaping	Silicon carbide and aluminum oxide dust.	Gr/CF exhaust	1.8 Gr/CF	2 cyclones in parallel.	51
Casting cleaning	Bronze and silica dust	Gr/CF exhaust	0.44 Gr/CF	Cloth filter	99+
	Iron scale and sand	Gr/CF exhaust	0.19–0.68 Gr/CF	Cloth filter	99+
Casting cleaning with metal shot, sandblasting and tumbling.	Metallic and silica dust.	Gr/CF exhaust	0.75–7.3 Gr/CF	Dynamic precipitator.	97–99+
Foundry tumbling	Dust	Gr/CF exhaust	0.28 Gr/CF	Dynamic precipitator.	99+
Steel scarfing	Iron oxide dust	Ton of process weight.	29 lb/ton	----------	----------
Truing and shaping abrasive products	Aluminum oxide and silica dust.	Gr/CF exhaust	0.1–3.65 Gr/CF	Cloth filter	99+
	Boron carbide dust	Gr/CF exhaust	0.075 Gr/CF	Cloth filter	97
	Silicon carbide and aluminum oxide dust.	Gr/CF exhaust	0.36 Gr/CF	Cyclone	58
	Silicon carbide and aluminum oxide dust.	Gr/CF exhaust	0.13–2.3 Gr/CF	Cloth filter	99+

Table 7-16 —*Continued*

ESTIMATED EMISSIONS FROM INDUSTRIAL OPERATIONS AND PROCESSES CLASSIFIED BY OPERATIONS AND/OR PROCESSES

(1) *Specific Operation or Process*	(2) *Air Contaminant*	(3) *Basis Used for Expressing Process Exhaust Loading*	(4) *Process Exhaust Loading Factors* **	(5) *Type of Air Cleaning Equipment, if any*	(6) *Collector Efficiency Wt. %*
WELDING, SOLDERING, BRAZING					
Electric arc welding	Iron oxide fume	Pounds of electrodes consumed.	0.01–0.02 lb/lb electrodes.	----------	----------
Soldering	Lead oxide fume	Pounds of solder consumed.	0.005 lb/lb solder	----------	----------

** Gr—grain (1 grain = 64.999 milligrams).

Table 7-17

AIR CONTAMINANTS RELEASED BY STRIPPING OPERATIONS

Coating to be Stripped	*Base Metal (Footnote)*	*Component of Bath Which May be Released to Atmosphere*	*Physical and Chemical Nature of Major Atmospheric Contaminant*	*Rate of Gassing*	*Usual Temp. Range-°F*
Anodized Coatings	1, 7	Chromic Acid	Acid Mist, Steam	Medium	120–200
Black Oxide Coatings	14	Hydrochloric Acid	Hydrogen Chloride Gas	Low-Medium	70–125
Brass and Bronze	8, 14 (a)	Sodium Hydroxide, Sodium Cyanide.	Alkaline, Cyanide Mists	Low-Medium	70– 90
Cadmium	8, 14 (a)	Sodium Hydroxide, Sodium Cyanide.	Alkaline, Cyanide Mists	Low-Medium	70– 90
	2, 4, 14	Hydrochloric Acid	Acid Mist, Hydrogen Chloride Gas.	Low-Medium	70– 90
Chromium	7, 8, 14 (a)	Sodium Hydroxide	Alkaline Mist, Steam	Low	70–150
	2, 4, 8, 14	Hydrochloric Acid	Hydrogen Chloride Gas	Medium	70–125
	2, 4, 8, 18 (a)	Sulfuric Acid	Acid Mist	Medium	70– 90
Copper	8, 14	Sodium Hydroxide, Sodium Cyanide.	Alkaline, Cyanide Mists	Low-Medium	70– 90
	7, 12, 14 (b)	None	None	Nil	70– 90
	14 (a)	Alkaline Cyanide	Cyanide Mist	Low-Medium	70–160
	1	Nitric Acid	Nitrogen Oxide Gases	High	70–120
	18 (a)	Sodium Hydroxide, Sodium Sulfide.	Alkaline Mist, Steam	Medium	185–195
Gold	4, 5, 6, 8, 9, 14 (a)	Sodium Hydroxide, Sodium Cyanide.	Alkaline, Cyanide Mists	Low-Medium	70– 90
	4, 5, 16 (a)	Sulfuric Acid	Acid Mist	Low-Medium	70–100
Lead	13 (c)	Acetic Acid, Hydrogen Peroxide	Oxygen Mist	Low	70– 90
	14 (a), (c)	Sodium Hydroxide	Alkaline Mist, Steam	Low-Medium	70–140
Nickel	2, 4	Sulfuric, Nitric Acids	Nitrogen Oxide Gases	Medium-High	70– 90
	2, 4 (a)	Hydrochloric Acid	Hydrogen Chloride Gas	Low	70– 90
	2, 4, 14 (a)	Sulfuric Acid	Acid Mist	Low	70– 90
	7	Hydrofluoric Acid	Hydrogen Fluoride Gas	Low-Medium	70– 90

Table 7-17—Continued

AIR CONTAMINANTS RELEASED BY STRIPPING OPERATIONS

Coating to be Stripped	*Base Metal (Footnote)*	*Component of Bath Which May be Released to Atmosphere*	*Physical and Chemical Nature of Major Atmospheric Contaminant*	*Rate of Gassing*	*Usual Range-°F Temp.*
Nickel—Continued					
	14	Fuming Nitric Acid	Nitrogen Oxide Gases	High	70– 90
	(a), (d)	Hot Water	Steam	Medium	200
	1, 18, 19 (a)	Sulfuric Acid	Acid Mist, Steam	Low-Medium	70–150
Phosphate Coatings	15	Chromic Acid	Acid Mist, Steam	Low	165
	16	Ammonium Hydroxide	Ammonia Gas	Low-Medium	70– 90
Rhodium	10	Sulfuric, Hydrochloric Acids	Acid Mist, Hydrogen Chloride Gas.	Low-Medium	70–100
Silver	1	Nitric Acid	Nitrogen Oxide Gases	High	70– 90
	2, 11	Sulfuric, Nitric Acids	Nitrogen Oxide Gases, Steam	High	180
	8, 14 (a)	Sodium Hydroxide, Sodium Cyanide.	Alkaline, Cyanide Mists	Low	70– 90
	17 (a)	Sodium Cyanide	Cyanide Mist	Low	70– 90
Tin	2, 3, 4	Ferric Chloride, Copper Sulfate, Acetic Acid.	Acid Mist	Nil-Low	70– 90
	(a)	Sodium Hydroxide	Alkaline Mist	Low	70– 90
	2, 4, 14	Hydrochloric Acid	Hydrogen Chloride Gas	Low-Medium	70– 90
	14 (a)	Sodium Hydroxide	Alkaline Mist, Steam	Medium	70–200
Zinc	1	Nitric Acid	Nitrogen Oxide Gases	High	70– 90
	8, 14	Sodium Hydroxide, Sodium Cyanide.	Alkaline, Cyanide Mists	Low	70– 90

FOOTNOTES

1. Aluminum
2. Brass
3. Bronze
4. Copper
5. Copper Alloys
6. Ferrous Metals
7. Magnesium
8. Nickel
9. Nickel Alloys
10. Nickel Plated Brass
11. Nickel Silver
12. Non Ferrous Metals
13. Silver
14. Steel
15. Steel (Manganese Type Coatings)
16. Steel (Zinc Type Coating)
17. White Metal
18. Zinc
19. Zinc Base Die Castings

(a) Electrolytic Process
(b) Refers only to Steel (14) when Chromic, Sulfuric Acids Bath is Used
(c) Also Lead Alloys
(d) Sodium Nitrate Bath

Table 7-18

AIR CONTAMINANTS RELEASED BY ELECTROPOLISHING, ELECTROPLATING, AND ETCHING OPERATIONS

Process	*Type*	*Footnote*	*Component of Bath Which May be Released to Atmosphere*	*Physical and Chemical Nature of Major Atmospheric Contaminant*	*Rate of Gassing*	*Usual Temp. Range-°F*
Electropolishing	Aluminum	(a)	Sulfuric, Hydrofluoric Acids	Acid Mist, Hydrogen Fluoride Gas, Steam.	Medium	140–200
	Brass, Bronze	(a)	Phosphoric Acid	Acid Mist	Low	68
	Copper	(a)	Phosphoric Acid	Acid Mist	Low	68
	Iron	(a)	Sulfuric, Hydrochloric, Perchloric Acids.	Acid Mist, Hydrogen Chloride Gas, Steam.	Medium	68–175
	Monel	(a)	Sulfuric Acid	Acid Mist, Steam	Medium	86–160
	Nickel	(a)	Sulfuric Acid	Acid Mist, Steam	Medium	85–160
	Stainless Steel	(a)	Sulfuric, Hydrofluoric, Chromic Acids.	Air Mist, Hydrogen Fluoride Gas, Steam.	Medium-High	70–300
	Steel	(a)	Sulfuric, Hydrochloric, Perchloric Acids.	Acid Mist, Hydrogen Chloride Gas, Steam.	Medium	68–175
Strike Solutions	Copper		Cyanide Salts	Cyanide Mist	Medium	70– 90
	Silver		Cyanide Salts	Cyanide Mist	Medium	70– 90
	Wood's Nickel		Nickel Chloride, Hydrochloric Acid.	Hydrogen Chloride Gas, Chloride Mist.	Medium	70– 90
Etching	Aluminum		Sodium Hydroxide-Soda Ash-Trisodium Phosphate.	Alkaline Mist, Steam	High	160–180
	Copper	(b)	Hydrochloric Acid	Hydrogen Chloride Gas	Medium	70– 90
	Copper	(c)	None	None	Nil	70
Electroplating Alkaline.	Platinum		Ammonium Phosphate, Ammonia Gas.	Ammonia Gas	Medium	158–203
	Tin		Sodium Stannate	Tin Salt Mist, Steam	Low	140–170
	Zinc	(d)	None	None	Nil-Low	170–180
Electroplating Fluoborate.	Cadmium		Fluoborate Salts	Fluoborate Mist, Steam	Low-Medium	70–170
	Copper		Copper Fluoborate	Fluoborate Mist, Steam	Low-Medium	70–170
	Indium		Fluoborate Salts	Fluoborate Mist, Steam	Low-Medium	70–170
	Lead		Lead Fluoborate-Fluoboric, Hydrofluoric Acids.	Fluoborate Mist, Hydrofluoric Acid.	Low	70– 90
	Lead-Tin Alloy		Lead Fluoborate, Fluoboric Acid	Fluoborate Mist	Low	70–100
	Tin		Stannous Fluoborate, Fluoboric Acid.	Fluoborate Mist	Low	70–100
	Zinc		Fluoborate Salts	Fluoborate Mist, Steam	Low-Medium	70–170

Table 7-18—Continued

AIR CONTAMINANTS RELEASED BY ELECTROPOLISHING, ELECTROPLATING, AND ETCHING OPERATIONS

Process	*Type*	*Footnote*	*Component of Bath Which May be Released to Atmosphere*	*Physical and Chemical Nature of Major Atmospheric Contaminant*	*Rate of Gassing*	*Usual Temp. Range-°F*
Electroplating Cyanide.	Brass, Bronze	(e), (f)	Cyanide Salts, Ammonium Hydroxide.	Cyanide Mist, Ammonia Gas	Nil-Low	60–100
	Black Nickel	(f)	None	None	Nil	70–100
	Bright Zinc	(f)	Cyanide Salts, Sodium Hydroxide.	Cyanide, Alkaline Mists	Low	70–120
	Cadmium	(f)	None	None	Nil	70–100
	Copper	(f), (g)	None	None	Low-Medium	70–160
	Copper	(f), (h)	Cyanide Salts, Sodium Hydroxide.	Cyanide Alkaline Mists, Steam	Low-Medium	110–160
	Gold	(f)	Cyanide Salts	Cyanide Mist, Steam	Nil-Low	75–214
	Indium	(f)	Cyanide Salts, Sodium Hydroxide.	Cyanide, Alkaline Mists	Low	70–120
	Silver	(f)	None	None	Nil-Low	72–120
	Tin Zinc Alloy	(f)	Cyanide Salts, Potassium Hydroxide.	Cyanide, Alkaline Mists, Steam	Low-Medium	120–140
	White Alloy	(f), (i)	Cyanide Salts, Sodium Stannate	Cyanide, Alkaline Mists	Low	120–150
	Zinc	(f), (j)	Cyanide Salts, Sodium Hydroxide.	Cyanide, Alkaline Mists	Low	70–120
Electroplating Acid.	Chromium		Chromic Acid	Chromic Acid Mist	High	90–140
	Copper	(k)	Copper Sulfate, Sulfuric Acid	Sulfuric Acid Mist	Nil-Low	75–120
	Indium	(l)	None	None	Nil	70–120
	Iron		Chloride Salts, Hydrochloric Acid	Hydrochloric Acid Mist, Steam	Medium	190–210
	Iron	(l)	None	None	Nil	70–120
	Nickel	(d)	Ammonium Fluoride, Hydrofluoric Acid.	Hydrofluoric Acid Mist	Low	102
	Nickel	(l), (m)	None	None	Nil	70–120
	Palladium	(m), (n)	None	None	Nil	70–120
	Rhodium	(l), (o)	None	None	Nil	70–120
	Tin		Tin Halide	Halide Mist	Medium	70– 90
	Tin	(l)	None	None	Nil	70–120
	Zinc		Zinc Chloride	Zinc Chloride Mist	Low	75–120
	Zinc	(l)	None	None	Nil	70–120

FOOTNOTES for Table 7-18

(a) Arsine may be produced due to the presence of arsenic in the metal or polishing bath
(b) Dull Finish
(c) Ferric Chloride Bath
(d) On Magnesium
(e) Also Copper-Cadmium Bronze
(f) HCN gas may be evolved due to the acidic action of CO_2 in the air at the surface of the bath
(g) Conventional Cyanide Bath
(h) Except Conventional Cyanide Bath
(i) Albaloy, Spakwhite, Bonwhite (Alloys of Copper, Tin, Zinc)
(j) Using insoluble anodes
(k) Over 90°F
(l) Sulfate Bath
(m) Chloride Bath
(n) Nitrite Bath
(o) Phosphate Bath

Table 7-19

AIR CONTAMINANTS RELEASED BY METALLIC SURFACE TREATMENT, PICKLING, ACID DIPPING, AND METAL CLEANING OPERATIONS

Process	*Type*	*Footnote*	*Component of Bath Which May be Released to Atmosphere*	*Physical and Chemical Nature of Major Atmospheric Contaminant*	*Rate of Gassing*	*Usual Temp. Range-°F*
Surface Treatment.	Anodizing Aluminum.		Chromic-Sulfuric Acids	Acid Mist	Medium	95 and up
	Black Magic		Conc. Sol. Alkaline Oxidizing Agents.	Alkaline Mist, Steam	High	260–350
	Bonderizing	(a)	Boiling Water	Steam	Medium-High	140–212
	Chemical Coloring		None	None	Nil	70– 90
	Descaling	(b)	Nitric-Sulfuric, Hydrofluoric Acids.	Acid Mist, Hydrogen Fluoride Gas, Steam.	Medium-High	70–150
	Dulite		Conc. Sol. Alkaline Oxidizing Agents.	Alkaline Mist, Steam	High	260–350
	Ebonol		Conc. Sol. Alkaline Oxidizing Agents.	Alkaline Mist, Steam	High	260–350
	Galvanic-Anodize.	(c)	Ammonium Hydroxide	Ammonia Gas, Steam	Low	140
	Hard Coating Aluminum.		Chromic-Sulfuric Acids	Acid Mist	High	120–180
	Jetal		Conc. Sol. Alkaline Oxidizing Agents.	Alkaline Mist, Steam	High	260–350
	Magcote	(d)	Sodium Hydroxide	Alkaline Mists, Steam	Low-Medium	105–212
	Magnesium Pre Dye Dip		Ammonium Hydroxide Ammonium Acetate	Ammonia Gas, Steam	Low	90–180
	Parkerizing	(a)	Boiling Water	Steam	Medium-High	140–212
	Zincate Immersion.	(e)	None	None	Nil	70– 90

Pickling	Aluminum		Nitric Acid	Nitrogen Oxide Gases	Medium	70– 90
	Aluminum		Chromic, Sulfuric Acids	Acid Mists	Low	140
	Aluminum		Sodium Hydroxide	Alkaline Mist	Medium	140
	Cast Iron		Hydrofluoric Nitric Acids	Hydrogen Fluoride-Nitrogen Oxide Gases.	Medium-High	70– 90
	Copper		Sulfuric Acid	Acid Mist, Steam	Low-Medium	125–175
	Copper	(f)	None	None	Nil	70–175
	Duralumin		**Sodium Fluoride, Sulfuric Acid**	**Hydrogen Fluoride Gas, Acid Mist.**	**Low**	**70**
	Inconel		**Nitric, Hydrofluoric-Sulfuric Acids.**	**Nitrogen Oxide HF Gases, Acid Mist, Steam.**	**Medium-High**	**150–180**
	Iron and Steel		**Hydrochloric-Sulfuric Acids**	**Hydrogen Chloride Gas-Acid Mist, Steam.**	**Medium-High**	**70–190**
	Magnesium		**Chromic-Sulfuric, Nitric Acids**	**Nitrogen Oxide Gas-Acid Mist, Steam.**	**Medium**	**70–160**
	Monel		**Hydrochloric-Sulfuric Acids**	**Hydrogen Chloride Gas-Acid Mist, Steam.**	**Medium-High**	**70-190**
	Nickel		**Hydrochloric-Sulfuric Acids**	**Hydrogen Chloride Gas-Acid Mist, Steam.**	**Medium-High**	**70–190**
	Nickel Silver		Sulfuric Acid	Acid Mist, Steam	Low-Medium	70–140
	Silver		Sodium Cyanide	Cyanide Mist, Steam	Low	70–210
	Stainless Steel	(g)	Nitric, Hydrofluoric Acids	Nitrogen Oxide, Hydrogen Fluoride Gases.	Medium	125–180
	Stainless Steel	(g), (h)	Hydrochloric-Sulfuric Acids	Hydrogen Chloride Gas-Acid Mist, Steam.	Medium-High	70–190
	Stainless Steel Immunization.		Nitric Acid	Nitrogen Oxide Gases	Medium	70–120
	Stainless Steel Passivation.		Nitric Acid	Nitrogen Oxide Gases	Medium	70–120
Acid Dipping	Aluminum Bright Dip.		Phosphoric, Nitric Acids	Acid Mist, Steam	High	200
	Aluminum Bright Dip.		Nitric, Sulfuric Acids	Nitrogen Oxide Gases	Medium-High	70– 90
	Acid Mist					
	Cadmium Bright Dip.		None	None	Nil	70
	Copper Bright Dip.		Nitric, Sulfuric Acids	Nitrogen Oxide Gases, Acid Mist	Medium-High	70– 90

Table 7-19—Continued

AIR CONTAMINANTS RELEASED BY METALLIC SURFACE TREATMENT, PICKLING, ACID DIPPING, AND METAL CLEANING OPERATIONS

Process	*Type*	*Footnote*	*Component of Bath Which May be Released to Atmosphere*	*Physical and Chemical Nature of Major Atmospheric Contaminant*	*Rate of Gassing*	*Usual Temp. Range-°F*
	Copper Semi Bright Dip.		Sulfuric Acid	Acid Mist	Low-Medium	70
	Copper Alloys Bright Dip.		Nitric, Sulfuric Acids	Nitrogen Oxide Gases, Acid Mist	Medium-High	70– 90
	Copper Matte Dip.		Nitric, Sulfuric Acids	Nitrogen Oxide Gases, Acid Mist	Medium-High	70– 90
	Magnesium Dip		Chromic Acid	Acid Mist Steam	Medium	190–212
	Magnesium Dip		Nitric, Sulfuric Acids	Nitrogen Oxide Gases, Acid Mist	Medium-High	70– 90
	Monel Dip		Nitric Sulfuric Acids	Nitrogen Oxide Gases, Acid Mist	Medium-High	70– 90
	Nickel and Nickel Alloys Dip.		Nitric, Sulfuric Acids	Nitrogen Oxide Gases, Acid Mist	Medium-High	70– 90
	Silver Dip		Nitric, Sulfuric Acids	Nitrogen Oxide Gases, Acid Mist	Medium-High	70– 90
	Zinc and Zinc Alloys Dip.		Chromic, Hydrochloric Acids	Hydrogen Chloride Gas (If HCl attacks Zn).	Nil-Low	70– 90
Metal Cleaning	Alkaline Cleaning	(i)	Alkaline Sodium Salts	Alkaline Mist, Steam	Medium-High	160–210
	Emulsion Cleaning.		Petroleum-Coal Tar Solvents	Petroleum-Cola Tar Vapors	Low-Medium	70–140
	Emulsion Cleaning.		Chlorinated Hydrocarbons	Chlorinated Hydrocarbon Vapors	Medium	70–140

FOOTNOTES

(a) Also Aluminum Seal, Magnesium Seal, Magnesium Dye Set, Dyeing Anodized Magnesium. Magnesium Alkaline Dichromate Soak. Coloring Anodized Aluminum

(b) Stainless Steel before Electropolishing

(c) On Magnesium

(d) Also Manodyz, Dow–12

(e) On Aluminum

(f) Sodium Dichromate, Sulfuric Acid Bath and Ferrous Sulfate, Sulfuric Acid Bath

(g) Scale Removal

(h) Scale loosening

(i) Soak and Electrocleaning

CHAPTER 8

AIR POLLUTION EFFECTS

Effects on Humans. The effects of air pollutants on health are generally respiratory in nature. These differ according to whether the exposures are acute or chronic. Acute exposures imply high concentrations of pollutants over relatively short periods of time, while chronic exposure implies low concentrations over relatively long periods of time.

a. Acute Exposures. Much more is known about acute exposures due to such disasters as occured in the Meuse Valley, Belgium, in 1930; in Donora, Pennsylvania, in 1948, and in London, England, in 1952. The seriousness of these incidents is amply illustrated by the Donora episode of 1948 which resulted in the death of 20 persons along with 6,000 illnesses out of a total population of 12,000.

(1) Most acute episodes are relatively short in duration (a few days to one week), occur during periods of atmospheric stagnation, produce effects which are mainly respiratory and cardiac in nature, and primarily affect older individuals and those already ill with cardiac and pulmonary diseases. Infants have also been found to be susceptible. These disasters resulted from the presence of high concentrations of a number of pollutants coming from many sources. None of the effects could be traced to individual contaminants.

(2) Air pollution episodes where one instead of many sources of pollutants were involved have been recorded. In such an instance the contaminant causing the health problems is easily identified. An example of such an acute exposure would be the Poza Rica episode in Mexico where a large quantity of hydrogen sulfide was accidentally discharged and caused the hospitalization of 320 of which 22 died. The one-source episodes have the same general features as the multi-source episodes with respect to durations, population affected, and existing meteorological conditions.

b. Chronic Exposures. Many more people are exposed to air pollutants on a chronic rather then an acute basis; however, little is known concerning the resulting health effects. In fact, a controversy exists as to whether such exposure even impairs health.

(1) Studies by the US Public Health Service have found a correlation between air pollution and deaths caused by cancer of the respiratory system, cancer of the gastro-intestinal tract, arteriosclerotic heart disease, and myocardial degeneration. In British studies a correlation has been shown between air pollution and chronic bronchitis. Air pollution has been accused of increasing the incidence of other specific diseases but, as in the studies mentioned, the effects of additional possible contributory factors such as crowded living conditions are hard to eliminate from the data.

(2) It is believed that the health effect of chronic exposure to low levels of air contaminants will be the intensification of existing diseases, particularly those of a cardiac or respiratory nature. Levels of air pollutants regularly found in cities and heavily industrialized areas will not be major causes of disease; however, the importance of such levels of contamination as secondary or aggravating factors in already ill or debilitated persons is of important consequence.

Effects on Animal Life:

a. It is well known that animals must be considered when evaluating exposures. The mechanism by which an animal can become poisoned in many instances is completely different from that by which humans are affected. As in humans, inhalation is an important route of entry in acute air pollution exposures such as occurred in the Meuse Valley and Donora incidents. However, probably the most common exposure to herbivorous animals grazing within a zone of pollution will be the ingestion of feed contaminated by air pollutants. In this case inhalation is of secondary importance.

b. Air pollutants that present a hazard to livestock, therefore, are those which are taken up by vegetation or deposited on the plants. only a few pollutants have been observed to cause harm to animals. These include arsenic, fluorides, lead, and molybdenum.

Effects on Vegetation:

a. The effects of air pollution on vegetation can appear as death, stunted growth, reduced crop yield, and degradation of color. It is interesting to note that in some cases of color damage such as the silvering of leafy vegetables by oxidants, the plant may still be used as a food without any danger to consumer; however, the consumer will not buy such vegetables on aesthetic grounds so the farmer still sustains a loss.

Effects on Materials:

a. **Discoloration.** Many air pollutants accumulate on and discolor buildings. Sooty material not only blackens buildings, but it can accumulate and become encrusted. This can hide lines and decorations and thereby disfigure structures and reduce their aesthetic appeal. Another common effect is the discoloration of paint by certain acid gases. A good example is the blackening of white paint with a lead base by hydrogen sulfide.

b. **Corrosion.** A more serious effect and one of great economic importance is the corrosive action of acid gases on building materials. Such gases can cause stone surfaces to blister and peel. Mortar can be reduced to powder. Metals are also damaged by corrosive action of some pollutants. Another common effect is the deterioration of tires and other rubber goods. Cracking and apparent "drying" occurs when these goods are exposed to ozone and other oxidants.

c. **Soiling of Goods.** Wearing apparel and household goods can easily be soiled by air contaminants. This can become expensive due to the necessity for more frequent cleaning. Also, more frequent cleaning often leads to a shorter life span for the materials involved and new goods will have to be purchased more often.

d. **Visibility.** The impairment of atmospheric visibility (i.e., decreased visual range through a polluted atmosphere) is caused by the scattering of sunlight by particles suspended in the air. It is not a result of sunlight being obscured by materials in the air. Since light scattering and not obscuration is the main cause of the reduction in visibility, reduced visibility due to the presence of air pollutants occurs primarily during bright days. On cloudy days or at night there may be no noticeable effect although the same particulate concentration may exist during these times as during the sunny days. Reduction in visibility creates several problems. The most significant are the adverse effects on aircraft, highway, and harbor operations. Reduced visibility can also cause adverse aesthetic impressions which can seriously restrict the growth and development of any area. Extreme conditions such as dust storms or sandstorm can actually cause physical damage in and by themselves.

The Cost of Air Pollution:

a. It is reasonably safe to assume that the direct economic loss to the United States is due to air pollution amounts to several billion dollars annually. It would be impossible to derive an exact figure.

b. Money is being lost daily due to the destructive nature of air pollution. This is not the only cost; millions of dollars are being spent to control air pollution. Local, state, and Federal agencies have been established to deal with the problem such as in New York City and Los Angeles County, the states of Colorado and Pennsylvania, and in the United States Public Health Service. Along with this activity comes research and training which is also expensive. Private industry is also bearing a major share of the cost of air pollution control by purchasing expensive control devices and sponsoring research activities.

CHAPTER 9

ATMOSPHERIC CARBON MONOXIDE

Carbon monoxide (CO) is the most widely distributed and the most commonly occurring air pollutant. Total emissions of CO to the atmosphere exceed those of all other pollutants combined.

Most atmospheric CO is produced by the incomplete combustion of carbonaceous materials used as fuels for vehicles, space heating, and industrial processing or burned as refuse. Man's activities are, therefore, largely responsible for CO contamination, and his technological advances have contributed to the present atmospheric concentrations.

No large natural source of CO has been positively identified, but a number of geophysical and biological sources of CO are known. Their ultimate contribution to urban atmospheric concentrations is thought to be relatively small.

CARBON MONOXIDE IN HISTORY

Human experience with CO probably began during prehistoric times when man first discovered fire. CO poisoning has been traced through Greek and Roman literature; in fact, this form of poisoning has been closely associated with the history of mankind.

Materials used for making fire in prehistoric times were probably wood, grasses, and other organic matter. Lewin cites many cases of CO poisoning due to the incomplete combustion of such fuels. He also includes references to the effects of CO on the health of man in the writings of Greek physicians of the Third Century.

With the increased use of coal for domestic heating in the Fifteenth Century, CO poisoning increased greatly. This increase was due to inhalation of CO formed in incomplete combustion in the heating of homes and to the exposure of men in mines where the deadly "white damp" was encountered after explosions and mine fires. The introduction of illuminating gas (a mixture of hydrogen, carbon monoxide, methane, and other hydrocarbons) for domestic heating (still used extensively in Europe but largely replaced by natural gas in the United States) further increased this hazard.

The introduction of the internal combustion engine for transportation and the development of a number of technological processes wherein CO is produced have greatly increased the production of CO and its release to the atmosphere.

Over the past few centuries, therefore, the problem of dealing with CO has spread from individual dwellings and work environments to include the ambient air in cities. Concern has now broadened from the acute and often lethal effects of high concentrations of the gas to encompass as well those effects that may occur as a result of considerably longer exposures to much lower concentrations.

OCCURRENCE

Technological Sources

Carbon monoxide is found among the combustion products of organic materials used as fuels. Transportation activities represent the largest source category. Other major technological sources of CO emissions are stationary heat-generating facilities, industrial processes, and solid-waste combustion. CO is also formed in explosions and in the firing of weapons. It occurs in high concentrations in cigarette smoke.

Natural Sources

a. Nonbiological

Some CO is reported to be produced in volcanos and is formed in natural gases (e.g., gases found in coal mines).[4] It has also been reported to be formed during electrical storms.[5] Forest fires caused by lightning sometimes make a considerable contribution to atmospheric CO.[6] A small amount of CO is also formed as a photochemical degradation product of various reactive organic compounds in the process involving the formation of photochemical smog.[7,8] Some CO is believed to be formed in the upper atmosphere above 70 kilometers (km) by photodissociation of carbon dioxide (CO_2).[9]

b. Biological

Small quantities of CO are formed by vegetation during seed germination and seedling growth of plants,[10] and it has been observed in injured, cut, or dried plants.[11] Some CO is found in marsh gases and it is also formed by certain brown algae (kelps).[12,13] Carbon monoxide concentrations of up to 900 milligrams per cubic meter (mg/m^3) (786 ppm) have been found in the floats of *Nereocyctis*, a familiar kelp or seaweed.[14] Mature leaves of green plants have been shown to produce CO.[11] Microorganisms also have shown to produce CO from plant flavonoids.[15]

Carbon-monoxide-producing colonies of marine hydrozoan jellyfish (siphonophores) are widespread and make up a large portion of the plankton in the warmer oceans of the world.[16,17] CO is also produced in the float cells of the surface-dwelling *Physalia physalis* (Portugese Man-of-War).[18] In addition to the generation of CO by ocean-dwelling biological specimens, other mechanisms may exist through which CO is generated in the ocean.

Another biological source of CO is endogenous CO, which is produced in measurable quantities in man and animals as a by-product of heme catabolism.[19-21] This endogenous Co is produced in larger amounts in hemolytic disease states.

PROPERTIES AND GASEOUS REACTIONS OF CARBON MONOXIDE

Physical Properties

Carbon monoxide is a colorless, odorless, tasteless gas, slightly lighter than air.[22] Quite flammable, it burns with a bright blue flame, but does not support combustion.

Physical properties of CO are outlined in Table **9-1.**

Gaseous Chemical Reactions of Carbon Monoxide

The likely gaseous reactions wherein CO might be oxidized to CO_2 have been reviewed by Bates and Witherspoon.[9] The reaction with molecular oxygen:

$$2\,CO + O_2 \rightarrow 2\,CO_2 \qquad (1)$$

is possible in the lower atmosphere, but is apparently unimportant. Both dry and moist experimental mixtures of CO and oxygen have remained unchanged after 7 years of exposure to sunlight.[23] The possibilities of reactions between CO and atomic oxygen have been reviewed by Leighton,[8] and he has concluded that they are unimportant; Bates and Witherspoon have come to a similar conclusion.[9]

a. Lower Atmospheric Reactions

Two oxidation reactions, although very slow, do occur in the lower atmosphere.[24-26]

$$CO + O_2 \rightarrow CO_2 + O \qquad (2)$$

and in the presence of moisture

$$CO + H_2O \rightarrow CO_2 + H_2. \qquad (3)$$

Both of these reactions have appreciable energy barriers, 51 and 56 kilocalories per mole (kcal/mole), respectively. Direct chemical reactions between CO and oxygen or water occur at a frequency of less than one reaction per 10^{15} molecular collisions at room temperature. These reactions become important as gas-phase processes primarily at temperatures above 500°C and even more so above 1000°C.

Table 9-1 **PHYSICAL PROPERTIES OF CARBON MONOXIDE**

Molecular weight	28.01
Melting point	-207°C
Boiling point	-192°C
Specific gravity relative to air	0.968
Density	
At 0°C, 760 mm Hg	1.25 g/liter
At 25°C, 760 mm Hg	1.15 g/liter
Explosive limits in air	12.5 to 74.2% (volume)
Solubility[a]	
At 0°C	3.54 ml/100 ml water
At 25°C	2.14 ml/100 ml water
Conversion factors	
At 0°C, 760 mm Hg	1 mg/m^3 = 0.800 ppm 1 ppm = 1.250 mg/m^3
At 25°C, 760 mm Hg	1 mg/m^3 = 0.874 ppm 1 ppm = 1.145 mg/m^3

[a]Volume of CO indicated is at 0°C, 760 mm Hg.

These reactions occur more readily and at lower temperatures on the surfaces of certain catalysts, usually metal oxides. For example, oxidation is catalyzed at room temperatures by metallic catalysts, such as Hopcalite, which is a mixture of the oxides of manganese and copper, and by palladium on silica gel.[27]

Ozone will oxidize CO to CO_2, but the rate of this reaction is extremely slow at atmospheric temperatures and concentrations.[28,29] A high activation energy of about 20 kcal for the oxidation of CO by ozone has been found.[30]

Oxidation by NO_2 in the reaction

$$NO_2 + CO \rightarrow CO_2 + NO \qquad (4)$$

has an even higher activation energy than does the oxidation of CO by ozone. The activation energy of 28 kcal essentially precludes the occurrence of this process in the atmosphere.[31]

Consideration has also been given to the possibility that some very rapid reactions may occur between CO and certain intermediates of photochemical smog reactions. One possible intermediate is the hydroxyl radical, which may occur when the photolysis of aldehydes produces perhydroxyl, which can then be reduced to the hydroxyl radical. Hydroxyl appears to react very rapidly with CO, and there is some indication that perhydroxyl also reacts with CO.[32] Such reactions involve chain-type mechanisms, and Doyle[33] has calculated that a global average hydroxyl concentration of only 10^{-9} to 10^{-8} part per million (ppm) would be sufficient to convert all emitted CO to CO_2.

Methane, which is found in the troposphere at concentrations one order of magnitude higher than CO concentrations, can also be oxidized by hydroxyl radicals. The rate of reaction is slow, however, and the extent to which methane competes with CO for the hydroxyl radicals in the atmosphere has not been defined.

Although future research with hydroxyl radicals may be rewarding, at present one must conclude that significant gaseous oxidation reactions of CO in the ambient atmosphere have not been proved.

b. *Upper Atmospheric Reactions*

Short-wavelength ultraviolet radiation (about 1700Å) dissociates CO_2 into CO and atomic oxygen[9] according to the reaction:

$$CO_2 + h\nu \rightarrow CO + O. \qquad (5)$$

The yield of CO from the photodissociation of CO_2 in the upper atmosphere, however, is considered to be relatively small at levels below 100 kilometers since the intensity of active ultraviolet radiation falls off rapidly at that level.

Carbon dioxide may be reformed by a three-body collision:

$$CO + O + M \rightarrow CO_2 + M \qquad (6)$$

where M represents the third body. The third body absorbs excess energy by collision from the energy-rich combination of a CO molecule and an oxygen atom. In the absence of such a collision, no stable CO_2 can be formed. The probability of this collision occurring is small, however, because of the low concentration of such third bodies at higher altitudes.

From the foregoing information, it must be concluded that no gaseous reactions have been shown to be important scavengers of CO in the atmosphere.

BACKGROUND LEVELS AND FATE OF ATMOSPHERIC CARBON MONOXIDE

Background Levels and Estimated Mean Lifetime of Carbon Monoxide

The amount of CO measurable in relatively unpolluted ("clean") air is small. Junge has estimated the background level of CO in the lower atmosphere to be in the range of 0.01 to 0.2 mg/m^3 (0.01 to 0.2 ppm).[34] Studies of background levels of CO using solar spectral techniques at Mt. Wilson, California; Ottawa, Canada; Jungfraujoch, Switzerland; and Columbus, Ohio, confirmed that such levels average about 0.1 mg/m^3 (0.1 ppm).[35-37] There appears to be no significant variation in such levels with geographical locations.[37] Robbins et al.,[38] using an experimental, continuous-measurement HgO method sensitive in parts per billion (ppb), have recently determined that North Pacific marine air mass concentrations may contain as little as 0.029 mg/m^3 (0.025 ppm) and the clean (nonurban) air mass over continental California contains 0.06 to 1.2 mg/m^3 (0.05 to 1.0 ppm). These investigators have also determined, on the basis of a 12-day sample, that typical background levels of CO at Inge Lehmann Station in northern Greenland range from 0.06 to 0.75 mg/m^3 (0.05 to 0.65 ppm).[39] The variability of CO in nonpolluted areas appears to be a characteristic of the air mass in transit and reflects its prior history. Robinson and Robbins estimate that arctic air masses passing over Greenland have CO concentrations of between 0.11 and 0.23 mg/m^3 (0.10 and 0.20 ppm) and that concentrations of 0.6 to 1.2 mg/m^3 (0.5 to 1.0 ppm) can occur when the air mass has recently traversed a heavily populated area.[39] Measurements at Point Barrow, Alaska, gave clear air CO levels, ranging from 0.063 to 0.299 mg/m^3 (0.055 to 0.260 ppm) and averaging 0.104 mg/m^3 (0.90 ppm).[40]

The exact turnover time of CO in the atmosphere is not known with certainty. A recent estimate, derived from radioactive measurements, of the residence time of CO in the lower atmosphere has a lower limit of 0.1 year.[41] Another estimate based on different technique is 3 years.[34] Robbins et al. estimate a mean residence time of about 5 years, and have tentatively concluded that the background levels of CO are not rising significantly at the persent time.[38]

The previously discussed relative inertness of CO in reactions with the normal gaseous atmospheric constituents and its photon transparency[43] effectively eliminate the possibility of chemical reactions as a mechanism for CO removal in the lower atmosphere with the possible exception of the still speculative reaction with the hydroxyl radical.

In spite of the above, calculations have shown that in the absence of removal processes, the estimated world-wide emissions of CO from technological sources (estimated to be on the order of 1.8 to 2.1 x 10^{11} kg or 198 to 231 million tons per year[13,40]) would be sufficient to raise the atmospheric background concentration of CO by 0.03 mg/m^3 (0.03 ppm) per year creating a current background level of 1 mg/m^3 (1 ppm).[9,38,44] It is therefore postulated that some "sink" or removal process for atmospheric CO exists.

Possible Processes for Carbon Monoxide Removal

The following discussion represents a review of possible CO removal processes. Some of the CO "sinks" discussed below are highly speculative, however; in fact, they may actually contribute CO rather than remove it from the atmosphere.

a. Atmospheric Migration (Upper Atmospheric Sink).

Conceivably CO in the lower atmosphere may eventually migrate by atmospheric mixing to a potential sink in the upper atmosphere, where it is oxidized to CO_2 in the presence of high-intensity ultraviolet solar radiation. A recent laboratory study by Harteck and Reeves[45] confirmed that CO, in the presence of NO_2 or other absorbing molecules, when subjected to high-intensity, ultraviolet radiation in an evacuated chamber, was oxidized to CO_2.

b. Biological Removal (Terrestrial and Marine Biosphere Sink).

Another possible removal agent of atmospheric CO is the presence, in significant numbers, of plants and microorganisms that can metabolize CO. The earth's surface is a possible vector for the removal of CO from the atmosphere. Carbon monoxide in contact with the soil may be oxidized to CO_2 and converted to methane (CH_4) by common specific anaerobic methane-producing soil microorganisms, *Methanosarcina barkerii* and *Methanobacterium formicum*, in the presence of moisture.

This action has been demonstrated in the laboratory by Schnellen,[46] who showed that pure cultures of these bacteria convert CO into methane. Schnellen has found that *Ms. barkerii* is capable of converting CO to CH_4 according to the equation:

$$4\,CO + 2\,H_2O \rightarrow CH_4 + 3\,CO_2 \qquad (7)$$

Stephenson,[47] however, indicates that CO, in the absence of H_2, reacts with water in these bacteria in two stages as follows:

$$4\,CO + 4\,H_2O \rightarrow 4\,CO_2 + H_2 \qquad (8)$$

and

$$CO_2 + 4\,H_2 \rightarrow CH_4 + 2\,H_2O. \qquad (9)$$

In the presence of H_2, these bacteria convert CO directly into methane and water:

$$CO + 3\,H_2 \rightarrow CH_4 + H_2O. \qquad (10)$$

An aerobic soil bacterium, *Bacillus oligocarbophilus (Carboxydomonas oligocarbophila)*, found in and isolated from arable soil, has also been demonstrated to oxidize CO to CO_2. When cultivated on simple organic media free from other carbon sources, this organism oxidizes CO to CO_2, which is then utilized as a source of energy.[48] Another bacterium, *Clostridium welchii*, when grown in the presence of CO, has been reported to produce lactic acid as a fermentation product.[49]

Although these soil bacteria utilize CO in their fermentation or metabolic processes, it is difficult to estimate global destruction rates of CO on the basis of such laboratory experiments.

Within the biosphere, the process of plant respiration may serve as a potential removal process; but this concept has not been firmly established, even though it is recognized that plants are scavengers for a wide variety of atmospheric materials.

c. Biochemical Removal (Biochemical Sink).

A potential biochemical removal process for CO is the binding of CO to the porphyrin-type compounds that are widely distributed in plants and animals. In particular, the heme compounds such as hemoglobin, which are analogous to porphyrin compounds found in plants are known to bind CO. It must be noted, however, that practically all of the CO absorbed by the heme compounds found in man and animals is eventually discharged from the blood, and only a small fraction is retained.[50] This type of process in vegetation may have an important potential for scavenging atmospheric CO.Permanent removal from the environment, however, would depend on whether CO subsequently enters into some reaction process to form CO_2 when the porphyrin compound is degraded.

d. Absorption in Oceans (Oceanic Sink).

While absorption in the world's oceans is a recognized sink process for atmospheric CO_2, there is no evidence at present that the oceans are a sink for CO because no process or reaction that would remove CO has been discovered.[1,14] The solubility of CO at 1 atmosphere (atm) pressure in sea water ranges from about 17 to 32 milliliters CO (0°C, 1 atm)/liter water over a water temperature and chlorinity range of -2° to 30°C and 15 to 21 grams of chlorine per kilogram of water, respectively.[51] This solubility is too low for the oceans to accumulate CO or for precipitation washout to be an effective scavenging agent. An apparent daily cycle for CO concentrations over ocean areas seems to indicate some involvement of the ocean in the fate of CO.[34] The high CO concentrations in some marine plants[14] may also be indicative of some complex system in the ocean involving CO.

Swinnerton et al. simultaneously sampled the CO content of the air and the surface waters at 29 different points along the path of an oceanographic cruise between Washington, D.C., and Puerto Rico.[52] These investigators found that the measured concentrations of CO in the water exceeded by 7 to 90 times the concentrations of CO theoretically expected to be present based on solubility calculations. Apparently, man-made pollution is not the principal source of CO in sea water, since the highest ratios were found in the open ocean. The ocean, therefore, may be a source rather than a sink for atmospheric CO, as discussed previously.

e. Adsorption on Surfaces.

Kummler et al. indicate that the gas-phase oxidation of CO by nitrous oxide (N_2O), considered too slow to be of importance in the atmosphere, is catalyzed in the presence of certain surfaces (such as charcoal, carbon black, and glass) at 300°C and above.[53] These investigators have extrapolated these reported high-temperature reaction rates to 27°C and consider that such a catalytic reaction is feasible at ambient temperatures. This conclusion is supported by laboratory studies of Gardner and Petrucci, where the chemisorption of CO on metallic films (such as copper, cobalt, and nickel oxides) at room temperatures have been measured by infrared spectroscopy.[54]

The necessary data for evaluation of the catalytic efficiency of common surfaces such as metals, soil, and atmospheric particles are presently unavailable; therefore, the possibility of such surfaces serving as a sink by adsorption of atmospheric CO is uncertain.

FORMATION OF CARBON MONOXIDE

CO arises primarily from incomplete or inefficient combustion of carbonaceous fuels. Oxygen concentration, flame temperature, gas residence time, and combustion chamber turbulence are important variables that affect the exhaust concentrations of CO.

Internal Combustion Engines

The two factors that effectively determine total CO emissions from internal combustion engines are the concentration of CO in the exhaust and the exhaust volume. The exhaust concentration depends mainly on the air-to-fuel (A/F) ratio entering the combustion cylinder; the exhaust volume depends on the power output.

Exhaust concentrations of CO increase with lower (richer) A/F ratios, decrease with higher (leaner) A/F ratios, but remain relatively constant with ratios above the stoichiometric ratio of about 14.5 to 1. The behavior of gasoline automobile engines before and after the imposition of pollutant control measures differs considerably. Depending on the mode of driving, the average precontrol engine operates at A/F ratios ranging from about 11 to a point slightly above the stoichiometric ratio. During the idling mode, at low speeds with light load (such as low-speed cruise), during the full-throttle mode until speed picks up, and during deceleration, the A/F ratio is low in precontrol cars and CO emissions are high. At higher speed cruise and during moderate acceleration, the reverse is true. Cars with exhaust controls generally remain much closer to stoichiometric A/F ratios in all modes, and thereby the CO emissions are kept lower. The relationship between CO concentration in exhaust and the A/F ratio is shown in Figure 9-1.

The exhaust flow rate increases with increasing engine power output. During idle the flow is minimum; during full throttle it is maximum. At cruise engine conditions (i.e., constant engine speed and power output) the exhaust flow remains constant. It increases during engine acceleration.

Correlations between total emissions of CO in pounds per vehicle mile and average route

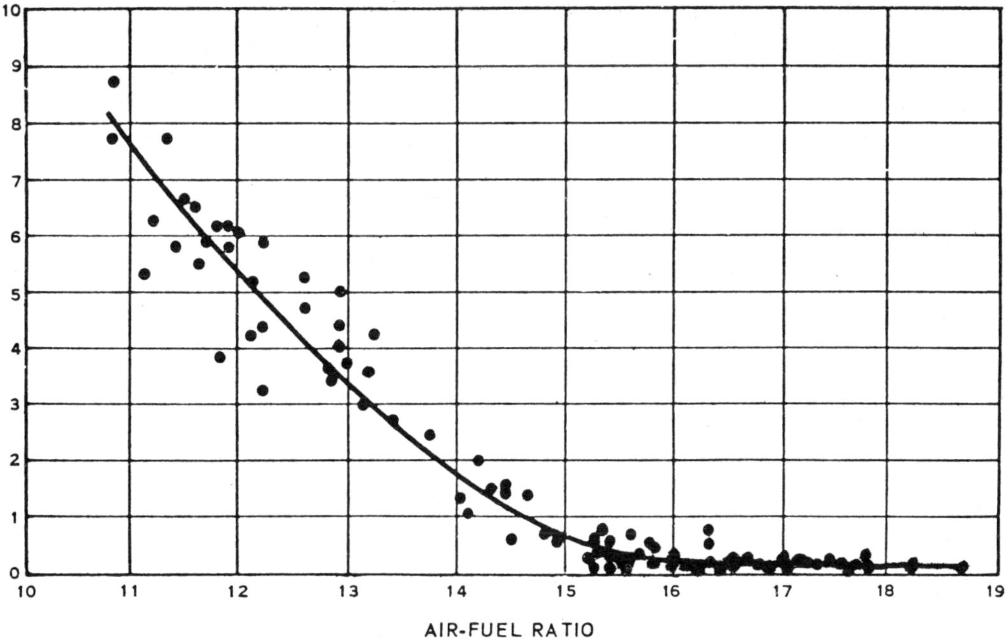

Fig. 9-1 Effect of air-fuel ratio on exhaust gas carbon monoxide concentrations from three test engines.

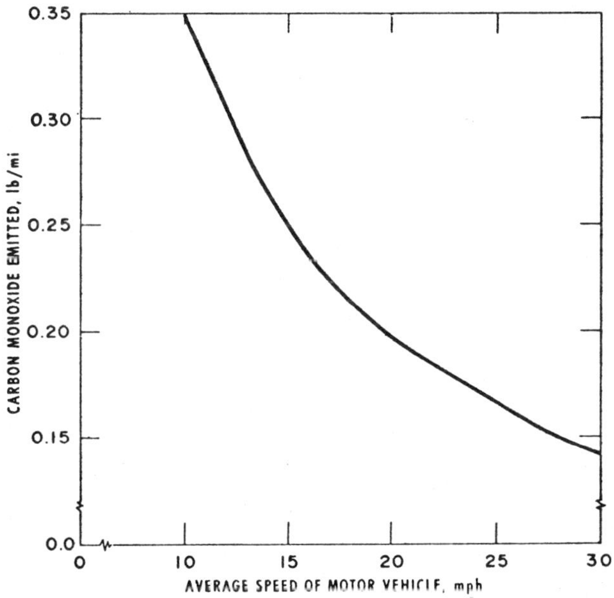

Fig. 9-2 Emissions of carbon monoxide vs. vehicle speed.

speed show decrease in emissions with increasing average speed. The greater emissions per mile during the low-speed conditions (below 20 miles per hour) are due to an increase in the frequency of the acceleration, deceleration, and idle stages of the driving cycle encountered in heavy traffic.

Stationary Combustion Sources

In stationary combustion systems, the concentration of CO is lowest near the stoichiometric ratio of air to fuel. At lower than stoichiometric A/F ratios, high CO concentrations reflect the relatively low oxygen concentration and the possibility of poor reactant mixing from low turbulence. These two factors can increase emissions even though flame temperatures and residence time are high. At higher than stoichiometric A/F ratios, increased CO emissions result from decreased flame temperatures and shorter residence times. These two factors control even though oxygen concentrations and turbulence increase. Minimal CO emissions and maximum thermal efficiency therefore require combustor designs that provide high turbulence, sufficient residence time, high temperatures, and near stoichiometric A/F ratios. Combustor design dictates the minimum CO emission that can be achieved, and operating practice dictates the actual approach to that minimum.

ESTIMATION OF CARBON MONOXIDE EMISSIONS

An estimated 92 x 10^9 kilograms (102 million tons) of CO were emitted in the United States in 1968. This amounts to 50 percent by weight of all major air pollutant emissions that year. Fuel combustion in mobile sources utilizing the internal combustion engine is the principal source category of CO (58 percent). Miscellaneous combustion sources, principally forest fires, and industrial process sources are the second (17 percent) and third (11 percent) largest categories, respectively. Disposal of solid wastes is the fourth (8 percent) greatest source category of CO emissions, and stationary fuel combustion is the fifth (2 percent).

REFERENCES

1. Jaffe, L. S. Ambient Carbon Monoxide and Its Fate in the Atmosphere. J. Air Pollution Control Assoc. *18*:534-540, August 1968.
2. Cooper, A. G. Carbon Monoxide. A Bibliography with Abstracts. U. S. DHEW, PHS. Division of Air Pollution. Washington, D.C. PHS Publication Number 1503; Bibliography No. 68. 1966. 440 p.
3. Lewin, L. Carbon Monoxide Poisoning: A Manual for Physicians, Engineers, and Accident Investigators. Berlin Julius Springer-Verlag, 1920.
4. Flury, F. and F. Zernik. Noxious Gases-Smoke, Fog, Fumes and Dust. Berlin, Julius Springer-Verlag 1931.
5. White, J. J. Carbon Monoxide and Its Relation to Aircraft. U. S. Naval Med. Bull. *30*(2):151-165, April 1932.
6. National Air Pollution Control Administration, Reference Book of Nationwide Emissions. U.S. DHEW, PHS, CPEHS, NAPCA. Durham, N.C.
7. Altshuller, A. P. and J. J. Bufalini. Photochemical Aspects of Air Pollution: A Review. Photochem. Photobiol. *4*(2):97-146, March 1965.
8. Leighton, P. A. Photochemistry of Air Pollution. New York, Academic Press, 1961. 300 p.
9. Bates, D. R. and A. E. Witherspoon. The Photochemistry of Some Minor Constiuents of the Earth's Atmosphere (CO_2, CO, CH_4, N_2O). Monthly Notices of the Roy. Astron. Soc. (London). *112*:101-124, January 1952.
10. Siegel, S. M., G. Renwick, and L. A. Rosen. Formation of Carbon Monoxide During Seed Germination and Seedling Growth. Science. *137*(3531):683-684, August 31, 1962.
11. Wilks, S. S. Carbon Monoxide in Green Plants. Science. *129*(3354):964-966, April 10, 1959.
12. Loewus, M. W. and C. C. Delwiche. Carbon Monoxide Production by Algae. Plant Physiol. *38*(4):371-374, July 1963.
13. Chapman, D. J. and R. D. Tocher. Occurrence and Production of Carbon Monoxide in Some Brown Algae. Can. J. Botany. *44*:1438-1442, October 1966.
14. Robinson, E. and R. C. Robbins. Sources, Abundance and Fate of Gaseous Atmospheric Pollutants. Stanford Research Institute., Menlo Park, Calif. Final Report. SRI Project PR-6755. February 1968. 124 p.
15. Westlake, D. W. S., J. M. Roxburgh, and G. Talbot. Microbial Production of Carbon Monoxide from Flavonoids. Nature. *189*(4763):510-511, February 11, 1961.
16. Barham, E. G. Siphonophores and the Deep Scattering Layer. Science. *140*(3568):826-828, May 17, 1963.
17. Barham, E. G. and J. W. Wilton. Carbon Monoxide Production by a Bathypelagic Siphonophore. Science. *144*(3620):860-862, May 15, 1964.
18. Wittenberg, J. B. The Source of Carbon Monoxide in the Float of the Portugese Man-of-War, *Physalia physalis L.* J. Exp. Biol. *37*(4):698-705, December 1960.
19. White, P. et al. Carbon Monoxide Production Associated with Ineffective Erythropoiesis. Blood. *24*(6):845, December 1964.
20. Middelton, V. et al. Carbon Monoxide Accumulation in Closed Circle Anesthesia Systems. Anesthesiology. *26*(6):715-719, November-December 1965.
21. Landaw, S. A. and H. S. Winchell. Endogenous Production of Carbon-14 Labeled Carbon Monoxide: An *In Vivo* Technique for the Study of Heme Catabolism. J. Nucl. Med. 7:696-707, September 1966.
22. Hygienic Guide Series: Carbon Monoxide. Amer. Ind. Hyg. Assoc. J. *26*(4):431-434, July-August 1965.
23. Mellor, J. W. Carbon. In: A Comprehensive

Treatise on Inorganic and Theoretical Chemistry, Vol. V. London, Longmans, Green and Co., 1924. p. 926-950.

24. Tunder, R. et al. Compilation of Reaction Rate Data for Non-Equilibrium Performance and Re-entry Calculation Programs. Aerospace Corp. El Segundo, Calif. Aerospace Report Number TR-1001(9210-02)-1. January 1967.

25. Fischer, E. R. and M. McCarty, Jr. Study of the Reaction of Electronically Excited Oxygen Molecules with Carbon Monoxide. J. Chem. Phys. *45*(3):781-784, August 1, 1966.

26. Graven, W. M. and F. J. Long. Kinetics and Mechanisms of the Two Opposing Reactions of the Equilibrium $CO + H_2O = CO_2 + H_2$. J. Amer. Chem. Soc. *76*(10):2602-2607, May 20, 1954.

27. Carbon Monoxide. In: The Merck Index of Chemicals and Drugs. Stecher, P. G. (ed.). Rahway, N. J., Merck and Co., 1960. p. 212.

28. Garvin, D. The Oxidation of Carbon Monoxide in the Presence of Ozone. J. Amer. Chem. Soc. *76*(6):1523-1527, March 20, 1954.

29. Harteck, P. and S. Dondes. Reaction of Carbon Monoxide and Ozone. J. Chem. Phys. *26*:1734-1737, June 1957.

30. Zatsiorskii, M., V. Kondratev, and S. Solnishkova. Radiation of the Flame of $CO + O_3$ and the Mechanism of this Reaction. Zh. Fiz. Khim. (Leningrad). 14(12):1521-1527. 1940.

31. Brown, F. B. and R. H. Crist. Further Studies on the Oxidation of Nitric Oxide; The Rate of the Reaction Between Carbon Monoxide and Nitrogen Dioxide. J. Chem. Phys. 9:840-846, December 1941.

32. Greiner, N. R. Hydroxyl-Radical Kinetics by Kinetic Spectroscopy. I. Reactions with H_2, CO, and CH_4 at 300°K. J. Chem. Phys. *46*(7):2795-2799, April 1, 1967.

33. Doyle, G. J. Unpublished Data. Stanford Research Institute. Menlo Park, Calif. 1968.

34. Junge, C. E. Air Chemistry and Radioactivity. New York, Academic Press, 1963. 382 p.

35. Migeotte, M. and L. Neven. Recent Progress in the Observation of the Solar Infrared Spectrum at the Scientific Station at Jungfraujoch, Switzerland [Recents Progres Dans L'Observation du Spectre Infrarouge du Soleil a la Station Scientifique du Jungfraujoch (Suisse)]. Mem. Soc. Roy. Sci. Liege. *12*(I-II):165-178, 1952.

36. Benesch, W., M. Migeotte, and L. Neven. Investigations of Atmospheric CO at the Jungfraujoch. J. Opt. Soc. Amer. *43*:1119-1123, November 1953.

37. Locke, J. L. and L. Herzberg. The Absorption Due to Carbon Monoxide in the Infrared Solar Spectrum. Can. J. Phys. *31*(4):504-516, May 1953.

38. Robbins, R. C., K. M. Borg, and E. Robinson. Carbon Monoxide in the Atmosphere. J. Air Pollution Control Assoc. *18*:106-110, February 1968.

39. Robinson, E. and R. C. Robbins. "Atmospheric CO Concentrations on the Greenland Ice Cap." Jour. Geophysical Research. 74(8):1968-1973, April 15, 1969.

40. Cavanagh, L. A. C. F. Schadt, and E. Robinson. Atmospheric Hydrocarbon and Carbon Monoxide Measurements at Point Barrow, Alaska. Environ. Sci. Technol. *3*:251-257, March 1969.

41. Weinstock, B. Carbon Monoxide: Residence Time in the Atmosphere. Science. *166*(3902):224-225, October 10, 1969.

42. Private communication with R. C. Robbins. Menlo Park, Calif. September 24, 1969.

43. Penndorf, R. The Vertical Distribution of Atomic Oxygen in the Upper Atmosphere. J. Geophys. Res. *54*(1):7-38, March 1949.

44. Haagen-Smit, A. J. and L. G. Wayne. Atmospheric Reactions and Scavenging Processes. In: Air Pollution, Stern, A. C. (ed.), Vol. I, 2d ed. New York, Academic Press, 1968, p. 149-186.

45. Harteck, P. and R. E. Reeves, Jr. Some Specific Photochemical Reactions in the Atmosphere. Presented at the Symposium on the Chemistry of the Natural Atmosphere, American Chemical Society Annual Meeting. Chicago. September 1967.

46. Schnellen, C. G. T. P. Research on Methane Fermentation [Onderzoekingen Over de Methaangisting]. Doctoral Thesis, Technische Wetenschap de Delft. Rotterdam, Drukkerij de Maasstad, 1947. 137 p.

47. Stephenson, M. Bacterial Metabolism. 3rd ed. New York, Longmans, Green and Co., 1949. 398 p.

48. Kaserer, H. The Oxidation of Hydrogen by Microorganisms [Die Oxydation des Wasserstoffes durch Mikroorganismen]. Zentr. Bakteriol. Parasitenk, Abt. II. *16*(14-16):681-696, 1906.

49. Waksman, S. A. Principles of Soil Microbiology. 2d ed. Baltimore, Williams & Wilkins Co., January 1932. 894 p.

50. Luomanmaki, K. Studies on the Metabolism of Carbon Monoxide. Ann. Med. Exp. Biol. Fenniae (Helsinki). *44*(Suppl. 2): 1-55, 1966.

51. Douglas E. Carbon Monoxide Solubilities in Sea Water. J. Phys. Chem. *71*(6):1931-1933, May 1967.

52. Swinnerton, J. W., V. J. Linnenbom, and C. H. Cheek. Distribution of Methane and Carbon Monoxide Between the Atmosphere and Natural Waters. Environ. Sci. Technol. *3*:836-838, September 1969.

53. Kummler, R. H. et al. Satellite Solution of the Carbon Monoxide Sink Anomaly. General Electric Co. and Barringer Research Ltd. Presented at the 50th Annual Meeting of the American Geophysical Union. Washington, D. C. April 21-25, 1969.

54. Gardner, R. A. and R. H. Petrucci The Chemisorption of Carbon Monoxide on Metals. J. Amer. Chem. Soc. *82*(19):5051-5053, October 5, 1960.

MEASUREMENT OF CARBON MONOXIDE CONCENTRATIONS IN AIR

Because of the need to measure the wide range of CO concentrations found in community air (from less than 1 to about 150 mg/m^3), numerous measurement techniques have been developed. The available methods for continuous measurement of CO include: (1) nondispersive infrared (NDIR) analysis, (2) electrochemical analysis, (3) mercury vapor analysis, (4) gas chromatography, and (5) catalytic and electrolytic analysis. The methods available for spot or integrated measurements include: (1) nondispersive infrared analysis, (2) infrared spectrophotometric analysis, (3) gas chromatographic analysis, and (4) colorimetric analysis.

Several tentative methods for the determination of CO in ambient air have been prepared for collaborative testing and evaluation as possible standards by the Intersociety Committee on Manual of Methods of Air Sampling and Analysis. Hopefully, one or more of these methods will qualify as a standard method against which all instruments and methods may be calibrated.[1]

PREPARATION OF CARBON MONOXIDE GAS STANDARDS

Volumetric Gas Dilution Techniques

Accurately prepared CO gas standards must be available for the calibration of methods and measuring instruments before CO can be quantified. Volumetric gas dilution is the technique used most frequently for the preparation of standards. If a large quantity of standard gas is needed, the dilution may be made in pressurized tanks.[2,3] Care should be exercised to assure accuracy of the standard because CO in pressurized tanks may be unstable at concentrations of less than 1 milligram per cubic meter (1 ppm).

A known volume of CO is placed in an evacuated tank of known volume, and the tank is subsequently filled to a known pressure with CO-free nitrogen or helium. The concentration of CO in the tank may then be calculated. Standard samples in 5- to 100-liter volumes also may be prepared in plastic bags.[4] The diluent gas is metered into an evacuated plastic bag with a suitable device: and as it flows into the bag, a measured volume of pure CO is injected by syringe into the line connecting the metering device to the bag. The concentration is calculated from the total volumes of diluent gas and CO in the bag.

It is important in these operations to verify that the diluting gas is free of CO. Some investigators suggest that pure helium, which is reliably free of CO, is preferable to nitrogen as a diluent. A less expensive method is to use helium periodically as a true "zero" gas to estimate the CO content in the nitrogen diluent. A correction factor is then used to account for the CO content in the nitrogen.

Gravimetric Methods for Standardizing Carbon Monoxide Gas Mixtures

Gravimetric methods are not used routinely in air pollution studies to measure CO because relatively large samples are needed to quantify CO concentrations. These methods are useful, however, for checking CO standards. McCullough et al.[5] made use of the reaction:

$$\mathrm{CO(gas)} + \mathrm{HgO\ (solid,\ red)} \xrightarrow{(175°\text{-}200°\mathrm{C})} \mathrm{Hg(gas)} + \mathrm{CO_2\ (gas)}.$$

During the passage of the gas sample, the red mercuric oxide is reduced and loses weight. This weight loss and the measured volume of the sample are used to calculate the concentration of CO present in the standard gas mixture.

Salsbury et al.[6] used a method in which the gas containing CO is drawn over Hopcalite at 195°C and the carbon dioxide formed is absorbed in tubes containing Ascarite. Gas volumes are determined using a flowmeter and a stopwatch. From the weight of the CO_2 absorbed in the tubes, the CO concentration is calculated. This method is accurate to about 2 percent for CO concentrations of from 25 to 1000 milligrams per cubic meter (22 to 870 ppm).

Lysyj et al.[7] oxidized CO to CO_2 with decomposed silver permanganate ($AgMnO_4$) as a catalyst and weighed the CO_2 absorbed in an Ascarite tube.

Chemical Assay of Carbon Monoxide Gas Standards

The reduction of iodine pentoxide may be used to measure CO in standard gas mixtures. Adams and Simmons[8] have given a detailed discussion of the method. The analysis is based on the reaction:

$$5\,CO + I_2O_5 \rightarrow 5\,CO_2 + I_2.$$

When the gas mixture containing CO is passed over heated I_2O_5, iodine is liberated. It is then collected and titrated with standard sodium thiosulfate.

MEASURING CARBON MONOXIDE IN ATMOSPHERE

1. Continuous Measurement of Carbon Monoxide

a. *Definitions of Terms Describing Instrument Performance*

Definitions of some terms commonly used to describe instrument performance are given below;

Sensitivity. The smallest change in pollutant concentration that can be detected.

Minimum detectable concentration. The sensitivity as the pollutant concentration approaches zero.

Response time. The amount of time taken by an instrument from the start of a change in input before the instrument output reaches a specified percentage of the ultimate change. (The specified percentage is usually 90 or 95). The response time is a measure of the minimum averaging time necessary to achieve valid output data.

Lag time. The time between the start of an input signal and the observed start of the output response corresponding to that signal. The lag time is a measure of the delay between input sample and output data, and is always less than or equal to the response time.

Accuracy. The degree of agreement of the true value with the indicated value or values.

True value. The number that is completely consistent with the (known) value and definition of the primary standard used.

b. *Nondispersive Infrared Analyzers*

Nondispersive infrared (NDIR) analyzers are the most commonly used continuous, automated devices for determining atmospheric CO and are generally accepted as being the most reliable reference method for the calibration of other instruments. An NDIR analyzer operates on the principle that CO has a sufficiently characteristic infrared absorption spectrum that the absorption of infrared radiation by the CO molecule can be used as a measure of CO concentration in the presence of other gases. Although the size, shape, sensitivity, and range of these instruments vary with manufacturer, basic components and configurations are similar. Most commercially available instruments include a hot filament source of infrared radiation, a rotating sector (chopper), a sample cell, a reference cell, and a detector. The detector senses pressure changes on either side of a diaphragm separating portions of gases being irradiated through the sample cell or the reference cell of the instrument. These pressure changes are converted to electrical signals corresponding to the difference between radiation received from the sample and reference cells. The signal is amplified and rectified, and then read on a meter or recorder calibrated to yield CO concentration.

Most commercially available instruments function on the double-beam principle, operate at atmospheric pressure, and are able to detect minimum CO concentrations of about 0.6 to 1.2 milligrams per cubic meter (0.5 to 1.0 ppm). The sensitivity of NDIR instruments and the minimum concentrations they can detect are proportional to the length of the cells, electronic amplification, and operating pressures. Measuring ranges usually extend from 1 to 58 mg/m^3 (1 to 50 ppm) CO or from 1 to 115 mg/m^3 (1 to 100 ppm) CO. Newer NDIR instruments, however, may be operated at from 1 to 29 mg/m^3 (1 to 25 ppm) CO at atmospheric pressure and with cell path lengths of less than 0.5 meter.

NDIR analyzer response times are determined by the physical dimensions of the system, the flow rate of the sample, and the response time of the meter or recorder. Response times may range from less than 1 to 5 minutes.

Carbon dioxide and water vapor interfere in the determination of CO by NDIR techniques. Filter cells are used to minimize these interferences. They may be placed in front of the sample cell or in front of both sample and reference cells. Filter cells, filled with carbon dioxide (CO_2) and water vapor, absorb radiation at interfering wavelengths so that normal atmospheric concentrations of CO_2 and water vapor in the sample have minimal effect on the radiation reaching the detector.[9]

Optical filters are also used successfully to limit the infrared wavelength and bandwidth to a range in which CO_2 and water vapor are transparent and thus invisible to the instrument.

Other techniques may be used to minimize or prevent CO_2 interference. Ascarite-filled tubes remove CO_2 from the entering gas stream; other systems add nominal quantities of CO_2 (300 to 400 ppm) to the zero and span gases used to calibrate the NDIR instrument.

Filter cells reduce water vapor interference; however, they do not adequately cope with the high absolute humidities frequently encountered in atmospheric monitoring. Air streams may be dried by using dehydrating agents such as silica gel, anhydrous calcium sulfate, magnesium perchlorate, or combinations of these media.[11,12] Jacobs et al.[13] found, however, anhydrous calcium sulfate to be unsatisfactory for the removal of water vapor from the gas stream. Satisfactory results were obtained by saturating the incoming air with water at a constant temperature, thus eliminating water as a variable. Some commercially available instrument systems use this technique to eliminate water vapor interference. Others reduce water vapor interference by passing the air sample through a refrigerator in front of the analytical system to condense moisture from the air. This system, however, cannot regulate the absolute humidity in the air stream undergoing analysis when the dew point of the sample air is lower than the refrigerator temperature (as is the case in cold weather or with pressurized calibration gases). It should be realized that most methods of reducing CO_2 and water vapor interferences also tend to reduce instrument sensitivity or increase instrument response time, or both.

NDIR systems have several advantages. They are not sensitive to flow rate, they require no wet chemicals, they are reasonably independent of ambient air temperature changes, they are sensitive over wide concentration ranges, and they have short response times. Further, NDIR systems may be operated by nontechnical personnel. Such systems also have some disadvantages, such as zero drift, span drift, nonlinearity, and high cost. Some newer instruments have minimum drift because good quality thermostats and solid-state electronics are used in their manufacture. Such instruments also have automatic zeroing, spanning, and recalibrating capability; they may also be obtained with essentially linear outputs. Features such as these should be considered when multiple-station networks are established.

c. *Electrochemical Analyzers*

(1). Galvanic analyzer. Galvanic cells employed in the manner described by Hersch[14,15] can be used to measure atmospheric CO continuously. When an air stream containing CO is passed into a chamber packed with I_2O_5 and heated to 150°C, the following reaction takes place:

$$5CO + I_2O_5 \xrightarrow{(150^\circ C)} 5CO_2 + I_2.$$

The liberated iodine is absorbed by an electrolyte and transferred to the cathode of a galvanic cell. At the cathode, the iodine is reduced, and the resulting current is measured by a galvanometer. Instruments using this detection system have been used successfully to measure CO levels in traffic along freeways.[16]

Mercaptans, hydrogen sulfide, hydrogen, olefins, acetylenes, and water vapor interfere. Water may be removed by sampling through a drying column; hydrogen, hydrogen sulfide, acetylene, and olefin interferences can be minimized by sampling through an absorption tube containing mercuric sulfate ($HgSO_4$) on silica gel.

(2). Coulometric analyzer. A coulometric method employing a modified Hersch-type cell has been used to measure CO in ambient air on a continuous basis.[12] The iodine pentoxide (I_2O_5) reaction with CO liberates iodine, which is then passed into a Ditte cell, and the current generated is measured by an electrometer-recorder combination. Interferences are the same as those discussed above for the galvanic analyzer.[17,18]

This technique may be used for a minimum detectable concentration of 1 milligram per cubic meter (1 ppm) with good reproducibility and accuracy if flow rates and temperatures are well controlled. This method requires careful column preparation and use of filters to remove interferences. Its relatively slow response time may be an added disadvantage in some work.

d. Mercury Vapor Analyzer

A continuous CO analyzer using the reaction of CO with hot mercuric oxide and photometric determination of the mercury vapor produced has been developed and used.[11] The reaction involved is:

$$CO + HgO\,(s) \xrightarrow{(210°C)} CO_2 + Hg(g).$$

Oxygenated hydrocarbons and olefins react quantitatively and, therefore, present a major interference; however, these are normally present in the ambient atmosphere in low concentrations relative to CO. Free hydrogen also interferes with the technique, but this interference is not serious because very little free hydrogen is present in the air. The effect of ozone is not reported, but ozone, too, is normally present in the atmosphere in very low concentrations relative to CO.

The mercury vapor instrument has a response time similar to that of the NDIR analyzer; but because of interferences and some electronic instability, it does not appear to be suitable for routine urban atmospheric monitoring. The instrument is, however, a useful, portable, continuous analyzer with a capability for analyzing concentrations of from 0.029 to 11.5 milligrams of CO per cubic meter (0.025 to 10.0 ppm). Changes of 0.002 mg/m^3 (0.002 ppm) are detectable. For this reason it has found application in determining geophysical "background" CO levels throughout the world.

e. Gas Chromatographic Analyzer

A prototype automated gas chromatographic system has been developed to measure both CO and methane.[19] A gas sampling valve, a back flush valve, a precolumn, a molecular sieve column, a catalytic reactor, and a flame ionization detector comprise the system. The precolumn prevents CO_2, water, and hydrocarbons other than methane from reaching the molecular sieve. The catalytic reactor quantitatively converts CO to methane, which is then measured by the flame ionization detector. The system is designed for semicontinuous operation with the capability of performing one analysis about every 5 minutes. It has a linear output for both CO and methane, and has a wide operating range suggesting its use in both heavily polluted areas and relatively clean locations. Simultaneous CO and methane concentrations of from 0.1 to 1,000 ppm may be measured with this instrument. Its semicontinuous characteristics suggest its use in special surveys and field studies rather than for routine air monitoring. The fact that the instrument must be operated by technically trained personnel may be considered a possible disadvantage.

f. Catalytic Analysis

Carbon monoxide may be oxidized catalytically to CO_2 using Hopcalite; the resulting temperature rise is recorded continuously and is used to indicate CO concentrations.[20,21] These systems are widely used in enclosed spaces; their applicability for ambient air monitoring is limited because they function best at high ambient concentrations.

2. Intermittent Analysis

a. Collection of Spot or Integrated Samples

Intermittent samples may be collected in the field and later analyzed in the laboratory. A detailed outline of this technique has been given by Mueller.[22] Sample containers may be rigid (glass cylinders or stainless steel tanks) or they may be non-rigid (plastic bags). Because of location or cost, intermittent sampling at times may be the only practical method for air monitoring. Samples can be taken over a few minutes or accumulated intermittently to obtain, after analysis, either "spot" or "integrated" results. Techniques for analyzing intermittent samples are described below.

b. Nondispersive Infrared Analysis

Field NDIR continuous monitoring analyzers discussed previously can be used for laboratory analysis of intermittent samples. NDIR instruments are designed by some manufacturers primarily for use in the laboratory and must be modified for continuous ambient air monitoring in the field.

c. Infrared Spectrophotometric Analysis

Carbon monoxide may be measured by its infrared absorption at 4.6 microns. Other gases may absorb at 4.6 microns, but they can be differentiated from CO by examining the complete spectrum from 2 to 15 microns. The sensitivity and range of the technique depends on the sophistication of the instrument used. A minimum detectable concentration of 2.9 milligrams per cubic meter (2.5 ppm) CO has been obtained by instruments using scale expansion and a 1-meter path-length cell.[23] It can be expected that a similar infrared spectrophotometer equipped with a 10-meter cell and ordinate scale expansion would have a 0.3-milligram-per-cubic-meter (0.3 ppm) CO detection capability. Water vapor and CO_2 must be removed from the sample before the infrared scan can be made; in most applications magnesium perchlorate and Ascarite are used to remove interfering substances. The technique is specific and accurate, but requires expensive, nonportable equipment.

d. Gas Chromatographic Analysis

After reduction of CO over a reduced nickel catalyst, a gas chromatograph equipped with a flame ionization detector, discussed earlier, becomes a satisfactory tool for determining low levels of CO.[10,19,24-26] One mole of CO produces 1 mole of methane. Since methane may be detected at concentrations of about 0.01 ppm, it is possible to detect such a concentration of CO.[27] The technique does not suffer from interferences because the measurement is made on a methane peak derived from a separated CO peak. Operating temperatures and other parameters must be controlled, however, before reliable results can be obtained.

e. Colorimetric Analysis

(1). Colored silver sol method. Carbon monoxide reacts in an alkaline solution with the silver salt of p-sulfamoylbenzoate to form a colored silver sol. Concentrations of from 12 to 23,000 milligrams per cubic meter (10 to 20,000 ppm) CO may be measured by this method.[28-32] The procedure has been modified to determine CO concentrations in incinerator effluents.[17] Samples are collected in an evacuated flask and reacted. The absorbance of the resulting colloidal solution is measured spectrophotometrically. Acetylene and formaldehyde interfere, but can be removed by passing the sample through mercuric sulfate on silica gel. Carbon monoxide concentrations of from 6 to 20,700 milligrams per cubic meter (5 to 18,000 ppm) may be measured with an accuracy of 90 to 100 percent of the true value.

(2). National Bureau of Standards colorimetric indicating gel. A National Bureau of Standards (NBS) colorimetric indicating gel (incorporating palladium and molybdenum salts) has been devised to measure CO in the laboratory and in the field.[33,34] The laboratory method involves colorimetric comparison with freshly prepared indicating gels exposed to known concentrations to CO. The method has an accuracy range of from 5 to 10 percent of the amount of CO involved, and the minimum detectable concentration is 0.1 milligram per cubic meter (0.1 ppm). This technique requires relatively simple and inexpensive equipment; but oxidizing and reducing gases interfere, and the preparation of the indicator tube is a tedious and time-consuming task.

(3). Length-of-stain indicator tube. An indicator tube method using potassium pallado-sulfite is a commonly used manual method.[35] Carbon monoxide reacts with the contents of the tube and produces a discoloration. The length of discoloration is an exponential function of the CO concentration. This method and other indicator tube manual methods are estimated to be accurate to within ± 25 percent of the amount present, particularly at CO concentrations of about 115 milligrams per cubic meter (100 ppm). Such indicator tube manual methods have been used frequently in air pollution studies. Ramsey[36] used the technique to measure CO at traffic intersections, and Brice and Roesler[37] used color-shade detector tubes to estimate CO concentrations with an accuracy of ±15 percent.

Colorimetric techniques and length-of-stain discoloration methods are recommended for use only when the other physiochemical monitoring systems are not available. They may be used in the field for gross mapping where accuracy is not required and might be of great value during emergencies.

ATMOSPHERIC CARBON MONOXIDE CONCENTRATIONS

Diurnal, weekly, and seasonal variations in CO concentrations can be observed. Diurnal and weekly variations correlate best with community traffic patterns; seasonal variations are most dependent on meteorologic variables.

Both macro- and micrometeorological factors play a role in the rate of dispersion of CO emissions. Micrometeorological factors, such as mechanical turbulence produced by automobiles and airflow around buildings, become important in determining street-side exposures. Macrometeorological factors can lead to air stagnation, which causes high community CO levels.

The concepts of averaging time and frequency of occurrence are important when describing ambient pollutant measurements. Because of physiological considerations, the averaging time of most interest for CO is 8 hours. Aerometric data from the Continuous Air Monitoring Program, the State of California, and Los Angeles County were analyzed for the 8-hour-averaging-time CO concentration exceeded 0.1 percent of the time at each available site. These values ranged from approximately 12 to 46 mg/m^3 (10 to 40 ppm).

To aid in analyzing aerometric data, a statistical model has been developed. While year-to-year mean and peak CO values may vary markedly within a community, the model can be used to calculate, based on any averaging time, a statistically probable annual maximum concentration.

Within a community, CO concentrations vary markedly with location. Calculated annual maximum concentrations in the most polluted 5 percent of the locations incorporated in a recent sample of a variety of sites showed that CO concentrations predicted inside the passenger compartment of motor vehicles in downtown traffic were almost 3 times those predicted in central urban areas and 5 times those expected in residential areas. Occupants of vehicles traveling on expressways and arterial routes were found to have CO exposures somewhere between those in central urban areas and in downtown traffic.

Concentrations exceeding 100 mg/m^3 (87 ppm) have been measured in underground garages, in tunnels, and in buildings constructed over highways.

Using emission and meteorological data, diffusion models can be used to estimate community air quality under a variety of conditions.

Carbon monoxide has not been shown to produce detrimental effects on the higher-type plant life at concentrations below 115 mg/m^3 (100 ppm) during exposures for 1 to 3 weeks. Nitrogen fixation by free-living bacteria was inhibited at exposures of 2,300 mg/m^3 (2,000 ppm) CO for 35 hours. Nitrogen fixation by efficient nitrogen-fixing bacteria in clover roots was also inhibited by 115 mg/m^3 (100 ppm) CO when exposed for 1 month.

Ambient CO levels rarely reach 115 mg/m^3 (100 ppm) even for very short periods of time. In view of this and the foregoing information concerning CO effects, a significant impact on vegetation and associated microorganisms seems improbable.

It should be pointed out that no information is available on CO concentrations in soils. Most higher plants grow with their roots in the soil, and nitrogen-fixing organisms as soil dwellers would be vulnerable to soil-borne CO. Hundreds of species of both flora and fauna, whose activities are important to soil development and to plant growth, may also be affected.

TOXICOLOGICAL APPRAISAL OF ATMOSPHERIC CARBON MONOXIDE

CO is absorbed by the lung and reacts primarily with hemoproteins and most notably with the hemoglobin of the circulating blood. The absorption of CO is associated with a reduction in the oxygen-carrying capacity of blood and in the readiness with which the blood gives up its available oxygen to the tissues. The affinity of hemoglobin for CO is over 200 times that for oxygen, indicating that carboxyhemoglobin (COHb) is a more stable compound than oxyhemoglobin (O_2Hb). About 20 percent of an absorbed dose of CO is found outside of the vascular system, presumably in combination with myoglobin and heme-containing enzymes. The magnitude of absorption of CO increases with the concentration, the duration of exposure, and the ventilatory rate. With fixed concentrations and with exposures of sufficient duration, an equilibrium is reached; the equilibrium is reasonably predictable from partial-pressure ratios of oxygen to CO.

Long-term exposures of animals to sufficiently high CO concentrations can produce structural changes in the heart and brain. It has not been shown that ordinary ambient exposures will produce this. The lowest exposure producing any such changes has been 58 mg/m^3 (50 ppm) continuously for 6 weeks.

The normal or "background" concentration of COHb in nonsmokers is about 0.5 percent and is attributed to endogenous sources such as heme catabolism. The body's uptake of exogenous CO increases blood COHb according to the concentration and length of exposure to CO as well as the respiratory rate of the individual.

In human exposure studies, continuous exposure to 35 mg/m^3 (30 ppm) CO has led to 80 percent of the equilibrium value of 5 percent COHb being approached in 4 hours, and the remaining 20 percent approached slowly over the next 8 hours. Theoretical calculations indicate a COHb equilibrium value of about 3.7 percent after continuous exposure to 23 mg/m^3 (20 ppm) and about 2 percent after continuous exposure to 12 mg/m^3 (10 ppm). The equilibrium values are generally reached after about 8 or more hours of exposure, although physical activity can shorten this time period.

Interference with the accurate estimation of time intervals has been demonstrated in nonsmokers with exposures to as low as 58 mg/m^3 (50 ppm) CO for 90 minutes. Such an exposure is likely to lead to COHb levels in the range of 2.5 percent. At a blood level of about 3 percent COHb, estimated by expired air analysis after exposure of nonsmokers to 58 mg/m^3 (50 ppm) CO for 50 minutes, significant changes in relative brightness threshold and visual acuity have been observed. Evidence of impairment in performance of other psychomotor tests has been associated with COHb levels of 5 percent in some instances. Experimental exposures of human subjects to CO leading to blood COHb levels above 5 percent have been associated with impairment in the oxidative metabolism of the myocardium in subjects with pulmonary emphysema and coronary heart disease. Persons in the latter group are unable to compensate for CO exposures by increasing coronary blood flow and are, therefore, particularly vulnerable. Persons with veno-arterial shunts in the circulation are also probably vulnerable, as are those with respiratory impairment. From physiologic considerations CO would be expected to have a greater effect with increasing altitude.

There is evidence that CO exposure increases the hematocrit of the blood and probably the circulating blood volume, although the significance of these changes is not clear. There is evidence that prolonged exposure to relatively high concentrations of CO increases the deposition of lipids in the major blood vessels of rabbits, and this could be a factor in the pathogenesis of arteriosclerosis.

Thus, in summary it may be stated that: (1) no human health effects have been demonstrated for COHb levels below 1 percent, since endogenous CO production makes this a physiological range; (2) the following effects on the central nervous system occur above 2 percent COHb: (a) at about 2.5 percent COHb in nonsmokers (from exposure to 58 mg/m^3 for 90 minutes), an impairment in time-interval discrimination, has been documented, (b) at about 3 percent COHb in nonsmokers (from exposure to 58 mg/m^3 for 50 minutes), an impairment in visual acuity and relative brightness threshold has been observed, (c) at about 5 percent COHb there is an impairment in performance of certain other psychomotor tests; (3) cardiovascular changes have been shown to occur at exposure sufficient to produce over 5 percent COHb; they include increased cardiac output,

Table 9-2 **EFFECTS OF CARBON MONOXIDE**

Environmental conditions		Effect	Comment	Reference
35 mg/m^3 (30 ppm)	for up to 12 hours	Equilibrium value of 5 percent blood COHb is reached in 8 to 12 hours; 80 percent of this equilibrium value, (4 percent COHb) is reached within 4 hours.	**Experimental exposure of nonsmokers. Theoretical calculations suggest exposure to 23 (20 ppm) and 12 mg/m^3 (10 ppm) would result in COHb levels of about 3.7 and 2 percent, respectively, if exposure was continuous for 8 or more hours.**	**Smith**
58 mg/m^3 (50 ppm)	for 90 minutes	Impairment of time-interval discrimination in nonsmokers.	Blood COHb levels not available, but anticipated to be about 2.5 percent. Similar blood COHb levels expected from exposure to 10 to 17 mg/m^3 (10 to 15 ppm) for 8 or more hours.	**Beard and Wertheim**
115 mg/m^3 (100 ppm) intermittently through a facial mask		Impairment in performance of some psychomotor tests at a COHb level of 5 percent.	Similar results may have been observed at lower COHb levels, but blood measurements were not accurate.	**Schulte**
High concentrations of CO were administered for 30 to 120 seconds, and then 10 minutes was allowed for washout of alveolar CO before blood COHb was measured.		Exposure sufficient to produce blood COHb levels above 5 percent has been shown to place a physiologic stress on patients with heart disease.	Data rely on COHb levels produced rapidly after short exposure to high levels of CO; this is not necessarily comparable to exposure over a longer time period or under equilibrium conditions.	**Ayres et al.**

increased arterial-venous oxygen difference, increased coronary blood flow in patients without coronary disease, decreased coronary sinus blood P_{O_2} in patients with coronary heart disease, impaired oxidative metabolism of the myocardium, and other related effects; these changes have been demonstrated to produce an exceptional burden on some patients with heart disease; and (4) adaptation to CO may occur through increasing blood volume, among other mechanisms.

EPIDEMIOLOGICAL APPRAISAL OF CARBON MONOXIDE

Those segments of the population most susceptible to the adverse effects associated with atmospheric CO can be predicted on a physiologic basis to include those people most sensitive to a decreased oxygen supply. These susceptible groups include, then, individuals with anemia, cardiovascular disease, anormal metabolic states such as thyrotoxicosis or fever, and chronic pulmonary disease and the developing fetus.

A major source of CO exposure is cigarette smoke; cigarette smokers generally have a COHb with a median value of 5 percent, whereas nonsmokers are usually found to have about 0.5 percent COHb. Community exposures of people who are cigarette smokers and already have an elevated COHb will either lead to an increase in COHb or will slow down the excretion of CO during intervals between cigarette smoking, depending upon the initial COHb level in the smoker and the magnitude and duration of the ambient CO exposure. Exposure of traffic policemen, in Paris, for 5 hours to between 12 and 14 mg/m^3 (10 and 12 ppm) has increased COHb levels in nonsmokers by about 0.7 percent. The extent of human CO exposure from sources other than smoking or the ambient outdoor air has not been well documented.

Several effects of long-term exposure to CO have been implicated. Recent data suggest that the mortality from myocardial infarction may be increased by exposure to average weekly CO concentrations of from 9 to 16 mg/m^3 (8 to 14 ppm), although these results are not conclusive and require replication. The probability of involvement in motor vehicle accidents may also be associated with CO exposure. This could be related to the influence of CO exposure on visual acuity, estimation of time intervals, or other psychomotor parameters.

REFERENCES

1. Analytical Methods of the Intersociety Committee on Methods for Ambient Air Sampling and Analysis. Health Lab. Sci., 7:1, in press, 1970.
2. Feldstein, M. Methods for the Determination of Carbon Monoxide. In: Progress in Chemical Toxicology, Stolman, A. (ed.), Vol, 3, New York, Academic Press, 1967. p. 99-119.
3. Kuley, C. J. Gas Chromatographic Analysis of C_1 to C_4 Hydrocarbons in the Parts Per Million Range in Air and Vaporized Liquid Oxygen. Anal. Chem. *35*(10): 1472-1475, September 1963.
4. Schuette, F. J. Plastic Bags for Collection of Gas Samples. Atmos. Environ. *1*(4):515-519, July 1967.
5. McCullough, J. D., R. A. Crane, and A. O. Beckman. Determination of CO in Air by Use of Red Mercuric Oxide. Anal. Chem. *19*:999-1002, December 1947.
6. Salsburg, J. M., J. W. Cole, and J. H. Yoe. Determination of Carbon Monoxide—A Microgravimetric Method. Anal. Chem. *19*:66-68, January 1947.
7. Lysyj, I., J. E. Zarembo, and A. Hanley. Rapid Method for Determination of Small Amounts of Carbon Monoxide in Gas Mixtures. Anal. Chem. *31*:902-904, May 1959.
8. Adams, E. G. and N. T. Simmons. The Determination of Carbon Monoxide by Means of Iodine Pentoxide. J. Appl. Chem. (London). *1*(Suppl. 1):20-40, 1951.
9. Information supplied by State of California. Dept. of Public Health. August 1968.
10. Moore, H. A Critical Evaluation of the Analysis of Carbon Monoxide with Nondispersive Infrared (NDIR). Presented at the 9th Conference on Methods in Air Pollution and Industrial Hygiene Studies. Pasadena. February 7-9, 1968.
11. Robbins, R. C., K. M. Borg, and E. Robinson. Carbon Monoxide in the Atmosphere. J. Air Pollution Control Assoc. *18*:106-110, February 1968.
12. Dubois, L., A. Zdrojewski, and J. R. Monkman. The Analysis of Carbon Monozide in Urban Air at the PPM Level, and the Normal Carbon Monoxide Value. J. Air Pollution Control Assoc. *16*:135-139, March 1966.
13. Jacobs, M. B., M. M. Braverman, and S. Hochheiser. Continuous Determination of Carbon Monoxide and Hydrocarbons in Air by a Modified Infrared Analyzer. J. Air Pollution Control Assoc. *9*(2):110-114, August 1959.

14. Hersch, P. Galvanic Analysis. In: Advances in Analytical Chemistry and Instrumentation, Reilley, C. A. (ed.), Vol 3. New York, Interscience Publishers, 1964. p. 183-249.

15. Hersch, P. Process for Measuring the Carbon Monoxide Content of a Gas Stream (Beckman Instruments, Inc., U.S. Patent No. 3, 258, 411). Official Gazette U.S. Patent Office. *827*(4): 1276-1277, June 1966.

16. Haagen-Smit, A. J. Carbon Monoxide Levels in City Driving. Arch. Environ. Health. *12*:548-551, May 1966.

17. Levaggi, D. A. and M. Feldstein. The Colorimetric Determination of Low Concentrations of Carbon Monoxide. Amer. Ind. Hyg. Assoc. J. *25*:64-66, January-February 1964.

18. Feldstein, M. The Colorimetric Determination of Blood and Breath Carbon Monoxide. J. Forensic Sci.*10*:35-42, January 1965.

19. Stevens, R. K., A. E. O'Keefe, and G. C. Ortman. A Gas Chromatographic Approach to the Semi-Continuous Monitoring of Atmospheric Carbon Monoxide and Methane. Presented at the 156th National Meeting of the American Chemical Society. Atlantic City. September 1968.

20. Linderholm. H. and T. Sjostrand. Determination of Carbon Monoxide in Small Gas Volumes. Acta Physiol. Scand. (Stockholm). *37*(2-3):240-250, September 26, 1956.

21. Lukaci, J. and J. Chriastel. Determination of Carbon Monoxide in Air. Anal. Abstr. *6*:751, February 1959.

22. Mueller, P. K. Detection and Analysis of Atmospheric Pollutants. Presented at the Engineering Extension Course of Combustion Generated Air Pollution. University of California. Berkeley. June 1965.

23. Kemmner, G., G. Nonnenmacher, and W. Wehling. Analytical Application of Infrared Spectroscopy with Gratings: Quantitative Analysis of Gas Tracings [Analytische Anwendung der Infrarot-Spektroskopie mit Gittern: Quantitative Analyse von Gasspuren]. Fresenius' Z. Anal. Chem. *222*(2):149-161, October 13, 1966.

24. Porter, K. and D. H. Volman. Flame Ionization Detection of Carbon Monixide for Gas Chromatographic Analysis. Anal. Chem. *34*(7): 748-749, June 1962.

25. Altshuller, A. P., I. R. Cohen, and T. C. Purcell. Photooxidation of Propionaldehyde at Low Partial Pressures of Aldehyde. Can. J. Chem. *44*(24):2973-2979, December 15, 1966.

26. Altshuller, A. P. and I. R. Cohen. Atmospheric Photooxidation of the Ethylene-Nirtic Oxide System. Int. J. Air Water Pollution. *8*(11/12):611-632, December 1964.

27. Recommended Methods in Air Pollution Measurements. California Dept. of Public Health. Berkeley. 1967.

28. Ciuhandu, G. New Method for the Determination of Carbon Monoxide in Air. Acad. Rep. Populare Romine, Baza Cercetari Stiint. Timisoara, Studii Cercetari Stiint., Ser. I (2):133-142, 1955.

29. Ciuhandu, G. Photometric Determination of Carbon Monoxide in Air [Photometrische Bestimmung von Kohlenmonoxyd in der Luft] Fresnius' Z. Anal. Chem. *155*(5):321-327, March 21, 1957.

30. Ciuhandu, G. Micromethod for the Photometric Determination of Carbon Monoxide in Air [Mikromethode fur die photometrische Kohlenmonoxydbestimmung in der Luft]. Fresenius' Z. Anal. Chem. *161*(5):345-348, May 3, 1958.

31. Ciuhandu, G. and G. Krall. Photometric Determination of Traces of Carbon Monoxide in Hydrogen [Photometrische Bestimmung von Kohlenmonoxydspuren in Wasserstoff]. Fresenius Z. Anal. Chem. *172*(2):81-87, January 8, 1960.

32. Ciuhandu, G., V. Rusu, and M. Diaconovici. Determination of Traces of Carbon Monoxide in Oxygen [Bestimmung von Kohlenmonoxydspuren in Sauerstoff]. Fresenius' Z. Anal. Chem. *208*(2):81-86, January 15, 1965.

33. Shepherd, M., S. Schuhmann, and M. V. Kilday. Determination of Carbon Monoxide in Air Pollution Studies. Anal. Chem.*27*(3):380-383, March 1955.

34. Shepherd, M. Rapid Determination of Small Amounts of Carbon Monoxide. Anal. Chem. *19*:77-81, February 1947.

35. Silverman, L. and G. R. Gardner. Potassium Pallado Sulfite Method for Carbon Monoxide Detection. Amer. Ind. Hyg. Assoc. J. *26*(2):97-105, March-April 1965.

36. Ramsey, J. Concentration of Carbon Monoxide at Traffic Intersections in Dayton, Ohio. Arch. Environ. Health *13*(1):44-46, July 1966.

37. Brice, R. M. and J. F. Roesler. The Exposure to Carbon Monoxide of Occupants of Vehicles Moving in Heavy Traffic. J. Air Pollution Control Assoc. *16*:597-600, November 1966.

CHAPTER 10

NITROGEN OXIDES

This document contains a consolidation and an assessment of the current state of knowledge regarding the group of air pollutants known as the oxides of nitrogen. This chapter provides a concise presentation of that information, including reasonable conclusions for evaluating the concentrations of nitrogen oxides (NO_x) and the accompanying situations that have an effect on either health or welfare. The studies and data cited comprise the best available basis for developing specific standards for NO_x in the ambient air, aimed at protecting public health and the environment.

Although the essential role of NO_x in the production of photochemical oxidants is treated from the physical-chemical standpoint in this document, little research has been done to demonstrate the significance of the indirect effects of NO_x on health, vegetation, and materials through the photochemical reaction mechanism; thus, only the direct effects of NO_x are treated here. A P C O publication AP-63, *Air Quality Criteria for Photochemical Oxidants*, provides a comprehensive review of photochemical oxidant effects.

Units of pollution concentration used in this document are expressed as both mass per unit volume (e.g., micrograms per cubic meter, $\mu g/m^3$) and as volume-ratios (e.g., parts per million, ppm). Conversion between these units requires a knowledge of the gas density, which varies with temperature and pressure measurement. In this document 25°C (77° F) has been taken as standard temperature, and 760 mm Hg (atmospheric pressure at sea level) as standard pressure. All references to NO_x are expressed in terms of NO_2 mass per unit volume on the basis of the conversion formula: ppm x 1880 = $\mu g/m^3$ at 25° C, 760 mm Hg, unless otherwise specified. Similarly, hydrocarbons and oxidant concentrations are expressed as mass of methane and ozone per unit volume, respectively.

PROPERTIES OF NITROGEN OXIDES AND PHYSICAL EFFECTS ON LIGHT TRANSMISSION

Of the oxides of nitrogen known to exist, only two, nitric oxide (NO) and nitrogen dioxide (NO_2) are emitted to the atmosphere in significant quantities. Nitric oxide is formed during all atmospheric combustion processes in a spontaneous chemical reaction between the nitrogen and oxygen in the air. The amount formed depends on the combustion temperature, the concentration of both reactants and products, and the length of time favorable conditions persist for the reaction.

Both NO and NO_2 are formed when combustion temperatures exceed approximately 1093°C (2000°F), but usually less than 0.5 percent is NO_2. More NO_2 is formed when atmospheric oxygen (O_2) reacts with NO, but at the dilute concentrations of NO characteristically found in ambient atmospheres, this reaction proceeds very slowly. During the initial phases of exhaust gas dilution, however, the concentration of NO is high, and forces the reaction to proceed more rapidly until the exhaust has been sufficiently diluted (to 1 ppm or less). At that time the major process for converting NO to NO_2 reverts to the photochemical cycle.

Visibility reduction is common in polluted atmospheres. Scattering and absorption of light rays by particles and gases reduce the brightness and contrast of distant objects. The degree of reduction depends on the concentration and properties of the pollutants. Nitrogen dioxide absorbs light energy over the entire visible spectrum, although primarily in the shorter, blue-wavelength regions; thus, NO_2 can by itself reduce visibility. At present, however, under most ambient conditions, aerosols make the major contributions to visibility reduction.

SOURCES AND CONTROL OF ATMOSPHERIC NITROGEN OXIDES

On a global basis, the total amount of nitrogen oxides generated by natural sources exceeds the amount from man-made, technological sources. Natural scavenging processes keep background levels in nonurban areas low on the order of 8 $\mu g/m^3$ (4 ppb) NO_2 and 2 $\mu g/m^3$ (2 ppb) NO. In urban areas, however, where 60 percent of the technological sources are located, the levels are frequently higher because pollutants are added faster than scavenging processes control them.

Fuel combustion is the major source of technological (nitrogen oxides) air pollution. Chemical processing is responsible for high, but localized emissions.

Control of NO_x emissions has been directed at both combustion sources and chemical processes. For stationary combustion sources, the control principle has been based on reducing either the flame temperature or the availability of oxygen, to prevent NO formation. Similar principles of control are applicable to motor vehicles. Catalytic principles, which have been applied to reduce NO_x emissions from chemical processes, are also being investigated for possible use in control of NO_x in motor-vehicle exhaust.

METHODS FOR MEASURING NITROGEN OXIDES

Research is still needed to develop and thoroughly evaluate more sensitive, reliable, and practical methods for measuring ambient levels of NO, NO_2, and NO_x. All of the field techniques in use at present can measure only NO_2 directly; NO must be oxidized to NO_2, then measured. NO_x can be determined either by summing NO and NO_2 concentrations that have been measured independently or by oxidizing NO to NO_2, then measuring the total as NO_2.

Any method used for measuring NO_2 in the ambient air should be calibrated against atmospheres containing known amounts of NO_2. The use of permeation tubes to generate the test atmospheres is recommended.

Two techniques are currently used in atmospheric monitoring programs. For sampling periods of 30 minutes or less, the most suitable currently available method for measuring NO_2 is the colorimetric Griess-Saltzman method. This method can also be automated for continuous measurement. The Jacobs-Hochheiser method is the most suitable of the available methods for long-term (up to 24 hours) sampling, or for situations requiring a delay of the analysis for more than 4 hours after sampling. The Griess-Saltzman and Jacobs-Hochheiser methods are not interchangeable, can yield different results, and must be chosen carefully, according to the purposes of the sampling to be done.

When used in conjunction with an oxidizing prescrubber to convert NO to NO_2, the continuous Griess-Saltzman method can be used to measure NO in ambient air in either a series or parallel mode, with the same or separate samples of air. Problems exist in obtaining complete NO to NO_2 oxidation, and researchers disagree as to which of the two modes is more satisfactory.

EFFECTS OF NITROGEN OXIDES ON MATERIALS

Significant effects of NO_x have been observed and studied on three classes of materials: textile dyes and additives, natural and synthetic textile fibers, and metals.

The most pronounced problem is associated with textile dyes and additives. Fading of sensitive disperse dyes used on cellulose acetate fibers has been attributed to NO_2 levels below 188 mg/m^3 (<100 ppm). Loss of color, particularly in blue- and green-dyed cotton and viscose rayon has occurred in gas dryers where NO_x concentrations range from 1.1 to 3.7 mg/m^3 (0.6 to 2 ppm). Yellow discoloration in undyed white and pastel-colored fabrics has recently been attributed to NO_x by controlled laboratory experiments.

Laboratory and field observations have shown that cotton and Nylon textile fibers can be deteriorated by the presence of NO_x, but specific reactants and threshold levels are undetermined at this time.

Failure of nickel-brass wire springs on relays has been related to high particulate nitrate levels. This type of stress corrosion has been observed when surface concentrations of particulate nitrates have exceeded 2.4

$\mu g/cm^2$ and relative humidity was greater than 50 percent. Another type of this corrosion has been associated with annual average particulate nitrate concentrations of 3.0 and 3.4 $\mu g/m^3$ with corresponding NO_x levels of 124 and 158 $\mu g/m^3$ (0.066 and 0.084 ppm).

EFFECTS OF NITROGEN OXIDES ON VEGETATION

The degree of injury occurring with the lower concentrations of NO_2 present in the atmosphere remains to be determined. Exposure of many kinds of plants to concentrations of NO_2 above 47 mg/m^3 (25 ppm) for any extended period causes acute necrotic leaf injury. Such lesions are usually characteristic for each plant, but their nonspecific character in relation to other toxicants renders these symptoms of little value in diagnosing NO_2 damage in the field.

The 1-hour visible-injury-threshold value for NO_2 can be achieved by exposing plants to 18.8 to 28.2 mg/m^3 (10 to 15 ppm). Increasing the exposure time, however, obviates the threshold level; 4.3 to 6.6 mg/m^3 (2.3 to 3.5 ppm) NO_2 administered for 8 to 21 hours and 1.9 mg/m^3 (1 ppm) NO_2 for 48 hours cause equivalent leaf injury. Continuous fumigation with 940 $\mu g/m^3$ (0.5 ppm) NO_2 for 35 days resulted in leaf drop and chlorosis in citrus, but no actual necrotic lesions developed.

The effects of exposure to low levels of NO_2 for extended periods are less evident. Recently completed studies suggested that 470 $\mu g/m^3$ (0.25 ppm) or less of NO_2, supplied continuously for 8 months will cause increased leaf drop and reduced yield in navel oranges.

The mechanism(s) by which NO_x causes direct injury to plants can only be postulated at this time. Evidence of diurnal fluctuation in sensitivity to NO_2 has been presented, and could indicate that the pollutant is reacting with a particular plant metabolite, which only accumulates at certain periods during the day. The absence of a protective metabolite within the plant at certain periods would also cause a diurnal sensitivity.

Limited information regarding the effect of nitric oxide on photosynthesis indicates that NO would reduce the growth of plants if concentrations in the range of 3.8 to 7.5 mg/m^3 (2.0 to 4.0 ppm) persisted continuously.

TOXICOLOGICAL EFFECTS OF NITROGEN OXIDES

Both of the prominent oxides of nitrogen present in ambient air are potential health hazards. At ambient concentrations, NO presents no direct threat to general health; NO_2 does, however. Effects of NO_2 determined in extensive studies are summarized in **Table 10-1**.

The toxicology of nitrous oxide (N_2O) and other oxides of nitrogen does not appear to be relevant to the problems of ambient air pollution at the present time.

1. Nitric Oxide

NO is not an irritant and is not considered to have adverse health effects at concentrations found in the atmosphere. Its greatest toxic potential at ambient concentrations is related to its tendency to undergo oxidation to NO_2. A 12-minute exposure to 3,075 mg/m^3 (2,500 ppm) NO has proved lethal to mice. In addition, NO has been observed to inhibit bacterial hydrogenase activity at lower concentrations - 24.6 mg/m^3 (20 ppm). This inhibition was reversible, however, until the exposure reached about 12,300 mg/m^3 (10,000 ppm).

2. Nitrogen Dioxide

NO_2 exerts its primary toxic effect on the lungs. High concentrations, greater than 188 mg/m^3 (100 ppm), are lethal to most animal species; 90 percent of the deaths are caused by pulmonary edema.

The concentration time product determines nonlethal morbidity effects of NO_2 exposures. At 940 $\mu g/m^3$ (0.5 ppm) for 4 hours or 1.9 mg/m^3 (1.0 ppm) for 1 hour, mast cells of rat lungs became degranulated, possibly signifying the onset of an acute inflammatory reaction. These cells returned to normal 24 hours after exposure was terminated. Lung proteins, collagen and elastin, were found to be altered structurally in rabbits exposed to 1.9 mg/m^3 (1 ppm) NO_2 for 1 to 4 hours. The condition was also reversible within 24 hours. Similar changes were observed in rabbits exposed to 470 $\mu g/m^3$ (0.25 ppm) NO_2,

Table 10-1 SUMMARY OF REPRESENTATIVE NO_2 EFFECTS

Effect	NO_2 concentration, ppm	NO_2 concentration, $\mu g/m^3$	Duration	Comment
Lowest level associated with reference oxidant production of 200 $\mu g/m^3$ (0.1 ppm)	0.04	80	3 hr (6 to 9 a.m.)	
Increased incidence of acute respiratory disease in families	0.062 to 0.109	117 to 205	2 to 3 yr	Chattanooga study - 6-mo mean concentration range
Increased incidence of acute bronchitis in infants and school children	0.063 to 0.083	118 to 156	2 to 3 yr	Chattanooga study - 6-mo mean concentration range
Human - olfactory threshold	0.12	225	- - -	Immediate perception
Rabbits - structural changes in lung collagen	0.25	470	4 hr/day for 6 days	Still apparent 7 days after final exposure
Navel orange - leaf abscission; decreased yield	0.25	470	8 mo, continuously	
Rats - morphological changes in lung mast cells characterized by degranulation	0.5	940	4 hr	Possibly precedes onset of acute inflammatory reaction
	1.0	1880	1 hr	
Mice - pneumonitis; alveolar distension	0.5	940	6 to 24 hr/day for 3 to 12 mo	Possibly emphysematous condition
Mice - increased susceptibility to respiratory infection	0.5	940	6 to 24 hr/day up to 12 mo	Based on mortality following challenge with *K. pneumoniae*
Navel orange - leaf abscission, chlorosis	0.5	940	35 days, continuously	
Rats - tachypnea, terminal bronchiolar hypertrophy	0.8	1504	Lifetime, continuously	

Table 10-1 (Continued). SUMMARY OF REPRESENTATIVE NO_2 EFFECTS

Effect	NO_2 concentration, ppm	NO_2 concentration, $\mu g/m^3$	Duration	Comment
Rats - bronchiolar epithelial changes, loss of cilia, reduced cytoplasmic blebbing, crystalloid inclusion bodies	0.8 to 2.0	1504 to 3760	Lifetime, continuously	Possibly pre-emphysematous lesion
Rabbits - structural changes in lung collagen	1.0	1880	1 hr	Denaturation of structural protein suggested
Sensitive plants - visible leaf damage	1.0	1880	21 to 48 hr	
Rats, monkeys - polycythemia	2.0	3760	3 wk, continuously	Approximate doubling of red cell number with lesser increases in hematocrit and hemoglobin
Man - increase in airway resistance	5	9,400	10 min	Transient
Monkeys - tissue changes in lungs, heart, liver, and kidneys	15 to 50	28,200 to 94,000	2 hr	Degree of damage directly related to concentration of NO_2

4 hours a day for 6 days, except that recovery was delayed and some denaturation was still apparent 7 days after the final exposure. Denaturation of collagen and elastin associated with repeated exposure to NO_2 has been suggested as a possible factor in the pathogenesis of pulmonary emphysema.

Early pulmonary emphysema-type lesions have been observed in dogs exposed continuously to 47.0 mg/m^3 (25 ppm) for 6 months. In lung tissue from monkeys exposed to 18.8 to 94.0 mg/m^3 (10 to 50 ppm) NO_2 for 2 hours, alveoli were expanded and had thin septal walls. This response involved increasing numbers of alveoli as the NO_2 concentration was increased. Hyperplasia has been observed in respiratory bronchiolar epithelium of hamsters exposed to 94.0 mg/m^3 (50 ppm) for 10 weeks, and a similar response was noted in major bronchi and distal portions of the respiratory tract of hamsters exposed to 18.8 mg/m^3 (100 ppm) for 6 hours.

Long-term exposures to NO_2 concentrations that do not produce acute inflammatory responses have a cumulative, sustained effect, suggestive of a pre-emphysematous condition. Examination of lung tissue from rats exposed to 3.8 mg/m^3 (2 ppm) for their natural lifetimes showed loss of cilia; decreased bronchiolar blebbing; and intercellular, crystalloid, rod-shaped, inclusion bodies. Similar effects have been seen in lungs of rats continuously exposed to 1.5 mg/m^3 (0.8 ppm). Alveoli in lungs of mice exposed to 940 $\mu g/m^3$ (0.5 ppm) for 3 to 12 months on 6-, 18-, and 24-hour daily schedules have shown increase in size from distension rather than from septal breakage. The accompanying inflammation of the bronchiolar epithelium and reduction in distal airway size suggested the development of early focal emphysema.

Rats chronically exposed to 18.8 to 47.0 mg/m^3 (10 to 25 ppm) NO_2 developed compensatory changes, such as polycythemia and thoracic kyphosis, with lateral flaring of the ribs.

Since certain pathological changes seen in animals after experimental NO_2 exposure are similar to changes that occur in the pathogenesis of chronic obstructive pulmonary disease in man, it is suggested that long-term, low-level exposures to NO_2 may play a significant role in the development of chronic lung disease.

Exposure of mice, hamsters, and squirrel monkeys to NO_2 increased susceptibility to bacterial pneumonia and influenza infection. The susceptibility has been demonstrated by a significantly increased mortality, decreased survival time, and a reduction in ability to clear infectious agents from the lungs. In mice, threshold for increased susceptibility to *Klebsiella pneumoniae* occurred after exposure to 6.6 mg/m^3 (3.5 ppm) NO_2 for 2 hours, if the infectious challenge was given within 1 hour after the NO_2 exposure. Squirrel monkeys exposed to 18.8 mg/m^3 (10 ppm) NO_2 for 2 hours and then challenged with *K. pneumonia* aerosol retained the infectious agent in their lungs for extended periods of time.

In long-term studies of mice, significantly increased susceptibility to infection occurred after continuous daily exposure to 940 μg/m^3 (0.5 ppm) NO_2 for 3 months, and after 6- and 18-hour daily exposures for 6 months. A significant increase in susceptibility to influenza virus or *K. pneumoniae* was also seen in squirrel monkeys continuously exposed to 18.8 and 9.4 mg/m^3 (10 and 5 ppm) NO_2 for 1 and 2 months, respectively. In addition, interferon formation has been impaired and resistance to viral infection has decreased following exposure of rabbits to 47.0 mg/m^3 (25 ppm) NO_2 for 3 hours. Researchers conjecture that such increased susceptibility to infection may also be significant in the pathogenesis of human lung disease.

Inhalation of NO_2 can produce other systemic effects, although these are generally secondary to the effects on the lungs. In monkeys exposed to 28.2 to 94.0 mg/m^3 (15 to 50 ppm) NO_2 for 2 hours, cellular changes appeared in heart, liver, and kidney tissue. A circulating substance, possibly a lung antibody, has been detected in the blood of guinea pigs exposed to 9.4 mg/m^3 (5.0 ppm) for 4 hours daily, 5 days per week for 5.5 months. Rats and monkeys continuously exposed to 3.8 mg/m^3 (2.0 ppm) NO_2 for 3 weeks developed marked polycythemia. Methemoglobin has been detected in the blood of several species exposed to NO_2 concentrations greater than 122 mg/m^3 (70 ppm) for 1 hour.

The small amount of information available concerning the toxicological effects of the oxides of nitrogen in man pertains to levels higher than those found in ambient air. Experimental exposure of volunteer subjects to 9.4 mg/m^3 (5 ppm) NO_2 for 10 minutes has produced a substantial, but transient, increase in airway resistance. Other data, derived from occupational exposure to high-concentration mixtures of NO and NO_2, are complicated by the presence of other pollutants. Impaired pulmonary function, evidenced by reduced maximal breathing capacity, increased expiratory resistance, and occasional decreased vital capacity, has been observed in patients accidentally exposed to high concentrations of nitrous fumes for a few minutes. Such evidence has persisted for more than 2 years after the exposure, in some cases. In one case, occupational exposure to 169 mg/m^3 (90 ppm) NO_2 for 30 minutes produced pulmonary edema and a vital capacity 50 percent lower than expected 18 hours later. Exposure to very high concentrations for about 5 minutes has produced death within 2 days to 5 weeks.

The threshold for odor perception of NO_2 is about 225 μg/m^3 (0.12 ppm).

EPIDEMIOLOGICAL APPRAISAL OF NITROGEN OXIDES

Nitrogen dioxide, the only oxide of nitrogen examined in epidemiological surveys, can be significantly correlated with increased respiratory disease at mean 24-hour concentrations between 117 and 205 μg/m^3 (0.062 and 0.109 ppm).

Effects of community exposure to NO_2 were studied in four residential areas of greater Chattanooga. The ventilatory performance ($FEV_{0.75}$) of children in a high-

NO_2 area was significantly reduced, when compared to the performance of children in control areas. In addition, an 18.8 percent relative excess of respiratory illness occurred among families exposed to high NO_2 concentrations. A 10.4 percent excess occurred among families in an elevated-particulate area. The increased incidence of acute respiratory disease was observed when the mean 24-hour NO_2 concentration, measured over a 6-month period, was between 117 and 205 $\mu g/m^3$ (0.062 and 0.109 ppm) and the mean suspended nitrate level was 3.8 $\mu g/m^3$ or greater.

In a retrospective study of the same Chattanooga area, exposure to intermediate and high levels of NO_2 in ambient air was associated with a significant increase in the frequency of acute bronchitis among infants exposed for 3 years and school children exposed for 2 and 3 years. When increase was observed, the mean 24-hour NO_2 concentration, measured over a 6-month period, had ranged between 118 and 156 $\mu g/m^3$ (0.063 and 0.083 ppm) and the mean suspended nitrate level had been 2.6 $\mu g/m^3$ or greater.

A report from Czechoslovakia indicates that NO_x has produced several alterations in the peripheral blood. Increased levels of methemoglobin were observed in school children residing in a town that had relatively high ambient levels of nitrogen oxides. The findings in that report require further clarifying investigation, however, before conclusions can be drawn.

CHAPTER 11

HYDROCARBONS

SOURCES, NATURE, AND PRINCIPLES OF CONTROL OF ATMOSPHERIC HYDROCARBONS

Most natural sources of hydrocarbon emissions are biological in nature. A conservative estimate made for the worldwide natural production rate of methane is 3 x 10^8 tons per year. A similar estimate of 4.4 x 10^8 tons per year has been made for volatile terpenes and isoprenes. It appears that nonurban air contains from 0.7 to 1.0 mg/m^3 (1.0 to 1.5 ppm) methane and less than 0.1 ppm each of other hydrocarbons.

Total nationwide technological emissions of hydrocarbons and related organic compounds for the year 1968 were estimated to be 32 x 10^6 tons. Transportation represented the largest source category and accounted for 52 percent of this estimate. The miscellaneous source category—principally organic solvent evaporation—was the second largest and represented 27 percent of the total emissions. Industrial processes (14 percent) was third; solid waste disposal (5 percent) was fourth; and fuel combustion in stationary sources (2 percent) was fifth.

Local emission estimates for 22 metropolitan areas ranged from about 0.05 to 1.3 million tons per year. Transportation sources accounted for from 37 to 99 percent of local emissions, and process losses accounted for from 1 to 63 percent. Hydrocarbon emissions, therefore, originate primarily from the inefficient combustion of volatile fuels and from their use as process raw materials.

The control of hydrocarbon emissions rests upon the basic principles of: (1) combustion process optimization, (2) recovery by mass transfer principles, (3) restriction of evaporative loss, and (4) process material and fuel substitution. The first three principles are all applied with varying degrees of success in the control of automobile emissions.

SAMPLING AND STANDARDIZATION METHODS FOR MEASUREMENT OF HYDROCARBONS

With few exceptions, atmospheric hydrocarbon measurements are made with relatively complex instruments that operate continuously. Continuous instrumentation demands dynamic calibration techniques. Standard gases for this purpose are available or may be generated by permeation tubes or dilution systems, or prepared in large containers.

Flame ionization analyzers are sensitive and reliable and are suitable for the continuous measurement of total hydrocarbons. They are generally accepted as the method of choice by NAPCA. They cannot, however, give the specific detailed information required for a thorough understanding of the atmospheric photochemical problem. Attempts to further define the hydrocarbon mixture by using pretreatment columns to measure only methane or various reactive classes have met with some success but limited application.

Spectrometric techniques both for total and specific analysis are complex and generally insensitive.

Gas chromatographic analysis provides the requisite sensitivity and specificity for the quantitation of individual hydrocarbons. Difficulties in qualitative analysis and data handling have limited the application of GC to short-term studies for the most part, and no continuous data are available.

Carbonyl compounds, specifically aldehydes and ketones, can be measured by several manual colorimetric techniques, but very little actual data are available.

EFFECTS OF HYDROCARBONS AND CERTAIN ALDEHYDES ON VEGETATION

Hydrocarbons were first recognized as

phytotoxic air pollutants about the turn of the century as a result of complaints of injury to greenhouse plants from illuminating gas. Ethylene was shown to be the injurious component. Renewed interest in hydrocarbons, and ethylene in particular, occurred in the mid-1950's when ethylene was found to be one of the primary pollutants in the photochemical smog complex. Research on several unsaturated and saturated hydrocarbons proved that only ethylene had adverse effects at known ambient concentrations. Acetylene and propylene more nearly approach the activity of ethylene than do other similar gases, but 60 to 500 times the concentration is needed for comparable effects.

In the absence of any other symptom, the principal effect of ethylene is to inhibit growth of plants, Unfortunately, this effect does not characterize ethylene because other pollutants at sublethal dosages, as well as some disease and environmental factors, will also inhibit growth.

Epinasty of leaves and abscission of leaves, flower buds, and flowers are somewhat more typical of the effects of ethylene, but the same effects may be associated with nutritional imbalance, disease, or early senescence. Perhaps the most characteristic ethylene effects are the dry sepal wilt of orchids and the closing of carnation flowers. Injury to sensitive plants has been reported at ethylene concentrations of 1.15 to 575 $\mu g/m^3$ (0.001 to 0.5 ppm) during time periods of 8 to 24 hours.

TOXICOLOGICAL APPRAISAL OF HYDROCARBONS AND ALDEHYDES

Hydrocarbon air pollutants enter into and promote the formation of photochemical smog and thus contribute to the development of eye irritation and other manifestations. They are intimately involved in the formation of formaldehyde and other aldehydes and ketones, and of various oxidants, including peroxyacetyl nitrate. In developing air quality criteria, there must be due consideration of the potential of the hydrocarbons under certain atmospheric conditions to form more hazardous derivatives.

Experimental data resulting from the exposure of animals and humans to various hydrocarbons compounds indicate that:

1. The aliphatic and alicyclic hydrocarbons are generally biochemically inert, though not biologically inert, and are only reactive at concentrations of 10^2 to 10^3 higher than those levels found in the ambient atmosphere. No effects have been reported at levels below 500 ppm.

2. The aromatic hydrocarbons are biochemically and biologically active. The vapors are more irritating to the mucous membranes than equivalent concentrations of the aliphatic or alicyclic groups. Systemic injury can result from the inhalation of vapors of the aromatic compounds; no effects, however, have been reported at levels below 25 ppm.

Pulmonary function and eye irritation have been evaluated by the exposure of experimental animals and humans to various mixed atmospheres that had the characteristic of providing the milieu necessary for hydrocarbon reactivity. Experimental data indicate:

1. Exposure of guinea pigs to ambient Los Angeles photochemical smog produces a temporary increase in pulmonary airflow resistance. Exposure of guinea pigs to laboratory-irradiated auto exhaust causes an increased tidal volume, minute volume, and flow resistance, with a decrease in the respiratory rate. These parameters return to normal immediately following exposure.

2. The precursors of the eye irritants are hydrocarbons (as well as other organic compounds) in combination with oxides of nitrogen; the alkylbenzenes and olefins are more effective precursors than paraffins, benzene, and acetylenes. The products of photochemical reactions that have been identified as effective eye irritants are formaldehyde, acrolein, peroxyacyl nitrates, and peroxybenzoyl nitrate. In general, the most characteristic and important effect of aldehydes for both humans and animals is primary irritation of the eyes, upper respiratory tract, and skin. The unsaturated aldehydes are several times more toxic than the corresponding aliphatic aldehydes, and toxicity generally decreases with increasing molecular weight within the unsaturated and

aliphatic aldehyde series. Animal experiments have shown that aldehydes can affect respiratory functions, causing such effects as an increase in flow resistance and tidal volume and a decrease in the respiratory rate. Animal populations, although many of the effects attributed to photochemical smog are indirectly related to ambient levels of these hydrocarbons.

Experimentai data on specific aldehydes have indicated:

1. Aldehyde concentrations have been correlated with the intensity of eye irritation during natural and laboratory-produced photochemical smog. Formaldehyde appears to be detectable by odor or physiological response (optical chronaxy) at concentrations in the order of 70 $\mu g/m^3$ (0.06 ppm). The threshold for eye irritation by formaldehyde has been estimated by various investigators to be between 12 and 1,230 $\mu g/m^3$ (0.0 1 to 1.0 ppm). Acrolein can be detected by both odor and eye irritation at concentrations as low as 600 $\mu g/m^3$ (0.25 ppm), but the threshold for optical chronaxy is reported as 1,750 $\mu g/m^3$ (0.75 ppm).

2. Formaldehyde at a concentration of 2,460 $\mu g/m^3$ (2 ppm) has been associated with a doubling of the total expiratory resistance of guinea pigs exposed to irradiated atmospheres of automobile exhaust gases.

3. Acetaldehyde is much less irritating than formaldehyde: its irritant threshold appears at concentrations of about 10,000 $\mu g/m^3$ (50 ppm). No physiological effects are known at levels that may be anticipated to occur in the ambient atmosphere.

CHAPTER 12

SULFUR OXIDES

This section presents criteria of air quality in terms of the effects empirically obtained and published for various concentrations of one family of pollutants, the sulfur oxides, their acids and acid salts. These effects do not, for the most part, derive solely from the presence of sulfur oxides in the atmosphere. They are the effects that have been observed when various concentrations of sulfur oxides, along with other pollutants, have been present in the atmosphere. Many of these effects are produced by a combination of sulfur oxides pollution and undifferentiated particulate matter; the contributions of each class are difficult to distinguish. Moreover, laboratory studies have shown that a combination of sulfur oxides and particulates may produce an effect that is greater than the sum of the effects caused by these pollutant classes individually. Because of the interactions between pollutants, and the reactions of pollutants with oxygen and with water in the atmosphere, and because of the influence of sunlight and temperature on these reactions, the criteria for sulfur oxides can not be presented as exact expressions of cause and effect that have been replicated from laboratory to laboratory. They are presented as useful statements of the effects that can be predicted when sulfur oxides are present in the atmosphere; they are derived from a careful evaluation of what has so far been reported.

The sulfur oxides are common atmospheric pollutants which arise mainly from the combustion of fuels. Solid and liquid fossil fuels contain sulfur, usually in the form of inorganic sulfides or sulfur-containing organic compounds. Combustion of the fuel forms about 25 to 30 parts of sulfur dioxide to 1 part of sulfur trioxide.

Sulfur dioxide is a non-flammable, non-explosive colorless gas that most people can taste at concentrations from 0.3 ppm to 1 ppm (about 0.9 mg/m^3 to 3 mg/m^3) in air. At concentrations above 3 ppm (about 8.6 mg/m^3), the gas has a pungent, irritating odor. In the atmosphere, sulfur dioxide is partly converted to sulfur trioxide or to sulfuric acid and its salts by photochemical or catalytic processes. Sulfur trioxide is immediately converted to sulfuric acid in the presence of moisture. The degree of oxidation of sulfur dioxide in the atmosphere is dependent on a number of factors, including residence time, amount of moisture present, and the intensity and duration of sunlight and its spectral distribution. The amounts of catalytic material, hydrocarbons and nitrogen oxides, and the amounts of sorptive and alkaline materials present, also affect the oxidation process.

In the United States, sulfur dioxide is most commonly measured by the colorimetric West-Gaeke (pararosaniline) and the conductometric methods. The West-Gaeke method is specific for sulfur dioxide and sulfite salts. The method has been modified to compensate for interferences produced by the presence of nitrogen oxides, ozone, or heavy metal salts in the sample, and the modified method is the method of choice of the National Air Pollution Control Administration. Conductometric methods measure sulfur dioxide concentrations as a function of change in the electroconductivity of a solution. These methods are general, in that they react to changes in electroconductivity brought on by other soluble gases, as well as by sulfur dioxide, and the indicated values for sulfur are sometimes very approximate.

The technique most frequently used in Europe to measure sulfur dioxide is the hydrogen peroxide acid titration method, or an automated conductometric version of the same technique. The presence of other acidic or alkaline gases in the sample may affect the results.

The lead peroxide candle is a widely used technique which determines a "sulfation rate." The method gives integrated values for relatively long periods, but provides no indication of short-term fluctuations. It provides only a rough indication of sulfur dioxide concentrations.

Recently, two long-path spectroscopic techniques have been introduced that sense sulfur dioxide concentrations remotely. Although these techniques are complex and expensive, they may eventually be developed to provide a sulfur dioxide "pollution contour" for large areas of a city, as well as pollution concentrations at different elevations.

Sulfuric acid aerosol in suspended particulate material may be measured by titration or by controlled decomposition to sulfur dioxide. The sulfur dioxide can be measured by a number of methods, including spectrophotometry, coulometry, and flame photometry. Particulate sulfate may be analyzed by spectrophotometric or turbidimetric methods.

Each method of measuring sulfur oxides pollution is unique in terms of measuring time resolution, operating costs and skills, time required for analysis, and the specificity of the technique. A single program may make use of both general and specific methods. In selecting the methods to be used in a sampling program, it is especially important to consider the degree to which data obtained from one method can be compared to data obtained from another.

An estimated 28.6 million tons of sulfur dioxide were emitted to the atmosphere of the United States in 1966, as compared with an estimated 23.4 million tons emitted in 1963. The principal share, 58.2 percent, of the 1966 tonnage came from the combustion of coal, primarily for the generation of electric power and for space heating. The combustion of residual fuel oil and other petroleum products, also primarily for power generation and space heating, accounted for 19.6 percent of the total, while the remainder came from the refining of petroleum (5.5 percent), the smelting of sulfur-containing ores (12.2 percent), the manufacturing of sulfuric acid (1.9 percent) the burning of refuse (0.4 percent), and the burning of coal refuse banks (0.4 percent). Paper-making and other industrial operations also contributed minor amounts to the total.

The National Air Pollution Control Administration operates two nationwide programs for surveying sulfur oxides pollution levels in the United States. The National Air Surveillance Network (NASN) takes 24-hour samples of sulfur dioxide from about 100 sites 26 times a year, and the Continuous Air Monitoring Project (CAMP) records 5-minute average concentrations of sulfur dioxide continuously in six large cities—Washington, Philadelphia, Cincinnati, Chicago, St. Louis, and Denver. The NASN program employs the colorimetric, West-Gaeke method of analysis, while the CAMP program uses the electroconductivity technique. Recently, continuous, colorimetric West-Gaeke monitoring devices were installed at the six CAMP locations.

Levels recorded in the CAMP cities over a 6-year period show mean annual concentrations ranging from 0.01 ppm, in San Francisco, to 0.18 ppm, in Chicago, with the averages exceeding, for 1 percent of the time, a concentration between 0.09 ppm and 0.68 ppm. The NASN annual average concentrations ranged from 0.002 ppm, in Kansas City, Missouri, to 0.17 ppm, in New York City. The highest 24-hour average concentration was 0.38 ppm, also in New York City, while the lowest 24-hour averages were below the minimum detectable range of the instruments—below approximately 0.001 ppm. Geographically, the highest values were recorded in the northeastern part of the United States, especially east of the Mississippi River and north of the Ohio River, where large quantities of sulfur-bearing fossil fuels are burned.

Relationship of Maximum Concentrations to Average Concentrations

Although it is convenient to discuss the various effects of sulfur dioxide in connection with the average concentrations of the gas over a long period, such as a year, some effects are thought to be associated with the peak concentrations that may occur during the period and to better define the kind and extent of effects that may be occurring in a given community, it is useful to know what these peak values may be.

For a given averaging time, measurements of sulfur dioxide concentrations follow a log-normal frequency distribution. The two statistical parameters used to describe this distribution are the geometric mean and the standard geometric deviation, which is an index of the deviation of the samples from the mean.

Once sufficient sampling data have been

gathered, the geometric mean and the standard geometric deviation can be used to calculate the expected maximum concentration, the minimum concentration, and the concentration at any percentile, provided that the averaging time is 1 hour or greater. Further, it is possible to calculate expected maxima for one averaging time from data obtained at another averaging time.

For a given typical, multiple source urban area, a fairly good approximation of the frequency distribution of hourly average concentrations for a year at a given station can be obtained from measurements taken for 24-hour periods on about 26 randomly selected days. The accuracy of the approximation will depend on the number of samples taken, compared to the total number of samples that could have been taken if air sampling had been continuous.

The CAMP data for various years in the period 1962 to 1967 appear to be fairly representative of the distribution of sulfur dioxide concentrations for large U.S. metropolitan areas. The range of standard geometric deviations for the CAMP cities is roughly from 2 to 2.5. These values correspond to hourly maxima that range from 10 to 20 times the annual mean, respectively. Similarly, the 8-hour maximum ranges from about 6 to 10 times the annual mean, and the 1-day maximum is between 4 and 7 times the annual mean.

The ratio of the maximum sulfur dioxide concentration to the average values may be greater for measurements made near a single point source than for a city as a whole. For averaging periods from 4 to 54 minutes, for example, the maximum concentrations encountered near a point source were, respectively, 30 to 160 times the 6-month average value.

Effects on Health

The current scientific literature indicates that, for the most part, the effects of the oxides of sulfur on health are related to irritation of the respiratory system. Such injury may be temporary or permanent.

Laboratory studies show that sulfur dioxide can produce bronchoconstriction in experimental animals such as the guinea pig, the dog, and the cat. Dose-response curves have been established for the guinea pig, the most susceptible laboratory animal studied to date. They relate the concentration of sulfur dioxide to the observed increase in pulmonary flow resistance produced by 1-hour exposures. Slight increases in resistance are detectable at 0.16 ppm (460 $\mu g/m^3$) and the changes are readily reversible.

Sulfuric acid and some, but not all, particulate sulfates also produce bronchoconstriction in the guinea pig. The response is highly dependent on particle size, with the smallest particles showing the greatest irritant potency. In this animal, as in man, sulfuric acid and irritant particulate sulfates have a greater irritant potency at a given concentration than does sulfur dioxide alone.

The potentiation by particulate matter of toxic responses to sulfur dioxide (synergism) has been observed under conditions which would promote the conversion of sulfur dioxide to sulfuric acid. The degree of potentiation is related to the concentration of particulate matter. A threefold to fourfold potentiation of the irritant response to sulfur dioxide is observed in the presence of particulate matter capable of oxidizing sulfur dioxide to sulfuric acid. Aerosols of soluble salts of ferrous iron, manganese, and vanadium have been observed to produce this potentiation, although the concentrations used (0.7 mg/m^3 to 1.0 mg/m^3) were considerably greater than any levels of the metals reported in urban air.

Generally speaking, the laboratory work that has been performed to date with animals has only partial relevance for air quality criteria. In most of the studies, the laboratory environment has not simulated very closely the actual environment. Exposures have been to high and constant concentrations, rather than to the low and fluctuating levels commonly found in the atmosphere. Other normally occurring stresses, such as fluctuating temperature, have not, in general, been applied. These studies do, however, provide valuable information on some of the bioenvironmental relationships that may be involved in the effects of the sulfur oxides on health. The data they provide on synergistic effects show very clearly that information derived from single substance exposures should be applied to ambient air situations only with great caution.

The response of bronchoconstriction in man may be assessed in terms of a slight increase in airway resistance. Normal individuals, exposed to sulfur dioxide via the mouth, exhibit small changes in airway resistance, which are often insufficient to produce any respiratory symptoms. The effects may be even smaller when the subject breathes through his nose. As in animals, sulfuric acid is a much more potent irritant in man than is sulfur dioxide. Again, the irritant effect is highly dependent on particle size.

Laboratory observations of respiratory irritations suggest that most individuals will show a response to sulfur dioxide when exposed for 30 minutes to concentrations of 5 ppm (about 14 mg/m^3) and above. Exposure of certain sensitive individuals to 1 ppm (about 3 mg/m^3) can produce detectable changes in pulmonary function. Similar exposure of these same individuals has, in some instances, produced severe bronchospasm. In most of the studies discussed, an increase in pulmonary flow resistance was the indicator of response employed.

Epidemiologic studies do not have the precision of laboratory studies, but they have the advantage of being carried out under ambient air conditions. In most epidemiologic studies, indices of air pollution level are obtained by measuring selected pollutants, most commonly particulates and sulfur compounds. To use these same studies to establish criteria for individual pollutants is justified by the experimental data on interaction of pollutants. However, in reviewing the results of epidemiologic investigations it should always be remembered that the specific pollutant under discussion is being used as an index of pollution, not as a physicochemical entity.

It has been suggested that industrial experience with sulfur oxides exposures may be relevant to ambient air quality criteria. In the absence of epidemiologic evidence, one might, as a rough approximation, select some fraction of the concentrations reported for industrial exposures. In selecting such a fraction, several factors should be taken into consideration. Industrial exposure, for example, is not continuous, and it may not include the synergistic effects which result from the presence of more than one class of pollutant. Further, the exposed population may not include the segments most susceptible to the effects.

From the epidemiologic studies available, it is easy to conclude that there is an effect of the oxides of sulfur in the ambient atmosphere on the health of the population, and that the degree of effect is related to the degree of pollution. Episodes of acute elevation of oxides of sulfur and other pollutant concentrations have been associated with a larger number of deaths than expected. Those predominantly affected were individuals with chronic pulmonary disease or cardiac disorders, or very young or old individuals. However, the general population has also been involved.

Studies of episodes occurring in London suggest that a rise in the daily death rate occurred when the concentrations of sulfur dioxide rose abruptly to levels at or about 715 $\mu g/m^3$ (0.25 ppm) (as measured by the hydrogen peroxide titrimetric method) in the presence of smoke at 750 $\mu g/m^3$. A more distinct rise in deaths has been noted generally when sulfur dioxide exceeded 1000 $\mu g/m^3$ (0.35 ppm) for one day, and particulate matter reached about 1200 $\mu g/m^3$ (measured by the British reflectometer method). Daily concentrations of sulfur dioxide in excess of 1500 $\mu g/m^3$ for one day (0.52 ppm) in conjunction with levels of suspended particles exceeding 2000 $\mu g/m^3$ appear to have been associated with an increase in the death rate of 20 percent or more over base line levels. This same effect has been observed at lower sulfur dioxide levels when the maximum pollution levels lasted for a longer period.

Air pollution episodes in New York City have been associated with exposures similar to those of the London episodes. In one case, for example, excess deaths were detected in New York following a 24-hour period during which sulfur dioxide concentrations exceeded 1500 $\mu g/m^3$ (>0.5 ppm) (as measured by the hydrogen peroxide titrimetric method) and suspended particulate matter was measured as a soiling index of 6 cohs or greater.

For Rotterdam, there have been indications of a positive association between total mortality and exposure for a few days to 24-hour mean concentrations of 500 $\mu g/m^3$ (0.19 ppm) sulfur dioxide. Further, it has been reported that: "There is a faint indication that this will happen somewhere between 300 and 500 μg SO_2 per m^3 per 24 hours" (0.11 ppm and 0.19 ppm).

A survey of emergency clinics at major New York City hospitals revealed a rise in visits for upper respiratory infections and cardiac diseases in both children and adults in all 4 hospital studies during a 10-day period of elevated pollution levels. Sulfur dioxide ranged between 200 $\mu g/m^3$ and 2460 $\mu g/m^3$ (0.07 ppm to 0.86 ppm) during the period studied; hospital admissions were clearly elevated at a time when concentrations had not yet exceeded 715 $\mu g/m^3$ (0.25 ppm). Smoke shade was close to 3 coh units.

In London, a one-day exposure to a daily average level of 600 $\mu g/m^3$ of sulfur dioxide (0.20 ppm) caused accentuation of symptoms in persons with chronic respiratory disease on the day following the high sulfur dioxide level if particulate matter at a substantial concentration was also a pollutant.

This finding in London was also observed in a Chicago study. The Chicago study noted a sharp rise in illness rates on the day following a one-day exposure to 715 $\mu g/m^3$ (0.25 ppm) of sulfur dioxide or more for patients 55 years of age and over with severe bronchitis. Particulate matter was also present.

Effects at lower levels of sulfur dioxides have also been noted. In Rotterdam, during a few days in which sulfur dioxide concentrations rose from about 300 to 500 $\mu g/m^3$ (0.11 ppm to 0.19 ppm), the number of hospital admissions for irritations of the respiratory system rose, particularly for older persons. Absenteeism from work under such conditions increased substantially, especially for those 45 years old and over.

The lowest levels at which effects are reported for short-time periods are those reported for the Ruhr area. An effect was noted on "functional disturbance" at an estimated daily indoor mean of 270 $\mu g/m^3$ (0.10 ppm) for 4 days and an increase in "symptoms, illnesses, or diseases" (breathlessness, throat and eye irritation, and "depression and apathy without further specification") was noted at a daily indoor mean of about 540 $\mu g/m^3$ (0.20 ppm).

Longer term exposure to lower levels than those found during an air pollution episode has also been associated with demonstrable health effects. It is for this reason that it must be emphasized that the levels of a pollutant at which effects are detected are not the concentrations at which the pollutant may begin to have an effect on health. The initiation of the deleterious effects presumably must take place before, and at a lower concentration, than that at which the existence of a strong association is accepted for statistical reasons. Repeated respiratory infections in early childhood, for example, which in one study appeared to be related to air pollution, may have contributed to the later development of the chronic bronchitis syndrome in the adult.

A major British study has found an association between mortality from bronchitis and lung cancer and levels of air pollution, after taking into consideration differences in age, smoking habits, social class, and occupational exposure. The sulfur dioxide values for a year averaged 116 $\mu g/m^3$ (0.040 ppm) for the polluted area, and 75 $\mu g/m^3$ (0.026 ppm) for the cleaner area. Corresponding smoke values were 160 $\mu g/m^3$ and 80 $\mu g/m^3$, respectively. It was not possible to separate the effects of sulfur dioxide from the effects of comparable amounts of smoke which were present. This is consistent with other British studies, which indicate an association of bronchitis death rates with air pollution concentrations that is apparently independent of social class differences, but which can not specify the specific role played by sulfur oxides.

Children and housewives appear to represent the most suitable subjects for determining the health effects of long-term exposures to routine levels of air pollution. A study in Genoa, Italy, included housewives 65 or more years of age, who were non-smokers, and who had lived for a long period in the same area without having any industrial experience. Sulfur dioxide was monitored for 10 years at 19 sites by a technique analagous to the volumetric method of the British. An increased frequency of cough, sputum, dyspnea, and bronchitis was noted in the moderately polluted area as compared to the relatively clean area. Differences were noted in the summer prevalence of respiratory diseases in the industrial area, with an annual mean of 265 $\mu g/m^3$ SO_2 (0.093 ppm) when compared to the middle [annual mean of 105 $\mu g/m^3$ SO_2 (0.037 ppm)] and low [annual mean of 80 $\mu g/m^3$ SO_2 (0.028 ppm)] pollution areas. The study also showed a very significant correlation between the frequency of bronchitis and the annual mean of sulfur dioxide levels for the seven districts of the city, whereas the correlation between the frequency of bronchitis and suspended matter and dustfall was not significant.

Another morbidity study which has included housewives is that conducted in Nashville. This showed a direct correlation between illnesses for all causes for housekeeping white females, 15 to 64 years of age, and sulfur dioxide levels. It also showed that the cardiovascular morbidity in the 55 and older age group, for both sexes, was twice as high in the most polluted area as compared with the least. Cigarette smoking was not taken into account, and the adjustment made for socioeconomic status may not have been wholly adequate.

Differences in respiratory symptoms have also been found in areas which had long-term values of 95 $\mu g/m^3$ sulfur dioxide (0.034 ppm) and 25 $\mu g/m^3$ (0.009 ppm) measured by hydrogen peroxide absorption, sulfuric acid titrations. The study is unusual in that the source of pollution was the single stack of a smelter, and daily averages were up to 17 times the annual mean. However, the findings that respiratory symptoms disappeared when pollution was locally reduced through installation of a higher stack, would appear to confirm an effect of sulfur dioxide at the concentrations originally measured, provided that no other change took place which could account for the disappearance of symptoms.

Studies of schoolchildren in Great Britain have indicated that increased frequency and severity of respiratory diseases occurred when long-term pollution levels exceeded averages of about 120 $\mu g/m^3$ (0.046 ppm) sulfur dioxide and 100 $\mu g/m^3$ for smoke.

Effects on Visibility

Particles suspended in the air reduce visibility, or visual range, by scattering and absorbing light coming from both an object and its background, thereby reducing the contrast between them. Moreover, suspended particles scatter light into the line of sight, illuminating the air between, to further degrade the contrast between an object and its background. This phenomenon is described in detail in the document *Air Quality Criteria for Particulate Matter.*

The scattering of light into and out of the line of viewing by particles in the narrow range of 0.1 μ to 1 μ in radius has the greatest effect on visibility. Of the total suspended particulate matter in urban air, commonly from 5 percent to 20 percent consists of sulfuric acid and other sulfates, and of these, 80 percent or more by weight are smaller than 1 μ in radius. Consequently, suspended sulfates in the air can contribute significantly to reduction in visibility.

Characteristic behavior of suspended particles in the size range mentioned makes it possible to relate visual range to concentrations of overall particulate matter. Since sulfur dioxide levels, in general, correlate with levels of overall suspended particulate matter, and since the ratio of sulfur dioxide to suspended sulfate can be estimated, given the relative humidity, it is possible to estimate visibility for various relative humidities from sulfur dioxide concentration. At a concentration of 285 $\mu g/m^3$ (0.10 ppm) of sulfur dioxide and with a relative humidity of 50 percent, visibility in New York City would typically be reduced to about 5 miles. At a visual range of less than 5 miles, operations are slowed at airports because of the need to maintain larger distances between aircraft. Federal Aviation Administration restrictions on aircraft operations become increasingly severe as the visual range decreases below 5 miles.

Effects on Materials

Laboratory and field studies underscore the importance of the combination of particulate and sulfur oxides pollution in a wide range of damage to materials. On the basis of present knowledge, it is difficult to evaluate precisely the relative contribution of each of the two classes of pollution; however, some general conclusions may be drawn.

Steel test panels, dusted with a number of active hygroscopic particles commonly found in polluted atmospheres, corroded at a low rate in clean air at relative humidities below 70 percent. The corrosion rate was higher at relative humidities above 70 percent. It greatly increased when traces of sulfur dioxide were added to the laboratory air.

It is apparent that corrosion rates of various metals are higher in urban and industrial atmospheres with relatively high levels of both particulate and sulfur oxides than they are in rural and other areas of low pollution. High humidity and temperature also play an important synergistic part in this corrosion reaction. Studies show increased corrosion rates in industrial areas where air pollution levels, including sulfur oxides and particulates, are higher. Further, corrosion rates are higher during the fall and winter seasons when particulate and sulfur oxides pollution is more severe, due to a greater consumption of fuel for heating. Depending on the kind of metal exposed as well as location and duration of exposure, corrosion rates were 1½ to 5 times greater in polluted atmospheres than in rural environments.

In Chicago and St. Louis, where steel panels were exposed at a number of sites, high correlations were found in each city between corrosion rates, as measured by weight loss, and sulfur dioxide concentrations, as measured by the West-Gaeke method. In St. Louis, except for one exceptionally polluted site, corrosion losses were 30 percent to 80 percent higher than losses measured in nonurban locations. Sulfation rates in St. Louis, measured by lead peroxide candle, also correlated well with weight loss due to corrosion. Measurements of dustfall in St. Louis, however, did not correlate significantly with corrosion rates. Over a 12-month period in Chicago, the corrosion rate at the most corrosive site (mean SO_2 level of 0.12 ppm) was about 50 percent higher than at the least corrosive site (mean SO_2 level of 0.03 ppm). Although suspended particulate levels measured in Chicago with high-volume samplers also correlated with corrosion rates, a co-variance analysis indicated that sulfur dioxide concentrations were the dominant influence on corrosion. Based on these data, it appears that considerable corrosion may take place (i.e., from 11 percent to 17 percent weight loss in steel panels) at annual average sulfur dioxide concentrations in the range of 0.03 ppm to 0.12 ppm, and although high particulate levels tend to accompany high sulfur dioxide levels, the sulfur dioxide concentration appears to have the more important influence.

Sulfur oxides pollution contributes to the damage of electrical equipment of all kinds. Studies have reported a one-third reduction in the life of overhead power line hardware and guy wires in heavily polluted areas. In some areas it has been found necessary to use more expensive, less corrodible metals, such as gold, for electrical contacts.

Sulfur oxides pollution attacks a wide variety of building materials—limestone, marble, roofing slate, and mortar—as well as statuary and other works of art, causing discoloration and deterioration. Certain textile fibers—such as cotton, rayon, and nylon—are harmed by atmospheric sulfur oxides. Dyed fabrics may fade in atmospheres containing sulfur oxides and other pollutants. Severe fading was noted for some dyes in fabrics exposed in Chicago, where annual average sulfur dioxide levels were 0.09 ppm. Leather exposed to sulfur oxides may lose much of its strength, and paper may become discolored and brittle.

Concentrations of 1 ppm sulfur dioxide can increase the drying time of some oil-based paints by 50 percent to 100 percent. Some films become softer and others more brittle, both developments adversely affecting durability. Sulfur dioxide also appears to render some paint films water sensitive, consequently reducing the film gloss. Under certain conditions sulfur dioxide levels of 0.1 ppm to 0.2 ppm cause the blueing of Brunswick green, and in the presence of ammonia produce a troublesome defect called crystalline bloom brought about by the formation of very small ammonium sulfate crystals.

Effects on Vegetation

Sulfur dioxide may cause acute or chronic leaf injury to plants. Acute injury, produced by high concentrations for relatively short periods, usually results in injured tissue drying to an ivory color; it sometimes results in a darkening of the tissue to a reddish-brown. Chronic injury, which results from lower concentrations over a number of days or weeks, leads to pigmentation of leaf tissue, or leads to a gradual yellowing, or chlorosis, in which the chlorophyll-making mechanism is impeded. Both acute and chronic injury may be accompanied by the suppression of growth and yield.

Acute injury apparently affects the plant's ability to transform absorbed sulfur dioxide into sulfuric acid, and then into sulfates. At high rates of absorption, sulfite is thought to accumulate, resulting in the formation of sulfurous acid, which attacks the cells. The amount of acute injury depends on the absorption rate, which is a function of the concentration. A given amount of gas at a high concentration will be absorbed in a shorter period and will cause more leaf destruction than the same amount of gas at a lower concentration. Mathematical expressions have been worked out which, for some plant species, relate concentration, time of exposure, and amount of damage.

Different varieties of plants vary widely in their susceptibility to acute sulfur dioxide injury. The threshold response of alfalfa to acute injury is 1.25 ppm over 1 hour, whereas privet requires 15 times this concentration for the same amount of injury to develop. Some species of trees and shrubs have shown injury at exposures of 0.5 ppm for 7 hours, while injury has been produced in other species at 3-hour sulfur dioxide exposures of 0.54 ppm and, in still others, at 8-hour exposures of 0.3 ppm. From such studies, it appears that acute symptoms will not occur if the 8-hour average concentration does not exceed 0.3 ppm. (From the data on the CAMP cities, a maximum 8-hour concentration of 0.3 ppm would correspond to a yearly average concentration of between 0.03 ppm and 0.05 ppm). However, sulfur dioxide concentrations from 0.05 to 0.25 ppm may react synergistically with either ozone or nitrogen dioxide in short-term exposures (e.g., 4 hours) to produce moderate-to-severe injury to certain sensitive plants.

Chronic plant injury results from the gradual accumulation of excessive amounts of sulfate in leaf tissue. Sulfate formed in the leaf is additive to sulfate absorbed through the roots, and when sufficiently high levels accumulate, chronic symptoms, accompanied by leaf drop, may occur. Chronic symptoms and excessive leaf drop have been reported in locations where the mean annual concentration is below approximately 0.03 ppm.

It has been suggested that sulfur dioxide might suppress growth and yield without causing visible injury. One investigator reported that yields of rye grass grown in unfiltered air were significantly lower than similar yields of plants grown in filtered air. No visible symptoms of injury were observed. Sulfur dioxide levels in the unfiltered air ranged from 0.01 ppm to 0.06 ppm, with exposure periods ranging from 46 to 81 days; other gaseous pollutants may have also been present. Usually, the suppression of growth and yield is accompanied by visible symptoms of injury—a linear relationship has been derived, for example, between the yield of alfalfa and the total area destroyed by acute symptoms, or the area covered by chlorosis.

Sulfuric acid mist, which may occur in polluted fogs and mists, also damages leaves. The acid droplets may cause a spotted injury to wet leaves at concentrations of 0.1 mg/m^3.

SOURCES AND METHODS OF MEASUREMENT OF SULFUR OXIDES IN THE ATMOSPHERE

Oxidized sulfur in the atmosphere exists in several different chemical and physical forms. Under normal conditions, the predominant state is gaseous sulfur dioxide, together with smaller amounts of non-volatile sulfuric acid mist or sulfate salts.

Since the proper and efficient establishment of criteria and standards ultimately depends upon reliable analytical data, it is particularly important that proper attention be given to (1) the sampling techniques, and (2) the analytical procedures employed. These have been reviewed recently by Hendrickson [1] and Katz.[2]

If gaseous sulfur dioxide were the sole pollutant in the atmosphere, its quantitative determination, even in the fractional ppm range, would not be unduly difficult. The presence of both sulfuric acid mist and sulfate salts adds complications to the analytical procedures. These are further complicated by real or potential interferences from a variety of additional air pollutants often found in conjunction with sulfur dioxide.

Some of the methods used for determining SO_2 are *specific*, while others are *general* and depend on a general property such as acidity or conductivity and are subject to errors because of interferences. This has led to some problems in the interpretation of analyses from different laboratories using different techniques.

In the following discussion, the man-made sources of atmospheric oxides of sulfur will be outlined, and the more important methods of their determination, including recent techniques involving remote sensing of SO_2, will be discussed.

Agreement between the various methods of measurement is not always good, because each method is subject to interference from differing causes which may lead to either high or low results.

Sulfuric acid mist is determined by the separation of the mist from sulfur dioxide and the subsequent measurement of the acid or of the sulfate content. Measurements of suspended particulate sulfate and sulfate in dustfall are obtained by means of conventional sulfate determinations.

SOURCES OF ATMOSPHERIC SULFUR OXIDES

Sulfur dioxide pollution results primarily from the combustion of fossil fuels, the refining of petroleum, the smelting of ores containing sulfur, the manufacture of sulfuric acid, the burning of refuse, paper making, and the burning or smoldering of coal refuse banks. In all of these processes, a small amount of sulfur trioxide or sulfuric acid also is emitted. Sulfur trioxide is normally present in the atmosphere in extremely small concentrations because it is converted to sulfuric acid soon after its entry into the atmosphere.

Specific reviews covering the coal industry, petroleum refineries, fuel oil combustion, burning coal mine refuse banks, sulfuric acid manufacture, the iron and steel industry, and sulfite pulping have been published.[3–12] Sources of sulfur oxides in the atmosphere also have been reviewed by Rohrman and Ludwig.[13]

Of considerable importance to the meteorological and chemical behavior of sulfur oxides in the atmosphere, as well as to their measurement, are the kinds of emitters, whether large or small and whether disperse or point sources. The trend of operations has been away from sulfur dioxide pollution by low-level disperse sources and toward large point sources, except for space heating and diesel trucks using fuels of high sulfur content. The large source emissions contain lower concentrations of polynuclear hydrocarbons and higher concentrations of nitrogen oxides and sulfuric acid. Particulate matter which interacts with the oxides of sulfur can be controlled to a greater extent when emitted from the larger sources. Further, emissions from large sources usually are emitted from higher stacks and although this practice appears to reduce average ground level concentrations and the frequency of air pollution episodes,[9 10 14–17] it may result in early morning fumigations.

Sources of atmospheric sulfur oxides can be considered from the viewpoint of their annual production. Obviously, sulfur-containing coal and fuel oil utilized for heating during the winter will produce a seasonal increase in atmospheric SO_2. This is in contrast with industrial sources, such as power stations and factories, whose effluents are more or less constant throughout the year.

Major sources of sulfur dioxide released to the atmosphere in 1963[13] and in 1966[18] are presented in Table 12-1.

MEASUREMENT OF GASEOUS SULFUR DIOXIDE CONCENTRATIONS

1. Sampling Techniques

Before discussing specific analytical meth-

Table 12-1—ATMOSPHERIC SULFUR DIOXIDE EMISSIONS IN 1963 AND 1966 BY SOURCE

Process	Sulfur dioxide [a] 1963 Tons	1963 Percent of total emissions	1966 Tons	1966 Percent of total emissions
Burning of coal:				
Power generation (211,189,000 tons, 1963 data)	9,580,000	41.0	11,925,000	41.6
Other combustion (112,630,000 tons, 1963 data)	4,449,000	19.0	4,700,000	16.6
Subtotal	14,029,000	60.0	16,625,000	58.2
Combustion of petroleum products:				
Residual oil	3,703,000	15.9	4,386,000	15.3
Other products	1,114,000	4.8	1,218,000	4.3
Subtotal	4,817,000	20.7	5,604,000	19.6
Refinery operations	1,583,000	6.8	1,583,000	5.5
Smelting of ores	1,735,000	7.4	3,500,000	12.2
Coke processing	462,000	2.0	500,000	1.8
Sulfuric acid manufacture	451,000	1.9	550,000	1.9
Coal refuse banks	183,000	0.8	100,000	0.4
Refuse incineration	100,000	0.4	100,000	0.4
Total Emissions	23,360,000	100.0	28,562,000	100.0

[a] A small amount of this tonnage is converted to sulfuric acid mist before discharge to the atmosphere. The rest is eventually oxidized and/or washed out. Only under unusual meteorologic conditions (Chapter 3) does accumulation occur. The increasing output of sulfur oxides due to increasing power demand is evident.

ods, it is important to recognize that regardless of how accurate the technique may be, the validity of the final results is dependent upon the sampling technique employed for the determination. For example, consideration must be given to factors such as adsorption on, and desorption from, inlet tubes utilized in the sampling apparatus.[19] Teflon *, Tygon *, glass, stainless steel and aluminum have been tested for various lengths and flow rates; in general, a conditioning period is required by Tygon and aluminum tubing. For relatively high flow rates (28.3 l/min), aluminu required a conditioning period of about 5 hours at 0.2 ppm and Tygon required a much longer period and should be avoided. Neither adsorption nor desorption was significant in glass or stainless steel tubing at various flow rates for tubing lengths as great as 30.5 meters. Teflon tubing falls into this same class. Any of these three materials can be used for sampling sulfur dioxide at sufficiently high flow velocities (greater than 3.7 m/sec) without prior conditioning and without effect by temperature or humidity.

Both continuous and intermittent sampling are commonly used for sulfur dioxide. To a large extent the type of instrument selected is dependent upon the measurement principle desired. The West-Gaeke and the Conductometric methods are most commonly employed in the United States. Many agencies in the country use intermittent or manual sampling where the collected sample is returned to the laboratory for analysis. Intermittent sampling can be used to provide integrated samples of from 1 hour to 24 hours and is less expensive than continuous monitoring. Continuous monitoring is necessary where it is important to show diurnal changes or the influence of local sources and meteorology.

2. Colorimetric Method: Pararosaniline

In the West-Gaeke method, sulfur dioxide is absorbed in dilute aqueous sodium tetrachloromercurate to form the nonvolatile dichlorosulfitomercurate ion, which then reacts with formaldehyde and bleached pararosaniline to form red-purple pararosaniline methylsulfonic acid. This reaction is specific for sulfur dioxide and sulfite salts. The color intensity of the dye, which is proportional to the concentration of sulfur dioxide, is measured at a wavelength of 560 mμ. The method can be used to determine concentrations in the air from 0.002 ppm to 5 ppm. Ozone and nitrogen dioxide reduce the apparent concentration by destroying some of the dye, although interference by nitrogen dioxide can be eliminated by adding sulfamic acid after sample collection or just prior to analysis. Heavy metal salts, especially iron salts, oxidize dichlorosulfitomercurate, which also results in a lowering of the apparent sulfur dioxide concentration. This effect can be eliminated by a membrane prefilter or by including the disodium ethylenediaminetetraacetic acid in the absorbing reagent to sequester metallic ions. Hydrogen sulfide precipitates mercuric sulfide from the collecting reagent, and such a precipitate must be removed by centrifugation from the sample before proceeding with color development.

Recently two improved West-Gaeke (pararosaniline) methods were developed for the determination of sulfur dioxide in ambient air.[20] These give greater sensitivity and reproducibility, as well as adherence to Beer's Law throughout a greater working range, than does the original West-Gaeke method. The improvements resulted from optimization of several important parameters. Specifically the pararosaniline dye was purified and standardized to reduce variability. Phosphoric acid was used in the final color development to control pH. The pararosaniline methylsulfonic acid produced in the reaction exhibits a hypsochromic spectral shift with increasing pH. Hence the reaction product behaves as a two-color pH indicator, and regardless of the pH condition under which the reaction is carried out, the spectra can be interchanged by addition of acid or base. The sharp peak with an absorbance maximum at $\lambda = 548$ mμ (rose-red) at the higher pH value (1.68) shifts to $\lambda = 575$ mμ (magenta) at a lower pH value (1.02). Interferences from nitrogen oxides, ozone, and heavy metals are minimal, and laboratory results are reproducible to within 4.9 percent (at 95 percent confidence level) if recommended precautions are taken. It is noted that these modifications, although specifically developed for sulfur dioxide in the atmosphere, can be applied to determination of sulfite in other materials. This method has been adopted as the standard or reference method by the NAPCA.

3. Conductometric Methods

The basis of these methods is the oxidation

* Mention of commercial products does not constitute endorsement by the National Air Pollution Control Administration.

of sulfur dioxide to sulfuric acid by aqueous hydrogen peroxide, and the subsequent measurement of the increased electrical conductivity of the solution. This is a *general* technique and one must take special precautions to eliminate other pollutants which could affect the conductivity of the solution. For example, acidic gases such as the hydrogen halides will increase the conductivity of the solution so that, in the presence of such gases, incorrectly high SO_2 values will be indicated. Weakly acidic gases such as hydrogen sulfide cause practically no intereference because of their slight solubility and poor conductivity, and nitrogen dioxide produces minimum interference because it is poorly absorbed. Sulfuric acid mist is not efficiently collected in the usual gas "scrubber" and therefore is not measured. Salt spray from ice and snow control may give high readings. Ammonia interferes with electroconductivity measurements by neutralizing the acid and forming the ammonium ion which has a high transport number of the opposite charge.

$$\underset{\text{Ammonia}}{2NH_3} + \underset{\text{Sulfuric Acid}}{H_2SO_4} \longrightarrow \underset{\text{Ammonium Sulfate}}{(NH_4)_2SO_4} \quad (2\text{-}1)$$

Conductivities that are too low are then recorded, since the transport number of all other cations is several times less than that of the hydrogen ion.

Conductometric methods can be automated readily so that one can obtain instantaneous readings. Air is drawn through either acidic hydrogen peroxide solution or through deionized water, and the sulfur dioxide concentration is estimated from the conductivity of the final solution. The interference effects of various gases on the conductivity have been described.[21, 22]

4. Acid Titration Method: Hydrogen Peroxide

In this method, the air sample is bubbled through 0.03*N* hydrogen peroxide solution adjusted to pH 5. Any sulfur dioxide present forms sulfuric acid, which is then titrated with standard alkali. The presence of other acidic gases in the sample will lead to erroneously high results, and alkaline gases or reactive basic solids give erroneously low results. Generally, a filter is placed in front of the sampling bottle so that no particulate matter or other aerosols are absorbed in the hydrogen peroxide solution. It is important that this be an efficient filter.

This technique is straightforward and the apparatus is inexpensive. It may be automated, if desired. Often, it is used to obtain 24-hour averages, particularly in Europe, where it is widely employed as the standard apparatus and procedure.

5. Spectroscopic Methods

In addition to techniques for making "on-the-spot" analyses for atmospheric sulfur dioxide, it would be a great advantage to have additional methods by which atmospheric concentrations of SO_2 could be determined by means of an instrument remote from the emission source or from the actual air mass being investigated. Recently, there have been developed several spectroscopic techniques which provide such remote sensing capability. Two of the most important are the multiple scan infrared interference spectrometer and the correlation spectrometer. Such instrumentation, though expensive, yields data in a uniquely useful form for a variety of applications.

A commercial model of a multiple-scan infrared interference spectrometer has been developed. Its utilization in the field has been discussed by Low.[23] With this instrument, remote detection of SO_2 and CO_2 in the stack effluent of a power plant has been reported by Low and Clancy.[24]

A correlation spectrometer mounted in an airplane recently has been used by Barringer and associates to estimate atmospheric sulfur dioxide "contours" for several U.S. cities.[25–27] The instrumental technique utilizes sunlight reflected from the earth's surface as a "light source" and, in a sophisticated fashion, measures the absorption of this reflected solar radiation by the sulfur dioxide in the air. The method depends on the fact that sulfur dioxide has a clearly structured absorption spectrum in the region near 3100Å and it gives the total sulfur dioxide concentration in the "line of sight" of the spectrometer. While work to date has been carried out primarily from aircraft, correlation spectrometry need not be limited to airborne applications.

6. Other Methods

The earliest methods for determining sulfur dioxide were based on its ability to reduce a starch-iodine solution. This method was considered to be reasonably accurate in the range 0.8 ppm to 3 ppm and it later was modified to cover the range 0.1 ppm to 60 ppm. This method has now been automated.[28] These and other methods are discussed by Katz.[29] He also reported a method of measuring the titratable acidity of the air by absorbing acid components in a water solution of hyperol, a compound formed by reacting hydrogen peroxide with urea. The oxidizable acid components react with the peroxide.

Sulfur dioxide can be determined by a fuchsin-formaldehyde method reviewed by Hochheiser [30] and reported by Alekseeva and Samorodiva.[31] The sample is collected in 0.1*N* sodium hydroxide in glycerol-water solution. Another procedure, reported by Bushtueva,[32] made use of collection in a potassium chlorate solution. The method of analysis was not stated, but it probably was by nephelometric analysis of barium sulfate or turbidimetric analysis of lead sulfate. Lyubimov [33] described a monitoring instrument in which sulfur dioxide is absorbed in a solution of barium chloride, and light transmission through the turbid solution, inversely related to the sulfur dioxide concentration, is recorded.

A further method for determining ozone and sulfur dioxide has been reported recently.[34] The measurement principle involves liberation of iodine by ozone from an iodide solution in one channel of the analyzer and the consumption of iodine by sulfur dioxide in a second analyzer channel.

7. Use and Comparison of Methods

The hydrogen peroxide acid titration method or an automatic conductometric version of the same method is used most frequently in Europe; in the United States, the colorimetric pararosaniline and automated conductometric (using either acidified peroxide or deionized water) methods are used most frequently.

The values obtained by the various methods of measuring gaseous sulfur dioxide in ambient air may not always correlate well with one another, since they do not measure the same thing and because interfering substances are present in varying amounts from time to time and from place to place. It is generally assumed that sulfur dioxide is the major constituent in air that will respond to any of the methods discussed; that the assumption may not be entirely valid is shown in some recent studies.

Recently the conductivity and colorimetric pararosaniline techniques have been evaluated in depth,[41, 42] and others have reported on comparisons between the methods.[41, 45] In one study, no consistent relationship between the two measurements was found, although the conductometric method usually gave higher values.[44]

There is evidence that hydrochloric acid may contribute significantly to the values obtained by the hydrogen peroxide method in England, but this may not be true for certain sections of the United States because of the negligible amounts of chloride found in the coal burned.[45–48]

Recently a flame photometric instrument was reported for the detection of total sulfur compounds in the atmosphere.[49] The instrument is based on the principle that when sulfur compounds in air are introduced into a hydrogen-rich flame, they emit light between 300 mμ to 425 mμ. A photomultiplier tube looks at the light of wavelength 394 mμ through a narrow-band optical filter. The quantity of light is proportional to the concentration of sulfur compounds. This instrument is readily automated for continuous monitoring.

Several commercial instruments are also available which are based on coulometry. These instruments measure the reducing properties of sulfur dioxide and, with the use of a suitable scrubber, can provide relatively specific and continuous measurements of sulfur dioxide. The principle of this method involves the titration of sulfur dioxide by electrogenerated bromine or iodine.

MEASUREMENT OF SULFURIC ACID AND SULFATES

These forms of sulfur are always found in the particulate phase in polluted air, either as suspended matter which can be removed by special measurement techniques or in dustfall. The acidity may be measured by titration or by a related procedure; alternatively, sulfate ion may be measured (frequently by conversion to the insoluble barium salt). Since the particle size is very small, the initial collection of the sample requires special techniques. Any technique for separating particulate matter from air may be used, but filtration and impaction usually have been employed.

A system capable of filtering 50 cubic feet

to 60 cubic feet of air in 1 hour through Whatman No. 4 filter paper (1-inch diameter) that had previously been washed with distilled water until the washing had a pH of 7 ± 0.10, is described by Mader.[50] The test filter holding the pollutants was macerated in 20 ml of distilled water, and the pH of the resulting solution was determined. Total acidity (sulfuric acid) was determined by titration of 0.002*N* sodium hydroxide to an end point equivalent to the pH of carbon-dioxide-free distilled water and corrected by a blank determination.

Commins[51] has described a method in which sulfuric acid in the air was collected by filtering air through Whatman No. 1 filter paper. The amount of acid was determined by immersing the collected sample in a known excess of 0.01*N* sodium tetraborate in deionized water of pH 7 and titrating back to pH 7 with 0.01*N* sulfuric acid.

Sulfuric acid aerosol has been collected by impaction or filtration to separate sulfuric acid from sulfur dioxide.[52] The acid is decomposed at controlled temperatures under a nitrogen stream, and the liberated sulfur trioxide is then reduced with hot copper to sulfur dioxide. Sulfur dioxide is determined spectrophotometrically, coulometrically, or by flame photometry. The method measures sulfuric acid in the presence of 10 to 100 times as much sulfur dioxide and inert sulfates. Ammonium sulfate, a primary lung irritant, is also measured as sulfuric acid; however, this sulfate probably originates in the atmosphere from the reaction of sulfuric acid and ammonia. The sensitivity is in the parts per billion range using 1 m^3 of air as a sample. DuBois has described a method in which sulfuric acid is collected on glass fiber filters and titrated following thermal separation at 200° C in a nitrogen atmosphere. This method measures free sulfuric acid and does not include ammonium sulfate or sulfuric acid which reacts with particles on the filter.

OTHER METHODS OF MEASURING POLLUTION BY SULFUR OXIDES

1. Sulfation Rates of Lead Peroxide Candles

A method for measuring sulfation involves use of the lead peroxide "candle." A paste made of lead peroxide in a gum tragacanth solution is applied to a cotton gauze wrapped around a glass or porcelain form; the product is the lead peroxide candle. The candle is exposed to the ambient air in a louvered shelter for an extended period of time, usually a month, and the lead sulfate formed in the candle is then determined. It is abvious that the method cannot give any information on short-term variations in sulfur oxides pollution levels and that it is only an empirical estimate of the average concentration. The method measures sulfuric acid, hydrogen sulfide, and other sulfur-containing compounds which can form sulfate. In spite of these limitations, sulfation measurements have correlated well with data on biological effects and the deterioration of materials.

A small, dish-type lead peroxide measuring device termed a "sulfation plate," which can be exposed without a protective shelter, has been described.[53] The plates are more reactive than candles, and it is proposed that exposure periods could be shortened to 1 day with more sensitive analytical methods. Turbidimetric determination of sulfate was found, not unexpectedly, to be less time-consuming than gravimetric analysis, though not as accurate.

The primary application of the lead peroxide candle is for mapping sulfur pollution in a given area as related to sources and meteorology. It is convenient since a simple shelter requiring no electrical power is used. A major problem with the lead peroxide candle is that results are influenced by wind movement and humidity. Another major drawback to the lead peroxide candle is that the laboratory preparation and analysis is quite time consuming. The cost of preparing and analyzing a lead peroxide candle is approximately the same as for collecting and analyzing a 24-hour integrated sulfur dioxide sample by the West-Gaeke procedure. Also, the lead peroxide candle provides intelligence on the oxidizable sulfur compounds in the atmosphere which seldom can be directly related to sulfur dioxide. It has been observed that in many cases monthly averages of sulfur dioxide rise and fall somewhat parallel to monthly sulfation values.

Sulfation rates in United States cities have been observed to range from a few hundredths of a milligram to about 8 mg of sulfur trioxide per 100 cm^2 of exposed lead peroxide candle surface per day (mg SO_3/100 cm^2-day).[35–37, 55]

2. Suspended Sulfate

Sulfate (sulfuric acid and sulfate salts) in suspended particulate matter has been extensively measured for a number of years by the National Air Surveillance Networks.[48,56] Sulfates are extracted from the particulate matter on the glass fiber filter by refluxing with distilled water. Sulfate in the extract is determined by the methyl thymol blue automated method or by the turbidimetric method. The maximum 24-hour average concentration estimated from 2,197 samples collected over the years 1957 to 1960 is 94 $\mu g/m^3$, and the national average for this period is 11.8 $\mu g/m^3$. There is considerable geographic variation. The observed average concentrations of 18.8 $\mu g/m^3$, 15.0 $\mu g/m^3$, 14.5 $\mu g/m^3$, and 13.3 $\mu g/m^3$ occur in the Mid-Atlantic, Midwest, Mideast, and New England areas, respectively. Lower average concentrations of 10.7 $\mu g/m^3$, 9.0 $\mu g/m^3$, 8.7 $\mu g/m^3$, 7.4 $\mu g/m^3$, and 5.8 $\mu g/m^3$ have been observed in the Southeast, Pacific coast, Great Plains, Gulf south, and Rocky Mountain areas, respectively. As with sulfur dioxide, the highest concentrations occur most frequently in the fall and winter months.

3. Sulfate in Dustfall

The register of Air Pollution Analyses[56] lists only four organizations which have measured and reported sulfate in dustfall in the United States. This measure may be expressed either as percent sulfate in total dustfall or as tons of sulfate per square mile per month. In a southwestern city, where 32 collecting stations were used, sulfate averaged 10.8 percent of the total dustfall, or 15.5 tons/mi^2-month over a 3-month period. In a northwestern city, sulfate averaged 27 percent of the total dustfall, or 7.4 tons/mi^2-month, which is similar to the 32 percent of the total dustfall and 7.8 tons/mile^2-month observed in a New England town.[54,57–59]

4. Sulfuric Acid Mist

Few attempts have been made to measure sulfuric acid mist in the United States. Concentrations measured in Los Angeles average about 25 $\mu g/m^3$, and a high concentration of 50 $\mu g/m^3$ has been observed.[50,60,61] In Chicago, during the months of November and December 1964, the average for the hours 10 a.m. to 4 p.m. was 9.2 $\mu g/m^3$.[62]

SUMMARY

In 1966, an estimated 28.6 million tons of sulfur dioxide were emitted to the atmosphere, as compared with 23.4 million tons in 1963. The principal share, i.e., 58.2 percent, came from the combustion of coal, most of which was used to generate electrical power. The combustion of residual fuel oil and other petroleum products accounted for 19.6 percent of the total, while the remainder came from the refining of petroleum (5.5 percent), the smelting of sulfur-containing ores (12.2 percent), the manufacturing of sulfuric acid (1.9 percent), the burning of refuse (0.4 percent), and the burning or smoldering of coal refuse banks (0.4 percent). Paper-making and some other industrial operations also contributed minor amounts to the total. In all of these processes, small amounts of sulfur trioxide or sulfuric acid are emitted also.

In the United States, the two most common methods of measurement are the colormetric West-Gaeke (pararosaniline) and the conductometric methods. The former is a specific method and the latter is a general method. The colorimetric West-Gaeke method employs a chemical reaction in which a red-purple dye is produced; the intensity of the dye, which is proportional to the concentration of sulfur dioxide, is then measured. This method is specific for sulfur dioxide and sulfite salts. While nitrogen oxides, ozone, and heavy metal salts in the sample can affect the measurement, there are ways to eliminate these interferences. For intermittent as well as continuous sampling, the modified West-Gaeke (pararosaniline) procedure is the most satisfactory, and it is the standard method of the National Air Pollution Control Administration.

Conductometric methods may employ an acidified peroxide solution in which the electrical conductivity changes directly with the concentration of sulfur dioxide in the absorbing solution. Although conductometric methods can be readily automated to obtain readings over long durations, they are *general* techniques and the user must be fully aware that other pollutants can affect the conductivity of the solution. Since often it is not known what other interfering pollutants may be present in the sample, the results are sometimes very approximate, particularly at high and very low sulfur dioxide concentrations.

The technique most frequently used in Europe is the hydrogen peroxide approach, in which sulfur dioxide forms sulfuric acid which is then titrated with standard alkali;

this approach may be considered a *general* method, since the presence of other acidic or alkaline gases in the sample may give erroneously high or low results.

An approach that has been widely used is the lead peroxide candle technique, which determines a "sulfation rate." The method gives integrated values for relatively long periods, but, unlike continuous monitoring instruments, provides no indication of short-term fluctuations. While the lead candle method is inexpensive in terms of equipment, the preparation and analysis costs are about the same as for collecting and analyzing 24-hour integrated sulfur dioxide samples by the West-Gaeke procedure. In general, the disadvantages of lead candles far outweigh their advantages. They are useful only in relatively localized studies where the meteorology and exposure do not vary, and then only if great care is exercised to keep the construction of the candles and the chemistry of the analyses constant. In addition, the lead candle provides only a rough indication of sulfur dioxide concentrations, since it is a general approach that may respond to a large number of oxidizable sulfur-containing compounds found in the atmosphere.

Recently, two long-path spectroscopic techniques have been introduced that sense sulfur dioxide concentrations remotely. A multiple-scan interference spectrometer detects from a distance the characteristic infrared sulfur dioxide emissions in heated plumes; a correlation spectrometer looks at the characteristic absorption by sulfur dioxide that occurs near 3100Å. Although these instruments are complex and expensive, they possess a potential for determining the sulfur dioxide "pollution contours" over large areas of the city, as well as the concentrations at different elevations.

Other techniques are still being perfected and may have much promise. Coulometric instruments involve the titration of sulfur dioxide by electrogenerated bromine or iodine. With the aid of suitable scrubbers, the technique may provide a method for determining sulfur dioxide that is both continuous and relatively specific. A newly developed flame photometric instrument measures total sulfur and is therefore not specific for any one sulfur-containing pollutant. In this method sulfur compounds undergo reduction in a hydrogen-rich flame; the light emitted at a characteristic wavelength is proportional to the total atmospheric sulfur.

Other methods are available for analyzing other sulfur-containing compounds. For example, sulfuric acid aerosol in suspended particulate material may be measured by titration or by controlled decomposition to sulfur dioxide. This then can be measured by a number of methods, including spectrophotometry, coulometry, and flame photometry. Particulate sulfate may be analyzed by spectrophotometric or turbidimetric methods.

Each method is unique in terms of the measuring time resolution, the costs, the skill and time required for the analysis, and the kinds of pollutants which may cause interference and thus impair the accuracy of the results. The techniques selected, therefore, are usually a compromise. Moreover, a single program often will make use of both general and specific methods. In selecting a technique, it is especially important to consider the degree to which the resulting data can be converted for comparison with results obtained from other instruments and at other time periods.

REFERENCES

1. Hendrickson, E. R. "Air Sampling and Quantity Measurement." Air Pollution, 2nd edition, Vol. II, A. C. Stern (ed.), Academic Press, New York, 1968, pp. 3–52.
2. Katz, M. "Analysis of Inorganic Gaseous Pollutants." Air Pollution, 2nd edition, Vol. II, A. C. Stern (ed.), Academic Press, New York, 1968, pp. 53–114.
3. Smith, W. S. "Atmospheric Emission from Fuel Oil Combustion." U.S. Dept. of Health, Education, and Welfare, Public Health Service, PHS–Pub–999–AP–2, Nov. 1962.
4. "Atmospheric Emissions from Petroleum Refineries." U.S. Dept. of Health, Education, and Welfare, Public Health Service, PHS–Pub–763, 1960.
5. Schueneman, J. J., High, M. D., and Bye, W. E. "Air Pollution Aspects of the Iron and Steel Industry." U.S. Dept. of Health, Education, and Welfare, Public Health Service, PHS–Pub–999–AP–1, June 1963.
6. Smith, W. S. and Gruber, C. W. "Atmospheric Emissions from Coal Combustion. An Inventory Guide." U.S. Dept. of Health, Education, and Welfare, Public Health Service, PHS–Pub–999–AP–24, 1966.
7. "Atmospheric Emissions from Sulfuric Acid Manufacturing Processes." Cooperative Study Project Manufacturing Chemists' Association, Inc., and U.S. Public Health Service, PHS–Pub–999–AP–13, 1965.
8. Stahl, R. W. "Survey of Burning Coal-Mine Refuse Banks." U.S. Bureau of Mines, Information Circular 8209, 1964.
9. Hangebrauck, R. P., von Lehmden, D. J., and Meeker, J. E. "Emissions of Polynuclear Hy-

drocarbons and Other Pollutants from Heat-Generation and Incineration Processes." J. Air Pollution Control Assoc., Vol. 14, pp. 267–278, July 1964.

10. Cuffe, S. T., Gerstle, R. W., Orning, A. A., and Schwartz, C. H. "Air Pollutant Emissions from Coal-Fired Power Plants, Report 1." J. Air Pollution Control Assoc., Vol. 14, pp. 353–362, Sept. 1964.
11. Hendrickson, E. R. (ed.) "Proceedings Conference on Atmospheric Emissions from Sulfite Pulping." Sanibel Island, April 28, 1966.
12. Cuffe, S. T. and Gerstle, R. W. "Emissions from Coal-Fired Power Plants: A Comprehensive Summary." U.S. Dept. of Health, Education, and Welfare, Public Health Service, PHS–Pub–999–AP–35, 1967.
13. Rohrman, F. A. and Ludwig, J. H. "Sources of Sulfur Dioxide Pollution." Chem. Eng. Prog., Vol. 61, pp. 59–63, Sept. 1965.
14. Engdahl, R. B. "Combustion in Furnaces, Incinerators and Open Fires." Air Pollution, Vol. II, A. C. Stern (ed.), Academic Press, New York, 1962, pp. 3–39. (See also 2nd edition, Vol. III, in press).
15. Lucas, D. H., Moore, D. J., and Spurr, G. "The Rise of Hot Plumes from Chimneys." Int. J. Air & Water Pollution, Vol. 7, pp. 473–500, Aug. 1963.
16. Lucas, D. H. "The Atmospheric Pollution of Cities." Int. J. Air Pollution, Vol. 1, pp. 71–86, Oct. 1958.
17. Scorer, R. S. and Barret, C. F. "Gaseous Pollution from Chimneys." Int. J. Air Pollution, Vol. 6, pp. 49–63, 1962.
18. Electrical World, *168*(20):101, Nov. 13, 1967.
19. Wohlers, W. C., Trieff, N. M., Newstein, H., and Stevens W. "Sulfur Dioxide Absorption on and Desorption from Teflon, Tygon, Glass, Stainless Steel, and Aluminum Tubings." Atmos. Environ., Vol. 1, pp. 121–130, March 1967.
20. Scaringelli, F. P., Saltzman, B. E., and Frey, S. A. "Spectrophotometric Determination of Atmospheric Sulfur Dioxide." Anal. Chem., Vol. 39, pp. 1709–1719, Dec. 1967.
21. Kuczynski, E. R. "Effects of Gaseous Air Pollutants and the Response of the Thomas SO_2 Autometer." Environ. Sci. Technol., Vol. 1, p. 69, Jan. 1967.
22. Rodes, C. F., Elfers, L. A., Palmer, H. F., and Norris, C. H. Sixty-First Annual Meeting of the Air Pollution Control Association, St. Paul, Minnesota, June 23–27, 1968.
23. Low, M. J. D. "Multiple-Scan Infrared Interference Spectroscopy." J. Chem. Ed., Vol. 43, p. 637, 1966.
24. Low, M. J. D. and Clancy, F. K. "Remote Sensing and Characterization of Stack Gases by Infrared Spectroscopy. An Approach Using Multiple-Scan Interferometry." Environ. Sci. Technol., Vol. 1, pp. 73–74, 1967.
25. Barringer, A. R. and Newberry, B. C. "Remote Sensing Correlation Spectrometry for Pollution Measurements." Ninth Conference on Methods in Air Pollution and Industrial Hygiene, Pasadena, California, Feb. 7–9, 1968.
26. Barringer, A. R. and Newberry, B. C. "Molecular Correlation Spectroscopy for Sensing Gaseous Pollutants." 60th Annual Air Pollution Control Association Meeting, Paper 67/196, 1967.
27. Barringer, A. R. and Shock, J. P. "Progress in the Remote Sensing of Vapors for Air Pollution, Geologic, and Oceanographic Applications." Proceedings of the 4th Symposium on Remote Sensing of the Environment, April 1966, pp. 779–791.
28. Bokhoven, C. and Niessen, H. G. L. "The Continuous Monitoring of Traces of Sulfur Dioxide in Air on the Basis of Discoloration of the Starch-Iodine Reagent with Prior Elimination of Interfering Compounds." Int. J. Air Pollution, Vol. 10, pp. 223–243, 1966.
29. Katz, M. "Sulfur Dioxide in the Atmosphere of Industrial Areas." In: Effect of Sulfur Dioxide on Vegetation, National Research Council, Ottawa, Canada, 1939, pp. 14–50.
30. Hochheiser, S. "Methods of Measuring and Monitoring Atmospheric Sulfur Dioxide." U.S. Dept. of Health, Education, and Welfare, Public Health Service, PHS–Pub–999–AP–6, Aug. 1964.
31. Alekseeva, M. V. and Samorodiva, R. Ya. S. "Determination of Sulfur Dioxide Colorimetrically with the Aid of Fuchsin-Formaldehyde Solution." Gig. i Sanit., Vol. 10, p. 42, 1953. In: Limits of Allowable Concentrations of Atmospheric Pollutants, Book 2, Office of Technical Services, Washington, D.C., 1955, pp. 90–95.
32. Bushtueva, K. A. "Ratio of Sulfur Dioxide and Sulfuric Acid Aerosol in Atmospheric Air, in Relation to Meteorological Conditions." Gig. i Sanit., Vol. 11, pp. 11–13, 1954. In: U.S.S.R. Literature on Air Pollution and Related Occupational Diseases. A Survey. Vol. 4. Translated by B. S. Levine, U.S. Dept. of Commerce, Office of Technical Services, Washington, D.C., Aug. 1960, pp. 193–196.
33. Lyubimov, N. A. "A Nephelometer with an Automatic 24-Hour Device for Continuous Recording of Sulfur Dioxide Concentrations in Atmospheric Air." In: Limits of Allowable Concentrations of Atmospheric Pollutants. Book 5. Translated by B. S. Levine, U.S. Dept. of Commerce, Office of Technical Services, Washington, D.C., 1962, pp. 120–127.
34. Schulze, F. "Versatile Combination Ozone and Sulfur Dioxide Analyzer." Anal. Chem., Vol. 38, pp. 298–752, May 1966.
35. McCaldin, R. O. and Bye, W. E. "Air Pollution Appraisal—Seward and New Florence, Pa." U.S. Public Health Service, Robert A. Taft Sanitary Engineering Center, 1961.
36. Stalker, W. W., Dickerson, R. C., and Kramer, G. D. "Atmospheric Sulfur Dioxide and Particulate Matter—A Comparison of Methods of Measurements." Am. Ind. Hyg. Assoc. J., Vol. 29, pp. 68–79, Jan. and Feb. 1963.
37. Farmer, J. R. and Williams, J. D. "Interstate Air Pollution Study, Phase II Project Report, Section III, Air Quality Measurements." Interstate Air Pollution Study St. Louis-East St. Louis Metropolitan Areas. U.S. Dept. of Health, Education, and Welfare, Public Health Service, Dec. 1966.
38. Hochheiser, S., Santner, J. T., and Ludman, W.

F. "The Effect of Analytical Method on Indicated Atmospheric SO_2 Concentration." J. Air Pollution Control Assoc., Vol. 16, pp. 266–270, May 1966. (Presented at Annual Meeting, Air Pollution Control Association, Toronto, June 1965.)

39. Greenburg, L. and Jacobs, M. B. "Sulfur Dioxide in New York City Atmosphere." Ind. Eng. Chem., Vol. 48, pp. 1517–1521, Sept. 1956.
40. Tabor, E. C. and Golden, C. C. "Results of Five Years' Operation of the National Gas Sampling Network." J. Air Pollution Control Assoc., Vol. 15, pp. 7–11, Jan. 1965.
41. Terabe, M., Comichi, S., Benson, F. B., Newill, V. A., and Thompson, J. E. "Relationships Between Sulfur Dioxide Concentration Determined by the West-Gaeke and Electroconductivity Methods." J. Air Pollution Control Assoc., Vol. 17, pp. 673–675, Oct. 1967.
42. Shikiya, J. M. and McPhee, R. D. "Multi-Instrument Performance Evaluation of Conductivity Type Sulfur Dioxide Analyzers." Presented at 61st Annual Meeting, Air Pollution Control Association, St. Paul, 1968.
43. Trieff, N. M., Wohlers, H. C., O'Malley, J. A., and Newstein, H. "Appraisal and Modification of West-Gaeke Method for Sulfur Dioxide Determination." J. Air Pollution Control Assoc., Vol. 18, pp. 329–331, May 1968.
44. Booras, S. G. and Zimmer, C. E. "A Comparison of Conductivity and West-Gaeke Analyses for Sulfur Dioxide." J. Air Pollution Control Assoc. Vol. 18, pp. 612–615, Sept. 1968.
45. "Air Quality Data (1963) National Air Sampling Network." U.S. Dept. of Health, Education, and Welfare, Public Health Service, 1965.
46. Gorham, E. "Atmospheric Pollution by Hydrochloric Acid." Quart. J. Roy. Meteorol. Soc., Vol. 84, pp. 274–276, July 1958.
47. Cummins, B. T. "Chemistry in Town Air." Research, Vol. 15, pp. 421–426, Oct. 1962.
48. "Air Pollution Measurements of the National Air Sampling Network, Vol. I. Analyses of Suspended Particulates, 1953–1957." U.S. Dept. of Health, Education, and Welfare, Public Health Service, PHS–Pub–637, 1958.
49. Stevens, R. K., O'Keeffe, A. E., and Ortman, G. C. Title not known, Environ. Sci. Technol. (In press, 1968.)
50. Mader, P. P., Hamming, W. J., and Bellin, A. "Determination of Small Amounts of Sulfuric Acid in the Atmosphere." Anal. Chem., Vol. 22, pp. 1181–1183, Sept. 1950.
51. Cummins, B. T. "Determination of Particulate Acid in Town Air." Analyst, Vol. 88, pp. 364–367, May 1963.
52. Scaringelli, F. P. and Rehme, K. A. "Determinations of Atmospheric Concentrations of Sulfuric Acid by Spectrophotometry, Coulometry, and Flame Photometry." (To be published in Anal. Chem.)
53. Huey, N. A. "The Lead Peroxide Estimation of Sulfur Dioxide Pollution." J. Air Pollution Control Assoc., Vol. 18, pp. 610–611, Sept. 1968.
54. Kenline, P. A. "In Quest of Clean Air for Berlin, New Hampshire." U.S. Public Health Service, Robert A. Taft Sanitary Engineering Center, Tech. Rept. A62-9, 1962.
55. "Air Pollution Measurements of the National Air Sampling Network, Vol. II. Analyses of Suspended Particulates, 1957-1961." U.S. Dept. of Health, Education, and Welfare, Public Health Service, PHS–Pub–978, 1962.
56. "Register of Air Pollution Analyses." U.S. Dept. of Health, Education, and Welfare, Public Health Service, PHS–Pub–610, Vol. I, 1958 and Vol. II, 1961.
57. "Air Pollution in the El Paso, Texas Area. Report on a Two-Year Study under a Community Air Pollution Demonstration Project Grant." El Paso City-County Health Unit, 1959.
58. Tyler, R. G. "Report on an Air Pollution Study for City of Seattle." Univ. of Washington, Environ. Res. Lab., March 15, 1952.
59. Anderson, D. M., Lieben, J., and Sussman, V. H. "Pure Air for Pennsylvania." Pennsylvania Dept. of Health and U.S. Dept. of Health, Education, and Welfare, Public Health Service, Nov. 1961.
60. Chaney, A. L. "Investigations to Detect the Atmospheric Conversion of Sulfur Dioxide to Sulfur Trioxide." Am. Petrol. Inst., Vol. 38, pp. 306–312, 1958.
61. Thomas, M. D. "Sulfur Dioxide, Sulfuric Acid Aerosol, and Visibility in Los Angeles." Int. J. Air Pollution, Vol. 6. pp. 443–454, Nov.-Dec. 1962.
62. Boone, R.E. and Brice, R. M. "Continuous Measurement of Acid Aerosol in the Atmosphere." APCA Paper 65–119. (Presented at Annual Meeting Air Pollution Control Association, Toronto, June 1965.)

CHAPTER 13

PHOTOCHEMICAL OXIDANTS

Photochemical oxidants result from a complex series of atmospheric reactions initiated by sunlight. When reactive organic substances and nitrogen oxides accumulate in the atmosphere and are exposed to the ultraviolet component of sunlight, the formation of new compounds, including ozone and peroxyacyl nitrates, takes place.

Absorption of ultraviolet light energy by nitrogen dioxide results in its dissociation into nitric oxide and an oxygen atom. These oxygen atoms for the most part react with air oxygen to form ozone. A small portion of the oxygen atoms and ozone react also with certain hydrocarbons to form free radical intermediates and various products. In some complex manner, the free radical intermediates and ozone react with the nitric oxide produced initially. One result of these reactions is the very rapid oxidation of the nitric oxide to nitrogen dioxide and an increased concentration of ozone.

The photochemical system generally is capable of duplication in the laboratory. For various reasons, however, laboratory results cannot be quantitatively extrapolated to the atmosphere. Theoretically generation of an atmospheric simulation model should be feasible, enabling the prediction of ambient oxidant concentrations from a knowledge of emission and meteorological data. The development of such a model, however, is dependent on the acquisition of more reliable and applicable quantitative information derived from direct atmospheric observations, as well as on the refinement of results obtained from irradiation chamber studies.

ATMOSPHERIC PHOTOCHEMICAL OXIDANT CONCENTRATIONS

The presence of photochemically formed oxidants has been indicated in all of the major U.S. cities for which aerometric data have been examined. On a concentration basis, ozone has been identified as the major component of the oxidant levels observed. Difficulties arise, however, in interpreting data obtained by the most commonly used oxidant measuring method; this method is nonspecific and subject to several interferences. Adjusted oxidant concentrations, obtained by correcting potassium iodide oxidant measurements for known interferences, have been found to be relatively close to concurrent measurements of ozone alone.

Since photochemical oxidants are the products of atmospheric chemical reactions, the relationship between precursor emissions and atmospheric oxidant concentrations is much less direct than is the case for primary pollutants. A further complicating situation is the dependence of these photochemical reactions on intensity and duration of sunlight, and on temperature.

In an analysis of oxidant concentration data for 4 years and 12 stations, the daily maximum 1-hour average concentration was equal to or exceeded 290 $\mu g/m^3$ (0.15 ppm) up to 41 percent of the time; maximum 1-hour average concentrations ranged from 250 to 1,140 $\mu g/m^3$ (0.13 to 0.58 ppm); short-term peaks were as high as 1,310 $\mu g/m^3$ (0.67 ppm). Yearly averages, commonly applied to other pollutants, are not representative of air quality with respect to oxidant pollution, because 1-hour average ozone concentrations will necessarily be at or about zero for approximately 75 percent of the time when photochemical reactions are minimal.

Peroxyacyl nitrates, through not routinely measured, have been identified in the atmosphere of several cities. These compounds may be assumed to be present whenever oxidant levels are elevated.

NATURAL SOURCES OF OZONE

Ozone can be formed naturally in the atmosphere by electrical discharge, and in the stratosphere by solar radiation, by processes which are not capable of producing significant urban concentrations of this pollutant. Maximum instantaneous ozone levels of from 20 to 100 $\mu g/m^3$ (0.01 to 0.05 ppm) have been recorded in nonurban areas.

MEASUREMENT OF PHOTOCHEMICAL OXIDANTS

The most widely used technique for the analysis of atmospheric total oxidants is based on the reaction of these compounds with potassium iodide to release iodine. The iodine may then be measured by either colorimetric or coulometric methods. Calibrating the oxidant measurement method used against a known quantity of ozone provides a measurement of the net oxidizing properties of the atmosphere in terms of an equivalent concentration of ozone. Most oxidant measurements are currently being made by the colorimetric method, although coulometric analyzers are used in a number of laboratory and field studies.

In order to generate comparable data, it is essential that all measurements be made by techniques which have been calibrated against the same standard or reference method. Since at the present time there is no standard method for the determination of total oxidants, the National Air Pollution Control Administration recommends use of the neutral-buffered 1 percent potassium iodide colorimetric technique as the method against which all instruments and other methods should be compared. In addition to serving as a manual procedure for determining oxidants, the reference method may be used in conjunction with a "dynamic calibration" technique for instrumental methods.

Reducing agents such as sulfur dioxide produce a negative interference in oxidant determination. Such interference can be reduced, however, by passing the air stream through a chromium trioxide scrubber prior to measurement. Unfortunately, a portion of the nitric oxide which may be present in the air stream is oxidized to nitrogen dioxide by the scrubber. This results in an apparent increase in the oxidant measurement of about 11 percent of the concentration of nitric oxide. Moreover, a portion of the atmospheric nitrogen dioxide concentration will also contribute to the oxidant measurement. Peroxyacyl nitrate concentrations are usually small and contribute only a very slight amount to the oxidant reading.

There are several means for the specific measurement of atmospheric ozone. Instrumental methods include chemiluminescent analysis based on the reaction of ozone with Rhodamine B, gas phase olefin titration, and ultraviolet and infrared spectroscopy. A semiquantitative method for ozone measurement is based on its ability to produce cracks in stretched rubber. Peroxyacyl nitrates can be measured in the atmosphere by gas chromatography with the use of an electron-capture detector.

For a better evaluation of the results of research on the effects of photochemical oxidants, it is essential that data be obtained for individual oxidants such as nitrogen dioxide, ozone, PAN, formaldehyde, acrolein, and organic peroxides. These data would either replace or complement data on total oxidants. Instrumentation currently available permits the accurate measurement of atmospheric ozone, nitrogen dioxide, and PAN. There exists, however, a further need to develop instruments capable of measuring other individual gaseous pollutants which have the properties of oxidants. Photochemical reactions and problems derived from oxidants can be much better defined using specific methods for measurement in preference to the traditional total oxidants determination.

EFFECTS OF PHOTOCHEMICAL OXIDANTS ON VEGETATION AND MICROORGANISMS

Injury to vegetation is one of the earliest manifestations of photochemical air pollution, and sensitive plants are useful biological indicators of this type of pollution. The visible symptoms of photochemical oxidant produced injury to plants may be classified as: (1) *acute injury*, identified by cell collapse with subsequent development of necrotic patterns; (2) *chronic injury*, identified by necrotic patterns with or without chlorotic or other pigmented patterns; and, (3) *phsyiological effects*, identified by growth alterations, reduced yields, and changes in the quality of plant products. The acute symptoms are generally characteristic of a specific pollutant; though highly characteristic, chronic injury patterns are not. Ozone injury to leaves is identified as a stippling or flecking. Such injury has occurred experimentally in the most sensitive species after exposure to 60 $\mu g/m^3$ (0.03 ppm) ozone for 8 hours. Injury will occur in shorter time periods when low levels of sulfur dioxide are present. PAN-produced injury is characterized by an under-surface glazing or bronzing of the leaf. Such injury has occurred experimentally in the most sensitive species after exposure to 50 $\mu g/m^3$ (0.01 ppm) PAN for 5 hours. Leaf injury has occurred in certain sensitive species after a 4-hour exposure to 100 $\mu g/m^3$ (0.05 ppm) total oxidant. Ozone appears to be the most important phytotoxicant in the photochemical complex.

There are a number of factors affecting the response of vegetation to photochemical air pollutants. Variability in response is known to exist between species of a given genus and between varieties within a given species; varietal variations have been most extensively studied with tobacco. The influence of light intensity on the sensitivity of plants to damage during growth appears to depend on the phytotoxicant. Plants are more sensitive to PAN when grown under high light intensities, but are more sensitive to ozone when grown under low light intensities. Reported findings are in general agreement that sensitivity of greenhouse-grown plants to oxidants increases with temperature, from 10° to 38° C (40° to 100° F), but this positive correlation may result from the overriding influence of light intensity on sensitivity. The effects of humidity on the sensitivity of plants has not been well documented. General trends indicate that plants grown and/or exposed under high humidities are more sensitive than those grown at low humidities. There has been little research in this direction, but there are indications that soil factors such as drought and total fertility influence the sensitivity of plants to phytotoxic air pollutants. The age of the leaf under exposure is important in determining its sensitivity to air pollutants. There is some evidence that oxidant or ozone injury may be reduced by pretreatment with the toxicant.

Identification of injury to a plant as being caused by air pollution is a difficult undertaking. Even when the markings on the leaves of a plant may be identified with an air pollutant, there is the further difficulty of evaluating the injury in terms of its effect on the intact plant. Additional problems arise in trying to evaluate the economic impact of air pollution damage to a plant.

The interrelations of time and concentration (dose) as they affect injury to plants are essential to air quality criteria. There are, however, only scant data relating concentrations and length of photochemical oxidant exposure to chronic injury and effects on reduction of plant growth, yield, or quality. There is also a dearth of information relating concentrations to acute injury. A larger body of information exists on the acute effects of ozone, but even in this instance, the information is far from complete. Sufficient data do exist, however, to tabularly present ozone concentrations which will produce 5 percent injury to sensitive, intermediate, and resistant plants after a given short-term exposure, as shown in Table **13-1**. Information available lists 20 species and/or varieties as sensitive, 55 as intermediate in sensitivity, and 64 as relatively resistant.

Bacteriostatic and bacteriocidal properties of photochemical oxidants in general have been demonstrated. The growth suppression of microorganisms by ozone is a well-known phenomenon, although the ozone concentrations for this activity are undesirable from a human standpoint. The bacteriocidal activity of ozone varies with its concentration, the relative humidity, and the species of bacteria.

EFFECT OF OZONE ON MATERIALS

The detailed, quantitative extent of damage to materials caused by atmospheric levels of ozone is unknown, but generally any organic material is adversely affected by concentrated ozone. Many polymers are extremely sensitive to even very small concentrations of ozone, this sensitivity increasing with the number of double bonds in the structure of the polymer.

Economically, rubber is probably the most

important material sensitive to ozone attack, particularly styrene-butadiene, natural, polybutadiene, and synthetic polyisoprene. Antiozonant additives have been developed and are capable of protecting elastomers from ozone degradation; synthetic rubbers with inherent resistance to ozone are also available. These additives are expensive, however, and add to the cost of the end product; in addition, increasing amounts of antiozonants are required as the amount of ozone which is to be encountered increases, and sometimes only temporary protection is provided.

Ozone attacks the cellulose in fabrics through both a free radical chain mechanism and an electrophilic attack on double bonds; light and humidity appear necessary for appreciable alterations to occur. The relative susceptibility of different fibers to ozone attack appears to be, in increasing order, cotton, acetate, nylon, and polyester.

Certain dyes are susceptible to fading during exposure to ozone. The rate and extent of fading is also dependent upon other environmental factors such as relative humidity and the presence of air pollutants other than ozone, as well as the length and concentration of ozone exposure and the type of material exposed.

TOXICOLOGICAL STUDIES OF PHOTOCHEMICAL OXIDANTS

1. Effects of Ozone in Animals

The major physiological effects of ozone are on the respiratory system. Inhalation of ozone at concentrations greater than about 5,900 $\mu g/m^3$ (3 ppm) for several hours produces hemorrhage and edema in the lungs. This reaction can be fatal to animals. Rats and mice appear to be more sensitive than rabbits, cats, and guinea pigs. The toxicity is greater for young animals and for exercising animals. It is abated by intermittency of exposure, by prophylactic administration of chemical reducing agents, or by introducing agents into the diet which reduce the activity of the thyroid gland. At exposures less than those which produce edema in the lungs, changes in the mechanical properties of the lung occur. These are accompanied by increased breathing rates and increased oxygen consumption. Repeated non-fatal exposures to concentrations greater than 15,700 $\mu g/m^3$ (8 ppm) for 30 minutes have produced fibrosis in the respiratory tract of rabbits, with the damage increasing in severity over the length of the respiratory tract from the trachea to the bronchioles.

Short-term exposures to ozone also produce chemical changes in the lung tissue elements of animals. A study conducted on a small number of rabbits showed that inhalation of 1,960 to 9,800 $\mu g/m^3$ (1 to 5 ppm) ozone for 1 hour can produce denaturation of the structural lung proteins. Ozone also appears to oxidize the sulfhydryl groups of amino acids in the lung.

Short-term exposures to ozone also produce changes in organs other than the lung. Concentrations of 5,900 $\mu g/m^3$ (3 ppm) for 20 hours can stimulate some adaptive liver enzymes. Inhalation of 390 to 490 $\mu g/m^3$ (0.2 to 0.25 ppm) ozone for 30 to 60 minutes makes the red blood cells of mice, rabbits, rats, and man more sensitive to the shape-altering effects of irradiation. Exposure of blood to ozone *in vitro* produces interference with the release of oxygen from red blood cells; this suggests that ozone exposure could impair the delivery of oxygen to the tissues.

Table 13-1 PROJECTED OZONE CONCENTRATIONS WHICH WILL PRODUCE, FOR SHORT-TERM EXPOSURES, 5 PERCENT INJURY TO ECONOMICALLY IMPORTANT VEGETATION GROWN UNDER SENSITIVE CONDITIONS

Time, hr	Ozone concentrations producing injury in three types of plants, ppm		
	Sensitive	Intermediate	Resistant
0.2	0.35–0.75	0.70–1.00	0.90 and up
0.5	0.15–0.30	0.25–0.60	0.50 and up
1.0	0.10–0.25	0.20–0.40	0.35 and up
2.0	0.07–0.20	0.15–0.30	0.25 and up
4.0	0.05–0.15	0.10–0.25	0.20 and up
8.0	0.03–0.10	0.08–0.20	0.15 and up

Ozone exposures at concentrations from 1,310 to 7,800 $\mu g/m^3$ (0.67 to 4.0 ppm) have been shown to reduce the *in vitro* phagocytic abilities of the pulmonary alveolar macrophages. A 3-hour exposure to 9,800 $\mu g/m^3$ (5 ppm) ozone has been shown to reduce the activity of bactericidal enzyme, presumably due to *in vivo* oxidation of the enzyme.

Ozone inhalation increases the vulnerability of animals to other agents. A single exposure to ozone at a concentration of 160 $\mu g/m^3$ (0.08 ppm) for 3 hours has increased the mortality among mice from inhalation of pathogenic bacteria. This occurred when the bacteria were administered both before and after exposure to ozone. Ozone also increases the toxicity of histamine in guinea pigs.

Long-term effects of ozone exposure include, in some species, the development of tolerance to biological effects of ozone, production of fibrotic changes in the lungs, and a possible increase in the rate of aging. While tolerance has been shown in rodents, it has not been shown in chickens, and it is not certain whether or not it occurs in man. In species where tolerance to ozone exposure has been demonstrated, information is not available concerning the duration and mechanism of tolerance following repeated exposure. The aging effect may be similar to the changes produced by exposure to free radicals or by irradiation.

2. Effects of Ozone in Humans

Some studies of human exposures to ozone have focused on the determination of the threshold level at which odor can be detected, and on the occurrence of changes in pulmonary function. Nine out of 10 subjects exposed to 40 $\mu g/m^3$ (0.02 ppm) ozone were able to detect the odor immediately, and it persisted for an average of 5 minutes. Thirteen of 14 subjects exposed to 100 $\mu g/m^3$ (0.05 ppm) ozone indicated the odor is considerably stronger at this concentration, and the odor persisted for an average of 13 minutes.

Occupational exposure of humans to ozone concentrations of up to 490 $\mu g/m^3$ (0.25 ppm) has not produced detectable changes in pulmonary function. Respiratory symptoms and a decrease in vital capacity in three out of seven smokers who had been occupationally exposed to ozone have occurred at concentrations greater than 590 $\mu g/m^3$ (0.3 ppm).

Experimental exposures of humans have been carried out at concentrations ranging from 200 to 7,800 $\mu g/m^3$ (0.1 to about 4 ppm) for periods of up to 2 hours. Exposure to 390 $\mu g/m^3$ (0.2 ppm) for 3 hours daily, 6 days a week, for 12 weeks has not produced any change in ventilatory function tests. Similar exposure to 980 $\mu g/m^3$ (0.5 ppm) produced a decrease in the forced expiratory volume during the last 4 weeks of exposure, with recovery taking place in a subsequent 6-week period. In each of 11 subjects, exposure to 1,180 to 1,570 $\mu g/m^3$ (0.6 to 0.8 ppm) for 2 hours resulted in an impairment of the diffusing capacity of the lung. Small decreases in vital capacity and forced expiratory volume were observed in some of these subjects. Resistance to flow of air in the respiratory tract increased slightly in some subjects after exposure to 200 to 1,180 $\mu g/m^3$ (0.1 to 0.6 ppm) for 1 hour, and increased consistently in each of four subjects after exposure to 1,960 $\mu g/m^3$ (1 ppm) for 1 hour.

Data obtained from animal experimentation cannot be used directly to define the ozone concentrations above which human health will be affected. Animal mortality studies, however, can be useful in determining the factors involved in toxicity. While the concentrations of ozone used in the determination of short-term non-fatal effects in animals are rarely found in ambient air, the changes in pulmonary function observed during and after exposure to these concentrations call attention to the possibility that similar effects may be observed in humans.

When interpreting the research conducted thus far using human subjects, it must be noted that occupational exposures differ from experimental exposures, because it is difficult in an occupational environment to define the exact nature and dose of the pollutants present.

3. Effects of Peroxyacetyl Nitrate

Experimental studies with peroxyacetyl nitrate (PAN) in animals indicate that mortality may be delayed for 7 to 14 days after exposure; however, the exposure levels required to produce this mortality never occur in ambient atmospheres.

A single experimental study of healthy human subjects exposed to 1,485 μg/m³ (0.3 ppm) peroxyacetyl nitrate indicated only that there may be a small increase in oxygen uptake with exercise. Sensitive pulmonary function tests were not obtained.

The data from animal and human studies are sparse and inadequate for determining the toxicological potential of peroxyacetyl nitrate. It would appear, however, that at the concentrations of this compound known to occur in ambient atmospheres, PAN does not present any recognized health hazard.

4. Effects of Mixtures Containing Photochemical Oxidants on Animals

Studies have been conducted on animals exposed to both synthetic and natural photochemical smog. Synthetic smog has been produced by the irradiation of diluted motor vehicle exhaust or by irradiation of air mixtures containing nitrogen oxides and certain hydrocarbons. Exposures to irradiated motor vehicle exhaust are complicated by the simultaneous presence of carbon monoxide and other non-oxidant substances which include high concentrations of formaldehyde. Guinea pigs show increased respiratory volume during a four-hour exposure to irradiated exhaust containing 1,570 μg/m³ (0.8 ppm) total oxidant.

Exposure of mice to both natural and synthetic smog for 3 hours, at concentrations greater than 780 μg/m³ (0.4 ppm) oxidants have produced changes in the fine structure of the lung. The nature and extent of the damage was the same after exposure to either type of smog with the same oxidant levels. The severity of the damage increased with age and became irreversible at age 21 months.

Chronic exposure of guinea pigs to ambient air with an average oxidant concentration of from 40 to 140 μg/m³ (0.02 to 0.07 ppm) leads to a significant increase in flow resistance when the peak oxidant concentrations exceed 980 μg/m³ (0.5 ppm).

When male mice, prior to mating, were given long-term exposures to irradiated auto exhaust containing from 200 to 1,960 μg/m³ (0.1 to 1.0 ppm) oxidant, a decrease in fertility and an increase in neonatal mortality of their offspring resulted; the irradiated mixture also contained varying concentrations of carbon monoxide, nitrogen oxides, and hydrocarbons. Similar exposures also cause a reduction in spontaneous running activity, which results in an adaptation response.

Thus a number of experimental studies have demonstrated that changes in lung tissue or lung function occur when animals are exposed for several hours to photo-oxidized mixtures containing 980 μg/m³ (0.5 ppm) or more of oxidants.

5. Effects of Mixtures Containing Photochemical Oxidants on Humans

Laboratory studies of human exposure to photochemical smog have involved primarily the measurement of eye irritation. Based on the existing data, it appears that: (1) the effective eye irritants are the products of photochemical reactions; (2) although oxidant concentrations may correlate with the severity of eye irritation, a direct cause-effect relationship has not been demonstrated since ozone, the principal contributor to ambient oxidant levels is not an eye irritant; (3) the precursors of the eye irritants are organic compounds in combination with oxides of nitrogen, the most potent being aromatic hydrocarbons; (4) the chemical identities of the effective irritants in synthetic systems are known as being formaldehyde, peroxybenzoyl nitrate (PBzN), peroxyacetyl nitrate (PAN), and acrolein, although the latter two contribute to only a minor extent; and (5) the substances causing eye irritation in the atmosphere have not been competely defined.

EPIDEMIOLOGICAL STUDIES OF PHOTOCHEMICAL OXIDANTS

Several studies have examined daily mortality rates in localities where photochemical air pollution occurs, to determine if a relationship exists with increased levels of oxidant. Such an association has not been shown. These studies, however, pose a number of unresolved questions. One of these is, what is the effect of temperature, either alone or in combination with oxidants? In some of the most severe episodes, there has been an associated increase in environmental temperature, sufficient to cause excess mortality by itself. Several studies of mortality among residents in nursing homes in Los Angeles showed such excess mortality. In recent heat wave and air pollution episodes, however, large proportions of the elderly and ill persons in nursing homes have been protected by air conditioning.

Evidence of increased morbidity has been sought through study of general hospital admissions, but no unequivocal association between photochemical air pollution and increased morbidity has been shown. Additional studies are indicated for improved definition. Peak oxidant values of 250 $\mu g/m^3$ (0.13 ppm), which might be expected in relation to maximum hourly average levels of 100 to 120 $\mu g/m^3$ (0.05 to 0.06 ppm), have been associated with aggravation of asthma. No association between ambient oxidant concentrations and changes in respiratory symptoms or function was shown, however, in two separate studies of subjects with preexisting chronic respiratory disease. Non-smoking subjects with chronic respiratory disease did, however, demonstrate less airway resistance when they were studied in a room where the ambient air of Los Angeles was filtered before entry. No acute or chronic effects of oxidant pollution on ventilatory performance of elementary schoolchildren were demonstrated in a study conducted in two communities within the Los Angeles basin.

Impairment of performance by high school athletes has been observed when photochemical oxidants ranged from 60 to 590 $\mu g/m^3$ (0.03 to 0.3 ppm) for 1 hour immediately prior to the start of activities. Significantly, more automobile accidents have also occurred on days of high oxidant concentrations, but no threshold level for this effect could be determined from the analysis.

Among the general community, eye irritation is a major effect of photochemical air pollution. In Southern California, it has affected more than three-fourths of the population. Eye irritation under conditions prevalent in Los Angeles is likely to occur in a large fraction of the population when oxidant concentrations in ambient air increases to about 200 $\mu g/m^3$ (0.10 ppm). This oxidant value might be expected to be associated with a maximum hourly average oxidant concentration of 50 to 100 $\mu g/m^3$ (0.025 to 0.50 ppm), depending on localized conditions. According to survey data gathered in 1956, asthma, cough, and nose and throat complaints were more frequent in Los Angeles, Orange, and San Diego counties than in the San Francisco Bay area or in the rest of the State.

Casual reports of the presence of the symptoms of eye irritation have been recorded in many cities in the United States. Epidemiologic studies have been inadequate, however, to relate these symptoms clearly to measured exposures to photochemical oxidants. In fact, one of the major photochemical oxidants, ozone, is not an eye irritant. That eye irritation is experienced whenever the oxidant level exceeds a certain value is an indication that oxidant concentrations correlate well with other aspects of the photochemical complex; oxidant levels are probably a measure of the photochemical activity which produces the eye irritants. On the other hand, it must be recognized that reactions of ozone with hydrocarbons do lead to hydrocarbon fragments which are eye irritants. Nor can the possibility be discounted that ozone in the photochemical complex may exert a synergistic effect on eye irritation. Because the oxidant reading measured only the net oxidizing property of the atmosphere, however, the same amount of eye irritation experienced in two different geographical locations from identical irritants could be associated with different levels of oxidant, if other pollutants differed in their concentration.

MEASUREMENT OF OXIDANTS

Total oxidants in the air may be defined as those compounds that will oxidize a reference material which is not capable of being oxidized by atmospheric oxygen. Methods for measuring oxidants, therefore, involve exposing an oxidizable substance to a sample of ambient air and determining the degree of oxidation which has occurred. There are many such methods, each one being dependent upon the reference substance chosen.

Two widely used methods of determining the degree of oxidation are colorimetric and coulometric analysis of the iodine released by atmospheric oxidants from a potassium iodide (KI) solution. Since these methods are the most widely used, they will be discussed in

Table 13-2 EFFECTS OF OZONE

Effect	Exposure ppm	Exposure $\mu g/m^3$	Duration	Comment	Reference
Vegetation damage[a]	0.03	60	8 hours	Sensitive species; laboratory conditions	Heck and Dunning
Cracking of stretched rubber	0.02	40	1 hour	Vulcanized natural rubber	Bradley and Haagen-Smit
Odor detection	0.02	40	<5 minutes	Odor detected in 9 of 10 subjects	Henschler et al.
Increased susceptibility of laboratory animals to bacterial infection	0.08 to 1.30	160 to 2,550	3 hours	Demonstrated in mice at 160 $\mu g/m^3$ and in mice at 2550 $\mu g/m^3$	Coffin et al. Miller et al.
Respiratory irritation (nose and throat), chest constriction	0.30	590	Continuous during working hours	Occupational exposure of welders, other pollutants probably also present	Kleinfeld et al.
Changes in pulmonary function:				Experimental exposure of 6 subjects.	Bennett
Diminished $FEV_{1.0}$ after 8 weeks	0.50	980	3 hours/day, 6 days/week, for 12 weeks	Change returns to normal 6 weeks after exposure. No changes observed at 390 $\mu g/m^3$ (0.2 ppm)	
Small decrements in VC, FRC, and DL_{CO} in, respectively, 3, 2, and 1 out of 7 subjects	0.20 to 0.30	390 to 590	Continuous during working hours	Occupational exposure. All 7 subjects smoked. Normal values for VC, FRC, and DL_{CO} based on predicted value.	Young et al.
Impaired diffusion capacity (DL_{CO})	0.60 to 0.80	1,180 to 1,570	2 hours	Experimental exposure of 11 subjects	Young et al.
Increased airway resistance	0.10 to 1.00	200 to 1,960	1 hour	Significant increase in 2 of 4 subjects at 200 $\mu g/m^3$ (0.1 ppm) and 4 of 4 subjects at 1960 $\mu g/m^3$ (1.0 ppm)	Goldsmith et al.
Reduced VC, severe cough, inability to concentrate	2.00	3,900	2 hours	High temperatures. One subject.	Griswold et al.
Acute pulmonary edema	9.00	17,600	Unknown	Refers to peak concentration of occupational exposure. Most of exposure was to lower level	Kleinfeld et al.

[a]Similar vegetation damage also occurs upon exposure to 0.01 ppm peroxyacetyl nitrate for 5 hours.

Table 13-3 EFFECTS ASSOCIATED WITH OXIDANT CONCENTRATIONS IN PHOTOCHEMICAL SMOG

Effect	Exposure, ppm	Exposure, $\mu g/m^3$	Duration	Comment	Reference
Vegetation damage	0.05	100	4 hours	Leaf injury to sensitive species	MacDowall et al
Eye irritation	Exceeding 0.1	200	Peak values	Result of panel response. Such a peak value would be expected to be associated with a maximum hourly average concentration of 50 to 100 $\mu g/m^3$ (0.025 to 0.05 ppm)	Renzetti and Gobran
Aggravation of respiratory diseases–asthma	0.13[a]	250	Maximum daily value	Patients exposed to ambient air. Value refers to oxidant level at which number of attacks increased Such a peak value would be expected to be associated with a maximum hourly average concentration of 100 to 120 $\mu g/m^3$ (0.05 to 0.06 ppm).	Schoettlin and Landau
Impaired performance of student athletes	0.03 to 0.30	60 to 590	1 hour	Exposure for 1 hour immediately prior to race	Wayne et al.

[a]Calculated from a measured value of 0.25 ppm (phenolphthalein method) which is equivalent to 0.13 ppm by the KI method.

greater detail.

It should be recalled that:

1. Ambient air contains a mixture of oxidizing *and* reducing compounds–ozone, nitrogen dioxide, peroxyacetyl nitrate (PAN), sulfur dioxide, hydrogen sulfide, aldehydes, unsaturated hydrocarbons, and others.
2. The concentrations of these compounds may be constantly varying within relatively wide limits–from 0 to 1 ppm.

Since reducing compounds in air have an opposite effect on the reference material - that is, cause a decrease in the degree of oxidation observed - the result obtained by these KI methods is a "net" oxidant value rather than a total one unless empirical corrections are made.[1] Thus the net value describes a condition of the air, rather than a specific compound concentration. For this reason, concentrated efforts are being made by researchers to obtain measurements for each specific oxidant present. Hopefully these efforts will enable air pollution scientists to better define this atmospheric "condition."

Most oxidant data which have been collected are net determinations, and a high correlation between these values and other pollutant levels has been observed.[2] Consequently until the more promising methods become routinely available net determination must serve as indicators of total oxidant levels.

It should be emphasized that in studies seeking a better evaluation of health effects, it is essential that data be obtained for individual oxidants, such as nitrogen dioxide, ozone, PAN, formaldehyde, acrolein, and organic peroxides, rather than data on total oxidants. Instrumentation currently available permits the accurate measurement of atmospheric ozone, nitrogen dioxide, and PAN. A further essential need exists to develop instrumentation capable of measuring the other individual gaseous pollutants which have the properties of oxidants.

Photochemical reactions and other problems derived from oxidants can be much better defined using specific methods for measurement of nitrogen dioxide and ozone, in preference to the traditional methods for determination of total oxidants.

REFERENCE METHOD FOR MEASUREMENT OF TOTAL OXIDANTS

Because of the complexity and variety of the methodology involved in oxidants determination, it has become necessary to adopt one reference method for use in comparisons between techniques. The most widely used method for this purpose is the neutral-phosphate-buffered KI colorimetric technique described by Saltzman.[3] In this method, oxidants are absorbed in neutral buffered 1 percent KI aqueous solution and the liberated tri-iodine ion is measured colorimetrically at 3,520 Å. Since one mole of ozone theoretically liberates one mole of iodine from this solution, the iodine measured is directly related to the concentration of ozone present in the air sample. Results may thus be expressed directly as equivalent volumes of ozone.

At present, there is no satisfactory means of testing this theoretical reaction. There is evidence that the amount of tri-iodine ions liberated by ozone is dependent on the design of the gas bubbler and on the phosphate concentration at neutral pH. Nevertheless, the error of measurement produced as a result of neglecting to account for the above variables is estimated to be no larger than ± 10 percent of the true value. Since ozone is believed to be the principal oxidant present in the atmosphere, this reference method has been adopted by NAPCA. A reference method in which the stoichiometry depends on fewer variables is, however, needed.

In addition to serving as a manual procedure for determining oxidants, the reference method may be used in conjunction with a "dynamic calibration" technique for instrumental methods, as follows: By passing scrubbed clean air over an ultraviolet lamp at a constant rate, ozone is generated at a constant rate. Analysis of the effluent with both the bubbler reference method and the particular system to be tested makes it possible to calibrate the various methods against this constant ozone source, and to express the results obtained as "ozone concentrations." Since some instruments use different colorimetric techniques which require static liquid calibrations, comparisons of these liquid

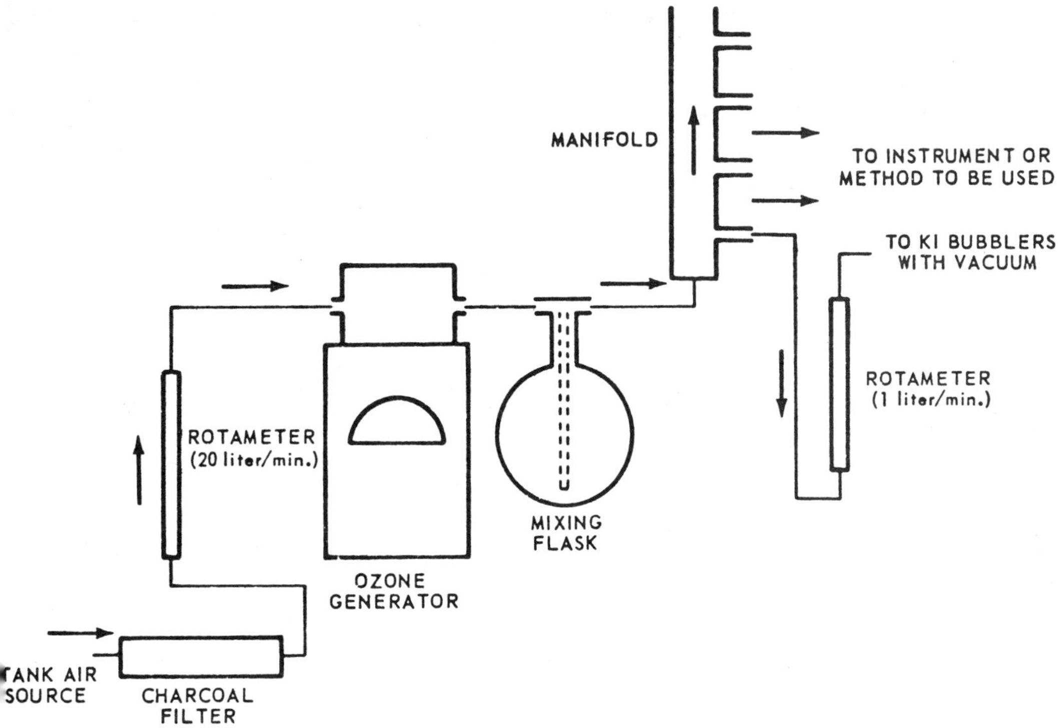

Fig. 13-1 Schematic of dynamic calibration of ozone measuring techniques.

standards may be made using such a dynamic system. A diagram of the procedure is shown in Figure 13-1.

As a manual procedure for the determination of oxidants in the atmosphere, the neutral-buffered KI technique is useful if analysis is completed immediately after sampling. Rapid analysis is necessary because the iodine-color deteriorates with time. Concentrations of oxidant in a range from about 20 to 19,600 $\mu g/m^3$ (0.01 to 10 ppm) can be determined using this method, although the accuracy for concentrations below 100 $\mu g/m^3$ (0.05 ppm) is compromised because of the cumulative errors of sampling and analysis.

METHODS FOR MEASUREMENT OF TOTAL OXIDANTS

1. Continuous Methods Utilizing Potassium Iodide

a. Colorimetric

Continuous analysis of total oxidants is most commonly accomplished with instrumentation using a 10 to 20 percent solution of neutral–buffered potassium iodide as the absorbing medium. In devices using a colorimetric method sample air is passed at a known rate through a liquid absorbing reagent with a vacuum pump arrangement. Oxidants in contact with the KI react to free iodine and the tri-iodine ion. These are determined colorimetrically in a continuous flow colorimeter at 3,520 Å. The continuous instrumental method is extremely sensitive to oxidizing substances and is suitable for determinations from 0 to 1,000 $\mu g/m^3$ (0 to 0.5 ppm) ozone with an accuracy similar to the KI reference method. The negative interference caused by sulfur dioxide is often reduced by passing the air sample over chromium trioxide (CrO_3) scrubbers, but this technique has not been entirely successful, primarily because the effects of humidity on scrubber systems often render them ineffective.[3] As discussed in Chapter 3, these scrubber systems also convert a portion of the ambient NO to NO_2, thus necessitating an additional correction factor. Despite these drawbacks, CrO_3 scrubbers remain the most generally used technique for removing SO_2 interferences; they are particularly useful for this purpose during periods of high SO_2 concentrations. Nevertheless, better methods for eliminating SO_2 interference are critically needed.

b. Coulometric

A coulometric method, commonly referred to as Mast-analysis, is also widely used for measurement of oxidants. In devices using this technique, sample air is passed in a manner similar to the colorimetric type of instrument into an electrolytic detector-cell containing KI.[4] The free iodine liberated by the oxidants is reduced at the cathode of the cell, causing a current flow through an external circuit. The current flow is proportional to the amount of iodine liberated, and, in turn, to the oxidants entering the solutions. The current is measured with a microammeter which is usually calibrated directly in pphm of ozone.

A method utilizing this technique for the continuous measurement of total oxidants has been prepared for collaborative testing by the Intersociety Committee on Manual of Air Sampling and Analysis.[5]

c. Colorimetric versus Coulometric Methods

Since these are the types of instruments which are the most widely used for oxidant measurements, there have been studies conducted comparing the results obtained with the colorimetric and coulometric approaches. Field comparisons of measurements from colorimetric and coulometric oxidant analyzers, where readings did not exceed 390 $\mu g/m^3$ (0.20 ppm) and corrections were made for NO_2 interferences, indicate that differ-

Table 13-4 EFFECT OF NITROGEN DIOXIDE ON OXIDANT DETERMINATION IN THE ABSENCE OF OZONE

Range of NO_2 concentration in air,[a] ppm	Method	NO_2 concentration (as ozone), %	Reference
> 0 to 36	1% KI, pH 7 bubbler	8 to 11	8
1.5 to 5.5	2% KI, pH 7 bubbler	6.4	9
2.0	10% KI, pH 7 contact column	21	7
> 0 to 1	20% KI, pH 7 contact column	30	8
0.24 to 0.43	20% KI, pH 7 contact column	12 to 47 (average: 25)	10
2.0	coulombic cell	10	7
2.0	U.V. photometer	2	7

[a]Based on 4-liter sample.

Table 13-5 CALIBRATION STABILITY OF COLORIMETRIC OXIDANT MONITORING STATIONS, STATEWIDE CONTINUOUS AIR-MONITORING NETWORK, CALIFORNIA,

Change from previous calibration, %	Number of stations	Occurrences	
		Number	% of total
± 0 to 10	7	20	62.3
±11 to 20	4	7	22.0
±21 to 30	2	3	9.4
±31 to 40	2	2	6.3

ences are no greater than would be expected if two colorimetric-type instruments were working side-by-side.[6] Comparative data, including readings greater than 390 $\mu g/m^3$ (0.20 ppm), show a correlation coefficient of 0.87 between readings of the two different instruments, both calibrated with respect to known ozone-streams.[7] In spite of the relatively good correlation obtained in this study, there were some instances when the agreement between the two methods was poor. To resolve some of these discrepancies, more comparative studies of this type are needed.

Perhaps the largest variations between the two methods are related to the differences in reagent formulation. Generally, 10 to 20 percent KI solutions are used in colorimetric instruments; at these concentrations, response to nitrogen dioxide is higher than at the 2 percent level used with coulometric instruments. Results of several studies are shown in Table 13-4. Analyzers utilizing other principles of detection vary in their sensitivity to nitrogen dioxide. As shown in Table 13-4, the coulometric cell analyzers register about 10 percent of the nitrogen dioxide concentration as ozone, and the U. V. photometer device shows the least interference from nitrogen dioxide.

Studies comparing the two methods indicate that both instruments must be checked and calibrated frequently and that variations are usually caused by changes occurring in the airflow or reagent flow and by interfering compounds. Table 13-5 gives some indication of the reliability of the continuous colorimetric oxidant analyzers used at seven California stations. These analyzers are dynamically and statically calibrated about twice a year. The percentage of change from the previous calibration indicates the uncertainty in measurements obtained during the time between calibrations. Table 13-5 shows that 62 percent of the time there was less than a 10 percent change, and 78 percent of the time there was less than a 20 percent change. Changes greater than 20 percent occurred at three of the seven stations, which indicates that some stations provide more reliable data than others.

2. Other Methods - Intermittent Sampling

a. *Ferrous Ammonium Sulfate*

The ferrous ammonium sulfate measurement technique is in use on an experimental basis by NAPCA because of its possible application to intermittent sampling by the gas sampling network. This method is based on the conversion of the ferrous ion by oxidation to the ferric ion. Samples of ambient air are drawn through an acidified ferrous ammonium sulfate solution. The ferric ion formed in the collection solution is complexed in the laboratory with ammonium thiocyanate, and the resultant color is measured on a spectrophotometer. The oxidized solution is stable after sampling and can be analyzed later. Depending on the sample flow rate and interval of collection, this method is useful in the range of 1 to 2,400 $\mu g/m^3$ (0.0005 to 1.2 ppm). As with the KI method, this method suffers from the drawback of interferences from reducing compounds in the atmosphere. Another weakness is the lack of agreement with results obtained by other methods.

b. *Alkaline Potassium Iodide*

This method[3] is intended for the manual determination of oxidants in a range from a few $\mu g/m^3$ to about 39,000 $\mu g/m^3$ (20 ppm). The advantage of this procedure over the neutral iodide, or reference, technique is that a delay between sampling and completion of analysis is allowed. Sampling is conducted in midget impingers containing 1 percent KI in a 1-normal sodium hydroxide solution. Because of the simplicity of the preferred reference neutral-buffered KI method, however, the alkaline procedure is not widely used.

c. *Phenolphthalin*

The phenolphthalin method is based on the oxidation of phenolpthalin to phenolphthalein by the oxidant present in the atmosphere. Phanolphthalein is pink in alkaline solution. Because the technique responds to pH and temperature changes, it also is not widely used at present, although it was used extensively to obtain oxidant data in the past. Measurement results obtained by this method are approximately twice those obtained by the KI method. Consequently, when data obtained by the two methods are compared, it is necessary to adjust for this difference.

METHODS SPECIFICALLY FOR MEASUREMENT OF OZONE

1. Chemiluminescence

The chemiluminescent method of measuring ozone was developed by Regener.[11,12] If air containing ozone is passed in the dark across the surface of a chemiluminescent or fluorescent substance such as Rhodamine B adsorbed on silica gel, light is emitted. The intensity of emitted light can be measured with a photomultiplier tube. The chemiluminescent reaction between the ozone and Rhodamine B is highly specific to ozone and concentrations less than 2 $\mu g/m^3$ (0.001 ppm) can be measured. Nitrogen dioxide, sulfur dioxide, and peroxyacyl nitrate give negligible, if any, interference. The sensitivity for ozone detection of the chemiluminescent surface gradually decays with time, and daily calibration with a constant ozone source is required. Although the method has not been widely used for routine ambient air monitoring, it shows considerable promise and potential for this application and represents a needed attempt to obtain ozone specificity in measurement. Devices of this type are currently being thoroughly evaluated in the field and are available commercially.

2. Ultraviolet Photometry

Two instruments have been developed based on ozone's absorption of ultraviolet radiation. Ozone is unique in having a high optical opacity within the wavelength band centered at 2,500 to 2,600 Å.[13] The first instrument, based on the researches of Stairs,[14] required an optical path-length of 250 feet between source and receiver units, which was a major difficulty.[15]

The second instrument designed to monitor ozone. It was of a more conventional design, using 10-inch cells, a single source of radiation, and a double path of fixed wavelength at 3,537 Å. One cell had ambient air passing through it, while the other cell sampled air with the ozone removed by catalytic decomposition.[15]

This photometer was available commercially for several years but, because of electronic instability and temperature dependence, it required frequent attention. As a result, it was not used extensively. It is not available presently.

3. Trans-butene-2 Gas Phase Titration

This method, which has been only recently described,[16] involves a periodic gas-phase titration of the air sample with trans-butene-2 prior to analysis for total oxidants. This reactive olefin selectively removes the ozone present in the air stream and provides a specific ozone concentration by difference. The technique is applicable to coulometric and colorimetric methods of iodine measurement but has been evaluated more fully with a colorimetric instrument.[17]

4. Rubber Cracking

One of the earliest techniques used to identify ozone in ambient air was based on the rapid cracking of stressed rubber strips. A strip of a suitable rubber is bent double, stretched, and exposed to the air. Either the time until the rubber begins to crack or the depth of crack after a specified time can be related to the ozone concentration. Although it does not provide a direct quantitative measurement, this technique is useful for survey work, as it involves inexpensive equipment and is a simple operation. It cannot, however, be automated to obtain continuous data and, although presumed specific for ozone, other reactive species such as free radicals may contribute to rubber cracking over extended periods.[18]

5. Other Chemical Methods

Hauser and Bradley[19] have described a specific ozone technique using 1,2-Di-(4 pyridyl) ethylene as the absorbing medium and colorimetric analysis of the resulting aldehyde by the 3-methyl-2-benzothiazolone hydrazone method.[3] This method can be applied in field analysis of atmospheric ozone because sampling and analysis need not be performed on the same day.

1-hexene, hydrogen peroxide, and PAN were found to interfere when a 24-hour sampling period was used. Glacial acetic acid, which is used in the collection solution, also presents a problem since it cannot be sent through the mails. Furthermore, the absorbing solution used freezes at 16° C and must be artificially warmed if used below this temperature. For a 30-minute sample collected in a bubbler at 0.5 liters/minute, the range of measurement for this technique is 100 to 2,300 $\mu g/m^3$ (0.05 to 1.2 ppm) ozone.

METHODS FOR DETERMINATION OF PEROXYACETYL NITRATE

1. Long-path Infrared Spectroscopy

Peroxyacyl nitrates were discovered using long-path infrared spectroscopy. Although water vapor interferes with the principal absorption band used in this determination, an alternate band may be used to avoid this effect. Strong bands are located at 5.75 and 8.62 μ and weaker bands at 7.3, 9.5, 10.0, 10.8, and 12.2 μ. A 500-meter-long-path infrared cell is used to detect PAN in polluted ambient air.[20]

2. Gas Chromatography

Increased sensitivity for the detection of PAN is provided by gas chromatography using an electron-capture detector.[21] The advantage of this detector is its relative insensitivity to other compounds which might interfere.

Short columns (9 to 18 inches in length by 1/8 inch in diameter) packed with carbowax coated on an inert substance are used for the separation. Retention time for PAN is 1 to 2 minutes. Two-milliliter samples containing 5 $\mu g/m^3$ (0.001 ppm) PAN give detectable peaks. This method has been automated, but currently it is used primarily as part of a manual sampling program.

REFERENCES

1. Renzetti, N. A. and J. C. Romanovsky. A Comparative Study of Oxidants and Ozone in the Los Angeles Atmosphere. Arch. Ind. Health. *14*(5):458-467, November 1956.
2. Renzetti, N. A. An Aerometric Survey of the Los Angeles Basin August-November 1954. Vol. 1. Air Pollution Foundation. Los Angeles, Calif. Report Number 9. July 1, 1955. 333 p.
3. Selected Methods for the Measurement of Air Pollutants. Division of Air Pollution. Cincinnati, Ohio. PHS Publication Number 999-AP-11. May 1965.
4. Mast, G. M. and H. E. Saunders. Research and Development of Instrumentation of Ozone Testing. Instr. Soc. Amer. Trans. *1*(4):325-328, October 1962.
5. Analytical Methods of the Intersociety Committee on Methods for Ambient Air Sampling and Analysis. Health Lab. Sci. In press, 1970.
6. Potter, L. and S. Duckworth. Field Experience with the Mast Ozone Recorder. J. Air Pollution Control Assoc. *15*:207-209, May 1965.
7. Cherniack, I. and R. J. Bryan. A Comparison Study of Various Types of Ozone and Oxidant Detectors which are Used for Atmospheric Air Sampling. J. Air Pollution Control Assoc. *15*:351-354, August 1965.
8. Byers, D. H. and B. E. Saltzman. Determination of Ozone in Air by Neutral and Alkaline Iodide Procedures. Amer. Ind. Hyg. Assoc. J. *19*(3):251-257, June 1958.
9. Information obtained from California Dept. of Public Health. June 1965.
10. Cholak, J. et al. Air Pollution in a Number of American Cities. Arch. Ind. Health. *11*:280-289, April 1955.
11. Regener, V. H. On a Sensitive Method for the Recording of Atmospheric Ozone. J. Geophys. Res. *65*:3975-3977, December 1960
12. Regener, V. H. Measurement of Atmospheric Ozone with the Chemiluminescence Method. J. Geophys. Res. *69*(18):3795-3800, September 15, 1964.
13. Stair, R. Measurement of Ozone in Terms of its Optical Absorption. *Ozone Chemistry and Technology*, American Chemical Society, Advances in Chemistry Series, No. 21, March, 1959 pp. 269-285.
14. Stair, R. G., T. C. Bagg, and R. G. Johnston, Continuous Measurement of Atmospheric Ozone by an Automatic Photoelectric Method. Journal of Research of the National Bureau of Standards, 52(133), March, 1954.
15. Romanovsky, J. C., et al., Air Monitoring of the Los Angeles Atmosphere with Automatic Instruments. Paper No. 49-22, presented at the 49th Annual Meeting of the Air Pollution Control Association. Buffalo. May 21, 1956.
16. Bufalini, J. J. Gas Phase Titration of Atmospheric Ozone. Environ. Sci. Technol. *2*:703-704, September 1968.
17. Dudley, J. E. et al. Continuous Monitoring for Atmospheric Total Oxidant and Simultaneous Determination of Ozone by Gas Phase Titration. Presented at Air Pollution Control Association Meeting. New York. June 24, 1969.
18. Vega, T. and C. Seymour. A Simplified Method for Determining Ozone Levels in Community Air Pollution Surveys. J. Air Pollution Control Assoc. 11:28-33, January 1961.
19. Hauser, T. R. and D. W. Bradley. Effects of Interfering Substances and Prolonged Sampling on the 1, 2-Di-(4-Pyridyl)Ethylene Method for Determination of Ozone in Air. Anal. Chem. *39*(10):1184-1186, August 1967.
20. Darley, E. F., K. A. Kettner, and E. R. Stephens. Analysis of Peroxyacyl Nitrates by Gas Chromatography with Electron Capture Detection. Anal. Chem. *35*:589-591, April 1963.
21. Stephens, E. R. Absorptivities for Infrared Determination of Peroxyacyl Nitrates. Anal. Chem. *36*:928-929, April 1964.

CHAPTER 14

PARTICULATE MATTER

The particulate matter commonly found dispersed in the atmosphere is composed of a large variety of substances. Some of these—fluorides, beryllium, lead, asbestos, for example—are known to be directly toxic, although not necessarily at levels routinely found in the atmosphere today. The evidence suggests that there may very well be others whose toxic effects have not yet been recognized. To evaluate fully the effects on health and welfare of the presence of each of these substances in the air requires that they be given individual attention, or attention as classes of similar substances. Such evaluations will be made in separate documents.

This document considers the effects on man and his environment of undifferentiated particulate matter. These effects oftentimes are produced by a combination of particulate and gaseous pollutants, the contributions of which are difficult to distinguish. Moreover, laboratory studies have shown that a combination of particulates and gases may produce an effect that is greater than the sum of the effects caused by these pollutants individually.

Particles in the atmosphere, whatever their individual characteristics, exhibit a number of similar properties, which are for the most part dependent on the particle size. Most of the available studies on the effects of particulate air pollution, however, do not specify particle size, and this document is limited to treating particulate matter as a whole, and to considering the effects which are generally associated with the presence of particles in the air.

Particulate air pollution, as used here, refers to any matter dispersed in the air, whether solid or liquid, in which the individual particles are larger than small molecules but smaller in diameter than 500μ. (One μ in one millionth of a meter.) Particles in this size range stay in the air anywhere from a few seconds to several months.

Generally speaking, particles smaller than 1 μ in diameter originate in the atmosphere principally through condensation and combustion, while larger particles, with the exception of rain, snow, hail, and sleet, arise principally from comminution. Particles larger than 10 μ in diameter result from mechanical processes such as wind erosion, grinding and spraying, and the pulverizing of materials by vehicles and pedestrians. Particles between 1 μ and 10 μ in diameter usually include local soil, process dusts and combustion products from local industries, and, for maritime locations, sea salt. Combustion products and photochemical aerosols make up a large fraction of the particles in the range 0.1 μ to 1 μ in diameter, and, although particles below 0.1 μ in diameter have not been extensively identified chemically, the typical urban increase over natural levels of particles in this size range seems to be entirely due to combustion.

Particles of a size less than 0.1 μ in diameter are characterized by random motions produced by collisions with gas molecules. They are highly concentrated, move rapidly, collide frequently, and through sorption and nucleation of gas molecules and adhesion with other particles grow larger quickly. Particles larger than 1 μ have significant settling velocities, and their motions may deviate significantly from the motion of the air.

Measurements of dustfall are commonly used to indicate the mass concentration of the settleable portion of particulate air pollution. Typical values for cities are 10 to 100 tons/mile^2-month; as high as 2,000 tons/mile^2-month have been measured in the vicinity of especially offensive sources. Levels of dustfall have apparently declined in some American cities, and dustfall measurements are probably not useful as an index of overall particulate air pollution. However, dustfall itself constitutes a nuisance, and its measurement can be used as an index of the dirtiness of air pollution.

Several methods are available for measuring suspended particulate concentrations. The most commonly used device is the high-volume sampler, which consists essentially of a blower and a filter, and which is usually operated in a standard shelter to collect a 24-hour sample. The sample is weighed to determine concentration, and is usually analyzed chemically. The hi-vol is considered a reliable instrument for measuring the weight of total particulate matter. Chemical analysis of the hi-vol sample, however, may be limited: the filter material may contaminate the sample; different substances in the sample may react with each other; and losses may occur through volatilization of material. Tape samplers, which collect suspended particulate matter on filters and analyze the sample optically, are also in common use. While these samplers are inexpensive and rugged, they yield data which cannot always be easily interpreted in terms of particulate mass concentration. Other techniques available for measuring particulate pollution include optical systems which provide an indication of concentration without requiring that a sample be taken.

The averaging time used for measuring suspended particulates is not as significant a factor as it is for gaseous pollutants. The basic unit of time is 24 hours. Values taken over this period may be combined into weekly, monthly, seasonal, and annual means as required. The relationships between daily and other longer time periods in the United States is known with some degree of precision, as data exist for a 10-year period.

Most of the data on mass concentrations of suspended particulates come from the National Air Surveillance Networks, (NASN), which uses the high-volume sampler. NASN currently operates some 200 urban and 300 nonurban stations, and is supplemented by State and local networks. From the NASN data, the annual geometric mean concentrations of suspended particulate matter in urban areas range from 60 $\mu g/m^3$ to about 200 $\mu g/m^3$. The maximum 24-hour average concentration is about three times the annual mean, but values of seven times the annual mean do occur. Mean particulate concentrations correlate, in general, with urban population class, but the range of concentrations for any class is broad, and many smaller communities have higher concentrations than larger ones. For nonurban areas the annual geometric mean is typically between 10 $\mu g/m^3$ and 60 $\mu g/m^3$.

Effects on Health

For the most part, the effects of particulate air pollution on health are related to injury to the surfaces of the respiratory system. Such injury may be permanent or temporary. It may be confined to the surface, or it may extend beyond, sometimes producing functional or other alterations. Particulate material in the respiratory tract may produce injury itself, or it may act in conjunction with gases, altering their sites or their modes of action.

Laboratory studies of man and other animals show clearly that the deposition, clearance, and retention of inhaled particles is a very complex process, which is only beginning to be understood. Particles cleared from the respiratory tract by transfer to the lymph, blood, or gastrointestinal tract may exert effects elsewhere. Few studies have investigated the possibility of eye injury by particles in the air; only transient eye irritation from large dust particles presently is known to be a problem.

The available data from laboratory experiments do not provide suitable quantitative relationships for establishing air quality criteria for particulates. The constancy of population exposure, the constancy of temperature and humidity, the use of young, normal, healthy animals, and the primary focus on short-term exposures in many laboratory studies make extrapolation from these studies of limited value for the general population, and singularly risky for special risk groups within the population. These studies do, however, provide valuable information on some of the bioenvironmental relationships that may be involved in the effects of particulate air pollution on health. The data they provide on synergistic effects show very clearly that information derived from single-substance exposures should be applied to ambient air situations only with great caution.

Epidemiological studies do not have the precision of laboratory studies, but they have the advantage of being carried out under ambient air conditions. In most epidemiological studies, indices of air pollution level are obtained by measuring selected pollutants, most commonly particulates and sulfur compounds. To use these same studies to establish criteria for individual pollutants is justified by the experimental data on interaction of pollutants. However, in reviewing the results of epidemiological investigations it should always be remembered that the specific pollutant under discussion is being used as an index of pollution, not as a physicochemical entity.

In epidemiologcial studies consistency of results at different times and places is important in determining the significance of observations. However, while polluted air has many similarities from place to place and from time to time, it is not identical in all communities or at all times, and complete consistency between epidemiological studies should not be expected. There are not a large number of suitable epidemiological studies available at present, but those that are available show some consistency in the levels at which effects were observed to occur.

Considerable data have been presented on a number of air pollution episodes in London and in New York City. In reviewing these data it should be remembered that British air pollution measurements are not entirely comparable with American measurements. The only published comparison indicates that the British method of measuring particulates tends to give somewhat lower readings than American methods.

Excess deaths and a considerable increase in illness have been observed in London at smoke levels above 750 $\mu g/m^3$ and in New York at a smokeshade index of 5–6 cohs. Sulfur oxides pollution levels were also high in both cases. These unusual short-term, massive exposures result in immediately apparent pathologic effects, and they represent the upper limits of the observed dose-response relationship between particulates and adverse effects on health.

Daily averages of smoke above 300 $\mu g/m^3$ to 400 $\mu g/m^3$ have been associated with acute worsening of chronic bronchitis patients in England. No comparable data are available in this country. Studies of British workmen found that increased absences due to illness occurred when smoke levels exceeded 200 $\mu g/m^3$.

Two recent British studies showed increases in selected respiratory illness in children to be associated with annual mean smoke levels above 120 $\mu g/m^3$. Additional health changes were associated with higher levels. These effects may be of substantial significance in the natural history of chronic bronchitis. Changes beginning in young children may culminate in bronchitis several decades later.

The lowest particulate levels at which health effects appear to have occurred in this country are reported in studies of Buffalo and Nashville. The Buffalo study clearly shows increased death rates from selected causes in males and females 50 to 69 years old at annual geometric means of 100 $\mu g/m^3$ and over. The study suggests that increased mortality may have been associated with residence in areas with 2-year geometric means of 80 $\mu g/m^3$ to 100 $\mu g/m^3$. The Nashville study suggests increased death rates for selected causes at levels above 1.1 cohs. Sulfur oxides pollution was also present during the periods studied. In neither study were the smoking habits of the decedents known.

Corroborating information is supplied from Fletcher's study of West London workers between the ages of 30 and 59. The data indicate that with a decrease of smoke pollution (yearly mean) from 140 $\mu g/m^3$ to 60 $\mu g/m^3$, there was an associated decrease in mean sputum volume. Fletcher noted that there may have been changes in the tar composition of cigarettes during the period studied; such a change could affect the findings. This study provides one of the rare opportunities to examine the apparent improvement in health that followed an improvement in the quality of the air.

Effects on Climate Near the Ground

Particles suspended in the air scatter and absorb sunlight, reducing the amount of solar energy reaching the earth, producing hazes, and reducing visibility. Suspended particulate matter plays a significant role in bringing about precipitation, and there is some evidence that rainfall in cities has increased as the cities have developed industrially.

Suspended particulate matter, in the concentrations routinely found in urban areas, considerably reduces the transmission of solar radiation to the ground, creating an increased demand for artificial light. The effect is more pronounced in the winter than in the summer, when particulate pollution loadings are higher, and sunlight must penetrate more air to reach the ground. For similar reasons the effect is also more pronounced during the workweek than on weekends, during industrial booms, and in higher latitudes. For a typical urban area in the United States, with a geometric mean annual particulate concentration of roughly 100 $\mu g/m^3$, the total sunlight, including that received directly from the sun and that reflected by the sky, is reduced five percent for every doubling of particle concentration. The reduction is most pronounced on ultraviolet radiation.

For urban areas in the middle and high latitudes, particulate air pollution may reduce direct sunlight by as much as one-third in the summer and as much as two-thirds in the winter. This effect may have implications for the delicate heat balance of the earth's atmospheric system. In spite of an increase in the carbon dioxide content of the atmosphere over the past several decades, which would presumably bring about an increase in atmospheric temperature, mean worldwide temperatures have been decreasing since the 1940's. Increased reflection of solar radiation back to outer space, brought about by increased concentrations of particulate air pollution, may be more than cancelling out the climatic effect of the increased carbon dioxide. That worldwide particulate air pollution has been increasing is evidenced by the fact that in the United States and in other countries, turbidity, a phenomenon produced by the back-scattering of direct sunlight by particles in the air, has increased significantly over the last several decades.

Effects on Visibility

Particles suspended in the air reduce visibility, or visual range, by scattering and absorbing light coming from both an object and its background, thereby reducing the contrast between them. Moreover, suspended particles scatter light into the line of sight, illuminating the air between, to further degrade the contrast between an object and its background.

The scattering of light into and out of the line of viewing by particles in the narrow range of 0.1 μ to 1 μ in radius has the greatest effect on visibility. Certain characteristics of behavior of these particles make it possible to formulate a useful approximate relationship between visual range and concentrations of particulate matter:

$$L_v \simeq \frac{A \times 10^3}{G'}$$

where G' = particle concentration ($\mu g/m^3$),

L_v = equivalent visual range, and

$A = 1.2_{0.6}^{2.4}$ for L_v expressed in kilometers and $0.75_{0.38}^{1.5}$

for L_v expressed in miles

The value 1.2 for A is the mid-range value empirically obtained from observations in a variety of air pollution situations. The data indicate that the range 0.6 to 2.4 covers virtually all cases studied. The relationship does not hold at relative humidities above 70 percent, nor does it apply to fresh plumes from stacks, and it may not hold for the products of photochemical reactions.

Within the limitations prescribed, the relationship provides a useful means of estimating approximate visual range from particulate concentrations. In addition to aesthetic degradation of the environment, reduced visibility has serious implications for safe operation of aircraft and motor vehicles. At a visual range of less than 5 miles, operations are slowed at airports because of the need to maintain larger distances between aircraft. Federal Aviation Administration restrictions on aircraft operations become increasingly severe as the visual range decreases below 5 miles. Using the upper and lower bounds of the relationship described above, visibility could be 5 miles at a particulate loading as high as 300 $\mu g/m^3$ or as low as 75 $\mu g/m^3$. However, on the average, visibility can be expected to be reduced to approximately 5 miles at a particulate concentration of 150 $\mu g/m^3$. At a level of 100 $\mu g/m^3$, visibility is reduced to 7½ miles. This limited distance, however, may be related to particulate concentrations as low as 50 $\mu g/m^3$ and as high as 200 $\mu g/m^3$.

Effects on Materials

Particulate air pollution causes a wide range of damage to materials. Particulate matter may chemically attack materials through its own instrinsic corrosivity, or through the corrosivity of substances absorbed or adsorbed on it. Merely by soiling materials, and thereby causing their more frequent cleaning, particulates can accelerate deterioration.

Laboratory and field studies underscore the importance of the combination of particulate matter and corrosive gases in the deterioration of materials. On the basis of present knowledge, it is difficult to evaluate precisely the relative contribution of each of the two classes of pollution; however, some general conclusions may be drawn.

Particulates play a role in the corrosion of metals. In laboratory studies, steel test panels that were dusted with a number of active hygroscopic particles commonly found in the atmosphere corroded even in clean air. Corrosion rates were low below a relative humidity of 70 percent; they increased at relative humidities above 70 percent; and they greatly increased when traces of sulfur dioxide were added to the laboratory air.

It is apparent that the accelerated corrosion rates of various metals in urban and industrial atmospheres are largely the result of relatively higher levels of particulate pollution and sulfur oxides pollution. High humidity and temperature also play an important synergistic part in this corrosion reaction. Studies show increased corrosion rates in industrial areas where air pollution levels, including sulfur oxides and particulates, are higher. Further, corrosion rates are higher during the fall and winter seasons when particulate and sulfur oxides pollution is more severe, due to a greater consumption of fuel for heating.

Steel samples corroded 3.1 times faster in the spring of the year in New York City, where annual particulate concentrations average 176 $\mu g/m^3$, than did similar samples in State College, Pennsylvania, where the average concentrations were estimated to range from 60 $\mu g/m^3$ to 65 $\mu g/m^3$. In the fall of the year, when particulate and sulfur oxide concentrations in New York were considerably higher than in the spring, the steel samples in New York corroded six times faster than the samples at State College. Similar findings were reported for zinc samples. Moisture may have contributed to the corrosion.

In Chicago and St. Louis, steel panels were exposed at a number of sites, and measurements taken of corrosion rates and of levels of sulfur dioxide and particulates. In St. Louis, except for one exceptionally polluted site, corrosion losses correlated well with sulfur dioxide levels, averaging 30 percent to 80 percent higher than losses measured in nonurban locations. Sulfation rates in St. Louis, measured by lead peroxide candle, also correlated well with weight loss due to corrosion. Measurements of dustfall in St. Louis, however, did not correlate significantly with corrosion rates. Over a 12-month period in Chicago, the corrosion rate at the most corrosive site (mean SO_2 level of 0.12 ppm) was about 50 percent higher than at the least corrosive site (mean SO_2 level of 0.03 ppm). Although suspended particulate levels measured in Chicago with high-volume samplers also correlated with corrosion rates, a covariance analysis indicated that sulfur dioxide concentrations were the dominant influence on corrosion. Based on these data, it appears that considerable corrosion may take place (i.e., from 11 percent to 17 percent weight loss in steel panels) at annual average sulfur dioxide concentrations in the range of 0.03 ppm to 0.12 ppm, and although high particulate levels tend to accompany high sulfur dioxide levels, the sulfur dioxide concentration appears to have the more important influence.

Particulate air pollution damages electrical equipment of all kinds. Oily or tarry particles, commonly found in urban and industrial areas, contribute to the corrosion and failure of electrical contacts and connectors. Dusts can interfere with contact closure, and can abrade contact surfaces. Hygroscopic dusts will absorb water and form thin electrolytic films which are corrosive.

Particulates soil and damage buildings, statuary, and other surfaces. The effects are especially severe in urban areas where large quantities of coal and sulfur-bearing fuel oils are burned. Particles may act as reservoirs of acids, and thereby sustain a chemical attack that will deteriorate even the more resistant kinds of masonary. Particles stick to surfaces, forming a film of tarry soot and grit which oftentimes is not washed away by rain. Considerable money and effort have been spent in many cities to sandblast the sooty layers that accumulate on buildings. Water-soluble salts, commonly found in urban atmospheres, can blister paint. Other particles may settle on newly painted surfaces, causing imperfections, thereby increasing the frequency with which a surface must be painted.

The soiling of textiles by the deposition of dust and soot on fabric fibers not only makes them unattractive, and thereby diminishes their use, but results in abrasive wear of the fabric when it is cleaned. Vegetable fibers, such as cotton and linen, and synthetic nylons are particularly susceptible to chemical attack by acid components of airborne particles.

Economic Effects of Atmospheric Particulate Matter

It is not possible at the present stage of knowledge to provide accurate measures of all the costs imposed on society by particulate air pollution. Selected categories of effects can be quantified; it is obvious that these estimates represent a significant understatement of the total cost.

Effects on Vegetation

Relatively little research has been carried out on the effects of particulate air pollution on vegetation, and much of the work that has been performed has dealt with specific dusts, rather than the conglomerate mixture normally encountered in the atmosphere. This document reports briefly on some of these specific particulate studies only to illustrate the possible mechanisms through which particulate matter may affect vegetation. This information is not presented for the purpose of establishing air quality standards on these specific pollutants.

There is considerable evidence that cement-kiln dusts can damage plants. A marked reduction in the growth of poplar trees 1 mile from a cement plant was observed after cement production was more than doubled. Plugging of stomates by the dust may have prevented the exchange of gases in leaf tissue that is necessary for growth and development. Moderate damage to bean plants occurred when the plant leaves were dusted at the rate of 0.47 mg/cm^2-day (400 $tons/m^2$-month) for 2 days and then exposed to natural dew. The mechanism through which the leaves are damaged is not entirely understood, but direct alkaline damage to tissues beneath the crust formed by the dust and moisture has been observed. The deposits may also plug stomates and block light needed for photosynthesis. Cement-kiln dusts may change the alkalinity of soils to benefit or harm vegetation, depending on the species.

Dust deposits may also eliminate predators, and thereby bring on increased insect injury to plants; they may interefere with pollen germination; and they may make plants more susceptible to pathogens.

Fluoride dusts apparently have a difficult time penetrating leaf tissue in physiologically active form, and they are much less damaging to vegetation than is gaseous fluoride. Soluble fluoride dusts may be absorbed by the plant, but the amount is relatively small compared to that which can enter the plant in gaseous form. The evidence suggests that there is little effect on vegetation at fluoride particulate concentrations below 2 $\mu g/m^3$. Concentrations of this magnitude and above can sometimes be found in the immediate vicinity of sources of fluoride particulate pollution; they are rarely found in urban atmospheres.

Ingestion of particles deposited on plants can be harmful to animal health. Fluorosis and arsenic poisoning have been brought on through this medium.

Soot may clog stomates and may produce necrotic spotting if it carries with it a soluble toxicant, such as one with excess acidity. Magnesium oxide deposits on soils have been shown to reduce plant growth, while iron oxide deposits on soils have been shown to reduce plant growth, while iron oxide deposits appear to have no harmful effects, and may be beneficial.

Effects on Public Concern

Several studies indicate that there is a relationship between levels of particulate pollution, used as an index of air pollution, and levels of public concern over the problem. A study conducted in 1963 in the St. Louis metropolitan region found a direct linear relationship between the fraction of a community's population who said air pollution was a nuisance, and the annual mean concentration of particulate air pollution in the community. The relationship, which was derived from data on communities in the St. Louis area whose annual concentrations ranged from 50 $\mu g/m^3$ to 200 $\mu g/m^3$, was formulated as:

$$y = 0.3x - 14$$

where y = population fraction (%) concerned, and

x = annual geometric mean particle concentration ($\mu g/m^3$).

It is thought that the reaction to suspended particulates as a nuisance probably occurs at peak concentrations, and not necessarily at the values representing annual means. However, the relationship provides a useful example of how the nuisance effect of air pollution relates to concentrations. Approximately 10 percent of the study population considered air pollution a nuisance in areas with suspended particulates at an annual geometric mean concentration of 80 $\mu g/m^3$. At this same level of pollution, 30 percent of the study population was "aware of" air pollution. In areas with 120 $\mu g/m^3$ (annual geometric mean), 20 percent were "bothered by" and 50 percent were "aware of" air pollution; in areas with an annual geometric mean of 160 $\mu g/m^3$, one-third of the population interviewed were "bothered by" and three-fourths were "aware of" air pollution.

Although data from other studies do not readily lend themselves to quantitative formulation, they do, in general, support the relationship reported by the St. Louis study. A study of communities in the Nashville, Tennessee, metropolitan area in 1957 found that at least 10 percent of the population expressed concern about the nuisance of air pollution at dustfall levels exceeding 10 tons/mi^2-month.

Suspended Particles as a Source of Odor

Particulate air pollution is not ordinarily considered a significant source of odors. However, there is evidence that liquid and even solid particles of some substances may be volatile enough to vaporize in the nasal cavity, and produce sufficient gaseous material to stimulate the sense of smell. Further, particles may carry absorbed odorants into the nasal cavity, and there transfer them to olfactory receptors. A survey of State and local air pollution control officials revealed that approximately one-fourth of the most frequently reported odors are those which are known to be, or are suspected to be, associated with particulate air pollution. The sources of these odorous particles are diverse, including diesel and gasoline engine exhausts, coffee-roasting operations, paint spraying, street paving, and the burning of trash.

CHAPTER 15

ODOROUS COMPOUNDS

Odorous compounds may have pleasant or unpleasant odors: an odor which is quite acceptable to one person may be unacceptable to another person. Although the quality of an odor is highly subjective, all healthy people are usually aware of odors and generally agree that some odorous compounds are obnoxious. Some of these offensive odors can be detected when the odorant is present in very low concentrations. For these reasons, malodors are one of the first manifestations of air pollution, and they frequently arouse extreme emotional reactions in people. Offensive odors are capable of producing nausea, vomiting, and headache; curbing the appetite, impairing nutrition, and curtailing water intake; disturbing sleep; upsetting the stomach; hampering proper breathing; offending the senses; and interfering with enjoyment of property. Most of all, bad odors can mar good dispositions and provoke emotional disturbances, mental depression, and irritability.[126]

Sociologically, such noxious odors can ruin personal and community pride, interfere with human relations in various ways, discourage capital improvements, lower socioeconomic status, and damage a community's reputation. Economically, they can stifle growth and development of a community. Both industry and labor prefer to locate in a desirable area in which to live, work, and play; and the natural tendency is to avoid communities with obvious odor problems. Tourists also shun such areas. The resulting decline in property values, tax revenues, payrolls, and sales can be disastrous to a community.[126]

No instrument has been developed with the sensitivity and versatility of the human nose for odor detection. Therefore, the methods currently used for odor measurement involve personal judgments by one or more people; results obtained are expensive and lack the desired precision.

The mode of expressing observation results is strictly of a qualitative nature. It is based largely upon the olfactory sense without any guide beyond human ability to associate and describe personal reaction. As a result, odors are often given such descriptive terms as dead-cat, wet-dog, manurial, rotten-egg, spoiled-fish, and others. In addition, the intensity of the odor is often rated on an arbitrary scale of 1 to 5. Such a system, of course, depends to a great degree upon the acuity of the observer's nose, his past experiences, and his ability to describe his reaction accurately. The most useful observation from an engineering point of view is to measure the number of dilutions which are necessary to reduce the odorant to the odor threshold concentration.

Odor may be defined as the sensation of smell perceived as a result of olfactory stimulus.[65] An odorant is a substance or mixture of substances that produces the sensation of smell.[65]

EFFECTS

Effects on Humans

To keep alive man must breathe. A single sniff of air may delight him with the perfume of vanillin, may nauseate him with a fecal odor, may warn him of the presence of toxic quantities of hydrogen sulfide, or may start his digestive juices flowing with the aroma of a broiling steak. Thus, odors may affect man in various ways, depending not only on the characteristics of the odor, but also on the particular man and his environment.

Characteristics of Odors

Odor Intensity

The human nose is an extremely sensitive gaseous detector. It can respond to thousands of different odor stimuli and detect both low and high concentrations of gaseous materials simultaneously. Moreover, the odorants may originate from sources at relatively great distances away. The intensity of the odor is defined as the numerical or verbal indication of the strength of an odor.[294]

Experimental findings on discerning odor intensity show that an average observer can distinguish between three intensities--weak, medium, and strong--whereas a trained observer may distinguish between five degrees of intensity,[294] and an expert can distinguish six.[190] Since 1920, experts have been rating the intensity of various odors by using the following scale:

Odor Intensity	Expert Description[140]
0	No odor
1	Very faint
2	Faint
3	Easily noticeable
4	Strong
5	Very strong

Trained observers have used[290,294] the following

scale to determine the odor intensity of tobacco smoke, as well as of domestic and industrial odors:

Odor Intensity	Odor Description[22]
0	A concentration of an odorant which produces no sensation.
1	Concentration which is just detectable (the threshold dilution).
2	A distinct and definite odor whose unpleasant characteristics are revealed or foreshadowed (the recognition threshold).
3	An odor strong enough to cause a person to attempt to avoid it completely.
4	An odor so strong as to be overpowering and intolerable for any length of time.

The sensation of intensity of an odor varies exponentially with the concentration of the odorant. This phenomenon is described by the well-known Weber-Fechner Psychophysical Law, which states that the intensity of the sensation is proportional to the logarithm of the strength of the stimulus. For the sensation of odor this may be expressed as

$$I = k \ln C$$

where I is the intensity of the odor sensation
k is a constant
and C is the concentration of odorant.

The data for three odorants, ethyl mercaptan, butyl thioether, and crotonaldehyde, follow this law over extremely large changes in concentrations. The range of intensity from 0 to 5 covers eight log cycles for ethyl mercaptan, six for butyl thioether, and four for crotonaldehyde.[36]

The concentration of odorant that just gives an intensity of zero may be defined as the detection threshold concentration.[127,217] However, this odor threshold concentration is more commonly defined as the minimum concentration which will result in the stimulation of the olfactory nerves. All people do not have the same sensitivity for detection of odors.[190] Therefore, an odor panel may be used to determine the odor threshold concentration.[26] As a result, the odor threshold may be reported as the "effective dosage" where 100 percent (ED_{100}), 50 percent (ED_{50}), or 0 percent (ED_0) of panelists perceive the odor. ED_{50} is the most commonly used. Tests are usually conducted to eliminate persons either highly sensitive or insensitive to odors from odor panels. (These tests are described in Section 6 .)

Two other bases of determining odor concentrations have been used: a recognition threshold concentration--the concentration at which the odor quality can be recognized; and the objectionability concentration--the concentration where the odor becomes objectionable. Leonardos _et al._[154] have argued that the recognition sensation is more reproducible than the detection sensation.

Unfortunately, odor threshold measurements depend largely on the purity of the odorant. Therefore, odor threshold concentrations of "pure" odorants vary widely, often overlapping both detection and recognition threshold concentrations. For example, the reported odor threshold concentrations of hydrogen sulfide vary from 0.65 to 1,400 $\mu g/m^3$.

The intensities of a mixture of odorants may be independent, counteractive, additive, or synergistic.[236] For example, if odorants A and B are mixed, the odor intensity (I) may be

Independence
$$I_{AB} = k \ln (C_A \text{ or } C_B)$$

Counteraction
$$I_{AB} < k \ln (C_A \text{ or } C_B)$$

Addition
$$I_{AB} = k \ln (C_A + C_B)$$

Synergism
$$I_{AB} > k \ln (C_A + C_B)$$

Mixtures of butanol and pyridine showed an additive effect on the odor intensity at the odor threshold, whereas the addition of p-cresol to the mixture showed a synergistic effect.

Horstman _et al._[116] have listed some of the factors which reportedly influence the olfactory sensitivity:

(1) The odor sensitivity of the individual observer varies from day to day, but the overall sensitivity of a group of observers is reasonably constant.

(2) The sense of smell becomes rapidly fatigued, though fatigue for one odor does not necessarily affect the perception of dissimilar odors.

(3) Responses to odors are not completely objective since psychological responses vary in different observers. What is unpleasant to one observer may be quite acceptable to another, and so may not be noted.

(4) The sensitivity of observers varies widely; some have extreme sensitivity while others are incapable of smelling an odor. The age of the observer seems to have an effect on sensitivity: sensitivity reaches a maximum at puberty and decreases with age.

(5) Meteorological factors influence reported odor levels; wind speed and vertical temperature gradient influence the dilution of odors. Temperature and humidity affect

odor perception, but there is considerable disagreement about their precise influence.

Odor Quality

Odor quality is a verbal description of the odor. The quality may be described in terms of such familiar odorants as coffee, onions, lemons (characteristic odors), or by associating an unfamiliar odor with a familiar odor. The observer often does not possess the vocabulary to describe the odor he smells.

The odor quality may change with dilution. In mixtures of odorants this may be because one odorant is more pervasive than the other odorant. Single component systems may also exhibit quality changes on dilution. The reason for this is not fully understood.[272]

Odor Acceptability

An odor may be either acceptable or unacceptable depending on its intensity and quality. The odors of new-mown hay or honeysuckle and roses are indicative of acceptable odors at normal concentrations. However, obnoxious odors may be unacceptable at much lower concentrations and become acceptable only at very low intensities. At high intensities the normally acceptable perfumes can be unacceptable.[272]

Physiological and Psychological Aspects of Odors

The influence of odors on the health and comfort of man is difficult to prove. Odors in themselves are usually not the cause of organic disease. However, the odorant may incite an allergic response. It is obvious that some highly toxic substances, such as hydrogen sulfide, are associated with offensive odors; but the dangerous properties of these types of substances do not derive from the odor itself. In fact, odor is valuable in serving as a warning of the presence of an injurious gas. Odor bears no relationship to toxicity, and some poisonous gases are odorless or have a rather pleasant odor.

The most frequent effects of odors on human health, according to McCord and Witheridge,[170] are insomnia and mental perturbation. They point out that many people regularly have difficulty sleeping and that any disturbance may prevent sleep, often after the person has been aroused from a deep sleep by an offensive odor. They admit that the extent to which odors contribute to loss of sleep cannot be proved; but they assert that odors do cause loss of sleep and, therefore, affect the health of a community. Long continuous exposure to offensive odors arouses a person to anger. Otherwise calm persons may become mildly maniacal, hysterical, and capable of carrying out acts entirely foreign to their usual natures. Thus odors may affect the mental health of a person.

Petri[213] states that malodorous substances may cause headaches, nausea, and similar phenomena. Even inherently fragrant substances, such as flavorings and chocolates, can cause considerable discomfort with protracted exposure, or at high concentrations. According to Petri, the psychological effects of an odor are highly subjective. Thus, the nuisance value of an odor depends on the attitude of the person, his disposition, and the time of day.

Air pollution in the form of malodors has been cited as the reason for certain lawsuits, picketing, rioting, and even forceful closure of plants.[124]

Allergies and Odors

Odors may cause attacks of asthma or other allergic conditions.

In many cases it is difficult to prove whether an allergic patient was affected by an odor or the odorant substance itself. Odorants from trees, shrubs, flowers, fabrics, animals, and household articles are ordinarily harmless to individuals not subject to allergies. However, these substances can, in small amounts under proper conditions, incite an allergic attack in sensitized individuals.[115]

Odorous substances have been cited as causing allergic symptoms or illnesses by various authorities as follows:

Odorous Substance

Cleaning fluid
Cooking odors
Feces
Fish
Food

Formaldehyde
Fresh paint
Furniture polish

Gasoline
Lighter fluid
Moth balls
Newspaper print
Oils
Perfume
Rubber
Spices
Tobacco smoke

Turpentine
Wood smoke

Food odorants in low concentrations also commonly excite allergic symptoms. Foods most frequently incriminated were eggs and fish, although chicken, pork, bacon, and cabbage were also mentioned. Atopic dermatitis was reported in a series of nine cases--the majority infants, but some, older children. In these patients the allergic signs and symptoms were provoked or aggravated by the mere presence of the foods in the patient's environment.

SOURCES

In 1955, the chief public officials responsible for control of air pollution in 67 major cities were surveyed.[211] The results of the survey showed that 78 percent received complaints of odors separately from other air pollution complaints, and 68 percent felt the public interest was increasing because of odor pollution. When asked to list the source of the odors in their communities, their replies were as follows:

Source	Percent*
Chemicals	62
Vehicles	52
Paint and varnish	49
Food processing	47
Domestic (homes, etc.)	45
Rendering plants	43
Plastics	33
Oil refineries	31
Coke works	31
Rubber	27
Steel	25
Insulation	21
Fish	21
Gas works	19
Pharmaceuticals	19
Soaps and detergents	17
Breweries	15

The odor pollution of an area depends on the odor strength and quality. The odor unit has been defined thus: one odor unit is the amount of odorant necessary to contaminate one cubic foot of clean air to the odor threshold.

*Percent of questionnaires in which the source was cited.

For any one odorant, the number of odor units can be calculated by knowing the volume of odor released and the odor threshold. For example, dimethyl amine has an odor threshold of approximately 0.5 ppm (1,000 μg/m^3). A release of 10 pounds of this substance per hour would result in the release of 2,800,000 odor units per minute.[37] This number can then be used in atmospheric diffusion equations to calculate the distance the odor may travel. Any value above one would be detectable. Similar calculations are also useful to engineers in designing systems which will avoid or abate odor pollution.

The mixture of two or more odorants may present complicating factors because the odorants may be additive, synergistic, counteractant, or independent.

ABATEMENT

Complete combustion is generally accepted as the best way to deodorize malodorous gases. However, it may not be the most economical method. Complete oxidation of odorants converts hydrocarbons to odorless carbon dioxide and water, and sulfur and nitrogen compounds to sulfur oxides and nitrogen oxides that usually have higher odor thresholds than the parent compounds. Partial oxidation may increase the odor problem by the formation of malodorous aldehydes.[292]

Oxidation at 1,200°F or above has been recommended and usually gives satisfactory results. The temperatures may be lower (500 to 800°F) when a catalyst is used. This will depend on the odorants and the possible catalytic poisons in the gas stream.[292]

Odor-laden smoke from the coffee roasting industry is most effectively controlled with afterburners, but fuel requirements are increased 100 to 150 percent over those for a conventional roaster.[210]

Where odorants are soluble in water or some other liquid or solution, absorption may be used. For example, ammonia may be removed by spraying water through a chamber containing the ammonia. Hundreds of methods have been devised for contact between vapor and liquid. Some of those used include simple vertical spray towers in single or multiple stages and cascade vertical towers packed with partition rings, Raschig rings, spiral rings, Berl saddles, hollow balls, helical packers, hexahelix blocks, double spirals, cyclohelix blocks, prismic packings, centrifugal or cyclone scrubbers, and bubble and sieve trays.[274]

Some solids will adsorb odorant compounds and thus remove them from the system. This process requires that adsorbent and odorant be matched. Activated charcoal has the particular advantage that it will adsorb all types of materials under almost any conditions. However, the efficiency of any adsorbent system is dependent on the temperature, pressure, and flow rate of the gas stream.[272]

Odor masking is the process of eliminating the perception of one odor or group of odors by superimposing another odor or group of odors on it to create a new odor sensation, preferably pleasant. Odor-masking chemicals are usually synthetic aromatic compounds or a mixture of these compounds.[292] Some examples of these are vanillin, methyl ionones, eugenols, benzyl acetate, phenyethyl alcohol, and heliotropin. The proper masking agent does not alter the composition of the preexisting odorant, but has a pleasant smell that is strong enough to overpower the offensive odorant.[272]

All smells cannot be masked. In particular, strong acids, even in traces, will defy masking because the agents used decompose under acid conditions.[272]

Masking agents may be applied directly to raw material (sewage, animal or vegetable waste, blood), drip-fed into process lines, added to scrubbing waters, injected into gas streams, soaked into covers for small leaks, or sprayed as a fog.[272]

The chief advantage of this method of control is that little or no capital costs are involved. Masking may be used as a temporary measure while other control methods are developed.[292] Odor counteractants are often used together with masking agents in a single application.[292]

The effect of odor counteraction is to reduce both the odor of the counteractant and the odor of the malodorant. When the two odors are sniffed together, both odors are diminished. This is often confused with odor masking, in which equal strengths of two odorants may both be distinguished and the masking agent concentration must be increased to overpower the malodors.[292]

When benzene, toluene, xylene, pseudo-cumene, and durene are mixed in small quantities, their odor strengths are additive. However, at higher concentrations, the odor becomes faint. Another example of counteraction is the pair, butyric acid and oil of juniper. When air is bubbled through a butyric acid solution, the characteristic strong, unpleasant odor is perceived. Oil of juniper also has an unpleasant odor, but a mixture of the vapors has a faintly pleasant odor.[190]

Odor dilution will obviously result in odor-free air as an odorant is diluted below the threshold concentration. Such a method is feasible to remove an odor problem from a plant area provided the weather conditions are favorable. However, unfavorable meteorological conditions may cause the odorant concentration to increase above the odor threshold. Thus, the odor emission rate (odor units per minute) must be weighed against the possibility of adverse meteorological conditions in order to prevent odor pollution.[272,292]

Another method of odor control is to eliminate or reduce the odor-producing substance. This type of control is being practiced by using low sulfur fuels.[272]

Many odorants are adsorbed on particulates, and the removal of these particulates may also reduce the odor.[49]

Chemical control of odors is possible in many industries. Chemical oxidation or combination may change an odorant to a nonodorous compound. Frequently, chlorine or potassium permanganate is added to a scrubbing solution to oxidize the odorants.[272] Other reactions result in ammonium acetate products that have no odor—for example, the reaction of acetic acid with ammonia which produces odorless ammonium acetate. Ozonation, catalytic chemical oxidation, silent electric discharge, and ultraviolet radiation will all result in chemical conversion of some odorants to compounds with a less offensive odor.[272]

Biological control may be possible in some operations. The biological degradation of sewage produces odorous gases. However, it is known that some microorganisms (*Beggiatoa alba*) oxidize hydrogen sulfide to water and sulfur, and this method has been suggested for sewage odor control.[272]

Containment offers a means of odor abatement in some situations. Covers on fuel tanks, sewage ponds, and other open storage areas will reduce the emissions of odor.[272]

ECONOMICS

Economically, the impact of odors is far-reaching. Because noxious and foul odors can ruin personal and community pride, interfere with human relations, discourage capital investment, lower socioeconomic status, and damage a community's reputation, the economics of a community may be closely related to any odor pollution problem. Both people and industry desire to locate in a desirable area in which to live, work, and play; the natural tendency is to avoid communities and localities with obvious odor problems. Tourists shun polluted areas. The resulting decline in market and rental property values, tax revenues, payrolls, and sales can be disastrous to a community.[126,166,263]

However, industries which cause odor pollution may be an economic advantage to a community, since they provide job opportunities both in the industry itself and in businesses which service the industry and its employees.

There are many socioeconomic aspects to odors which are difficult to assess. However, some incidents are easily evaluated. For example, a downtown theater in Washington D.C., was once evacuated because of some odor which penetrated the theater, requiring the manager to refund the price of admission. The exact source of the odor was not reported, but sewer gas was suspected.[302]

The cost of odor control by an industry is economically important. Often the cost is offset by economic benefits gained through the control methods or from recovery of waste products. For example, odorous compounds are often controlled by incineration, and the heat generated by incineration used to provide heat for some industrial process. Good examples are the heat from the recovery furnace in kraft pulp mills, heat recovered from incineration of odorous gases in rendering plants, and heat produced from sewage gas burners.

SUMMARY AND CONCLUSIONS

Offensive odors in the air are a major air pollution problem in some areas. These malodors cause many complaints, provoking emotional disturbances, mental depression, and irritability. In some instances health effects such as nausea, vomiting, headache, loss of sleep, loss of appetite, and impaired breathing are induced. Contact with odorants may cause varying degrees of reactions in allergic individuals, particularly children.

Sociologically, odor pollution can interfere with human relations in many ways. It can damage personal and community pride, discourage capital investment, and lower the socioeconomic status of both the individual and the community. Some State, county, and city governments have enacted laws that prohibit the emission of air pollutants which unreasonably interfere with the enjoyment of life and property. However, no odor standard has been established.

No information has been found on the effects of odor air pollution on animals. Odors *per se* have no effect on plants or materials. However, some odorants such as hydrogen sulfide and sulfur dioxide may affect animals, plants, and materials.

The sources of odors are numerous and include pulp and paper mills, animal rendering plants, sewers and sewage treatment plants, garbage dumps and incinerators, chemical plants, petroleum refineries, metallurgical plants, and internal combustion engines, particularly diesel and aircraft engines. The most offensive odors come from plants or processes which produce low molecular weight sulfur and nitrogen compounds, such as ethyl- and methyl-mercaptans, hydrogen sulfide, ammonia, and dimethylamine. Environmental air concentrations of obnoxious odorants frequently exceed the odor threshold concentration in some local areas, and the odor has on occasions been recognized 20 miles from the source.

The most generally accepted method of abatement of odors is incineration at the source. However, improper incineration may in itself be a source. Other abatement methods include adsorption, absorption, particulate removal, source elimination, process changes, chemical control, containment, odor masking, odor counteraction, biological control, and dilution.

Economically, noxious odors may stifle the development and growth of a community. Both people and industry desire to locate in a place where it is pleasant to work, live, and play. Tourists shun polluted areas, and rental and real estate property values may decrease. The control of odor pollution is often very costly to an industry, depending on the odor problem and the type of industry. This cost may be reduced by economic benefits derived from recovered heat or waste products. About $75 million have been spent for air pollution control in the kraft paper industry alone.

Both laboratory and field methods have been developed for measurement of odors at the source and in the ambient air. The human nose is the only valid odor detector, and all methods rely on the judgment of one or more people who make up the odor panel. Only gas chromatography has been developed to measure hydrogen sulfide, methyl mercaptan, and sulfur dioxide at concentrations near the odor threshold.

REFERENCES

The publications referred to in Chapter 15 may be found in "Air Pollution Aspects of Odorous Compounds" by R. J. Sullivan, Litton Systems, Bethesda, Md., Sept. 1969, pp. 115-152

CHAPTER 16

PESTICIDES

Pesticides, or economic poisons, include a spectrum of chemicals used to control or destroy pests that cause economic losses or adverse human health effects. These chemicals can be grouped as insecticides, herbicides (weed and brush killers, defoliants, and desiccants), fungicides, rodenticides, ascaricides, nematocides, molluscacides, algaecides, repellents, attractants, and plant growth regulators. There are presently approximately 45,000 registered pesticide formulations using some 900 chemicals. These are employed in agriculture, forestry, food storage, urban sanitation, and the home.

The pesticides are meant to be toxic to certain forms of life--the pests which are the targets of control. However, only a small portion of the total quantity of the pesticide used affects the target pest; the remainder enters the environment as a contaminant and may affect other forms of life directly or indirectly. Depending upon the target of control, the application of a pesticide may contaminate soil, water, air, plants, animal life, and humans. Pesticides in the air may settle in other parts of the environment close to the site of application or be carried some distance by air currents. Pesticides in the soil may remain on the surface, be washed off by rain, be blown into the air, or be absorbed by plants. In water, pesticides may be absorbed by aquatic forms of life. Pesticides on plant surfaces may wash off into the soil or water or be ingested as a residue by animals or humans when these plants are utilized as a food. Pesticides applied directly to animals or humans may be washed off into the environment or be absorbed. Thus, a complex cycle exists for the distribution and fate of a pesticide in the environment; the extent of distribution is dependent upon the ease with which the specific pesticide is degraded and the volume and method of its use.

Effects on Humans

Pesticides are intended to be especially toxic only for certain forms of life: insecticides for insects, herbicides for plants, fungicides for fungi, etc. In general, however, these specific uses have proved to be more quantitative than qualitative. At a sufficiently high dosage, these materials can cause adverse effects in humans as well. Pesticides can enter the body by ingestion, absorption through the intact skin, or by inhalation; and if the dosage is sufficiently high for each means of exposure, the effects will be the same.

Cases of poisoning, both accidental and occupational, have been reported for practically every known insecticide in all countries where they are used.[24] In general, chemicals given at equivalent dosages are absorbed more rapidly and more completely through the respiratory tract than through the skin. In cases of accidental occupational poisonings, it has usually been impossible to determine if the exposure was predominantly respiratory or dermal.[27]

Insecticide Effects and Toxicities

The chlorinated hydrocarbons and the organophosphates are the two most important groups of synthetic organic insecticides because of their wide use and toxicity to humans. On the average, the acute toxicity of the organophosphate compounds is somewhat greater than that of the chlorinated hydrocarbon compounds. However, the chlorinated hydrocarbons, because of their greater stability, may present a greater residue problem than do the organophosphates.

As a group, the chlorinated hydrocarbons are more chemically stable in the environment than are the organophosphates. They are not easily biodegradable and persist for long periods of time. When absorbed into the body, some of the chlorinated hydrocarbons are not metabolized rapidly but are stored in the fat tissue. DDT, BHC, and dieldrin have been found in the body fat of people in the general population whose only significant exposure to these insecticides has been through foods.[41] The consequences of such storage are presently under extensive investigation by many researchers. The organophosphates do not present a great residue problem in the body since they are metabolized much more rapidly than the chlorinated hydrocarbons.

Effects on Animals

Pesticides are generally introduced into the environment of commercial and domestic animals by direct treatment of the animal for pest control and by pesticide residues in feed. Usually, such animal exposures are carefully controlled and no ill effects are noted. However, accidents have occurred, and Street[90] cites a personal communication from Radeleff in 1963 stating that pesticide accidents account for 0.5 to 1.0 percent of all domestic animal losses from disease. These acute poisoning accidents usually were due to carelessness or misuse of the pesticides and involved the more toxic organophosphates in addition to endrin and dieldrin. As in humans, chronic effects have been difficult to identify in domestic and wild mammals.[90] The chlorinated hydrocarbons generally can penetrate the skin if applied in an oil solution or emulsion, and dieldrin can be absorbed in the dry powder form. Except for methoxychlor, all are stored in the body fat at various levels.[18] Again as with humans, the significance of this storage on health is not completely understood.

The presence of DDT and other stored pesticide residues in the secreted milk of dairy animals has been of concern since it relates to humans and it also provides a method for study of these residues. In addition to DDT, small amounts of BHC, chlordane, and dieldrin are also in the secreted milk. Only insignificant amounts of Toxaphene are present, and practically no methoxychlor. After ingestion of as little as 7 to 8 ppm of DDT on hay (a normal residue following spraying), approximately 3 ppm will be secreted in cow's milk, and butter made from such milk will contain about 65 ppm.[108] Witt _et al_.[108] in their studies found that DDT present in cow's milk in Arizona fluctuated seasonally with a low of about 1.0 ppm in the milk fat in the late spring and early summer and a high of 3.0 to 3.5 ppm in the late fall and early winter. This fluctuation was correlated with ingestion of residue in the feed at the higher dosage levels but not at the lower levels. They also studied the relative importance of respiratory (intratracheal) exposure of the cows as compared with other means of exposure to DDT. Their results showed that intratracheal exposure resulted in slightly less total excretion of DDT and its metabolites in milk than either of the two alimentary modes of exposure.

Fowl, fish, and many lower forms of wildlife have been extremely susceptible to low dosages of DDT and related insecticides. DDT residues in the bodies of such animals have been reported in all parts of the world, including penguins in the Antarctic.[83] Birds are particularly affected by such residues because they interfere with calcium deposition in eggs. Thin-shelled eggs are laid thus causing a loss in reproduction.

Cattle have been poisoned by drifting dust of 1 percent TEPP. Three incidences occurring in Washington in August 1963 were observed and documented.[73] Each involved the drift of TEPP dust from a hop vineyard to adjacent areas, and a prolonged static condition of the air resulted in an extra high exposure of the animals. In one instance, 15 cattle showed symptons of staggering, gasping for air, drooling, and some degree of cyanosis; two young cows died within the second hour after onset of the symptoms. Two geese, a cat, and chickens in the same dust cloud were not harmed. In another incident, a heifer in an open barn began coughing and developed symptoms of respiratory stress within 30 minutes of initial exposure. It was given one-fourth grain of atropine and was normal by the next morning. In the third incident, 17 head of cattle became severely ill after exposure to TEPP dust, with symptoms of diarrhea, excessive urination, wobbly gait, and loss of appetite. All recovered without special treatment.

The phenoxyacetic acid herbicide derivatives (2,4-D and 2,4,5-T) are comparatively harmless to most animals. All of the alleged cases of herbicidal poisoning of livestock and wildlife have been diagnosed as due to things other than herbicides. Inhalation of these herbicide dusts and sprays is relatively harmless, and percutaneous absorption is negligible. Even when administered to cows in large doses by ingestion, 2,4-D has not been found in the secreted milk. In cattle, a daily dose rate of 3,000 μg/kg body weight of 2,4-D or 2,4,5-T can be tolerated for long periods. However, at higher dosages of 100,000 μg/kg, pathological changes occur if feeding is continued for over a week, and at repeated doses of 1 gm/kg, illness and death result. Horses, cattle, sheep, pigs, and chickens have foraged on pastures freshly sprayed with 2,4-D or 2,4,5-T at two to four times the normal application with no ill effects.

Exposure of poultry to abnormally high residues of these herbicides led to reduced egg production but did not affect fertility or hatchability.[18]

The main hazard to livestock from fungicides is likely to arise from their use as seed-dressings for the protection of stored grain, potatoes, and other crops. However, the newer fungicides such as dodine, zineb, and maneb appear to present little toxic hazard.[18]

Effects on Plants

Many of the insecticides and fungicides have been developed for use against pests on or near plant life. Therefore, these materials generally have no adverse effect on plant growth. Herbicides, however, have been developed specifically to prevent the growth of plants. They can be classed into two categories, selective (affecting only certain types of plants) and nonselective (affecting all types of plants). However, there are only quantitative differences for many herbicides in the two categories; many selective herbicides, when used in higher dosages, become nonselective.[66] Many of the insecticides, although they do not affect the plant *per se*, have affected plant taste and flavor when used for food. Also, translocation of insecticides into the plant from the soil and the plant surface has been observed and studied.

Effect on Materials

There have been no reports of damage to inanimate materials from the pesticides *per se*, but some of the solvents used in spray application could have a damaging effect on paint and other surfaces.

ABATEMENT

The abatement of pesticidal contamination of the ambient air is a complex problem but it is being attempted. It appears that the contamination arising from the production processes can be controlled. Although incidences of occupational poisonings have been reported, proper protective measures are available. Precautions similar to those used in general chemical industries are taken to prevent the dusts and fumes from leaving the production plant into the outside environment. Bag packers, barrel fillers, blenders, mixing tanks, and grinding operations are generally completely enclosed or hooded and the air is vented through baghouses or cyclone separators. Similar control procedures are used when liquids are involved; liquid scrubbers are used however, instead of baghouses.[5,23] Although Tabor[95] has monitored the air near a formulating plant and found air levels similar to that observed in earlier agricultural samplings in the same area, too little air monitoring data are available at the present time to properly evaluate the production air control measures.

The physical state of the pesticide is quite important in drift. The drift potential from pesticide dust is very high because of particle size. Dust materials are generally screened to incorporate only particles ranging from 1 to 25 μ in size; on the average, 80 to 90 percent of the particles in the formulation are under 25 μ. Spray droplets of 50 u in size show less drift than dusts of smaller particle size. Therefore, the use of dusts has been decreasing in recent years, and the Federal Government has banned the use of 2,4-D dust.[6]

Evaporating the spray droplet in spraying operations has been considered as a drift control measure. Water is the most frequently used diluent because of its availability, its low cost, and its nonphytotoxic effects on agricultural crops. However, water droplets may experience considerable evaporation and reduction in size while airborne in the spray, resulting in a sufficiently smaller droplet size and therefore more drift than had been anticipated. As a control measure, oils which are relatively nonevaporative are frequently used as the pesticide carrier in sprays. They are especially used as carriers for low volume (1 to 3 gal/acre) applications of very fine sprays for forests and range lands.[6]

Improvement in delivery equipment and techniques is another partial solution to the abatement problem. Equipment has been developed for producing invert emulsions (water-in-oil emulsions) of which greater than 90 percent is in the water phase. In contrast to conventional oil-in-water emulsions, preparations containing a high water content are quite viscous. When using the invert emulsions, sprays consisting of large droplets can be delivered aerially, minimizing both drift and evaporation. Trials indicated that smaller quantities of pesticide could be used in the invert emulsion with equivalent results in terms of insect kill or herbicidal efficiency. At the same time, the accuracy of delivery was improved so that the invert emulsions could be applied under more adverse meteorological conditions than conventional sprays.[33]

Various spray nozzles have been designed and used with varied pesticide formulations having different viscosity, density, and surface tension in attempts to control drift during application. Additional factors such as the angle of the nozzle with airstream or the use of screens or discs at the nozzle also contribute to the characteristic of the spray.

Meteorological conditions are extremely important parameters that are considered in the application of pesticides and the control of potential drifts. Wind direction and velocity, humidity and temperature at ground and higher levels, and the amount of sunshine or rain are all interrelated factors that are considered. Because of the importance of such meteorological information, the U.S. Weather Bureau provides this specialized data as part of their service. Dusting and spraying advisories are sent out on a teletypewriter circuit 24 hours a day emphasizing local weather conditions for aerial and ground applications for various agricultural chemicals. In addition, these advisories include information relating certain insect and other pest activities with weather conditions, so that pesticides can be applied at the proper time to produce maximum pest control.[81]

Volatilization of pesticides into the air from soil, water, plants, and other treated surfaces is known to occur. However, the control of this is complicated by the fact that the extent to which volatilization occurs is not known. Some control over volatilization from soil can be effected by the use of cover crops. It has been observed that two to three times more insecticidal residues were recovered from alfalfa-covered plots than from fallow ones.[56]

ECONOMICS

The economic and social benefits gained by the use of pesticides are great, as noted in a number of discussions. The President's Science Advisory Committee[100] reported that modern agricultural efficiency is maintained only through the use of pesticides. It has been estimated that without the use of pesticides, 50 percent of the agricultural and forestry crops of California would be destroyed, and that the present loss in spite of pesticides is 21 percent.[75] In the United States, agricultural losses from insect damage have been estimated at approximately 3.5 billion dollars, losses from plant diseases at 2.9 billion dollars, and losses from weeds at 3.7 billion dollars; the annual total worldwide damage by pests to agriculture has been estimated to be about 80 billion dollars.[62] In addition, pesticides have contributed to the eradication or reduction of a number of human diseases such as malaria, typhus, and yellow fever in less developed countries of the world.[44,100] However, the estimate for worldwide malaria is still placed at about 200 million cases and 2.5 million deaths a year.[62]

Insect-borne diseases have been highly prevalent in the past in the United States. However, largely because of the introduction of DDT and other synthetic organic insecticides, the number of deaths and number of reported cases for some of these diseases have declined considerably. Malaria at one time had a severe economic impact on the Southern States in the United States, but today it is almost nonexistent.[44]

The economic benefits derived from the use of pesticides by agriculture in the United States have been quite significant. These benefits have been not only in increased yields of production, but also in increased quality of the product. In many cases, the improvement in quality has been such that a high percentage of the crop would not have been otherwise marketable. In California, for example, in crops not treated with insecticides the percent of wormy fruit in 1956 was 21 to 23 percent, but it was only 0.5 percent following the use of guthion.[44]

Although the value may vary for different crops and regions of the country, it has been estimated that nationally about five dollars are saved for every dollar invested in chemical pesticide usage. The pesticide cost includes research, development, and price of the material.[75]

The costs of pesticide damage to wildlife resulting from air pollution is impossible to estimate. The direct effects of such pollution on wildlife, if compared with those on livestock, are probably small. However, the indirect effects on the entire ecosystem resulting from the presence of pesticides in the air might be greater. However, at the present time, insufficient information exists regarding the translocation of pesticides in the ecosystem to make any valid cost estimates or conclusions.

The economic costs of pesticide contamination of the air with respect to human health have not been estimated. While there have been reported episodes of death or illnesses resulting from respiratory exposure to pesticides, it is not known how many unreported illnesses may occur. The percentage of the body burden of DDT resulting directly from respiratory exposure probably is low (Campbell et al.[16] have estimated it to be about 0.06 percent) but the percentage that is indirectly due to overall environmental pesticide pollution is not known.

Furthermore, outside of those costs estimated for agricultural damage, there appears to be no cost figures available for controlling air pollution by pesticides from manufacturing or formulating plants.

REFERENCES

The publications referred to in Chapter 16 may be found in "Air Pollution Aspects of Pesticides" by H. Finkelstein, Litton Systems, Bethesda, Md. Sept. 1969, pp.78-87

CHAPTER 17

MEASURING AIR QUALITY

General. Some of the purposes for which an air quality monitoring program may be undertaken are as follows:

a. To determine the nature and extent of air pollution.

b. To relate air pollution levels to factors such as topography, population density, industrial activity, climate, and meteorological conditions.

c. To relate air pollution levels to human health effects, plant and animal damage, property damage, and other economic and esthetic losses.

d. To determine the effectiveness of control programs.

e. To determine air pollution sources that require the greatest amount of curtailment. In any evaluation of air quality, techniques of sampling will vary according to the putpose intended. The purpose will determine site selection, number of samples, frequency of sampling and the duration and time periods selected for sampling. Before undertaking air quality sampling, the agency to provide analytical services should be contacted regarding instrumentation, analytical techniques and other aspects of the sampling program.

f. To compare with a standard.

Measurement of Particulate Pollutants

a. General:

(1) For the purposes of this discussion particulate pollutants will be classified as either settleable or suspended particulate matter.

(2) The first thing that should be considered when sampling for atmospheric particulate pollutants is the type of analyses to be performed on the collected material. Some types of analyses that could be conducted on collected particulate matter are gravimetric, microscopic, chemical (organic and inorganic), spectrographic and radiological. In order to coordinate analytical and sampling procedures, one must understand the methodology and limitations of each type of analysis in addition to the interferences and limitations of the sampling procedure.

b. Settleable Particulate Matter (Dustfall):

(1) *Equipment Description.* The dustfall jar is an apparatus which collects atmospheric particulates. It is nothing more than an open mouthed container usually constructed of polyethelene, glass or some other inert material.

(2) *Operation:*

(a) Large particles settle quite rapidly under the influence of gravitational forces, depositing themselves upon some horizontal surface. The dustfall jar takes advantage of this settling phenomenon to collect the particles. Due to atmospheric disturbances, very small particles have a difficult time reaching the ground. This means than the vast majority of particles collected by a sedimentation method will be large particles.

(b) The dustfall jar may be used with a liquid, usually water, or dry. The introduction of water into the dustfall jar introduces difficulties not encountered in dry operation. During the winter, an antifreeze must be added and during the summer an algae suppressant may be added if conditions warrant.

(c) *Analysis.* The easiest analysis that can be performed on the collected particles is a determination of the weight. The sampling period for this method is 30 days. Results are reported as weight per unit area per unit time (usually tons/square mile/month). A determination of the organic and inorganic composition of the material may be performed. However, if an antifreeze or algae suppressant was added during the sampling period, its composition must be considered in the chemical analyses.

(d) *Advantages.* The obvious advantages of the dustfall jar are its relatively low cost and ease of operation.

c. Suspended Particulate Matter:

(1) *High Volume Sampler:*

(a) *Equipment description.* The high volume sampler, more commonly known as the "Hi-Vol," is a vacuum cleaner motor which has been modified to take filter paper.

(b) *Operation.* Average sampling rates of Hi-Vols range from 40 to 60 cfm. with a normal sampling period of 24 hours. Due to the high flow rates and the length of the sampling period, large quantities of particulates are collected. The Hi-Vol collects most of the fine particulates and some of the coarse particulates.

(c) *Analysis.* The usual analyses performed on samples collected by this method are a mass determination and chemical analyses, particularly organic analyses. The microscope can be used in identifying the collected particles. Results of both mass and chemical determinations are expressed in units of mass per volume (usually micrograms per cubic meter) and represent an average consideration.

Sample Calculation of Mass Determination
Weight of filter:
before—3.182 gm
after —3.550 gm
Flow rate:
start —64 cfm
finish —56 cfm
Time:
start —3 PM 2-26-63
finish —3 PM 2-27-63

$$\text{Average sampling rate} = \frac{64+56}{2} = 60 \text{ cfm}$$

$$\text{Sample volume} = \frac{60 \text{ ft}^3}{\text{min.}} \times 1440 \text{ min.}$$

$$\times 0.0283 \frac{m^3}{\text{ft}^3} = 2450 m^3$$

$$\text{Mass of collected material} = 3.550 - 3.182 = 0.368 \text{ gm}$$

$$\text{Concentration} = \frac{0.368 \text{ gm}}{2450 m^3} = 150 \times 10^{-6} \text{gm}/m^3 = 150 ug/m^3$$

(d) *Limitations.* The interferences of the filtering medium must be taken into account in a chemical analysis. A complete inorganic analysis is difficult due to the presence of many inorganic compounds (iron, aluminum, magnesium, sodium, and potassium) present in the filter material. However, if the exact amounts of these compounds present in the filter are known, a complete inorganic analysis is possible.

(2) *Tape Sampler:*

(a) *Equipment description.* The tape sampler has a pump, a filtering mechanism, and a timing device. The timing device, is the unique feature of the tape sampler. It allows the sampling period to be varied automatically from about 5 minutes to 3 hours. This makes it possible to more accurately measure peak periods of pollution instead of an average value over a long sampling period.

(b) *Operation.* Air is pulled through the filter tape at a rate of 0.1 to 0.5 cfm. for a pre-selected period of time. At the end of this sampling period, the tape is automatically advanced to a new spot. The tape sampler collects the fine particles. It does not collect particles as large as those collected with a Hi-Vol.

(c) *Analysis.* The usual analysis of samples collected with the tape sampler is a measurement of the amount of light that is absorbed by the particles. This type of analysis is performed by passing a beam of light through a clean portion of the tape and is directed upon a photoelectric cell which registers the intensity of the light beams. The intensity of the light is adjusted until a reading of 100% light transmittance is attained. The portion of the tape with the sample is then placed between the light source and the photoelectric cell with the corresponding intensity being noted. It will always be less than 100 because some of the light will be absorbed by the particulate material on the filter.

The results of this type of analysis are reported in *cohs per 1000 lineal feet.* The coh (coefficient of haze) is a measure of light absorption by particles and is defined as that quantity of light scattering solids producing an optical density of 0.01 when measured by light transmittance. Optical density is defined as follows:

$$\text{optical density} = \log \frac{I_{100}}{I_{meas}}$$

where:

I_{100} is the intensity reading *without* interfering particles present

meas is the intensity reading *with* interfering particles present (always less than I_{100})

Sample Calculation
average sampling rate = 0.3 cfm
sampling time = 100 min.
I meas = 50
Diameter of tap shot = 1.0 in.

(d) *Limitations.* Due to the small quantity of material collected with a tape sampler, other types of analyses, particularly chemical, are not usually conducted on samples.

d. Special Techniques. Equipment which use membrane filters can be classified as special technique because due to operational difficulties, they are limited in application. The unique feature of this equipment is the filtering medium. These filters are made of cellulosic esters and have very tiny pore openings. This results in high efficiencies of collection of small particles and surface deposition of the particles. The surface deposition of particles permits complete microscopic analyses. The membrane filter offers high resistance to the passage of air and can be used only for the collection of small quantities of pollutant.

Measurement of Gaseous Pollutants:

a. Sampling Methods. The methods normally

$$\text{Volume sampled} = 100 \text{ min.} \times 0.3 \frac{ft^3}{min} = 30 \text{ ft}^3$$

$$\text{O.D.} = \log \frac{I_{100}}{I_{mean}} = \log \frac{100}{50} = \log 2 = 0.301$$

$$\text{cohs} = \frac{0.301}{0.01} = 30.1 \text{ cohs}$$

The entire volume of air sampled could be visualized as being contained in a cylinder whose cross-sectional area was that of the spot on the tape and some height.

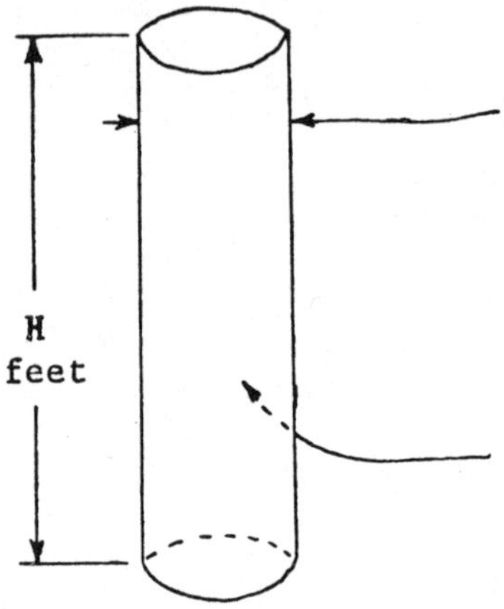

1.0 in. $$H = \frac{Vol}{A} = \frac{30 \text{ ft}^3}{\left[\frac{\pi \times 1}{4}\right] \text{in}^2 \left[\frac{1 \text{ ft}^2}{144 \text{ in}^2}\right]}$$

$$H = 5500 \text{ ft} = 5.5 \text{ (1000 ft)}$$

Vol. = 30 ft^3 $$\text{cohs/1000 lineal feet} = \frac{30.1}{5.5}$$

$$\text{cohs/1000 lineal feet} = 5.48$$

used for sampling gases in the atmosphere are absorption, adsorption, and freeze out. There may be some instances where grab sampling techniques may be applicable. Absorption employs impingers or bubblers filled with the media appropriate for the pollutant being sampled. Adsorption involves collecting the gas on the surface of solid materials such as activated charcoal, silica gel, activated alumina, or molecular sieve. Freeze out techniques involve condensation of the gaseous pollutants using flasks cooled by ice and sale (−16° C), dry ice and acetone (−80° C), liquid air (−147° C) or liquid nitrogen (−195° C).

b. **Recommended Analytical Methods for Specific Pollutants:**

(1) *Sulfur Dioxide* (*West and Gaeke Method*) The West and Gaeke method is applicable to the determination of sulfur dioxide in outside ambient air in the concentration range from about 0.005 to 5 parts per million (ppm). Sulfur dioxide in the air sample is absorbed in 0.1 M sodium tetrachloromercurate. Nonvolatile dichlorosulfitomercurate ion is formed in this process. Addition of acid-bleached pararosaniline and formaldehyde to the complex ion produces red-purple pararosaniline methylsulfonic acid, which is determined spectrophotometrically. The system obeys Beer's Law up to about 10 l of sulfur dioxide per 10 ml of absorbing solution. This method is more sensitive than the hydrogen peroxide method and is not subject to interference from other acidic or basic gases or solids such as SO_3, H_2SO_4, NH_3, or CaO; however, the analysis should be completed within 1 week after sample collection, and the concentrations of ozone and nitrogen dioxide should be less than that of the sulfur dioxide.

(2) *Nitrogen Dioxide and Nitric Oxide* (*Saltzman Method*). The Saltzman method is intended for the manul determination of nitrogen dioxide in the atmosphere in the range of a few parts per billion (ppb) to about 5 ppm. Sampling is conducted in fritted bubblers. The method is also applicable to the determination of nitric oxide after it is converted to an equivalent amount of nitrogen dioxide by passage through a permanganate bubbler. Concentrations of both gases of 5 to 100 ppm may be sampled in evacuated bottles. The nitrogen dioxide is absorbed in Griess-Saltzman reagent. A stable pink color is produced within 15 minutes and may be read visually or in an appropriate instrument. Only slight interfering effects occur from other gases.

(3) *Oxidants, Including Ozone* (*Alkaline Potassium Iodine Method*). This method is intended for the manual determination of oxidants (including ozone) in the range of a few pphm to about 20 ppm. Ozone, chlorine, hydrogen

peroxide, organic peroxides, and various other oxidants will liberate iodine by this method. The response to nitrogen dioxide is limited to 10% by the use of sulfamic acid in the procedure to destroy nitrate. It is customary, for convenience, to express the results as ozone. The advantage of this procedure over a neutral iodide procedure is that a delay is permissible between sampling and completion of the analysis. Sampling is conducted in midget impingers containing 1% potassium iodide in 1 N sodium hydroxide. A stable product is formed that can be stored with little loss for several days. The analysis is completed in a laboratory by addition of phosphoricsulfamic acid reagent, which liberates the iodine. The yellow iodine color is read in an appropriate instrument. Serious interfering effects occur from reducing gases and dusts.

CHAPTER 18

AIR QUALITY MONITORING INSTRUMENTATION

· GAS AND VAPOR ANALYSES

Gases and vapors probably represent the largest fraction of total pollutants in the atmosphere. Hydrocarbons from automobiles and sulfur oxides from thermal power generation are responsible for much atmospheric pollution. Local concentrations of materials that are hazardous to health probably represent the most serious monitoring problem because the levels at which health hazards occur are not known for specific pollutants. The analytical methods must be capable of providing meaningful data for gas concentrations that may be in the parts-per-billion range and for a variety of materials.

Wet-Chemical Analysis

Some of the most common pollutants are analyzed by wet-chemical analysis. Sulfur dioxide (SO_2) is usually analyzed by a colorimetric method known as the West-Gaeke method. A color change is observed with a photometer. Concentrations to 0.01 ppm can be analyzed, but interferences sometimes occur from nitrogen oxides. A minimum gas sample of 2.5 liters is usually required, and an analytical time allowing hourly sampling intervals can be obtained. An analog readout is delivered from the photometer. An automatic colorimetric device that requires only periodic addition of reagents can be purchased for approximately $5000 (ref. 1).

SO_2 in ambient air can be titrimetrically determined in a concentration range from about 0.01 to 10 ppm by oxidizing the SO_2 in hydrogen peroxide (H_2O_2) to sulfuric acid (H_2SO_4) and titrating. A sample size of approximately 1 cubic foot of air is required. Analytical time is approximately 10 minutes, and manual operation is involved (ref. 2).

Automated conductometric analysis is often used for SO_2 measurements from the atmosphere. In this method SO_2 is collected in an acid solution and oxidized to H_2SO_4 by H_2O_2. The method is sensitive in the range of 0.01 to 2 ppm. Materials that form strong electrolytes in solution interfere. The sample size is approximately 10 cubic feet of air and the analytical time is approximately 10 minutes. The data readout is analog, and the cost of a commercially available instrument is approximately \$5000 (ref. 3). There are many additional methods of measuring SO_2, but the above are the major analytical methods for this gas.

Hydrogen sulfide (H_2S) can also be analyzed by a wet-chemical colorimetric method. Methylene blue is produced by a reaction with suitable reagents, and concentration of H_2S is determined by measurements of optical absorption (ref. 4). Sensitivities in the parts-per-billion range have been reported. Sample sizes of some 30 cubic feet are required with an analytical time of approximately 2 hours. Data are obtained by photometric readout; although analog data are obtained, this system has not been successfully adapted for automatic instrumentation.

The existence of nitrogen oxides is determined by dye formation using the nitrite ion (NO_2^-) as part of the reaction (ref. 5). The sensitivity of this method ranges from a few parts per billion to about 5 ppm. Sample sizes up to ½ cubic foot produce sufficient color change for meaningful results. The analytical time is approximately 1 hour. Photometric readout is obtained, but no automated instruments have yet been developed for this procedure.

Oxidizing gases, such as ozone, are analyzed by observing color change after absorption in potassium iodide (KI) (ref. 6). The oxidants can be analyzed in the range of a few parts per hundred million to about 10 ppm. A sample size of approximately 2 cubic feet of air is suitable, and analytical time of up to 1 hour is required. The photometric readout produces analog data, but the overall procedure is manual.

Hydrogen fluoride (HF) and other volatile fluorides are analyzed colorimetrically after titration or spectrophotometrically (ref. 7). Concentrations below 1/100 μg/cu ft can be analyzed. An air-sample size of approximately 10 cubic feet is the minimum and, in some cases, 100 cubic feet is required. Analytical time is approximately 7 hours because a detailed procedure is required to eliminate interferences that are caused by a large variety of materials. These are removed by double distillation with perchloric acid ($HClO_4$). Analog data are produced by the photometer or colorimeter, but the method is manual.

A number of materials including SO_2, nitrogen oxides, oxidants, chlorine (Cl_2) or chlorides, and fluorides or cyanides may be measured by modified wet-chemical procedures in a system of gas absorption followed by colorimetric analysis. Continuous analyzers using programed sampling valves, proportioning pumps for reagent delivery, heating baths, distillation columns, temperature control, colorimeters, and so forth, can be purchased (ref. 8). Sensitivities and sample sizes for these colorimetric instruments are equivalent to those given for the manual methods, and analytical time is of the same order as that for the manual methods. Reliability, of course, is much less, and analog data are produced for the colorimetric readout. The cost for a complete system capable of analyzing a number of gases may run as high as \$70 000 to \$80 000.

Most organic gases are analyzed by instrumental methods. Volatile organic substances containing oxygen are usually analyzed by colorimetric methods. Analyses are reported, for example, as "total aldehydes," in which a number of individual aldehydes and ketones may be present (ref. 9). Sensitivities of several parts per million are reported with sample sizes of a few cubic feet. Analytical time is of the order of 1 hour with analog readout using colorimetric analysis.

Spectrographic Analytical Method

Spectrographic and spectroscopic analytical methods are used primarily for particulate materials that have been collected and can be placed in a spectrograph or an emission spectrometer. A number of gases and vapors, however, can be analyzed by such methods as emission and fluorescence spectroscopy. An example of one of these methods is the use of detector elements for chromatographic columns; these elements utilize flame emission to characterize the materials passing through the column.

Organic vapors can be detected by a flame-ionization analyzer (ref. 10). The flame-ionization analyzer responds in proportion to the number of carbon atoms in a gas sample. Sensitivity below 1 ppm can be obtained depending on the gas being analyzed. Sample sizes down in the cubic-centimeter range can be used, and analytical time is essentially zero with real-time data production. The type of data produced is analog. Depending on the gas concentration and type of analyzer, costs in the thousands-of-dollars range can be expected.

Both organic and inorganic vapors have been analyzed by mass spectroscopy (ref. 11). Monopole, quadrupole, and time-of-flight instruments have been used for gas analysis. Concentrations to several parts per million can be analyzed. Sample sizes of a few hundred cubic centimeters are adequate, and analytical time is of the order of 5 to 20 minutes, depending on the complexity of the analysis desired. An instrument can be set to scan for mass numbers from one to several thousand. The scan rate is usually of the order of 20 minutes to 1 hour, and printout is digital. The cost of such an instrument can be several tens of thousands of dollars.

The infrared absorption characteristics of several gases and vapors make it possible to detect and analyze their presence in the atmosphere with continuous analyzer systems. CO can be analyzed with a sensitivity of approximately 1 ppm in a nondispersive infrared analyzer (ref. 12). A sample size of approximately one-half cubic foot can be used, and analytical time is of the order of seconds. Data produced are analog, and the cost for a typical analyzer is several thousand dollars. Similar instruments of the same design have been used for analysis of organic materials. Dispersive infrared spectrometers have also been used to determine hydrocarbon emissions based on absorption at the 3.4-micro wavelength (ref. 13). Interferences occur that can be avoided in the nondispersive infrared-sensitized systems, but other characteristics of this analytical procedure are the same as for a nondispersive infrared analyzer.

A third type of infrared analysis has been recently described in which correlation spectroscopy is used (ref. 14). In this system, a specific compound is analyzed by comparing the infrared spectrum of an unknown gas to the infrared spectrum of a known material. Sensitivities of approximately 20 ppm are claimed. The device has the advantage of being suitable for use as a remote sensor using the Sun or other radiant-energy source to pass energy through a particular pollutant plume. Recently a passive infrared-analysis system has been studied in which a heated stack plume emitted enough infrared energy over the background so that it served as its own infrared energy source.

Ultraviolet absorption can be used to detect unsaturated hydrocarbons. A recent development of a Geiger counter for detection of low energy photons has permitted development of a field analyzer that is capable of detecting hydrocarbons in the atmosphere. The sensitivity of this device is 2 ppm. Sample flow rates of several hundred cubic centimeters per minute are used, and real-time analysis is possible. Analog data are produced, and the cost of the device is in the $1000 range.

Thermal Conductivity Measurement

The concentrations of several gases can be determined by measuring the thermal conductivity of the sample gas as compared to that of a known reference gas. Four filaments, arranged as an electrical resistance bridge, are heated, and imbalance caused by changing gas compositions is reported as a temperature difference. Sensitivity to approximately 0.1 percent can be obtained with a gas-flow rate of approximately 1/4 cu ft/min. Analytical time in the minutes range is normal for data in terms of relative concentrations based on a calibrating or standard gas. Cost for such an instrument is approximately $500.

Chromatographic Analysis

Chromatographic analysis is probably the most generally used method of measuring and detecting gases in the atmosphere. It is capable of separating, in suitable columns, essentially all of the polluting gases, both organic and inorganic, that may be present in the atmosphere. In brief, the method utilizes a column through which a mixture of gases is passed. The difference in the mobility or the diffusion rate of the component gases in the mixture passing through the column is a function of the molecular structure of each gas. Selection of a suitable detector from the six or more available detecting devices is based on the composition of the gases to be analyzed. Detectors measuring changes in thermal conductivity have been used along with detection devices involving flame-ionization and electron-capture methods.

In recent years a vast amount of work has been done on gas chromatography for air pollution analysis. A meaningful discussion of all the results of this work is beyond the scope of this survey. To summarize briefly, however, sensitivities in the parts-per-billion range can be obtained, and sample sizes from a few cubic centimeters to several cubic meters of gas are usable. Fractionating freezeout methods along with representative absorption sampling can be used to obtain gas samples for injection into the chromatographic column. The analytical time is normally of the order of 20 minutes. The data obtained are digital. The cost of a usable gas chromatographic system ranges from several thousand to several tens of thousands of dollars, depending on the number of columns, degree of automation, and the readout devices used.

GAS AND VAPOR CONTROL METHODS

There are four principal methods of controlling the emissions of vapors to the atmosphere. These are absorption, adsorption, incineration, and catalytic combustion.

Absorption

Absorption is a basic chemical engineering unit operation, which is frequently referred to in the air-pollution-control field as "scrubbing." It is a diffusional process involving the transfer of gas molecules to a liquid phase. Concentration differentials at the interfaces serve as driving forces, and the process is favored by high-interfacial surface areas, turbulence, and high diffusional coefficients. Because some soluble component of a gas mixture is to be dissolved in a liquid, the absorbent chosen should have high solubility for, or react irreversibly with, the absorbed gas.

Equipment used for the absorption of gases can be placed in five categories. These are packed tower, plate tower, spray chamber, venturi scrubber, and jet scrubber.

In a packed tower, crushed rock, Raschig rings, burl saddles, and various other materials and shapes are used for packing. The liquid is usually introduced at the top of the packing and trickles through the layers of packing that provide the necessary large surface area. The contaminated gas flows upward through the packing, contacts the absorbing liquid, and leaves cleansed at the top of the packed tower. The liquid commonly used is water, but solutions of neutralizing chemicals, organic solvents, oils, and emulsifiers are also employed.

A plate tower is usually cylindrical with several sections of perforated horizontal plates; the two most common are termed "bubble caps" and "sieves." In a plate tower the gas-liquid contact takes place in a series of steps (at the plates) rather than continuously as in a packed tower.

The simplest spray chamber consists of a cylinder in which spray nozzles are mounted in the upper section and droplets are forced downward by a liquid flowing down the interior walls of the chamber. Although the liquid and gas flows may be cocurrent, the gas stream is usually introduced tangentially at the lower section of the tower, so that a spiral motion is imparted to it, and the liquid sprays are directed radially. In this way, the spray is directed centrifugally to the wall from which it drains as a film flow. Simple spray towers can be either countercurrent or cocurrent in operation, have a relatively low pressure drop, and are not subject to plugging, as are packed columns and plate towers. Some spray towers have been successful in the absorption of relatively insoluble gases; but because the mixing of spray and gas is not as vigorous in the countercurrent towers as in other equipment, such as a venturi scrubber, spray towers are perhaps best suited to applications involving easily absorbed gases.

The venturi scrubber is essentially a spray tower in which an effort is made to maximize the mixing of the gas stream passing through the main section of the venturi by introducing a liquid at the throat of the venturi. In the venturi scrubber the gas stream passes at a high velocity through the venturi, with the scrubbing liquid being introduced at relatively low pressure at the venturi throat where the liquid is sheared and droplets are formed. Contacting is essentially cocurrent with the flow energies being separately supplied for the gas and for the liquid. The energy required for forming droplets and for intimate mixing with the gas is furnished by the gas stream. The liquid is introduced at conventional line pressures. The resultant liquid spray mixes with the gas, a degree of absorption takes place, and an entrainment separator—usually of the simple centrifugal or spray type—must be provided at the venturi discharge to remove liquid from the washed gases. The pressure drop across the venturi scrubber ranges from 10 to 100 inches of water with gas velocities at the throat exceeding 5000 ft/min. The high pressure drop in the gas stream results in a high power cost, and necessitates high tip speed on the main fan with an increase in problems of erosion, noise, or rotor imbalance.

The liquid jet scrubber is essentially an ejector designed as a fume scrubber. Energy is imparted to the liquid spray, which, through an aspiration effect, produces the suction necessary to move a waste gas and also succeeds through intermixing or surface-area contact in removing an amount of absorbable gases. The capacity of such scrubbing equipment can be varied by changing the liquid flow rate or designed characteristics at the venturi. A droplet entrainment is required at the discharge end. In such scrubbers, water at about 100 lb/in. generates pressure differentials of the order of a few inches of water, and gas-flow rates do not usually exceed 1000 cubic feet min/unit scrubber.

Generally, the two-film absorption theory is used for the design of absorbers and interpretation of performance. In this design approach, the assumptions are that (1) the transfer rate is proportional to concentration differentials at the interfaces, (2) equilibrium conditions exist between the gas and liquid at the interface, and (3) no holdup or interfacial resistance to flow exists.

In general, packed towers are prepared for small installations, for use with corrosive gases, high-forming liquids and high liquid/gas ratios, and for low-pressure-drop requirements. Plate columns are frequently more economical than packed towers because their ability to tolerate higher gas velocities allows use of smaller diameter columns. They can handle liquor rates that would flood typical packing, are less susceptible to plugging, and can be equipped to cool the liquor to remove heat of dilution, either with cooling coils on the plates or with external coolers. Spray towers are of particular importance when the pressure drop is a major consideration and when dust particles are present in the waste-gas stream.

Adsorption

In atmospheric pollution control, the principle of adsorption is employed primarily to prevent highly odorous or offensive organic vapors from escaping into populated areas. Adsorption systems are widely used in industry to remove gases and vapors that can produce human irritants, obnoxious odors, fire hazards, explosive reactions, unhealthy conditions, hazardous concentrations, unsafe operations, undesirable products, corrosive atmospheres, and inadequate processing. Solvent recovery, odor control, removal of gaseous contaminants, protection against gas-warfare and hazardous chemical reactions are suggested candidates for adsorption systems. Food services, hospitals, laboratories, petrochemical manufacturers, chemical users, printing plants, fallout shelters, and atomic energy installations are some of the users of adsorption systems to prevent air pollution or to eliminate pollution of the premises and processes by intake air. Because adsorbents are versatile enough to solve these problems, the list of possible air pollution applications is extremely long.

In air pollution control, the principal use of adsorption is for the physical adsorption of gaseous matter on solids in dynamic systems. Physical adsorption is defined as the capture and retention of molecules from the gas phase by the surfaces and capillaries of solid absorbents. Physical adsorption is completely and easily reversible; the adsorbed material is removed by high temperature and/or low pressures. Thus, adsorption systems are capable of being regenerated.

Although a great variety of clays, chars, gels, oxides, silicates, and activated carbons have been used as sorbates, activated carbons have been the subject of the most effort and application in air pollution abatement. Because physical adsorption has basically been a surface phenomenon, the more surface area per unit volume that is obtained, the more effective the adsorbent. Adsorbents display a specificity, or differential affinity, for various materials. Fortunately, many of the specificity relationships have been established by experience in the field. For example, activated carbons specifically adsorb hydrocarbon vapors in preference to water vapor, while silica gel specifically adsorbs water in preference to hydrocarbons.

One-pass nonregenerative dynamic systems are used when average contaminant concentrations are of the order of a few parts per million or less. Although the systems are designed to take contaminant slugs up to 1000 ppm, it is not economical to use this type of adsorber on concentrated contaminant gases. Holding capacity in one-pass dynamic adsorber systems is generally 5 to 20 percent of adsorbent weight; that is, in the case of an easily adsorbed vapor, the contaminant breakthrough does not occur until total vapor pickup exceeds 20 percent of the adsorbent weight. Because one-pass nonregenerative adsorber elements usually contain highly active and relatively expensive adsorbents, the replacement cost or factory reactivation becomes burdensome if the system is used to remove high concentrations of contaminants.

One-pass regenerative adsorber systems are used when the concentration of contaminant exceeds a few parts per million. Two adsorber vessels are usually employed with one adsorbing while the other is being regenerated. In reactivation, low-pressure steam is normally used to raise the temperature of the adsorbent bed and drive off the adsorbate. Steam and adsorbate are usually condensed together for recovery or disposal. Steam also acts as a stripping vapor; because a portion of it condenses in the adsorbent, the bed temperature is reduced by evaporation of the steam during the adsorption cycle. Bed-temperature increases created by heat of adsorption must be dissipated when handling heavy concentrations of contaminants; otherwise the efficiency and/or holding capacity of the bed is seriously reduced.

The original equipment cost for adsorption systems, excluding the erection cost, is approximately $32.50 per pound of contaminant. Operating costs on a nonregenerative system include replacement or factory reactivation of the adsorbent, maintenance of the bed holders and frames, air movement, space occupied, transportation to and from the factory, and labor to remove and replace elements. Replacement or factory reactivation and factory maintenance charges on bed holders are estimated to be of the order of $10 per pound of contaminant retained. Assuming a typical bed pressure drop of 1.5 inches of water, 0.75 brake horsepower (hp) is required for every 1000 cu ft/min of capacity installed. For a regenerative system, additional equipment is required to generate steam, such as a cooling-water system, dry-air heaters, and additional fans or air pumps. Rough estimates of equipment capacities for the costs above are:

Equipment	Capacity
Steam generating	3 to 5 lb/lb of vapor retained
Cooling water system	7 to 10 gal/lb of vapor retained
Dry heat	0.10 to 0.15 kw hr/lb of vapor retained
Air moving	5.0 brake hp/1000 cu ft/min of air

Incineration of Gases and Vapors

Incineration is rapid, high-temperature oxidation in the gas phase. Contaminated gases and liquid wastes are incinerated for odor control, reduction in opacity of plumes, reduction in emission of reactive hydrocarbons, and reduction of explosion hazards.

Many of the waste materials disposed of by flame afterburners are liquid aerosols, and the fuel for the afterburner can be a liquid such as oil. In the afterburner the actual combustion occurs between oxygen and the vapors, or gaseous decomposition products, of the waste and fuel liquids. In the combustion of hydrocarbons, a complex series of chain reactions plays a predominant role. At the high temperature of flames, hydrocarbons disassociate into radicals that in turn recombine. There is evidence that the radicals—OH, HO_2, H, and O—produced in the oxidation of hydrogen are important chain carriers. Thus, aldehydes and other oxygenated organic compounds are present in the stepwise oxidation of hydrocarbons, and these intermediates will appear in the end products if anything disrupts the chain reactions and prevents their ultimate combustion to form CO_2 and H_2O. These oxygenated materials tend to be irritating and odorous and are unwelcome products in the effluent.

Important factors in the combustion process are reaction time; temperature, which increases reaction rate; and turbulence, which provides thorough mixing of oxygen, intermediate products, and chain carriers.

Flame combustion, as a method of air pollution control, has been applied to a variety of processes and effluents including flares and furnaces for disposal of waste gases and liquids from coffee roasters, smokehouses, rubber curing, cupola gases, rendering plants, varnish kettles, enamel baking ovens, asphalt blowing, and printing-press vents.

The yearly capital and operating costs for a flame incinerator on a continuous roaster is about $22 000. Approximately three-fourths of the cost is for fuel for the disposal of halogenated hydrocarbons. A flame incinerator capable of handling 4 to 7 gal/min of waste is reported to cost $250 000, with operating costs about 6 cents per gallon of waste. This cost is compared to 10 cents per gallon for disposal by drum burial.

Catalytic Combustion

Direct-flame incineration, as a method for the disposal of organic vapors and gases, is often limited in application because of the excessive cost of the high temperature required. The high-ignition-temperature requirement can be overcome by passing the waste gas stream through a catalyst bed, because the catalyst causes the reaction to occur below the normal combustion temperature and does not become part of the end product. The process lessens the temperature necessary for the exothermic reaction of organic vapors or gases with oxygen by as much as 500° F. The chemical union of oxygen with the organic fumes occurs without flame on the surface of the catalyst. The temperature rise of the waste gas stream after passage through the catalyst bed depends on the Btu value of the fume.

Many substances have catalytic properties and are used for this purpose. Metals of the platinum family are generally chosen for effluent control because they produce the lowest catalytic ignition temperature. Sufficient surface area must be available to permit completion of the oxidation reaction within the catalyst bed. Normally, catalyst surface areas of 0.2 sq ft/cu ft/min are utilized, although certain difficult applications may require as much as 0.5 sq ft/cu ft/min.

The catalytic combustion method of fume disposal has a wide range of industrial applications in such diversified manufacturing activities as foundry-core baking; fish-oil and animal-fat rendering; solvent evaporating; oil refining; paper printing; and varnish, chemical, and plastics manufacturing. The method can be usefully applied to paint-baking ovens, if the paint does not contain volatile catalytic poisons. Satisfactory results have been obtained in the combustion of low-boiling hydrocarbons; of heavy vapors liberated in the cooking of oils, tars, and asphalt; and in the chemical reduction of nitrogen oxides under reducing conditions.

The catalytic method has been found to be unsuitable when the exhaust fumes contain such particulate material as fly ash, inorganic solids, or vaporized metals. These materials cause a rapid deterioration of the catalyst with resultant excessive maintenance and replacement costs. For this reason the exhaust gases from foundry cupolas, blast furnaces, and coal-fire boilers cannot be handled by catalytic methods. Some halogenated hydrocarbons and silicone paint additives are also harmful to the catalyst.

Properly designed catalytic combustion units have been operated with very acceptable efficiencies; disposal of various hydrocarbons and other organic vapors has exceeded 98 percent efficiency in some installations. A distinct advantage of this method over furnace combustion is that practically no nitrogen oxides are produced. Catalysts have been used, in fact, to reduce nitrogen oxides to nitrogen in waste-gas streams.

The initial costs of catalytic combustion units, including installation, vary widely with each installation requiring a separate design study. In general, a cost in the range of $3 to $5.50/cu ft/min covers most installations, and the catalyst element is about $1/cu ft/min. The operation cost of this disposal method covers maintaining the catalyst and system and preheating the fume, which is dependent on the nature of the fume. Maintenance costs include a periodic cleaning to insure proper efficiency and the infrequent replacement of the element. Under normal operation a platinum catalyst element should be inspected every 6 months; immediate inspection should be made if abnormal operation is observed. Dust and dirt may be removed from the element by immersion and agitation in hot water containing detergent. Elements coated with oils or resins should be heated to 1100° F, cooled, and treated in a nitric acid (HNO_3) bath. With proper maintenance, the catalyst may last several years with no loss in efficiency.

PARTICLE-COLLECTION AND MONITORING SYSTEMS

The qualitative nature of the hazard from airborne and settled particles has long been known. However, as in many cases involving chronic exposures, quantitative information is not available concerning the tolerable dose before damage to health or property occurs.

Atmospheric particulate materials, which are often generated by industrial processes, may contain or be composed of toxic, corrosive, and erosive compounds. Their presence in the atmosphere or on sensitive surfaces may not only reduce visibility but also cause damage to health, to appearance, and to plant life. In London, a man was executed in 1306 for burning coal while Parliament was in session.

Present monitoring systems are designed either to meet legal limitations or to obtain sufficient information about the nature of the particulate air pollution to permit remedial action. Thus, the present systems are used to observe the particulate debris that is deposited on horizontal or vertical surfaces, to observe the particles suspended in the ambient atmosphere, or to monitor the vents and stacks associated with a process that may cause emission of particulate debris.

It should be emphasized that scientifically justified limitations on tolerable levels of particulate air pollution are not quantitatively known. The complicating factors such as environmental conditions, contributing pollutants, synergistic effects, collection site status, and health deterioration or damage cannot be specifically stated for any single air pollutant or for a simple combination of air pollutants. Limitations have been set by industrial hygienists concerned with such aspects as control of radioactive hazards. Maximum-acceptable-concentration (MAC) limits have been set for 8-hour exposures of healthy working male individuals in reasonably controlled environments. These limits, however, are not useful for generalized air pollution control.

Air pollution results in exposure of the entire cross section of population from strong, young men to newborn infants under environmental conditions that are widely variable and uncontrolled. The best MAC limits are only one part of the input for finding the tolerable air pollutant levels. Combinations of materials may act in a synergistic manner not completely defined. Even more important, long exposure times at low levels have not been adequately studied.

Esthetic considerations involved in seeing a clear view as compared to a smoky view are also of real, but unmeasurable, importance in setting air-pollutant-concentration limitations.

When a qualitative analysis of particulate contamination is required, a quantity of particles is usually collected and analyzed by procedures similar to those used in any analytical laboratory. Physical measurements range from direct observation of the particles in the atmosphere and of their effects as they deposit or interact with various substrates to indirect observation of interactions between the particles and the means of measurement. The techniques ranging from routine to conceptual are in varied states of development.

Present instrumental techniques are capable of very high sensitivity for analysis of particulate contamination. An excellent sensitivity is required because of the lack of quantitative information for tolerable contamination levels. In general, the analysis of particulate contamination is based on physical measurements.

In categorizing the instruments used for monitoring particulate air pollutants, the methods in terms of instrument types, of monitored pollutants, of effects of monitored pollutants, or of instrumental applications may be considered. The authors have chosen instrumental applications as the most useful way to categorize the instruments used in the examination of settled particles, of airborne particles, and of the emitting sources considered.

The settled dust particles from the atmosphere may have irritating and corrosive effects and are usually very obviously present. They are usually measured and reported in terms of mass of material of several types per unit area. Sizes of airborne particles permit inhalation and retention in the respiratory system; thus, health hazards occur. Airborne particles also cause reduced visibility in the atmosphere. These particles are measured in terms of mass per unit gas volume, number of particles per unit gas volume, or, indirectly, as a visibility through the atmosphere. Emissions are of interest because they are normally the source of both settled and airborne particles and are a direct indication of the processes that cause air pollution. They are usually measured in terms of mass per unit volume of emission, visibility through the emission, or as mass of materials emitted per specific process or process subdivision.

Instruments used in one monitoring application can often be used in another. However, correlations from one application to another are usually difficult and often impossible. For example, settled dust must come from the atmosphere. In a fixed volume covering a given area, a direct correlation should be found. However, the differences in settling rates for particles of different Stokes' diameters and the presence of random eddy currents through the atmosphere make such a direct correlation impossible.

Some of the instrumental techniques, methods, and devices used to analyze particulate air pollution will be discussed. Particle-collection techniques are applicable to both particle analysis and particle control (collection of particles removes them from the environment). In situ measurement techniques are applicable to analysis only.

Settled-Particle Analysis

Instruments for settled-particle analysis are perhaps the simplest devices. Interpretation of results may be complex because of the diverse nature of the deposition of particulate material in the settled-particle collectors, the so-called "dust jars."

The dust jar is usually a glass, metal, or plastic container, 6 inches in diameter and 8 inches in height. It is placed in a stand at a level where reentrained dust from the normal traffic is not lifted to its interior. A layer of liquid is placed in the bottom of the jar. During winter or inclement weather, antifreeze may be added. A fungicide or algicide should be included to prevent growth of cultures that could change the reported results. Bird guards are usually used to prevent birds from perching on the edge of the jar and adding deposits to the fluid in the jar. Jars are usually left out for a period of 1 month.

Settled particles are analyzed by weight. An aliquot of the liquid is taken after the settled material is thoroughly dispersed. The liquid is evaporated, and the settled material is analyzed in terms of weight per unit area in the jar; the result then is extrapolated to unit weight per square meter or, in some cases, per square mile. It is also possible to extract, with suitable solvents, the organic-soluble and water-soluble components to determine combustible materials, and to report each component separately (ref. 15).

Another device used for determination of settled particulate material is the tacky or adhesive sheet. The adhesive-coated sheet is exposed to the atmosphere for a fixed period of time, ranging from 1 hour to 1 week. The sheet is then examined visually for specific particulate contamination. In some cases, magnification with a low-powered microscope may be necessary (ref. 16).

A variation on the above technique allows the use of nutrient plates for analysis of bacteriological debris, especially in the interiors of hospital rooms. The nutrient plates are exposed for a suitable time period, and viable colonies are visually counted to indicate the level of bacteriological contamination in the atmosphere.

In general, measurement of settled particulate material is helpful in determining trends in air pollution. Trends in airborne particles often follow trends in settled particles and, therefore, long-term changes in air pollution can usually be followed by observing the levels of settled debris. However, quantitative correlations between ambient-air-pollution levels in terms of mass per unit volume in the atmosphere and the total number or mass of settled particles are difficult, if not impossible, to obtain.

Airborne-Particle Analysis

Adequate sampling from the atmosphere is often difficult because the particles suspended in the atmosphere are not uniformly dispersed, do not follow the gas-stream lines in the atmosphere, and are subjected to forces other than purely aerodynamic ones. Before any analytical system is used for monitoring, a suitable representative sample must be obtained from the atmosphere. This sample must: (1) be representative of all particles in the environment, (2) indicate the magnitude of the environment from which the sample has been obtained, (3) be of sufficient size to permit a reasonable analysis, and (4) not be degraded during collection or sample handling.

Adequate sampling of particulate material usually involves isokinetic sampling in which the material to be sampled enters the sampling orifice without being disturbed or accelerated. Normally this means that the sampling orifice aspirates the sample at the same velocity as the sample stream. Anisokinetic sampling produces errors proportional to particle sizes because large particles do not follow changes in the direction and velocity of airflow (ref. 17).

Errors resulting from anisokinetic sampling of suspended dust are tolerable in the open atmosphere but are of extreme concern in sampling from stacks or other emitting sources. Errors resulting from anisokinetic sampling increase with particle diameter squared according to a complex function.

Some of the collecting devices that obtain a sample of particles for a subsequent analysis are considered in airborne-particle-analysis systems.

Filtration.—In the broadest sense, a filter is any coarse structure composed of granular or fibrous materials that will remove suspended matter from a fluid passed through it. When the filtration process is applied in the field of air pollution, the problem always involves fluid flow through a porous medium; the particles deposited on the medium; a resultant decrease in void volume; and a proportional increase in resistance to flow, or pressure drop, as filtration progresses. Some means of periodically removing solids must be provided when the percentage of void volume is rapidly reduced with time. The utility of a filter is governed by how efficiently it collects particulate material without developing excessive resistance to flow.

The filter medium used in any particular application is selected on the basis of the type of analysis to be performed on the collected particles. For example, for visual observation, a surface collection that can be obtained by membrane filter media is desirable. If weight analysis is required, high-flow, high-capacity systems of paper or glass-fiber medium-retaining particles larger than a fraction of a micron are usually used. Fabric filters are one of the oldest and most reliable of the high-efficiency devices available for the collection of dry particulate materials. Their wide acceptance is attributable to a number of advantages, the most important being that they are capable of maintaining 99 percent or better collection efficiency, despite wide variations in properties of the carrier gas. Minor process disturbances or process cycling do not result in a discharge of particulate material to the atmosphere. Fabric filters are simple to operate and maintain, do not require skilled labor, the initial cost is generally lower than that of competitive control equipment with equally high efficiency, and power consumption is moderate.

Major disadvantages of fabric filters are that the temperature is limited to that withstood by the fabric (500° F maximum for fabrics currently available), and that improper application or operation may result in diminishing the flow of gas through the device. The fabric filter must be operated above the dew point of the gas. Major considerations in the selection of a fabric, besides temperature, are resistance to abrasion and resistance to flexing.

Fabric filters operate so that flow through both the fabric and the deposited dust layer is viscous. Thus the fundamental characteristics embodied in Darcy's law apply. As previously stated, the dust collected on a fabric filter must be periodically removed if excessive resistance to flow is to be avoided. The built-up dust cake is usually removed from the fabric by one of three methods: mechanical cleaning using shaking or rapping devices, collapse cleaning accomplished by introducing a slight negative pressure on the interior of the bags, or reverse-jet cleaning in which air jets are used to blow the dust cake from the filter.

The three most important costs associated with fabric filtration are the first equipment cost, power cost, and maintenance cost. Generally, \$2.50 per square foot of cloth area is a good figure for estimating installation cost (exclusive of dust-disposal equipment) for woven fabrics that are intermittently or periodically cleaned. Because this figure is based on cloth area, the cost in a given situation depends on the filtration velocity to be used. Normally, filtration velocities range from ½ to 1 ft/min. Costs for specialized equipment, such as reverse-jet or pulse-air equipment, run higher per square foot of cloth; but because higher filtration velocities are possible, this equipment may be economically advantageous. Cost for the equipment is about \$14 per square foot, but filtration velocities generally range from 3 to 25 ft/min.

For chemical analysis, suitably soluble filter media are selected; but when sampling from high-temperature environments, alundum (porous aluminum oxide, Al_2O_3) and/or glass-fiber filters are used. A wide variety of commercially available filter media is available. After the particles have been collected on the filter media, the technique of analysis can be selected at the discretion of the analyst.

The filter assembly may contain a sequential sampling device in addition to the filter holder, filter media, and vacuum source. These sampling devices, which are commercially available, can be used to obtain information about the time variation in levels of particulate debris in the atmosphere. They are generally paper-tape samplers in which a roll of filter tape is used to collect the debris. The tape is advanced sequentially at preset intervals to permit sampling through a single section of tape for a given time period.[1]

The density of a filter-tape deposit can be expressed in terms of light transmission through the deposit. To determine the density of the deposit, the volume of the gas sample is divided by the tape-deposit area, giving the length of a sample-gas column having a specific uniform haze density that would produce the same reduction of light transmission as the deposit. The density is expressed as a coefficient of haze per 1000 feet of air (ref. 18).

Inertial Collection Devices.—An inertial particle collector takes advantage of the tendency of particles moving in an airstream line to continue in the original direction when the airstream line is deflected by an obstacle. The particles impact on the obstacle and can be retained by a suitable adhesive coating for subsequent analysis. Several instruments exist for collecting particulate material by virtue of particle inertia.

A single-stage impactor consists of a substrate for collection and subsequent examination of particulate material. The substrate faces an orifice through which the gas stream that may contain particles for analysis is drawn. This instrument is usually sensitive to particles from approximately 0.5 micron to approximately 50 microns in diameter. High collection efficiency results from the gas particles being accelerated through the orifice; however, the acceleration causes adiabatic expansion. This expansion causes water vapor to condense on the particles being collected, resulting in erratic operation. Instruments of this type range in cost up to approximately $750.[2]

The wet impinger is essentially a single-stage impactor. Its efficiency is maintained at a constant rate by continuously washing off the collecting substrate with a stream of liquid. Thus the collected dust particles are moistened and retained in the liquid. The sample is examined under a microscope by wet-chemical or colorimetric analysis or by depositing an aliquot of the sample on nutrient plates for culturing for bacteriological measurements.[3]

The single-stage impactors require that sufficient energy be imparted to the particles so that the smallest particle of interest is collected. If a large-size distribution is present, the larger particles may be impacted

[1] Instruments for filter-tape sampling can be obtained from the Gelman Instrument Co., Ann Arbor, Mich.; Instrument Development Co., Riverside, Calif.; Research Appliance Co., Allison Park, Pa.; and VonBrand Filtering Recorders, New York.

[2] Some typical units are the Casella sampler from Union Industrial Equipment Co., Fall River, Mass.; airborne bacteria monitor from Anderson Samplers, Provo, Utah; annular impactor from the Staplex Co., Brooklyn, N.Y.; Konimeter from W. Watson & Sons, Ltd., Barnett, England; and the rotorod collector from Metronics Associates, Inc., Palo Alto, Calif. The rotorod sampler, an example of a single-stage impactor, reverses the usual process of pulling air through the collector by rotating a collecting rod through the air. Particles are impacted on the collecting-rod face for subsequent examination under a microscope.

with enough force to cause damage. For this reason, multiple-stage or cascade impactors consist of a series of single-stage impactors, each operating at a higher velocity. The largest particles are collected in the first stages during which the velocity through the orifice is the lowest. The smallest particles are collected at the highest velocity in the final stage.[4]

The cyclone typifies centrifugal collectors. A cyclone separator is a structure, without moving parts, that removes particulate matter from gas streams by causing the fluid stream to move in a confined vortex within which inertial forces continuously produce a tangential motion of particles toward the collector wall. The inertial separating force in a cyclone is the radial component of simple centrifugal force. Because of the relative design simplicity, cyclones are inexpensive and can be constructed of materials capable of meeting many temperatures and pressures. Furthermore, cyclones may be designed to handle nearly any dust load or air volume. These factors lead to an extensive application of cyclones, especially when only moderate collection efficiency, 80 percent, is required.

Because of their simplicity of design, cyclones are among the most trouble-free particulate collectors available. Other than erosion and corrosion, plugging of the dust outlet and cake buildup on walls are the two most common operating problems. The former problem can be minimized by proper design, and the latter problem can be minimized by any of several methods, depending on the circumstances. Some of the methods are eliminating condensation through insulation, increasing inlet velocity, electropolishing inner walls, removing precleaners to allow scouring of walls by coarse material, and flushing or grounding.

Cost estimates for cyclones have varied from 7 to 50 cents/cu ft/min of capacity. A typical cyclone of moderate capacity, therefore, costs about 10 cents/cu ft/min uninstalled. Installation costs vary depending on the location. A comparative study of annual operating costs for various dust collectors including power, maintenance, and water (when used), but not amortization, shows for dry cyclones an average of 6 cents/cu ft/min/yr and for wet cyclones 16 cents/cu ft/min/yr. When 10-year amortization is included, the total, average annual costs are, respectively, 8 and 20 cents.[5]

Electrostatic Precipitators.—Air-sampling electrostatic precipitation of airborne particles is useful for collecting relatively large samples with a high collection efficiency for all particle sizes. Electrostatic

[3] Sources for wet impingers include MSA midget impinger from Mine Safety Appliances, Pittsburgh, Pa.; Greenburg-Smith impinger from Willson Products Division, Rayovac Co., Reading, Pa.; Unico microimpinger from Union Industrial Equipment Co., Fall River, Mass.; and Gelman midget impinger from Gelman Instrument Co., Ann Arbor, Mich.

[4] Sources for such multiple-stage impactors include: Casella cascade impactors from Brinkman Instruments, Inc., Westbury, N.Y.; Unico cascade impactor from Union Industrial Equipment Co., Fall River, Mass.; Anderson sampler from Anderson Samplers, Provo, Utah; and Battelle cascade impactors from Scientific Advances, Inc., Columbus, Ohio.

[5] Sources of supply for such devices include: Union Industrial Equipment Co., Fall River, Mass.; Zimeny Corp., Monrovia, Calif.; and Mine Safety Appliances Co., Pittsburgh, Pa.

precipitation is also useful for depositing small-quantity samples with very small particle sizes directly onto an electron-microscope grid. The sampling rate is not affected by the mass loading, or by the total collected material in the sample, and the sample is in a readily recoverable form.

Collection by electrostatic precipitation requires that the particle to be collected have an electrical charge and that it be introduced into an electrical field and directed toward a collecting substrate maintained at a polarity opposite to that of the charged particle.

For cleaning industrial gases by electrostatic precipitation, the high-voltage direct-current corona method of particle charging is universally used. In this process, a high-voltage discharge electrode placed at a proper distance from a grounded electrode ionizes the gas between the electrodes, and current passes through the ionized air. This ionization occurs only in a finite temperature and voltage range, and is dependent on the physical properties of the gas. Below the lower limit of the range there is no current flow, and the upper limit is defined by the dielectric breakdown of the gas when sparking takes place. The rate of ion generation must be sufficient to saturate the field so that the dust particles pass through. The ions attach themselves to the dust particles and these ionized particles are then subject to the electric forces of collection. With sufficient power generation, the corona discharge generates enough ions to accomplish this charging almost instantaneously. When a negative-charge discharge wire is used, a measurable amount of ozone (O_3) is formed. A positive-charge discharge wire system forms very little O_3. For this reason, positive corona systems are used for space ventilation.

Once the dust particles have been transported to the collector electrode, they must be removed with a minimum of reentrainment during the time the precipitator is operating. For the designer of dry precipitators (collection of dry dust), reentrainment is a very serious problem. Liquid films in wet precipitators self-drain and usually present no reentrainment problem.

Although the fundamentals of electrostatic precipitation are understood, theoretical efficiency is never attained in practice. Thus, despite years of experience, the prediction of precipitator performance is largely empirical. Maintaining the design efficiency in day-in-and-day-out operation is more difficult with precipitators than with many other types of equipment. All of the parameters affecting efficiency—gas volume and density, dust concentration, particle-size distribution, resistivity, gas distribution, and power settings—must be at favorable design conditions to yield design efficiency. It is not uncommon to find precipitators in apparently normal operation having efficiencies 5 to 10 percent below design efficiency. A precipitator efficiency of 75 to 90 percent, depending on application, is relatively easy to obtain. Above these levels and especially above approximately 98 percent, sectionalization and automatic power controls are necessary.

The purchase cost of precipitators for rates over 400 000 cu ft/min is approximately $1/cu ft/min. This does not include duct work, dust-handling equipment, or insulation. Total installation costs are quite dependent on the collection efficiency required. Operation cost is the sum of the power cost required for precipitation and the cost required to move the gas through the precipitator. Precipitation power, the major power-cost variable, is dependent on the desired efficiency. Corona power can range up to 500 w/1000 cu ft/min to which the energy conversion efficiency of the power supply must be added. Maintenance requirements for a precipitator installation generally require the service of an engineer or a trained operator in addition to the usual personnel for mechanical maintenance.[6]

Thermal Precipitators.—Thermal precipitators deposit dust particles from a small volume of air onto a surface from which they may be examined, counted, or weighed. Deposition is accomplished by the thermophoresis force that an airborne particle experiences in a temperature gradient. This force causes particles to be driven toward a cold substrate. The thermal precipitator quantitatively collects all particles from the smallest size that can be observed in an electron microscope up to approximately 3 to 5 microns in diameter. Most thermal precipitators operate at a sampling rate of a few centimeters per minute. High-capacity precipitators that sample up to 1 cu ft/min have recently been developed. Costs range from $300 or $400 up to $1,000 per unit.[7]

In Situ Analyzers

The collectors described above employ physical methods of removing particles from the atmosphere and depositing them on suitable substrates for examination, which may involve determining particle concentration, particle mass, or particle composition. In situ analyzers, on the other hand, use physical processes to observe the particles at their positions in the atmosphere. These analyzers are used mainly for determining particle concentration or mass. The composition of particulate contamination is seldom described by the physical means available for in situ analysis. These physical means depend on observation of interaction between a particle and some physical process. Instruments used for air pollution monitoring of particulate contamination in situ are described briefly below.

Photometric Analysis.—The dimensions of airborne particles are usually within an order of magnitude of the wavelength of the light used to illuminate them. When a particle is illuminated with a beam of light, the amount of light scattered by that particle varies roughly with the projected area of the particle. By causing particles to pass singly through a light beam, it is possible to collect pulses of light scattered by each particle at a particular angle, analyze the total light flux in each pulse, and obtain a particle-size distribution from the individual particles.

Alternatively, an aerosol photometer collects the light scattered from a volume of aerosols containing many particles and produces data that

[6] Supply sources of such precipitators include the Bendix Corp., Cincinnati, Ohio; Bell Electronics Corp., Mount Vernon, N.Y.; Mine Safety Appliances Co., Pittsburgh, Pa.; and Gardner Associates, Inc., Schenectady, N.Y.

[7] Electrical thermal precipitator sources include: Casella standard thermal precipitator, Union Industrial Equipment Co., Fall River, Mass.; Strong-Ficklen thermal precipitator, Joseph B. Strickland, Pasadena, Calif.; and thermopositor, American Instrument Co., Silver Spring, Md.

indicate the total area of all the particles in the volume of gas. By careful optical design, it is possible to measure light transmission through the aerosol; for example, no scattered light is permitted to enter the collecting optics. This means that all the undeviated light in the beam transmitted through the suspension and none of the scattered light reaches the collecting optics.

To summarize, the photometer is used to determine the total number of particles in a given volume of aerosol. The light-scattering, single-particle counter describes the particle-size distribution by the number of particles in a given volume of aerosol.

Light-scattering, single-particle counters detect single particles of at least 2/10 micron. Scattering angles approaching 90° are used. Concentration up to 1 million particles/cu ft can be analyzed without significant coincidence. Costs range from approximately $3,000 to $20,000, depending on desired accessories.[8]

An extinction meter indicates the total number of particles in a given volume of aerosol. The extinction meter can be used with an artificial light source to determine the concentration of particles in a given volume of aerosol, or the Sun can be used as a light source to determine atmospheric turbidity or plume transmittance from an emitting air pollution source. Plume transmittance can be obtained by the photometer sighting a pair of contrasting targets, one through and the other clear of the plume.

Typical optical device costs range from $1,000 up.[9]

In recent years high-power laser systems and high-sensitivity photometers have made it possible to collect light scattered back from discrete aerosol clouds at long distances. These devices involve a pulse-laser system producing a collimated beam and viewing optics, also collimated, to intersect the laser beam at a given distance within the aerosol cloud. The amount of light scattered back from the aerosol into the receiving optics gives an indication of the concentration of particles within the aerosol cloud at the intersection of the laser beam and the axis of the receiving optics.

No commercial devices of this type are presently available. Experimental systems have been investigated primarily under Government sponsorship and show great promise for remote detection and analysis of aerosol clouds.

Condensation Nuclei Counters.—If submicron aerosol particles are injected into a volume in which the relative humidity is near saturation or supersaturation with respect to liquid water, the vapor in the environment condenses on the particles and produces droplets that can be easily seen with simple optical systems. The submicron particles

[8] Sources of supply include the Royco Instrument Co., Menlo Park, Calif.; Bausch & Lomb, Inc., Rochester, N.Y.; Particle Technology, Inc., Sunnyvale, Calif.; and Climet, Inc., Sunnyvale, Calif.

[9] Such devices include the Brice-Phoenix photometer from Phoenix Instrument Co., Philadelphia, Pa.; Aminco angular scattering photometer from American Instrument Co., Silver Spring, Md.; Royco aerosol photometer from Royco Instrument Co., Menlo Park, Calif.; Leitz Tyndalloscope from Ernst-Leitz, Inc., New York; Sinclair-Phoenix photometer from Phoenix Instrument Co., Philadelphia, Pa.; and the particle detection unit from T-D Associates, Baltimore, Md.

act as heterogeneous condensation nuclei. By controlling the degree of supersaturation, the sizes of particles activated to nucleate the water vapor can be selected. This phenomenon also occurs with gas-molecule clusters, such as SO_2, SO_3, and NH_3. Such gas molecules are commonly present in environments in which air pollution occurs. Thus, the number of activated nuclei is in direct relation to the total level of particulate pollution present in the environment.

It is possible to achieve the required supersaturation by adiabatic expansion of a compressed air sample. The droplets that form when the air sample is expanded are viewed with a light-extinction meter. The concentration of droplets is directly proportional to the amount of light obscured. The concentration of droplets is also directly equivalent to the total concentration of nuclei.[10]

Experimental systems.—The aerosol analyzing systems described above are generally available commercially. In addition to these systems, however, several other systems in various stages of development apply physical principles. These include:

(1) *Spectral emission.*—If aerosol particles are passed through a flame or through a plasma, the particles are vaporized, heated to incandescence, and the light pulse of a particular wavelength is emitted. This light pulse can be analyzed on a suitable spectrophotometer for both concentration and composition. It is possible to detect particles as small as 0.1 micron.

(2) *Electrical mobility.*—Small particles can be charged unipolarly and their mobility determined by measuring their collection in a given electrical field. Single-electron charging of submicron particles, combined with current measurement, gives data on total quantity of submicron particles. Saturation charging of larger particles and current collection in any preset field give mobility data (ref. 19).

(3) *Acoustical techniques.*—If particles in air are passed through a sensor in which they are gradually accelerated to about 100 m/sec and suddenly projected into a wide hexacavity, the particles produce a pressure pulse that results in an audible click (ref. 20).

(4) If particles are removed from an airstream by a probe that is connected to the grid of an electrometer, the electrical charge of each particle normally produces a significant deflection in the electrometer reading (ref. 21).

This review of new experimental systems is by no means complete. One-of-a-kind prototypes of such systems are being evaluated. Additional concepts indicate the variety of physical phenomena that can be used to detect particles in the ambient air.

Emission-Source Analysis

Essentially all of the collection devices described in the previous section can be used with certain precautions for stacks and other emission sources.

In many municipal ordinances, allowable emissions from process vents or from power plant stacks are limited in terms of plume

[10] Two commercial sources for condensation nuclei counters are General Electric Co., Schenectady, N.Y., and Singco Associates, a division of Bolt, Beranek and Newman, Cambridge, Mass.

opacity. Opacity may be estimated by visually scanning the plume against the sky background with or without a comparison chart. The opacity of the plume can also be determined by photometric observation across a fixed distance within the stack or vent itself. A light source and photometer, acting as an extinction meter, are used. It is necessary, however, that precautions be taken to insure that deposits on the lens system of the light source and extinction meter are eliminated or prevented. A stream of clean air across the lens system is often adequate.

In addition to the optical visibility limitations, many stack emissions are limited in terms of the mass of particles that can be emitted per unit volume of gas or, in terms of the mass that can be emitted, per unit weight of input process material or per unit of power generated. The mass of emitted material is determined as a function of gas flow or of time. The gas velocity within the stack is measured, usually with a pitot tube, to insure isokinetic sampling. An isokinetic sample probe is inserted and is traversed across the stack. Materials are collected on a suitable filter medium or in wet or dry collectors, considering the temperature and dew point of the gases in the stack. Usually, manual methods are used to collect the sample and to adjust sample velocity to insure isokinetic sampling. In the last few years, however, modifications of this technique have been developed. In some systems the sample particles are collected on a suitable filter tape, and the mass of particles per unit area of filter tape (correlative to unit volume samples) is determined by beta-ray transmission (ref. 22).

In addition to these essentially standard techniques, a number of one-of-a-kind developments have been reported; for example, triboelectric measurements of charge deposited on a collecting cone grounded through an electrometer depend on the quantity of charge particles impacting on the cone.

In general, measuring stack or emission pollutants requires the same types of devices as ambient-air monitoring. Isokinetic sampling must be carried out within the stack, and the sample system must not be affected by the high temperature and/or high dew point of water or acid gases within the stack. Corrections for temperature and gas compositions must be included, because combustion products with densities other than that of air may be present.

NASA CONTRIBUTIONS

SPACECRAFT-CABIN ATMOSPHERE MONITORS

A spacecraft cabin is a closed ecologic environment, and that a nontoxic atmosphere must be maintained within the cabin is a requisite for manned spaceflight. The National Aeronautics and Space Administration has, therefore, invested a great deal of effort in identifying, measuring, and controlling the components of a sealed-spacecraft-cabin atmosphere. Much of this investment has been channeled into the design and production of monitoring equipment, and some of this equipment is functionally suited to air pollution monitoring on a broader scale. NASA developments in monitoring devices that represent novel or improved features applicable to air pollution monitoring have been selected for review here.

Tests with unmanned and manned cabin simulators have shown that nearly 150 vapors are generated in a closed system (ref. 23). These vapors, some of them potentially toxic, are primarily produced by three sources: (1) the metabolic processes of the crew, (2) the supplies and food stored for the use of the crew, and (3) the operation of the spacecraft systems (refs. 24 and 25). Additional sources are the materials of which the cabin is constructed (ref. 26) and the reactions of products from other sources. The studies that have been conducted to identify and measure these contaminants utilized commercially available analytical instrumentation, such as gas chromatographs, infrared spectrometers, mass spectrometers, and biological sampling devices (refs. 27, 28, and 29). This instrumentation has also been used for laboratory analysis in air pollution studies. However, the emphasis in air pollution instrumentation is shifting toward portability, durability, and simplicity of operation (manual or automatic). The development of flight-rated instrumentation by the National Aeronautics and Space Administration has achieved some of the goals of air pollution investigators. Eight NASA-developed instruments originally designed for cabin monitoring, which we feel are of interest to the field of air pollution, are described in detail below.

Small General Purpose Cycloidal-Focusing Mass Spectrometer

The NASA Flight Research Center at Edwards Air Force Base, Calif., developed this mass spectrometer (refs. 30 through 33) in response to a need for a flight-rated monitor for respiratory gases. The original use of the instrument was to monitor continuously the respiration of a test pilot for correlation of quantities of various respiration products with stress conditions being undergone by the pilot.

The mass spectrometer monitors vapors with molecular weights of 3 to 100 atomic mass units (amu)(fig. 18-1). For flight operation, 12 gases are monitored simultaneously with a continuous input. The delay time (determined by input system) is 300 milliseconds, and the instrument response time is 30 to 50 milliseconds. The accuracy of measurement varies with the concentration of gases monitored. For high concentrations, the accuracy is ±3.5 percent of full scale, ranging to ±50 percent for low concentrations of 600 ppm.

This compact instrument is 10 by 11 by 11 inches and weighs 28 pounds, exclusive of the vacuum system. It has a high reliability rating. Tests at the Flight Research Center have shown that zero drift is approximately 1 percent in 6 months, and that output with a constant input varies ±2.5 percent. The instrument has an automatic calibration circuit that recalibrates every 2 minutes. The life of the ion source is approximately 250 hours, and the solid state electronic circuitry has a life expectancy measured in years.

This instrument is suitable in its present form for use as an air pollution monitor for gases within the 3- to 100-amu range. Because

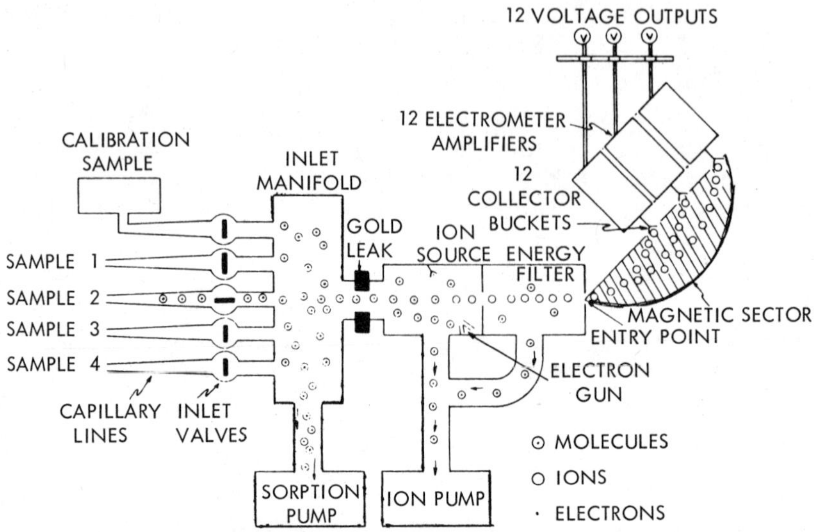

Fig. 18-1—Schematic diagram of operation of cycloidal focusing mass spectrometer.

of its small size and durability, it can be used in any type of vehicle as well as in a stationary location. There are no commercially available mass spectrometers that meet the size and performance specifications of this device. No additional developmental costs would be required for using this instrument in air pollution monitoring. The estimated cost of the instrument, \$50 000, is difficult to justify for an enforcement agency but would be acceptable for a research group.

Nonmagnetic Flight Mass Spectrometer

This flight-weight monopole mass spectrometer (refs. 33 and 34) was developed by the General Electric Co., Apollo Systems Department, Daytona Beach, Fla., for the NASA George C. Marshall Space Flight Center. The developers took a commercial GE Monopole 600 partial-pressure analyzer and made it smaller and more rugged and reduced its power requirement, thus designing a feasibility model for instrumentation to be flown in Saturn launch vehicles. Three applications were sought:

(1) Detecting hydrogen, oxygen, and hazardous-gas mixtures from gas systems and monitoring light atmospheric gas constituents with mass ranges of 2 to 50 amu.

(2) Measuring ambient atmospheric constituents, hydrocarbon fuels, toxic gases, products of offgassing, and so forth, with mass ranges 2 to 300 amu.

(3) Detecting heavier hydrocarbon contaminants and toxic gases with mass ranges 2 to 600 amu.

The development and testing thus far has been restricted to the 2- to 300-amu range. However, the instrument as designed is capable of detecting materials with mass unit numbers up to 600. Sensitivity is high—partial-pressure measurements can be made up to 10^{-10} torr. The instrument provides unit resolutions over the entire mass range with a scan rate of 60 seconds for the entire range. Manual scan is also provided. The output is suitable for driving a 0- to 5-volt recorder. Peak position on the recorder shows mass number, and peak height indicates quantity at that number.

This instrument is nonmagnetic. The samples are drawn through a small leak into the ion-source section where the atoms and molecules are ionized. Ion separation is performed by a combined radiofrequency (rf) and direct-current (dc) field. Under any given set of applied voltages, only ions of one mass stream to the end plate of the chamber. These pass through an aperture and strike the first stage of an electron multiplier. The rf and dc voltages can be systematically varied to cover a desired range of ion masses. Monitoring these voltages provides, then, a measure of the mass at any given moment. A plot of mass number versus partial pressure can then be used to identify the chemical nature of the constituents of the sample drawn into the tube. After the initial analysis, succeeding identifications can be partially automated.

The instrument's dimensions are 8 by 6 by 16 inches, and it weighs 16.8 pounds, exclusive of a vacuum system. Power requirement is less than 30 watts. Because this device was developed as a feasibility study, the commercial quality and reliability requirements were not considered, but an operating flight unit would be required to meet standards and procedures such as MSFC–FTB–154. The feasibility study demonstrated that these requirements could, in fact, be met by a flight unit designed according to this prototype.

An instrument built to the specifications of this prototype would be suitable for detection and analysis of the gaseous constituents of the atmosphere with mass numbers in the range of 2 to 600 amu. Although the gas sampling rate is very, very low—a fraction of a cubic centimeter per second, the large number of molecules that can be analyzed in a relatively short time for both composition and concentration makes the device useful for atmospheric problems. Both the GE Monopole 600 and the Consolidated Electrodynamics Corp. analytical mass spectrometer perform essentially the same functions. However, both are several orders of magnitude larger and require considerably more power. The modified instrument is admirably suited for mobile use, and its relatively low estimated manufacturing cost of $15,000 makes it attractive for civic and research groups. The estimate for design, checkout, and field testing for a production device is approximately 3 man-years of effort.

Trace-Gas Analysis Procedures

In 1966 the National Aeronautics and Space Administration and the Air Force School of Aerospace Medicine jointly conducted a series of experiments on contaminants in a closed ecologic environment. Aerojet-General (refs. 35 and 36) was under contract to the Air Force to perform analytical analyses of the contaminants, and they developed procedures for collecting and analyzing the gases. Personnel from Aerojet-General, working with a NASA team and Southern Research Institute personnel, are continuing the development of this procedure at the John F. Kennedy Space Center to analyze contaminants offgassed by materials and systems used in the Apollo cabin.

This procedure represents a well-defined, highly reliable technique for the use of commercially available equipment in an application intimately related to air pollution research. The salient features of this procedure are in sample treatment and data handling. Analytical systems used are gas chromatography and mass spectrometry with commercially available equipment. The procedure permits analysis of some 100 compounds contained in samples from the Apollo cabin in approximately 90 minutes, although manual planimetry of chromatograms now requires more time. The procedure is as follows:

A sample is taken by cryogenically fractionating the spacecraft-cabin effluent by freezeout at 0° C, −78° C, and −175° C. Each of the three fractionated samples is passed, in turn, through four chromatographic columns, the packing material in each column having been selected to obtain maximum separation of the compounds with slightly different fractionalities. The gas-flow system is arranged to direct samples to the mass spectrometer either from the fractionated sample bottles or from one of the gas chromatographic columns.

The method has demonstrated high reliability for a large number of samples but highly skilled personnel are required to maintain this level of reliability. Thus, the possibility of transferring this technique to an automated or routine laboratory analysis is remote.

This system is specifically designed for repetitive analysis of gas samples that may contain hazardous materials. In the field of air pollution it would be most suitable for monitoring special manufacturing processes or in research work. If the gas chromatographic and mass spectroscopic equipment is available, the gas-handling system and cryogenic storage facility would cost approximately $5000. Although variations on this procedure exist in many analytical laboratories, there are no commercial systems of this nature. The proven reliability and the documentation of this system recommend its use to air-pollution-research facilities.

Colorimetric Glycol Detector

This device is a specific detector for glycol vapor and is to be used in spacecraft-cabin atmospheres.

The glycol detector is currently in the developmental stage, and its limits of operation are being sought at the NASA Manned Spacecraft Center. At this time the device seems sensitive to 1 ppm of glycol vapor. Interferences with other materials are being investigated. About 40 have already been tested, including alcohols, ketones, esters, and aldehydes. Formaldehyde produces the same reaction (color) as glycol does, and the other aldehydes (acetaldehyde, butyraldehyde, and so forth) produce different colors. The other materials tested have no effect on the detector.

The detector consists of a glass tube 6½ inches in length and ¼ inch in diameter(fig.**18-2**).The interior is divided into three reagent sections by porous spacers. The first section contains periodic acid (HIO_4), the second contains catechol dimethyl ether, and the third contains H_2SO_4 stored on fused silica gel. The tube is sealed after being packed with these reagents, and the ends are opened for use. If glycol is present in the sample introduced into an open tube, the HIO_4 acts as an oxidizing agent and produces an aldehyde. The aldehyde is condensed by the

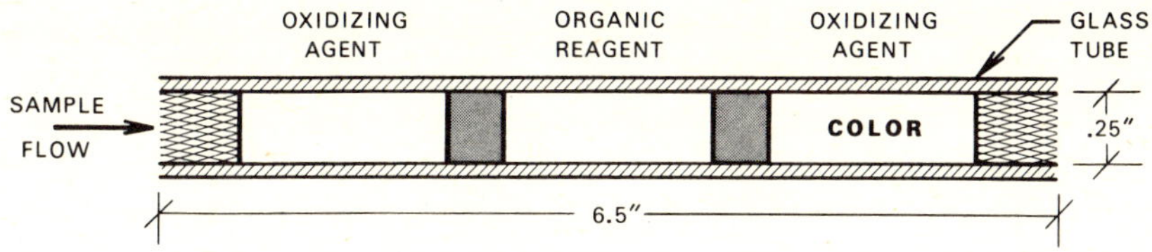

Fig. 18-2—Colorimetric glycol detector.

organic reagent in the second section to a methylene compound, which in turn is oxidized in the third section to produce a color— the intensity of which is proportional to the concentration of glycol in the sample.

No reliability has been established for this device because it is still experimental. Proper sealing is necessary, because atmospheric moisture collected on the silica gel in the last section will degrade operation. Storage life and shelf life are expected to exceed 12 months.

This device is designed only for detection of glycol vapors with manual use. The cost is very low, approximately $1 for each detector. Sampling costs are also low, because a real-time indication of the presence or absence of glycol is obtained during the sampling process. Other colorimetric detectors exist, such as the Kitagawa precision gas detector marketed by the Union Industrial Corp. and the Universal testing kit sold by the Mine Safety Appliances Corp. However, neither is specific for glycol. The glycol detector represents a device suitable for detection and analysis of glycol in the atmosphere at a very low per sample cost (the development of a production item would require about a ½-man-year effort).

The principle utilized suggests two other possible applications:

(1). Use of the operating principles of sensitization, reaction, and indication in tubes specific for vapors other than glycol.

(2). Production of an automated device using these operating principles. It appears at least feasible that tapes could be manufactured in a process analogous to that used in producing Polaroid film and used with a filtered tape-air sampler for continuous monitoring (ref. 37).

Hot-Wire Detector for Chemically Active Materials

A hot-wire detector for chemically active materials (ref. 38) used in gas chromatography was developed at the NASA Manned Spacecraft Center. Gas chromatography, as used for cabin-contaminant identification and analysis, has been limited because the filament in available hot-filament detectors reacted with certain gases and vapors. Thus, only materials that did not react with the filament could be measured.

The new detector can measure gases and vapors that are reactive with the detector filament. The filament is designed to change its electrical resistance as a function of the effluent composition when it reacts chemically with the effluent vapors in the gas chromatographic apparatus.

A titanium or zirconium filament is used for analysis of vapors that are evolved at high temperatures. Other reactive metals may be used for vapors that are evolved at low temperatures. Before sample introduction an inert gas such as helium is passed at a controlled rate over the filament that is heated to the desired operating temperature by a constant-current source. When equilibrium is reached, the electrical resistance of the filament is measured. The sample is then introduced, and the resultant resistance change of the filament with time is measured and plotted on a chromatogram. The quantity of each component in the sample is determined, as in conventional gas chromatography, by measuring peak heights on the chromatogram and multiplying by calibration factors. Because the filament is chemically reactive, its characteristics change, and frequent recalibration is necessary.

Microwave Spectroscopy

For several years the NASA Langley Research Center has been developing microwave spectroscopy into a tool for gas analysis of trace contaminants (refs. 39 through 47). Several years ago, Langley personnel built a microwave spectrometer, but now use a commercially available device manufactured by the Hewlett-Packard Co. The techniques developed at Langley regarding the theory and use of microwave spectroscopy are of prime interest in air pollution monitoring.

Microwave spectroscopy is an analytical technique similar to mass spectroscopy and has several potential advantages. Highly specific identifications can be made by measuring the frequencies of the microwave absorption lines of molecules. Compounds may be unambiguously identified, even in mixtures. Molecules are not fractured by ionization (as in mass spectroscopy) nor is a carrier gas needed (as in gas chromatography). Because the technique is all electronic, it is very amenable to automation.

Detection of molecules is based on the absorption of various microwave wavelengths by different gases. Only polar molecules absorb microwave energy, so nonpolar gases and vapors such as methane, benzene, nitrogen (N_2), and carbon tetrachloride (CCl_4) cannot be detected by this method. The average number of lines in the absorption spectrum of a polar molecule is 400. The number indicates highly unique spectra, and this uniqueness, combined with high-resolution and highly accurate frequency measurements of the microwave spectrometer, usually permits identification of a gas by measurement of only a few of the 400 lines.

The Langley Research Center program is concerned with establishing useful standards for microwave spectroscopy. Mississippi State University is cataloging spectral lines under a NASA grant. A computer-operated spectrometer will soon be used to generate identification spectra for a large number of compounds, and Langley is currently discussing the maintenance of these spectra by the National Bureau of Standards. Work is also being done in the development of a dual-resonance microwave spectrometer that will have even greater precision in identifying the components of a mixture and in establishing techniques that can be used to make quantitative measurement more accurate.

Langley Research Center is leading the development and application of this relatively new analytical technique. Microwave spectroscopy should prove to be a valuable tool for air pollution control centers and laboratories. Because of the cost and size of the equipment (comparable to commercial mass spectrometers), this technique is definitely laboratory oriented.

Contaminant Sensor

A laboratory prototype of a contaminant sensor (ref. 48) was developed by Perkin-Elmer under contract to NASA Langley Research Center. This prototype is suitable for adaptation to a flight instrument to be used in a spacecraft cabin for continuous monitoring of the atmosphere.

The prototype contaminant sensor was built and tested with samples of known gases and unknown mixtures. Concentrations in the range of 0.5 to 5 ppm can be detected, and concentrations over 100 ppm can be quantitatively determined.

This device consists of a mass spectrometer coupled with an accumulator cell containing a sorbent material on which the contaminants from the air are concentrated. Contaminants collected in the accumulator cell at room temperature are subsequently desorbed by heating the cell or by a vacuum system and are passed to the inlet system of the mass spectrometer. An initial precut removes residual air from the accumulator cell to provide pressures compatible with the mass spectrometer input leak and to increase the relative concentration of the contaminants.

The mass spectrometer used in this sensor is a CEC 21–614–1 with a scan range of 2 to 200 amu, and is available commercially. The collection system is the novel portion of this instrument. The system has three accumulator cells, each containing a different sorbent so that a wide range of contaminants could be recovered. Each cell has a Teflon-lined steel tube 3/8 inch in diameter. The sorbents used were Porapak Q, charcoal, and molecular sieve no. 5.

The techniques used in the contaminant sensor are applicable to air-pollution monitoring stations. A wide range of contaminants can be collected and measured, which is useful for laboratory applications.

Photoionization Source for a Mass Spectrometer

This instrument was developed by the GCA Corp. for the Langley Research Center (ref. 49). After the prototype instrument was made and tested, certain limitations were found in its discrimination between ions with differing ionization potentials and coinciding mass numbers, for example, CO and N_2. The instrument was modified by using an argon filter that reduces scattering and the higher order ultraviolet spectra. The modified instrument is capable of detecting CO and N_2 to a lower limit of 100 ppm.

This instrument differs from other mass spectrometers in the method by which the atoms or molecules to be analyzed are ionized. The usual technique is bombardment by electrons or ions. This instrument ionizes the particles with ultraviolet light. Although ionization is slower this way, much less dissociation of complex molecules results in simpler spectra and an easier determination of the composition of gas mixtures containing high-molecular-weight gases.

The instrument can detect gas traces in the 10- to 100-ppm range. Work to date has indicated that detection in the 1-ppm range should be possible with this technique, and that an automated scan of the 2- to 50-amu range would take about 5 minutes for this degree of precision.

The applicability of this device to air pollution monitoring is tenuous at present, because only one prototype instrument has been produced. However, the technique appears promising for use in laboratory analysis. The possibility of ionizing high-molecular-weight gases without extensive fragmentation indicates specific applicability to analysis of photochemical smog, hydrocarbons, and so forth.

HAZARDOUS-GAS AND TRACE-GAS ANALYSIS

The identification and analysis of possible toxic concentrations of gases and vapors within a spacecraft cabin are not the only areas of gas monitoring of interest to the National Aeronautics and Space Administration (refs. 50, 51, and 52). Several other aerospace systems require analytical work with gases. These include:

(1) *Leak detection.*—The detection of leaks in storage vessels, transfer lines, and spacecraft is highly important when hazardous fuels are used. For instance, hydrogen is highly combustible, and a leak in an enclosed space, such as the interstage area of a spacecraft, must be detected before an explosive concentration is reached.

(2) *Purity tests.*—Gases used in certain aerospace applications, such as fuels and low-friction gas bearings, must have a high degree of purity; and highly sensitive analytical methods are required.

(3) *Trace-gas analysis.*—The products offgassed from a pyrolyzing surface, such as an ablative heat shield, can be used to give insight into its mode of decomposition and hence provide a measure of its utility.

The National Aeronautics and Space Administration has developed and sponsored the development of novel and improved techniques of gas analysis toward these ends. As in the case of cabin-atmosphere monitoring, some of these techniques have applicability to the monitoring of gaseous air pollutants, and we describe such developments in detail below.

Gas Chromatograph to Mass Spectrometer Transfer System

This system was developed at the NASA Ames Research Center by Dr. John Parker and Marius Kendall. They were studying ablative heat shields for reentry vehicles, and required a method of analyzing the material offgassed as the shield pyrolyzes. Standard analytical techniques call for the use of infrared spectroscopy, gas chromatography, and mass spectroscopy. Shield materials are pyrolyzed in helium in specific time increments and the offgassed products sampled with a pyrolysis sampler. The same sample should be run in both a gas chromatograph for high resolution and a mass spectrometer for high speed.

Coupling of comercially available gas chromatographic and mass spectroscopic equipment is the unique feature of the system developed at Ames. The gas-transfer chamber of the system takes the gases that are separated by the gas chromatograph and delivers them to the mass spectrometer in a continuous operation. There is also a provision for transfer to an infrared spectrometer. The chamber is a molecular separator and removes 95 percent of the helium carrier gas with a loss of only 5 percent of the sample gases. The gas chromatograph provides readout on strip chart and magnetic tape, and the mass spectrometer reads out on oscilloscope and strip chart. A computer program for reduction of data recorded on magnetic tape has also been generated at Ames.

Similar procedures are used in several laboratories engaged in gas analysis, including some doing air pollution research. The transfer mechanisms used are "home built" by each laboratory, and comparison with the Ames device is difficult. However, the high efficiency, reliability, and record of use speak well for the Ames procedure. It is not applicable to remote or mobile monitoring, but provides improved laboratory capability for air pollution research. The cost of the entire system used by Ames is $80,000, but the transfer mechanism alone would cost only about $2,000.

Technique for Measuring Nitrogen in Argon

This technique, developed for the NASA Kennedy Space Center, permits the measurement of nitrogen in argon gas in the parts-per-million range (ref. 53). NASA required such a technique because of the need for ultrapure argon as a welding gas, an inert atmosphere gas, and so forth. The technique is also applicable to detection of nitrogen in oxygen.

The technique requires material and equipment usually available in a gas-analysis laboratory and permits accurate measurements in a wide sensitivity range with only a small sample. Two gas-sampling chambers, a molecular sieve column, and a thermal-conductivity detector cell are used. The two chambers, arranged in series, are purged with a common carrier gas into the molecular sieve column that elutes into the detector cell. Gas samples to be analyzed are injected into a chamber (using a 30-cc sample) to determine concentrations of 5 ppm and up or the second chamber (using larger size samples) to determine concentrations in the parts-per-billion range. Determinations are made with a standard gas chromatographic column and a thermocouple detector.

The inherent reliability of this technique is limited only by the skill of the personnel who carry out the necessary manipulations. NASA has no plans to automate the procedure.

Because this system is designed for the analysis of trace gases in a single-component high-purity carrier gas, it is not applicable to general air pollution monitoring. However, in some instances a stack-gas analysis problem occurs in which only two components are present; in this case, the procedure might be suitably modified. The relatively low cost (approximately $500 for the sampling chambers and controls, exclusive of the gas chromatographic and detection systems) for implementation of this procedure would make it worthwhile for simple or easily definable two-gas systems.

Hydrogen-Gas-Detection System

This instrument system, which was developed at the Kennedy Space Center, permits continuous, electrical monitoring from a remote location (ref. 54). The areas monitored are potential gaseous hydrogen (H_2) leak points in liquid and gaseous H_2 transfer and storage areas, launch-pad towers, and pad aprons. Continuous atmospheric sampling is accomplished by air diffusion through the perforated cover and convection through the heated elements in the detector. The bridge output signal is a function of the amount of combustible gaseous H_2 in the atmosphere being sampled over a range of 0.5 to 4.0 percent in air, and is produced by an amount of gaseous H_2 in excess of a predetermined level (1 percent), actuating an alarm circuit at the control module and display panel.

The system consists of a commercially developed detector (sensor), a control module (General Monitors, Inc., model M135), and related alarm-transmission equipment and display panels. The detector head is basically two legs of a Wheatstone bridge; one leg provides the environmental temperature reference, and the other leg is covered with a H_2-sensitive catalyst. The control module provides individual control and readout for its detector; contains adjustment circuitry for setting parameters, a meter for measuring the percentage of H_2 gas concentration, and an alarm lamp for indicating when a specific H_2 concentration is exceeded. The display panel contains an alarm-condition lamp and confidence indicator (to monitor the excitation current to the detector). A self-test capability verifies that the detector is operable by electrically simulating a hazard condition at the detector.

The system has been in continuous use for more than a year at the Kennedy Space Center, and no serious difficulties have occurred. This is indicative of a high degree of system and equipment reliability.

This instrument system was specifically designed for the detection of H_2 in air; however, it is adaptable (with proper catalyst selection) for detecting any combustible gas. At a cost of less than $500 per complete module, it is eminently suited for use in both industrial hygiene and air-pollution-monitoring systems (especially when storage facilities for combustible gases and/or vaporized liquids exist).[1]

Other similar devices use platinum catalyst detectors for all combustible gases. With the proper catalyst selection, several combustible gases could be monitored, and the concentration level of each gas could be obtained on a readout meter.

Hazardous-Gas-Sampling System

This sampling system, developed by John Barnes at the Marshall Space Flight Center, has been installed at launch pads at Cape Kennedy. It provides multipoint sampling for the detection of leaks

---

[1] Similar devices are on the market. The Erdco Engineering Co. produces a combustible gas detector, using a platinum-coated catalyst as an element in a Wheatstone bridge circuit to detect combustible gases; the Mine Safety Appliances Co. produces a combustibie gas indicator, using a platinum filament as one element in a Wheatstone bridge circuit for indicating explosive and combustible limits of gases; the Teledyne Inc. produces a series 100 combustible analyzer, using catalytic combustion on a heated platinum filament as a detector for combustible gases in process streams; and the Davis Instruments produces a model 11–3400 portable combustion detector, using a coiled platinum filament as one element of a Wheatstone bridge circuit and reading out any imbalance as percent combustible gas.

in any fuel or gas system in the interstage areas of Saturn rockets. Because of the large volumes involved, several points must be monitored to determine concentrations of hazardous gases or vapors before explosive levels are reached.

Monitoring is done with a commercially available Consolidated Electrodynamics Corp. residual gas analyzer. Only tubing and sampling orifices are within the spacecraft; the rest of the analyzer is outside, approximately 500 feet from the sampling locations. The ¼-inch stainless steel tubing runs through the umbilical cable to the sampling orifices in the interstage area and is disconnected just prior to launch. The purpose of having most of the monitoring equipment outside the spacecraft is to minimize size and weight problems in the spacecraft.

A valve sequencer, operated by a computer, is used to collect gas from each orifice in succession. The four hazardous gases to be detected are H_2, O_2, monomethylhydrazine (CH_3NHNH_2) vapor, and nitrogen tetroxide (N_2O_4) vapor. Because only these four gases are of interest, an automatic peak selector rather than a spectrum scan is used. The coupling device on the tubing at the umbilical disconnect point restricts gas flow, so response time is approximately 50 seconds. The gold leak detector of the CEC mass spectrometer cannot be used in this system because the oil-pumped vacuum system of this detector introduces too much dirt into the lines. The gold leak has been replaced by a Granville-Phillips servocontrolled capillary, which regulates gas flow as a function of pressure.

This system has been very reliable and has survived launch conditions in its protected position under the launch pad. Continuous monitoring has been performed for periods in excess of 24 hours. The system is not sensitive to low concentrations in the parts-per-million range, the lower level of sensitivity being about 0.1 percent.

The cost of the system, exclusive of the computer and the mass spectrometer, is approximately $20 000. As an air pollution monitor it would be best suited for fixed-station monitoring, such as stack-gas monitoring, in an industrial process system. Remotely located instrumentation in a fixed installation provides a new approach to monitoring inaccessible locations.

Gaseous Hydrocarbon Monitor

This instrument was developed by the IIT Research Institute for the Marshall Space Flight Center to detect gaseous hydrocarbons (refs. 55 through 58). It was developed to sample gaseous systems rapidly (air, nitrogen (N_2), O_2, and helium (He)) used in Saturn and other spacecraft for life-support systems and for gas-lubricated bearings. Traces of hydrocarbons in O_2 systems could introduce chemical-reaction hazards, and hydrocarbons in a gas used for bearing lubrication could condense, increase friction, and impair the performance of the guidance system.

The gaseous hydrocarbon monitor is based on chromatographic separation and determination by flame ionization(fig. **18-3**). It automatically samples gas at a variety of pressures and separates all hydrocarbon constituents, in a specially packed short chromatographic column. Each hydrocarbon is introduced into a flame-ionization detector, and the detector's output is integrated and recorded by digital methods.

This instrument is calibrated for heavy hydrocarbons—heavier than $C_{14}Hn$—by using a known mixture of heavy hydrocarbons as a reference standard and for light hydrocarbons by using hexane as a reference standard. The lower limit of detection of hydrocarbons in gaseous systems is approximately 0.05 ppm (by weight).

This instrumental procedure replaces the standard analytical method of bubbling the gas through a solvent, evaporating it, and analyzing it with infrared spectroscopy. It greatly reduces the time required for analysis—each analysis taking 60 to 80 seconds—and increases sampling frequency.

This instrument would be useful for air pollution research in determining hydrocarbon content in air from automobile and aircraft exhausts and industrial production. It is not suited for mobile use, but would be a useful research tool.

Hydrogen Detector

This instrument is used as a portable detector for hydrogen in air, oxygen, nitrogen, or helium in the atmosphere or in gas supply systems (ref. 59).

The sniffer is a portable monitor that indicates the presence of H_2 in contact with activated palladium black by a change in color of a thermochromic paint and indicates the quantity of H_2 present by means of a sensor probe and continuous readout.

It consists of a short, thin-walled glass or metal tube, packed with 0.5 gram of powdered, activated palladium black held in place with compressed glass fibers(fig.**18-4).** The outside of the tube adjacent to the palladium black is coated with a thermochromic paint having two successive color-change points at 55° and 85° C. A temperature-sensing probe is inserted into the mass of palladium black and sealed through one end of the tube to give a continuous, direct indication of thermal changes. The gas to be sampled is drawn through the gas inlet and passes through the palladium black. If H_2 is present, it releases heat when it contacts the palladium black, the amount of heat being directly proportional to the amount of H_2 present. When the temperature of the tube reaches 55° C, the thermochromic paint changes color to give a visual indication of the presence of H_2. Suitably calibrated, the temperature-sensing probe can be used for direct, quantitative data.

This instrument can be used "as is" for industrial hygiene and air-pollution-monitoring applications as a detector for hazardous concentrations of H_2. The developers have also used this principle of activated palladium black and thermochromic paint to prepare a continuous, automatic sampling device. In this application a finely woven, but porous, glass-fiber tape is impregnated with the palladium black and coated with a pattern of thermochromic paint. An automatic H_2 detection system could be used in a continuous tape sampler with a color-change monitor.

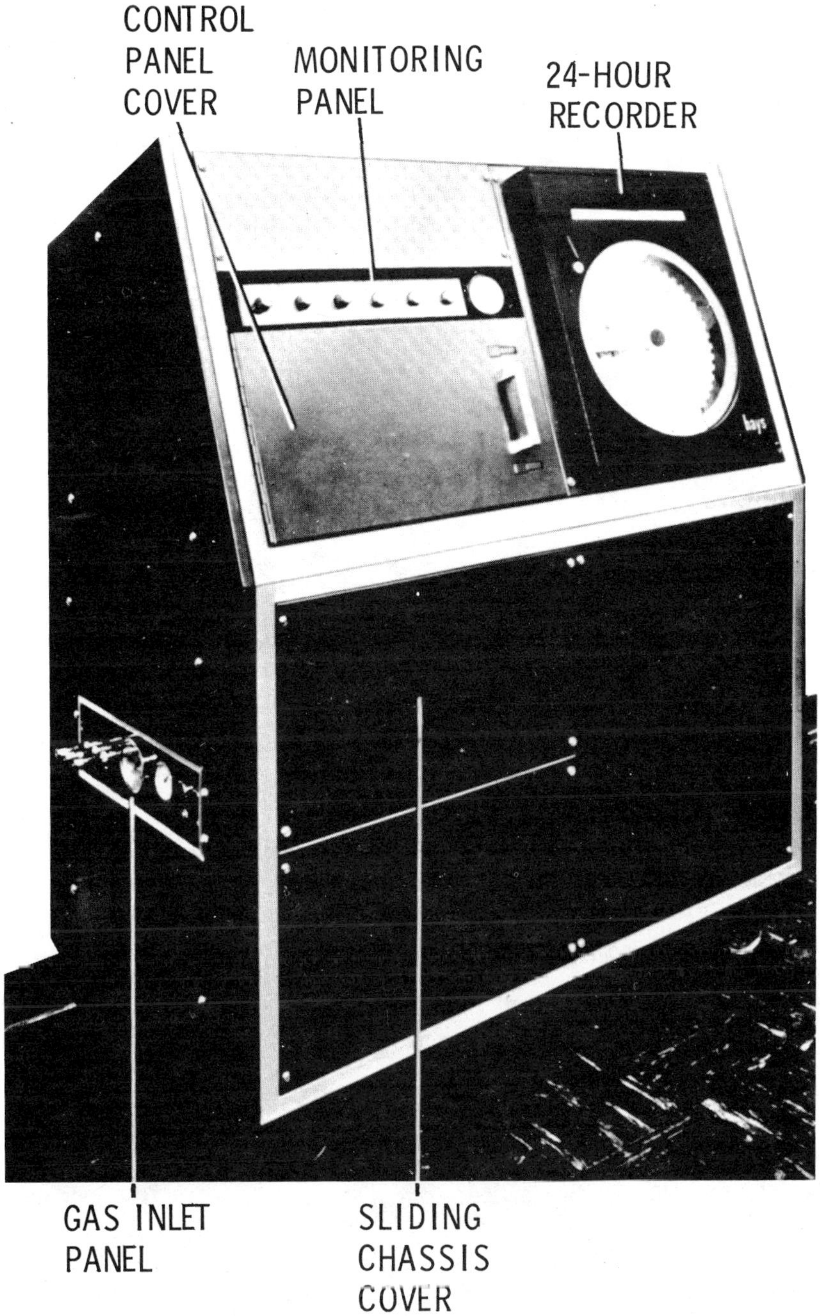

Fig. 18-3—Gaseous hydrocarbon monitor.

WELDING GAS MONITORS, WELDING SPECTROMETER ANALYZER

Innovations in welding technology have been among the significant NASA contributions to industrial processing. NASA sets extremely stringent requirements on welds in terms of materials, strength of weld, lack of porosity, minimal weight, and so forth. Heliarc welding is used extensively by NASA for joining aluminum and increasingly

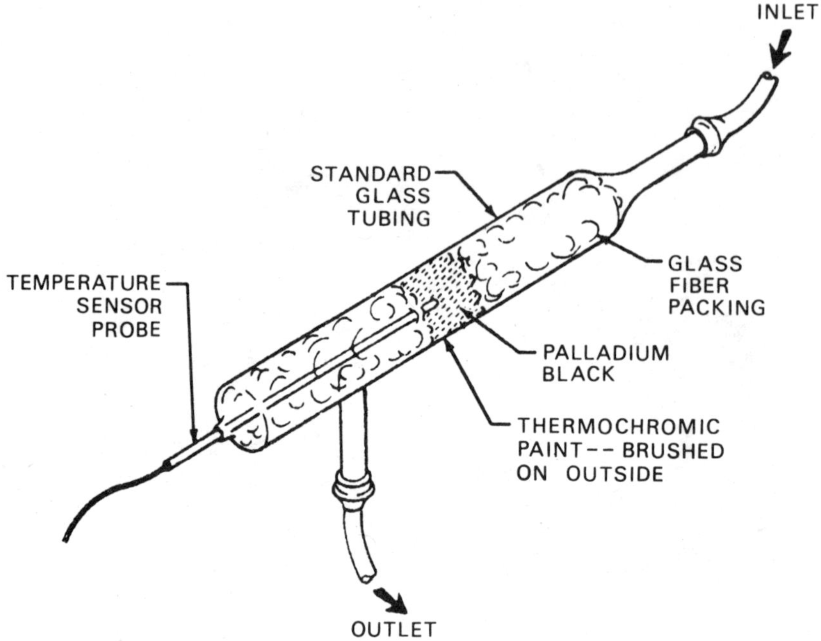

Fig. 18-4—Portable hydrogen detector.

more frequently by industry. In the process the weld is made under an inert gas such as helium (He) or argon (A) to avoid oxidation of aluminum. The amounts of H_2 and O_2 in the inert gas affect the quality of the weld, so a monitor was developed to measure these trace constituents. The monitor developed is applicable to air pollution monitoring.

This instrument (refs. 60 and 61) was developed for the Marshall Space Flight Center to monitor the O_2, H_2, and A constituents of the inert gas shield used in heliarc welding of aluminum. Even trace amounts of O_2 and/ or H_2 in the inert gas can result in an unsatisfactory weld.

This device is a spectrometer that uses photosensitive devices to monitor the emitted light from the inert gas shield(fig.**18-5)**.The light is collected through a lens system that incorporates a fiber optics bundle to follow the arc. The light is split in a diffraction grating, and three components (at 6562.8, 7771.9, and 8521.4 Å) go to three photomultipliers (for H_2, O_2, and A, respectively). The amounts of H_2 and O_2 are directly read out as ratios to A, which is used as an internal standard. The problem of blackbody radiation (background noise) entering the spectrometer is eliminated by the use of an oscillating rectangular prism in the nondispersed light beam. The oscillation causes the emission line to oscillate across the exit slit, which effectively produces an ac signal over the dc blackbody signal.

The instrument is sensitive to less than 5 ppm of O_2 or H_2 in A and is highly reliable. The electronic circuitry is solid state, and circuits have been included that allow use of an oscilloscope for rapid, simple optical alinement of the exit slits.

As constructed, this device detects only H_2 and O_2, but the design

Fig. 18-5—Welding spectrometer analyzer.

principles can be used to construct a monitor for virtually any gas, A heliarc welding instrument is basically a dc arc, and thus a dc arc could be used as the light source for an air pollution monitor. Photomultipliers responsive to characteristic emission lines of desired gases could then be employed to create an air pollution monitor based on these principles. An automatic continuous monitor is possible, because the gas stream to be measured could be passed continuously through the dc arc, and the output of the photomultipliers could be recorded on a strip-chart recorder or similar readout device. The cost of such an instrument would vary depending on the number of different constituents to be monitored, because each requires a separate photomultiplier and associated electronic circuitry.

CLEAN ASSEMBLY-AREA MONITORING

Contaminants in or on spacecraft and their components have been of great concern for two reasons. First, in a device as complex as a spacecraft, with literally thousands of functioning parts, contamination of even microscopic proportions can result in the failure of a mission, the loss of millions of dollars worth of equipment and effort, and hazards to personnel. An oxide film on a switch, a metallic particle in a valve, or dust motes on an optical surface can hinder the operation of the spacecraft, cause incorrect data to be returned from a sensor, or mean failure of a critical component.

Second, the effect of the experimenter's influence on his experiment is a problem in aerospace research. For example, the detection of the absence or presence of life on Mars will have far-reaching consequences on man's knowledge of the formation of the solar system and, indeed, his total body of knowledge. But if the space vehicle sent to Mars to detect life is already covered, internally and externally, with terrestrial organisms, the experiment will prove a failure.

These considerations have resulted in a great deal of NASA effort in the clean-room field—the use of noncontaminated areas for the manufacture and assembly of spacecraft and their components (refs. 62 through 71). Much work has also been done in sterilization for removal of contaminants. Our prime interest with respect to air pollution monitoring, however, is the NASA-developed technology for measuring contaminants in clean rooms. To determine the efficiency of the various designs and the uses of clean rooms, NASA has sponsored the development of sophisticated devices for measuring small amounts of contaminants. These devices are directly applicable to measurement of pollutants in the atmosphere. They are described in detail below.

Automatic Aerosol Particle Counter

This particle counter, developed by the Royco Instrument Co. and modified by the IIT Research Institute for the NASA Marshall Space Flight Center, monitors particulate contamination in a clean room (refs. 72 and 73). This includes counting the particles, categorizing them in terms of size, and recording the data. For adequate sampling information, counts were necessary at a minimum of three locations within the clean-room area.

The instrument is a light-scattering device; that is, a device in which the number and size of particles present in the sample stream cause a given degree of scattering of a light beam. A transducer head transforms this light signal into electrical signals for data recording.

The particle counter consists of a light source, an optical lens arrangement, and a photomultiplier tube(fig.**18-6).** The filament of the lamp is focused on a slit of known size in a light-shield plate. The lens system images this illuminated slit in the center of a sample stream. A similar lens system and a slit of the same size are used as a receiving system to carry the information to the photomultiplier tube. This arrangement defines an illuminated cube in space. The photomultiplier tube is excited only when a particle passes through this viewing volume. The scattered light intensity is directly proportional to the size of the particle, and thus the signal emitted by the photomultiplier can be calibrated according to particle size. Measurement of the size and number of pulses constitutes the counter. A pumping system conveys the sample stream through the transducer. A concentric air sheath surrounds the sample tube, and filtered air travels through the annular space at the same rate as the sample travels in the inner tube. This arrangement prevents loss of particles into the sample chamber, insuring that particles are counted only once and that the lenses remain clear. The sampling rate is 1 cu ft/min.

This instrument has high reliability and has been used both for tests and as a continuous monitoring device at the Marshall Space Flight Center. It is applicable in its present form for air pollution monitoring. Tests by NASA have shown that the count obtained with this instrument is directly proportional to that obtained in visual counting with a microscope—the standard technique. However, this device has

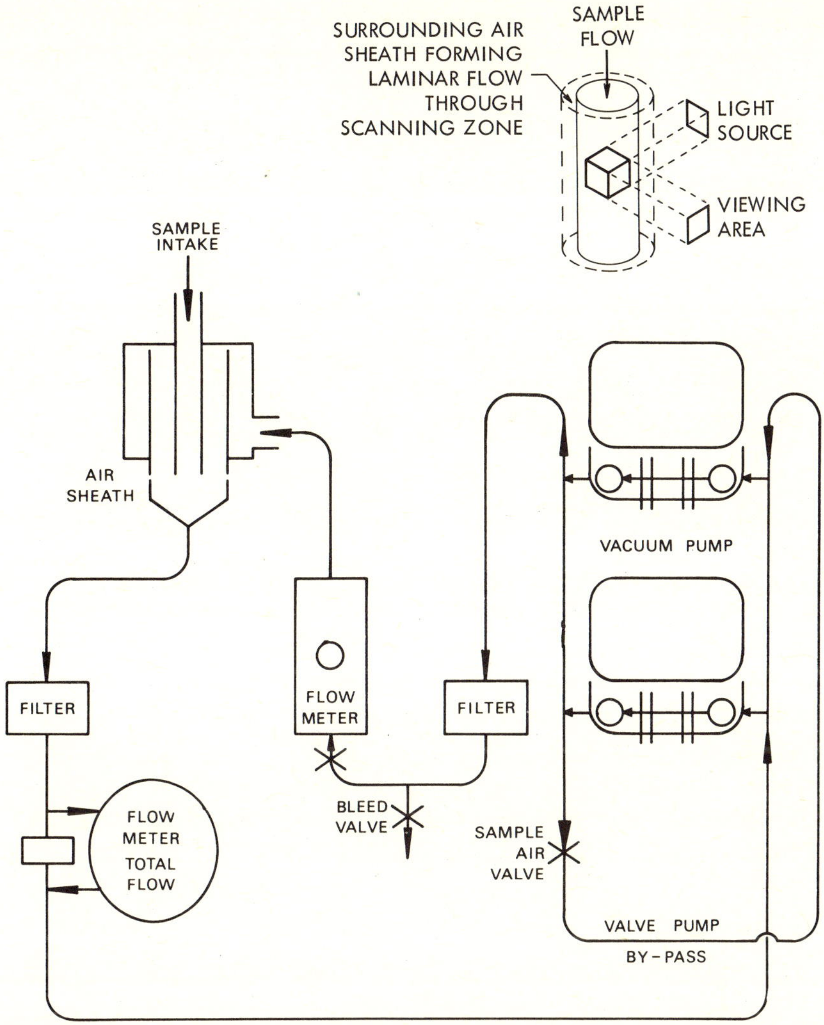

Fig. 18-6—Flow diagram of operation of automatic aerosol particle counter.

distinct advantages over microscopy:

(1) A microscopist is usually limited to counting particles 5 micrometers or larger in diameter, whereas this instrument can count particles with diameters as low as 0.6 micrometer.

(2) The instrument requires only minutes per count as opposed to at least 1 hour for microscopy.

(3) The instrument monitors continuously, providing a printed record for analysis of trends and/or association with specific events. It also has an alarm signal when a predetermined limit is exceeded.

The cost of this instrument on a production basis would be approximately $5000. It can be used by civic and industrial groups concerned with particulate air pollution.

Increased Sampling-Rate Monitor System

This instrument was developed for continuous monitoring of clean rooms (ref. 74). It samples at least 1 cu ft/min of air and counts particles 5 micrometers or larger in diameter.

Standard particle counters, such as the Royco PC200A, have a sampling rate of 0.01 cu ft/min. For extremely clean areas, such as a class 100 clean room (class 100 clean rooms have a maximum of 100 particles/cu ft), this sampling rate is insufficient for adequate measurement.

Sandia Corp. produced this prototype particle counter, which has a sampling rate of 1 cu ft/min. The device uses forward light scattering rather than the 90° scattering used in the Royco-IITRI counter. Adequate particle sizing was not possible because the light uniformity of the optics system was inadequate. However, the particle-count correlation was good.

This is a prototype instrument and no reliability figure has been established. However, the prototype was successfully used for a series of tests for monitoring a clean room, monitoring work locations within a clean room, and checking HEPA (high-efficiency particulate air) filters for leaks.

This device has the same advantages over microscopy that were discussed in regard to the automatic aerosol particle counter, and it is also suitable for civic or research use in air pollution monitoring. It appears to be quite similar to the IITRI-modified device (one of the modifications by IITRI increased the sampling rate of the Royco instrument 140-fold to approximately 1 cu ft/min), with the exception of inadequate particle sizing. This problem can be overcome by modifications to the optical system of the prototype.

Flight-Qualified Aerosol Particle Counter

This instrument was developed by the NASA Electronics Research Center (ERC) in cooperation with Block Engineering, Inc., and the Harvard University School of Public Health to obtain information on aerosol particle buildup in the cabin of an orbiting spacecraft (ref. 75). Because a flight-qualified instrument was needed, weight and power consumption were minimized, and the unit was strengthened and supplied with its own battery power.

The instrument weighs 5½ pounds, its envelope dimensions are 3¾ by 5 by 7½ inches, and its power consumption is less than 4 watts. It operates on the principle of light scattering by single particles and has an air-sampling rate of 1/60 cu ft/min.(fig. **18-7**). The amount of light scattered is a function of the particle size. This is a multichannel instrument, and sorts the information on sizes and counts into one of five channels: 0.5 to 1 micrometer, 1 to 2 micrometers, 2 to 3 micrometers, 3 to 5 micrometers, and 5 to 10 micrometers. This is not a continuous monitor; rather, it is designed for intermittent use for reasons of power consumption. It has built-in sequence timers, and in use it performs the following series of operations after the start button is pushed: a 10-second warmup, a 60-second sampling time, sequential display (via digital Nixie tubes) of the total count in each of the five particle-size ranges, and automatic termination of cycles.

This instrument has been thoroughly tested by ERC and has very high reliability. Battery life is sufficient for about 150 measurement cycles. ERC is currently engaged in refinement of the instrument to lower the minimum particle size measured to 0.3 micrometer (this is possible with an increase in light output), to improve the optical system, and to further miniaturize the electronic circuitry.

This instrument is suitable for use as an air pollution monitor. Its portability, low power consumption, and capability to measure particles simultaneously in five size ranges are not found in commercially available instruments. ERC personnel estimate that the instruments could be produced in volume at a price that would make them acceptable for use by civic and law enforcement groups.

CONTAMINANTS IN FLUIDS

Many fluids are used in spacecraft, and two of these, hydraulic fluid and fuel, must be of exceptionally high purity. Impurities in these fluids are of three forms: dissolved gases, other fluids, and particulate matter.

NASA has sponsored the development of monitoring instruments for dissolved gases and particulate impurities in fluids (refs. 76 and 77). We are including descriptions of these two instruments because they could be useful in air pollution monitoring after pretreatment of the air sample places the impurities in a liquid stream.

Ultrasonic Particle Counter

This instrument (refs. 55, 78, and 79) was designed and fabricated for the NASA Marshall Space Flight Center to detect, count, and size solid particles in a fluid stream. It has been used for monitoring hydraulic fluids, both in continuous and discrete sampling.

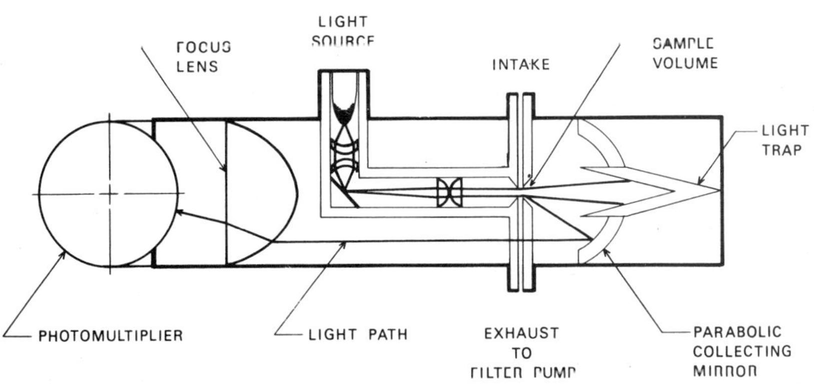

Fig. 18-7—Schematic diagram of flight-qualified aerosol particle counter.

This instrument uses the ultrasonic-reflectoscope method of analyzing echo amplitude from discrete reflectors (particles in the fluid). Two thousand pulses per second of a 5 megahertz frequency signal are passed through the fluid. If the ultrasonic wave hits a particle in the fluid, it is reflected back to the source, and the amplitude of the reflected signal is a function of the particle size. The device detects particles with a diameter of at least 25 micrometers. The maximum flow rate of the fluid is limited by the concentration of impurity and the percentage of the fluid sampled. For essentially 100 percent examination of each unit volume, the maximum flow rate in gallons per minute is made inversely proportional to particle count per volume. Thus, with a concentration of 10 particles/100 milliliters, the flow rate is 500 gal/min, and with a concentration of 100 particles/100 milliliters, the flow rate is 50 gal/min. The particle counting rate is 4000/second.

Skilled technicians are needed to maintain and calibrate the electronic components. Once calibrated, the instrument produces counts that can be correlated with a microscopic count, although the relationship of the correlation is not linear

This instrument is not suited for use as a remote monitor or mobile monitor for air pollution, but could be used in air pollution research. It would be most useful in counting particles captured in a filter or impactor and subsequently collected in an aqueous solution.

Nonvolatile Residue Nephelometer

This instrument (ref. 55), designed and built by the IIT Research Institute under contract to NASA Marshall Space Flight Center, automatically determines the nonvolatile soluble residue in solvents as a step in the purification of liquid-oxygen systems. Although pure liquid oxygen (LOX), the most widely used oxidizer in space vehicles, is stable and not subject to deterioration, mixtures of LOX with most organic materials explode under impact. Even the grease from fingerprints left on aluminum valves and transport lines during assembly has combined with LOX, and the mixture has caused explosions.

Thus, exceptional measures are taken to insure removal of incompatible organic materials and traces of grease from all LOX systems. After cleaning and degreasing, the systems are inspected by rinsing them with highly purified trichloroethylene or Freon, and the rinsings are examined for traces of contamination to verify the thoroughness of the cleaning. In the standard laboratory method, the traces are extracted by evaporating the rinsings just to dryness and then weighing the residue to determine the nonvolatile, soluble residue (NVR). The nephelometer is an automatic device for determining the presence of NVR in the solvent without going through this long, tedious procedure.

The instrument is based on the light-scattering characteristics of an aerosol spray of the solvent. Solvent is continuously fed to a nebulizer, in which it is atomized with a large volume of air to produce micrometer-size liquid particles of solvent and soluble contaminant. The large volume of air evaporates the volatile solvent, and the NVR is left behind as an aerosol in trichloroethylene vapor and air. The aerosol particle size depends on contaminant level. The aerosol, surrounded by a clean air sheath, is fed to a photometer, and the scattered light from a defined field of view is measured. The amount of scattered light is a measure of the contaminant level. The NVR nephelometer has two modes of operation; one in which a signal is given from a single sample, and one in which the contaminated sample is compared to a reference sample.

The device is quite reliable. Tests at Marshall Space Flight Center have shown that once calibrated, the instrument provides accurate analysis of the NVR in chlorinated solvents within 40 seconds per sample. These tests have been compared with the standard laboratory procedure, and a high degree of correlation has been found. The device will detect down to 0.1 to 0.5 milligram of degreasing NVR/1 solvent.

This device is useful in air pollution research for determining products of industrial processes, and for checking solvent purity and decreasing efficiency in general industrial application. It is not a general air-pollution-monitoring device but could be useful in research or industry where there is a requirement to measure NVR that might be released into the atmosphere. For such an application, it is much faster and as reliable as the standard laboratory method.

BIOSPHERE STUDIES

The interest of NASA in microbiological content of the Earth's atmosphere is threefold. First, NASA research vehicles (spacecraft) permit men to measure biological life at high altitudes for the first time. Second, techniques for detecting life on other planets are being tested in the Earth's atmosphere. Third, as mentioned in the section "Clean Assembly-Area Monitoring," spacecraft sent to other worlds must not carry terrestrial organisms with them. Thus, NASA must know to what extent a spacecraft can become biologically contaminated as it passes through the atmosphere.

These studies (refs. 80 through 90) and the techniques and instrumentation developed in their course are important to air pollution monitoring. Most obvious, of course, are the collection techniques that have been developed. Several of these are applicable to collection of any type of particulate matter from the atmosphere. Also of interest are sampling and counting techniques, because microbiological contamination is a form of air pollution (i.e., in hospitals), and also because air pollution research makes use of vegetative and spore cultures for particles in aerosols.

Accordingly, some of these NASA developments that are novel and/or improved instruments or techniques applicable to air pollution monitoring will be described below.

Air Sampler for Minute Particles

This instrument (ref. 91) was designed and fabricated for collection of extraterrestrial dust greater than 0.1 micrometer in diameter. The particles were to be both collected and protected for use in analysis.

The device consists of an air-ejector impactor sampler that causes impaction of particle-laden air onto several collection surfaces within a collection cylinder (fig. 18-8). When not in operation, the collection cylinder is retracted within a protective envelope.

An air-ejector pump, comprised of a compressed gas source and associated tubing, valve, and flowmeter, simultaneously sends compressed gas through valves to a nozzle in a diffuser and into a pneumatic actuator. The gas flow into the actuator extends the collector from the retracted position to the exposed position. Gas flow into the nozzle and diffuser entrains surrounding air and produces a reduced pressure within the collection cylinder, causing air to flow through its impactor slits and particles to impact on the collection surfaces within. When the gas flow is shut off, a spring draws the collection cylinder back into the protected position.

This device has been shown to have high reliability because it uses a compressed gas for power (the only electrical power required is for

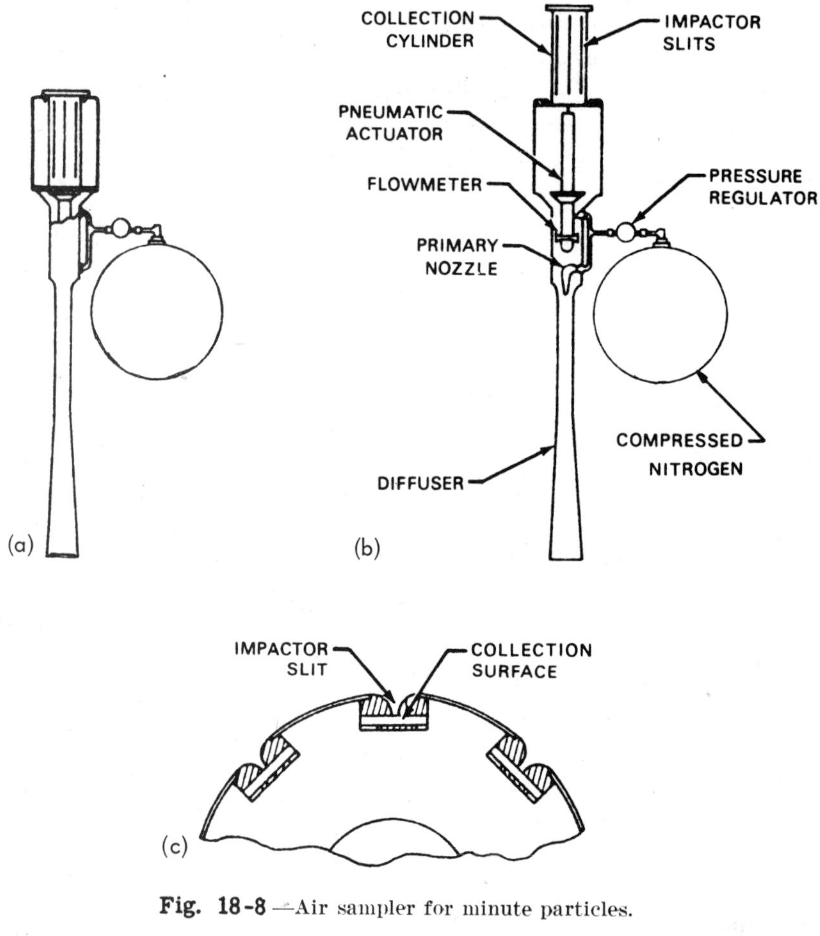

Fig. 18-8—Air sampler for minute particles.

the valve, and that can be provided by batteries) and is very simple to operate.

This is an excellent device for airborne collection of particulate matter at any altitude. Its primary application would be for aircraft monitoring. However, the device can operate in a stationary position, because it produces its own airflow.

Several instruments for sampling biological species at high altitudes have been developed by NASA efforts. These are relatively sophisticated units using motor-driven, air-movement devices for sampling at high altitude and eliminating microbiological contamination while reaching sampling altitude. For these reasons, these devices appear to be overly complex for the requirements of air pollution monitoring. Nevertheless, the Litton Mark II and Mark III sampling systems (ref. 92), the NCAR Corp. enclosed cryocondenser for air recovery (ref. 93), the University of California microbe sampler (ref. 94), the NASA Jet Propulsion Laboratory work conducted by Gerald A. Soffen (ref. 95), and the X–15 mounted micrometeorite collector may be of interest to air pollution researchers.

Wolf Trap Mars Microbe Detector

This instrument (ref. 96) was designed and built by the University of Rochester for NASA. It was designed to detect microbial life on the surface of Mars and has features necessary for surviving launch and Martian environments.

The device collects soil samples and places them in five separate enrichment cultures. It then optically monitors the cultures for evidences of turgidity caused by multiplication of microbiological organisms. A sensitive pH monitor measures the acidity changes in each culture. The device has been tested in the Mojave Desert and successfully detected life. Engineering for space-flight requirements insures a very rugged device. Standard laboratory culturing and counting techniques are usually less expensive and more reliable. However, in cases where remote monitoring might be necessary, this device would be applicable.

A device operating on the principle of microcalorimetry, that is, detection of a very small amount of heat, has been produced by Beckman Instruments for NASA Ames Research Center.

Another instrument for remote detection of life employs the "firefly reaction," that is, the generation of light that occurs when the firefly enzyme, luciferin, reacts with adenosinetriphosphate (ATP). An instrument to extract ATP from a soil sample, introduce luciferin, and monitor the light produced (if any) has been produced by Hazleton Laboratories for the NASA Goddard Space Flight Center (refs. 97 through 103).

These instruments were designed for life detection on Mars and, though operating on different principles, have the same utility for air pollution monitoring as the Wolf trap.

References

1. Gaeke, G. C.; and West, P. W.: Fixation of Sulfur Dioxide as D sulfitomercurate (II) and Subsequent Colorimetric Estimation. Anal. Chem., vol. 28, 1956, pp. 1816–1819.
2. Greenburg, L.; and Jacobs, M. B.: Sulfur Dioxide in New York City Atmosphere. Indus. Eng. Chem., vol. 48, 1956, pp. 1517–1521.
3. Thomas, M. B.; et al.: Automatic Apparatus for Determination of Small Concentrations of Sulfur Dioxide in Air. New Counter-current Absorber for Rapid Recording of Low and High Concentrations. Indus. Eng. Chem., vol. 18, 1946, pp. 383–387.
4. Jacobs, M. B.: Ultramicrodetermination of Sulfides in Air. Anal. Chem., vol. 29, 1957, pp. 1349–1351.
5. Saltzman, B. E.: Colorimetric Microdetermination of NO_2 in the Atmosphere. Anal. Chem., vol. 26, 1954, pp. 1949–1955.
6. Saltzman, B. E.; and Wartborg, A. F.: Absorption Tube for Removal of Interfering Sulfur Dioxide in Analysis of Atmospheric Oxidant. Anal. Chem., vol. 37, no. 6, 1965, pp. 779–782.
7. Sikes, J. H.; and Yamamora, S. S.: Direct Spectrophotometric Fluoride Determination. Anal. Chem., vol. 34, no. 10, Sept. 1962, pp. 1308–1312.
8. Zaleiko, N. S.: Automation of Colorimetric Chemical Procedures as Applied to Analytical Requirements in Air Pollution Programs. J. Air Pollution Control Assoc., vol. 13, no. 11, Nov. 1963, pp. 531–534.
9. Altshuller, A. P.; and McPherson, S. P.: Spectrophotometric Analysis of Aldehydes in the Los Angeles Atmosphere. J. Air Pollution Control Assoc., vol. 13, no. 3, Mar. 1963, pp. 109–111.
10. Guerrant, G. O.: Portable Instruments Based on Flame Ionization Detectors for Analysis of Air for Trace Organic Constituents. Anal. Chem., vol. 37, no. 4, Apr. 1965, pp. 516–520.
11. Matoi, H. J.; and Rounds, G. L.: Electrostatic Sampler for Dust-Laden Gases. Anal. Chem., vol. 27, no. 5, 1955, pp. 829–830.
12. Jacobs, M. B.; et al.: Continuous Determination of CO and Hydrocarbons in Air by a Modified Infrared Analyzer. J. Air Pollution Control Assoc., vol. 9, 1959, pp. 110–114.
13. Ping, A. Y.; Clayton, L. R.; and McEwen, T.: The Application of Silica Gel in Source Testing, Part 1: Collection of Samples (paper no. 66–79). Proc. 59th Annual Meeting, Air Pollution Control Assoc. (San Francisco, Calif.), June 20–24, 1966.
14. Barringer, R.; and Newbury, B. C.: Molecular Correlation, Spectrometer for Sensing Gaseous Pollutants (paper no. 67–196). Proc. 60th Annual Meeting, Air Pollution Control Assoc. (Cleveland, Ohio), June 20–24, 1967.
15. Paxton, R. R.: Measuring Rate of Dust Fall. Rock Products J., vol. 54, no. 2, Feb. 1951, pp. 114–118.
17. Watson, H. H.: Errors Due to Anisokinetic Sampling of Aerosols. Am. Ind. Hyg. Assoc. Quart., vol. 15, no. 1, Mar. 1954, pp. 21–25.
18. Gruber, C. W.: Recommended Standard Methods for Continuing Air Monitoring for Fine Particulate Matter. J. Air Pollution Control Assoc., vol. 13, no. 9, Sept. 1963, pp. 397–404.
19. Clark, W. E.; and Whitby, K. T.: Electrical Aerosol Particle Counting and Size Distribution Measuring System for the 0.015 to 1 Micron Size Range. Tellus, vol. 18, no. 2, 1966, pp. 573–586.
20. Langer, G.: An Acoustic Particle Counter, Preliminary Results. J. Coll. Sci., vol. 20, no. 6, 1965, pp. 602–609.
21. Keily, D. T.: An Airborne Cloud Drop Size Distribution Meter. J. Meteorol., vol. 17, 1960, p. 349.
22. Bulba, E.; and Silverman, L.: A Mass Recording Stack Monitoring System for Particulates (paper no. 65–141). Proc. 58th Annual Meeting, Air

Pollution Control Assoc. (Toronto, Can.), June 20–24, 1965.

23. CHAUDET, JULIAN H.; AND DESCHMERTZING, HANNIBAL: Utilization of Infrared Spectrophotometry in Microcontaminant Studies in Sealed Environments, Final Report. Rept. no. SAM–TR–67–2, June 2–29, 1965, Melpar Inc., Falls Church, Va., Jan. 1967. AF 41(609)–1962; AD 650 000.
24. ANDERSON, I. C.; BETHEA, R. M.; AND BRUCE, R. A.: Contaminant Collection and Identification (N67–14247). Anon.: Selected Papers on Environmental and Attitude Control of Manned Spacecraft. NASA TM X–1325, 1966.
25. ADAMS, J. D.; CONKLE, J. P.; MABSON, W. E.; WELCH, B. E.; AND ZEFT, H. J.: A Detailed Study of Contaminants Produced by Man in a Space Cabin Simulator at 760 mm Hg. NASA CR–87451, 1967.
26. HODGSON, F. NEIL; PUTSINGER, JOHN V., JR.; AND ROSS, WILLIAM D.: Identification of Volatile Contaminants of Space Cabin Materials. Final Report. Rept. no. AMRL–TR–66–53, Monsanto Research Corp., Dayton, Ohio, June 1966. AF 33(615)–1779; AD 642 054.
27. HULTZMAN, WILLIAM W.: Characteristics, Selection and Use of Residual Gas Analyzers. NASA TM X–1281, 1964.
28. CONKLE, JAMES P.: Sampling and Analysis of Atmospheric Contaminants (N66–24078). Anon.: Proc. Conference on Atmospheric Contamination in Confined Spaces. Wright-Patterson AFB, Dayton, Ohio, Mar. 30–Apr. 1, 1965. Rept. no. AMRL–TR–65–230, 1965. AF 33(657)–11305; AD 629 622.
29. FLOWERS, O. L.; HONMA, M.; NYGREN, R. E.; RINEHART, R. W.; TUTTLE, W. N.; ET AL.: Program for Delineation of Trace Constituents of a Closed Ecologic System, Final Report. Rept. no. SAM–TR–67–4, June 2–29, 1965. Lockheed Missiles & Space Co., Sunnyvale, Calif., Jan. 1967. AF 41(609)–2788; AD 650 003.
30. BRIGDEN, WAYNE H.; AND ROMAN, JAMES: Flight Research Program: IV. A Small Gas Analyzer for Aerospace. Aerospace Med., vol. 37, no. 10, Oct. 1966, pp. 1037–1040.
31. BRIDGEN, WAYNE H.; AND ROMAN, JAMES: Flight Research Program: V. Mass Spectrometer in Medical Monitoring. Aerospace Med., vol. 37, 12, Dec. 1966, pp. 1213–1217.
32. HALL, L. G.; AND MOBINSON, C. F.: Small General Purpose Cycloidal-Focusing Mass Spectrometer. Rev. Sci. Instr., vol. 27, no. 7, July 1956, pp. 504–508.
33. MCCLEARY, U., JR.: Mass Spectrometry. Ind. Res., vol. 7, no. 5, May 1965.
34. HEDMAN, JOHN; LITTLE, W. SHANNON; SHAFFER, LARRY W.; AND WAKEFIELD, WAYBERN L.: Development of a Non-Magnetic Flight Type Mass Spectrometer, Final Report, General Electric Co., Daytona, Beach, Fla.
35. MOBERG, M. L.: Analysis of Trace Contaminants in Closed Ecologic Atmospheres. Rept. No. SAM–TR–66–99, Aerojet-General Corp., Azusa, Calif., Dec. 1966. AF 41(609)–2783; AD 647 618.
36. MOBERG, M. L.: Analysis of Trace Contaminants Contained in Samples From a Closed Environment at 258 mm Hg, Final Report. Rept. no. SAM–TR–67–6, Mar. 16–Apr. 12, 1966. Aerojet-General Corp., Azusa, Calif., Feb. 1967. AF 41(609)–2783; AD 651 287.
37. GIBSON, A. DAVID: A Tape Programmed Sequential Sampler. Rept. no. PB–175925, Natl. Center for Atmospheric Res., Boulder, Colo., 1967.
38. ANON.: Hot-Wire Detector for Chemically Active Materials Used in Gas Chromatography. NASA Tech. Brief B66–10139, MSC–269, 1966.
39. JONES, G. E.: Microwave Spectroscopic Identification of Atmospheric Contaminants, Status Report. NASA CR–79490, 1966.
40. EASLEY, W. C.; AND WHITE, W. F.: Practical Use of Microwave Spectroscopy for Trace Contaminant Detection. Presented at the Fourth Annual Conf. on Atmospheric Contamination in Confined Spaces. Wright-Patterson AFB, Dayton, Ohio, Sept. 10–12, 1968.
41. WHITE, W. F.: Qualitative Analysis of Gas Mixtures by Microwave Spec-

troscopy. Presented at the 155th Natl. Meeting, Am. Chem. Soc. (San Francisco, Calif.), Mar. 31–Apr. 5, 1968.

42. Harrington, H. W.: On the Separation of the Broadening-Relaxation Time and Molecular Concentration From Pure-Rotational Spectroscopic Intensity Data. J. Chem. Phys., vol. 46, no. 10, May 15, 1967, pp. 3698–3707.

43. Jones, G. E.: Microwave Spectroscopic Identification of Atmospheric Contaminants, Status rept., Apr.–Aug. 1968. Dept. of Physics, Miss. State Univ., 1968.

44. Rinehart, E. A.: Investigation of Available Methods of Microwave Spectral Line Intensity Measurements. Preliminary Report, Univ. Wyoming.

45. Lee, C.: Investigation of a Dual Resonant Cavity Absorption Cell Using the Double-Resonance Microwave Spectroscopy Modulation Technique. M.S. Thesis, Univ. Va., 1968.

46. Volpicelli, R. J.; Stiefvater, O. L.; and Flynn, G. W.: A Study of Trace Contaminant Identification by Microwave Double Resonance Spectroscopy. NASA CR–967, 1967.

47. White, W. F.: The Application of Microwave Spectroscopy to Contaminant Analysis. Chem. Eng. Prog. Sym. Ser., vol. 62, no. 63, 1966, pp. 70–75.

48. Rotheram, M. A.: Laboratory Contaminant Sensor, Final Report. NASA CR–66606, 1968. (Addendum, NASA CR–66606–A.)

49. Poschenrieder, W. P.; and Warneck, P.: Improvement and Optimization of a Mass Spectrometer Employing a Photoionization Source. NASA CR–1018, 1968.

50. Johnson, E. T.; Miller, R. R.; and Williams, D. D.: A Method for the Determination of Strong-Acid Vapors in the ppm Range, Final Report. Rept. no. NRL–6332, Naval Research Laboratory, Washington, D.C., Oct. 14, 1965. AD 624 296.

51. Davies, P. J.; and Hamlin, A. G.: A Sorption Analyser for the Determination of Trace Components in Gas Streams. Rept. no. PG–700(CA), United Kingdom Atomic Energy Authority, Capenhurst, England, 1966.

52. Young, Robert A.: Measurement of Nitric Oxide in the Earth's Atmosphere, Final Report. Rept. no. DASA–1887, Stanford Research Inst., Menlo Park, Calif., Mar. 23, 1967. DA–49–146–XZ–112; AD 649 829.

53. Fickey, E. W.: Gas Analysis Technique for Measuring Nitrogen in Argon. Rept. no. 67–5, The Bendix Corp., Launch Support Div., Kennedy Space Center, Fla., Feb. 8, 1967.

54. Wojtasinski, R. J.: Hydrogen Gas Detection System. Flash Sheet No. KSC–10065, Kennedy Space Center, Fla., May 10, 1967.

55. Riehl, W. A.: Monitoring of Impurities in Fluids and Gases. Marshall Space Flight Center, Huntsville, Ala. (N66–17710) Anon.: Symposium on Technology Status and Trends. Apr. 21–23, 1965. NASA SP–5030, 1966. N66–17706.

56. Stockham, John D.: Supplemental Operations Manual for the IITRI Continuous Gas Analyzer. Report no. IITRI–U6000–40, Suppl., IIT Res. Inst., Chicago, 1968.

57. Scholz, Robert G.: Development of Analytical Methods and Test Equipment for Determination of Hydrocarbon Contamination, Final Report. Rept. no. IITRI U6000–40, IIT Res. Inst., Chicago, Nov. 29, 1965.

58. Scholz, Robert G.: Operation Manual for IITRI Continuous Gas Analyzer. IIT Res. Inst., Chicago, Mar. 22, 1966.

59. Dayan, V. H.; and Rommel, M. A.: "Sniffer" Used as Portable Hydrogen Leak Detector. NASA Tech Brief B66–10356, M–FS–846, M–FS–806, Aug. 1966.

60. Grove, E. L.: Portable Spectrometer Monitors Inert Gas Shield in Welding Process. NASA Tech Brief B67–10326, M–FS–12144, Sept. 1967.

61. Grove, E. L.: Design, Development, and Fabrication of a Portable Spectrometer-Type Monitor. Rept. no. IITRI–U6010–18, June 29, 1964–Jan. 31, 1966. IIT Res. Inst., Chicago, 1966.

62. POWERS, EDMUND M.: Microbial Profile of Laminar Flow Clean Rooms. NASA TM X–55320, 1965.

63. ETTINGER, H. J.; MOORE, R.; POSNER, S.; AND PRATT, O. R.: Evaluation of Particle Sizing Techniques; Comparison of Computer Particle Sizing Programs, Technical Progress Report. Rept. no. LF–30, July 1966. Lovelace Foundation for Medical Education and Research, Albuquerque, N. Mex., AT (29–2)–1013; N67–22326.

64. ANON.: Better Room Climate—"White Rooms"—Better Performance. Staub, vol. 26, no. 8, Aug. 1966, pp. 37–40. Israel Program for Sci. Trans., Jerusalem.

65. FAVERO, MARTIN S.: Microbial Contamination in Conventional and Laminar Flow Clean Rooms. NASA CR–81751, 1966.

66. MICHAELSEN, G. S.; RUSCHMEYER, O. R.; AND VESLEY, D.: The Bacteriology of Clean Rooms. NASA CR–890, 1967.

67. ANON.: Contamination Control Principles. NASA SP–5045, 1967.

68. POSTMA, A. K.; AND SCHWENDIMAN, L. C.: Current Research on Aerosols at Pacific Northwest Laboratory. Presented at the 60th Annual Meeting of the Air Pollution Control Assoc., Cleveland, Ohio, June 11–16, 1967. Rept. no. BNWL–SA–1255, Battelle-Northwest, Pacific Northwest Laboratory, Richland, Wash., May 10, 1967.

69. COLLINS, R. H.; THOMAS, A. L.; AND MUMMA, V. R.: A Particle Size Analyzer for Aerosols. Ann. N.Y. Acad. Sci., vol. 99, June 29, 1962, pp. 298–308.

70. CRIDER, W. L.; AND STRONG, A. A.: Flame Ionization-Pulse Aerosol Particle Analyzer (FIPAPA). Rev. Sci. Instr., vol. 38, no. 12, Dec. 1967, pp. 1772–1775.

71. CRAWFORD, A.; AND RICH, E.: Engineering Guidelines for Clean Assembly and Sterilization of Spaceflight Hardware. Rept. no. X–624–67–578, Goddard Space Flight Center, Greenbelt, Md., Nov. 1967.

72. LEWIS, THEODORE W.: Evaluation of an Automatic Aerosol Particle Counter for Measuring the Airborne Contamination Level in a Controlled Environment. NASA TM X–53416, 1966.

73. NELSON, M. B.: Clean Room Air Sampler Counts, Categorizes, and Records Particle Data. NASA Tech Brief B67–10076, M–FS–2221, June 1967.

74. MASHBURN, J. C.; AND WHITFIELD, W. J.: Development of an Increased Sampling Rate Monitoring System. NASA CR–79537, 1966.

75. ANON.: Development and Fabrication of a Flight Qualified Aerosol Particle Analyzer. Final Rept., July 1–Sept. 30, 1965, Block Engineering, Inc., Cambridge, Mass., Feb. 7, 1967.

76. ANTHONY, H. V.; DREW, J. G.; LARUE, E. F.; AND MCDONALD, J. P.: Study of Contamination Sensors. Vol. I: Executive Summary Report. NASA CR–78437, 1966.

77. GAYLE, J. B.; AND ROMINE, J. O.: Evaluation of High Accuracy Products Corporation Model PC 202 Automatic Particle Counter. NASA TM X–53145, 1964.

78. BEAKLEY, JOHN WILLIAM; MASHBURN, J. C.; AND WHITFIELD, W. J.: Microbiological Studies Relating to Clean Environments. Part I: Evaluation of the Efficiency of a Class 100 Laminar Flow Clean Room for Viable Contamination Cleanup. NASA CR–83246, 1966.

79. ANON.: Equipment Specification: Contamination Level Indicator for Subsieve Particles—Style 58E003. Specification SS 50–264, Howe Sound, Inc., Sperry Products Co. Div., Brookfield, Ill., July 12, 1960.

80. ANDRUS, PAUL G.; DUFFEE, RICHARD A.; AND SHULZ, EARL J.: Investigation of Methods to Measure Size Distribution of Fog Droplets and Condensation Nuclei, Final Report. Rept. no. AFCRL–65–724, Apr. 1, 1962–Apr. 17, 1965, Battelle Memorial Inst., Columbus, Ohio. May 16, 1966. AF 19(628)–344; AD 633 232.

81. BISCAYE, PIERRE E.; FEELY, HERBERT W.; FRENCH, EUGENE R.; KATZMAN, DAVID; AND PANACCIONE, JOHN B.: Flight Data and Results of Radiochemical Analyses of Filter Samples Collected During 1965 Under Project

Stardust. Rept. no. HASL–176, Isotopes, Inc., Westwood, N.J., May 1, 1966. DA–49–146–XZ–079.

82. JUNGE, CHRISTIAN E.: Studies of the Chemistry of Unpolluted Atmospheres, Final Report. Stanford Res. Inst., Menlo Park, Calif., Nov. 10, 1966. Cwb–11151, AD 646 465.

83. AMICK, JAMES, L.; AND GILL, GERALD C.: A Stationary Cylindrical Wind Speed and Direction Sensor for Use in Model Testing. No. 130, Dept. of Meteorology and Oceanography, Univ. Mich., Sept. 1967.

84. GILL, GERALD C.: Wind Speed Sensors—Basic Principles and Dynamic Response. Seminar No. 2 from a Seminar Series at White Sands Missile Range, Oct. 24–28, 1966, Temperature and Wind Measuring Systems—Basic Principles and Dynamic Response, Dept. of Meteorology, Univ. Mich.

85. ANON.: Product Specification Brochure for Model 1500 Moving Slide Impactor. Meteorology Research, Inc., P.O. Box 637, Altadena, Calif.

86. LEDOUX, F. N.: Clean-Room Facilities for Explorer XXXV Spacecraft. NASA TN D–4577, 1968.

87. LEDOUX, F. N.: Microbial Decontamination and Sampling Program for Anchored Interplanetary Monitoring Platform (AIMP–E) Spacecraft. Rept. no X–723–67–375, Goddard Space Flight Center, Greenbelt, Md., July 1967.

88. LEDOUX, F. N.: Decontamination of the AIMP–D Spacecraft. Rept. no. X–723–67–171, Goddard Space Flight Center, Greenbelt, Md., Apr. 1967.

89. PICCIOLO, G. L.; POWERS, E. M.; AND RICH, E.: An Experiment to Detect Microorganisms in the Upper Atmosphere Flown on Aerobee NASA 4.150. Rept. no. X–624–66–359, Goddard Space Flight Center, Greenbelt, Md., Dec. 1966.

90. POWERS, E. M.: Microbiological Burden on the Surfaces of the AIMP Spacecraft, Part 1. Rept. no. X–624–66–342, Goddard Space Flight Center, Greenbelt, Md., May 1966.

91. WOOD, R. C.: Air Sampler Collects and Protects Minute Particles. NASA Tech Brief B67–10661, HQ–10037, Dec. 1967.

92. PEDERSON, PAUL D., JR.: Microbiological Exploration of the Stratosphere. NASA CR–1101, 1967.

93. MARTELL, E. A.: Sixteenth Quarterly Status Report. Mar. 15–June 15, 1968, Natl. Center for Atmospheric Res., Boulder, Colo., June 15, 1968.

94. COOPER, ROBERT C.; NICHOLSON, DAVID; AND OSWALD, WILLIAM J.: Detection and Study of Microbial Population in Upper Atmosphere. Simulated High Altitude Microbial Sampling With an Electrostatic Precipitator. Univ. Calif. Sukalo, Lucille H., ed. Proc. Atmospheric Biology Conference, Apr. 13–15, 1964. NASA CR—62786, 1965.

95. SOFFEN, GERALD A.: Atmospheric Collection at 130,000 feet. Calif. Inst. Technol. (N65–24000). Sukalo, Lucille H., ed. Proc. Atmospheric Biology Conference, Apr. 13–15, 1964. NASA CR–62786, 1965.

96. RIED, L., JR.: Wolftrap Mars Microbe Detection Device, Final Report, NASA CR–85589, 1967.

97. ANON.: Fireflies for Space Research. Goddard Space Flight Center, Greenbelt, Md.

98. CHAPPELLE, E. W.; PICCIOLO, G. L., AND ALTLAND, R. H.: A Sensitivity Assay for Flavin Mononucleotide Using the Bacterial Bioluminescence Reaction. Biochem. Med., vol. 1, no. 3, Dec. 1967, pp. 252–260.

99. PICCIOLO, G. L.; CHAPPELLE, E. W.; AND RICH, E.: Use of an Assay for Flavin Nucleotides as a Technique for Detecting Extraterrestrial Life. Bioscience, vol. 18, no. 5, May 1968, pp. 427–428.

100. CHAPPELLE, E. W.; RICH, E.; AND MACLEOD, N. H.: Prevention of Protein Denaturation During Exposure to Sterilization Temperatures. Science, vol. 155, no. 3767, Mar. 10, 1967, pp. 1287–1288.

101. ANON.: Bioluminescence. Rept. no. X–624–68–43, Goddard Space Flight Center, Greenbelt, Md. 1968.

CHAPTER 19

CONTROL OF AIR POLLUTION

Principles of Control:

a. There are two fundamental approaches to control:

(1) Control of the pollutant at the source so that excessive amounts are not emitted to the atmosphere.

(2) Natural dilution of the pollutant to such a concentration that man, animals, vegetation, and materials will not be harmed.

b. Control of the Pollutant at the Source:

(1) Control of the pollutant at the source may be accomplished by:

(a) Keeping the pollutant from coming into existence.

(b) If it comes into existence, destroy, alter, or trap it before it can get out into the atmosphere.

(2) The most positive way to abate air pollution is to keep the pollutant from coming into existence or, if this is impossible, to keep the quantity that does form to a bare minimum.

(a) For industries already in operation, this may be done by:

Raw material changes.
Operational changes.
Modification or replacement of process equipment.
Adopting an alternate method to produce the same product.
More efficient operation of existing equipment.

(b) For industries developing entirely new products or processes, it is important to think about pollution problems in the research laboratory. With such an approach:

It may be found that, when looked upon from an air pollution angle, other lines of research differing from those being followed, may prove more promising.

Or, by developing a more expensive process which reduces or eliminates the need for air pollution control equipment, the overall production costs may be cheaper.

If the pollutant cannot be prevented from forming, and its uncontrolled mass rate of emission to the atmosphere is excessive, equipment which destroys, alters, or traps the pollutant will be required.

(c) There are several basic methods of reducing a pollutant to tolerable levels of concentration before it is emitted.

Destroying the pollutant. In destroying the pollutant, two general methods are used: use of fire and use of catalytic burners.

Masking the pollutant. Odor masking is the process of eliminating the human nasal perception of one odor by superimposing another odor to create a brand new odor sensation, preferably pleasant. In no way is the pollutant destroyed or transformed. The odor is merely "covered up."

Counteracting the pollutant. This method is also limited to those pollutants which cause odor. It actually "cancels out" the odor. It has been found that certain pairs of odors are antagonistic, and when two such odors are intermingled, both odors are diminished when sniffed.

Collecting the pollutant. This most widely used method of abating air pollution is to trap the pollutant by use of collection equipment. Thus, the pollutant is prevented from escaping in excessive quantities into the atmosphere.

c. **Control of Dilution in the Atmosphere.** Natural Dilution is accomplished by the use of tall stacks and meteorological control of the process operations.

(1) *Use of Tall Stacks.* In employing tall stacks, it is hoped that the discharge is high enough so that natural movement of the air currents will dilute and disperse the pollutants in such a way that they will not be harmful when they reach ground level.

(2) *Meteorological Control of Process Operations.* In this method, operations are curtailed or completely shut down during periods of adverse meteorological conditions.

Control of Particulate Emissions:

a. **Principles of Particle Collection.** Dust particles evolving from known sources and confined to well-defined gas streams can be removed from a carrier gas by various collection devices. Com-

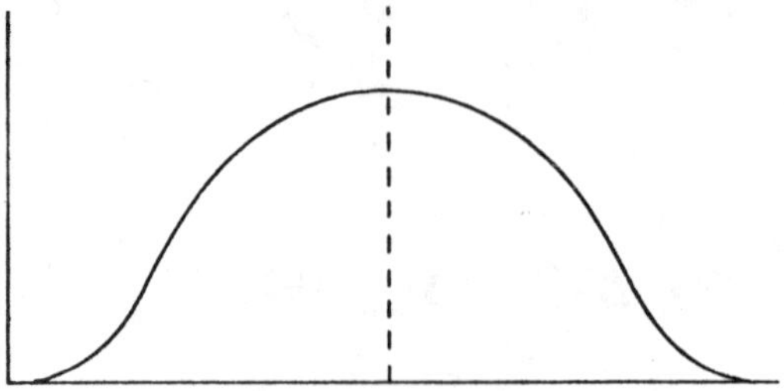

Fig. 19-1 Typical Size-Distribution Curve for an Industrial Dust

mon reasons for particle removal include:

(1) Recovering material for re-use in manufacturing processes or for sale as by-product.

(2) Desire to reduce quantity of particulates discharged to the atmosphere.

b. **The Removal Process.** Particle removal can usually be broken down into three distinct steps:

(1) Deposition of dust particles on a collecting surface by application of directional forces other than that of the motion of the gas stream.

(2) Retention of the particles on the collecting surface without allowing them to be reentrained in the gas stream.

(3) Intentional removal of the particles from the collecting surface by cleaning mechanisms, either intermittent or continuous.

c. **Collection Mechanisms.** The basic mechanisms for separating particles from a gas stream can be reduced to the following:

(1) *Gravity Settling.* The horizontal carrier gas velocity is reduced sufficiently to allow the particles to settle by force of gravity.

(2) *Inertial Forces.* By suddenly changing the direction of the gas flow, the greater momentum of the larger particles causes them to exit the gas stream.

(3) *Filtration.* Dust-laden gas passes through a porous medium upon which dust particles are retained leaving a cleaner gas to be discharged.

(4) *Electrostatic Attraction.* Electrically charged particles are attracted to objects having the opposite charge.

(5) *Particle Conditioning.* By causing intimate contact of dust particles and water, a heavier water-particle agglomerate is formed that can be more easily separated from the gas stream by one of the other collection mechanisms.

d. **Selecting a Dust Collector.** Selection requires balancing of the following factors:

(1) *Particle Characteristics:*
- (a) Size distribution.
- (b) Shape.
- (c) Density.
- (d) Stickiness.
- (e) Hygroscopicity.
- (f) Electrical properties.

(2) *Carrier Gas properties:*
- (a) Temperature.
- (b) Moisture content.
- (c) Corrosiveness.
- (d) Flammability.

(3) *Process Factors:*
- (a) Gas flow rate.
- (b) Particle concentration.
- (c) Allowable pressure drop.
- (d) Continuous or intermittent operation.
- (e) Desired efficiency.

(4) *Economic:*
- (a) Installation cost.
- (b) Operating cost.
- (c) Maintenance cost.

e. **Collection Efficiency:**

(1) Any dust collector operating under specific flow conditions and with a given gas will have a collection efficiency corresponding to each particle size or particle size-distribution passing through the collector. Thus, there is a definate "particle size-collection efficiency" relationship for every dust collector for given flow conditions.

(2) The term "particle size," in this instance, refers to an apparent size determined from measuring the terminal settling velocity of a given particle in the carrier gas. Apparent size can be calculated from Stokes Law which expresses terminal settling velocities for spherical particles under laminar flow conditions. The resulting "apparent size" would probably not be identical to particle size as determined by sieving or microscopic methods since it includes the effect of par-

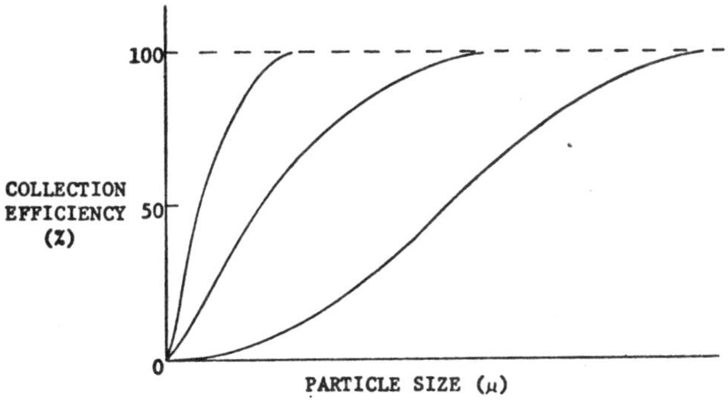

Fig. 19-2 Typical Particle Size-Efficiency Curves

ticle shape and density as well as size. Particle diameter determined in this manner is the equivalent spherical diameter the particle would have if its density were equal to the average density of the aggregate dust being analyzed.

(3) When speaking of overall collection efficiency, particle size-distribution is the most important and often the least considered parameter. A typical size-distribution curve for dusts found in air pollution work is shown in figure **19-1.** If the curve is normal (symmetrical about a vertical axis) when plotted on semi-logarithmic paper, the curve is said to be "log-normal." This situation occurs frequently for industrial dusts normally encountered in waste gas streams.

(4) The effect of particle size on efficiency of a collector can be demonstrated by collecting dust samples on the inlet and outlet of the operating collector. Knowing the inlet and outlet dust concentrations and the inlet and outlet size distributions, collection efficiency can be plotted as a function of particle size on arithmetic coordinates. The resulting plot, known as a "size-efficiency" curve, describes the performance capabilities of the collector in question when operating at stated gas flow conditions The collector would have a different "size-efficiency" curve for each new gas flow condition, but the curve would remain unchanged for different dust concentrations and different size distributions at the same flow. Thus, for any particle size passing through a given collector, a definite and predictable collection efficiency will be observed, flow conditions remaining constant. "Size-efficiency" curves are especially useful in predicting overall collection efficiency for a specific dust having known size-distribution characteristics. Some typical "size-efficiency" curves are shown in **Fig. 19-2.** Steeper curves indicate "more efficient" collectors or those having greater collection efficiency for small particles.

f. **Classification of Particulate Collection Equipment:**

(1) Collection equipment may be divided into two broad classifications:

(a) Dry collectors.

(b) Wet collectors.

(2) Dry collectors:

(a) Settling chambers.

(b) Centrifugal collectors:

Dynamic precipitators.

Cyclones:

simple cyclone (large diameter).

high efficiency (long cone).

multicyclone.

(c) Inertial separators:

Baffle chamber.

Impingement type.

Louver type.

(d) Fabric collectors.

(e) Electrical precipitators:

Wet collectors.

Gravity spray tower.

Wet impingement scrubber.

Self-induced spray deduster (orifice type).

Disintegrator.

Wet dynamic precipitator.

Venturi scrubber.

g. **Selecton of Control Equipment. Fig. 19-3** and table **19-1** illustrate and describe characteristics of typical particulate collection equipment.

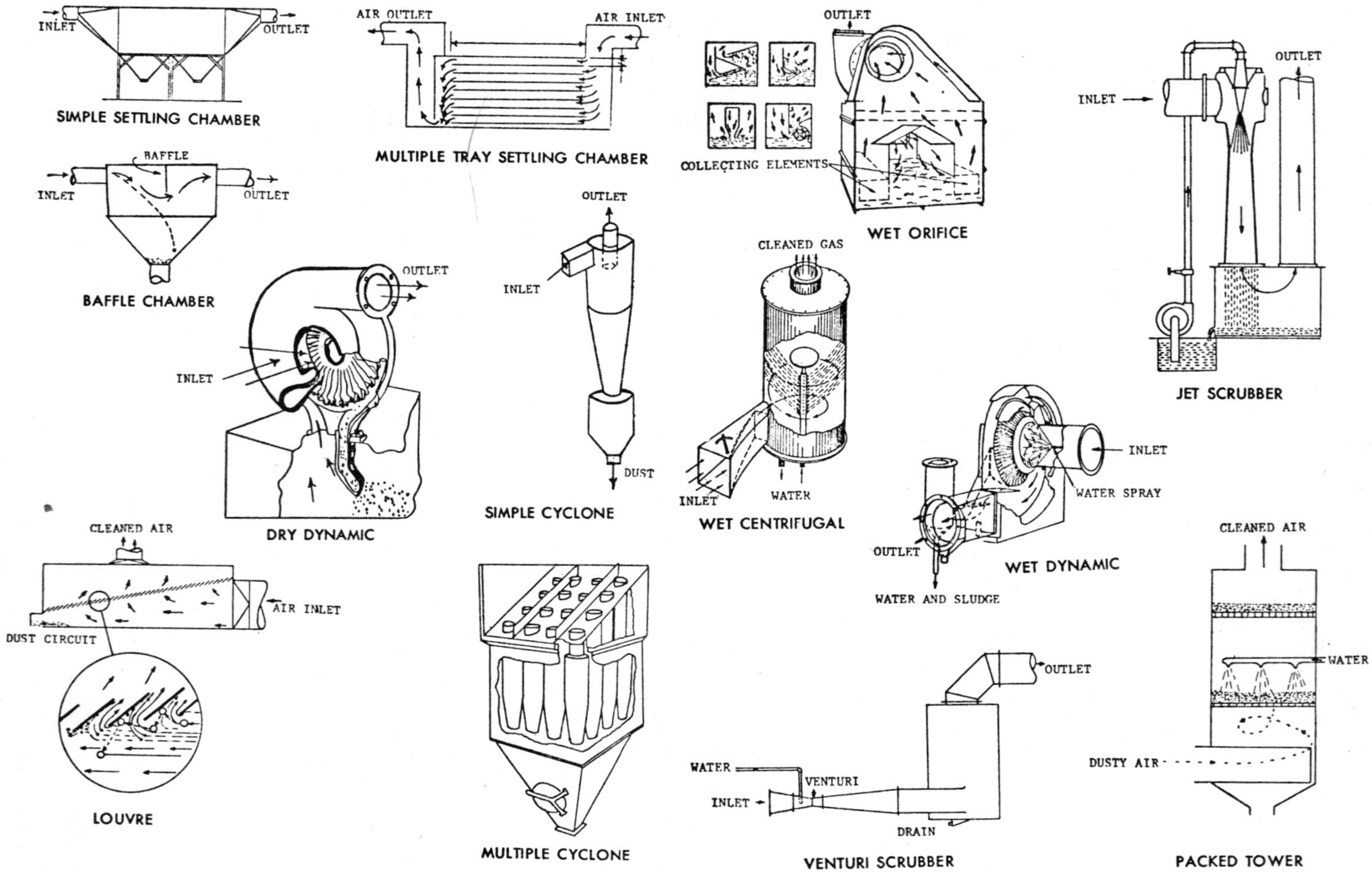

Fig. 19-3 Types of Particulate Collection Equipment

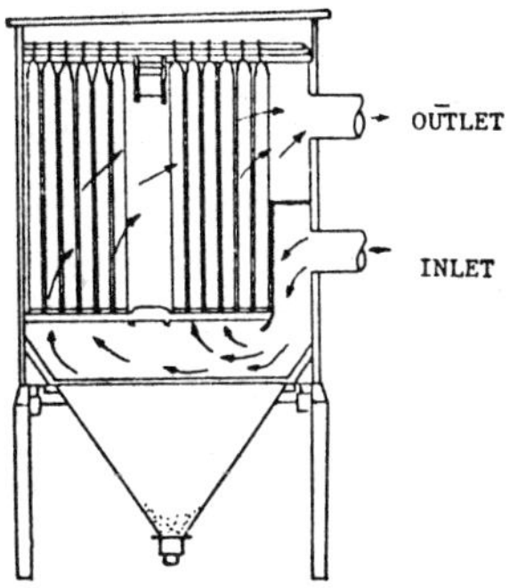

• TUBE OR STOCKING TYPE FABRIC FILTER

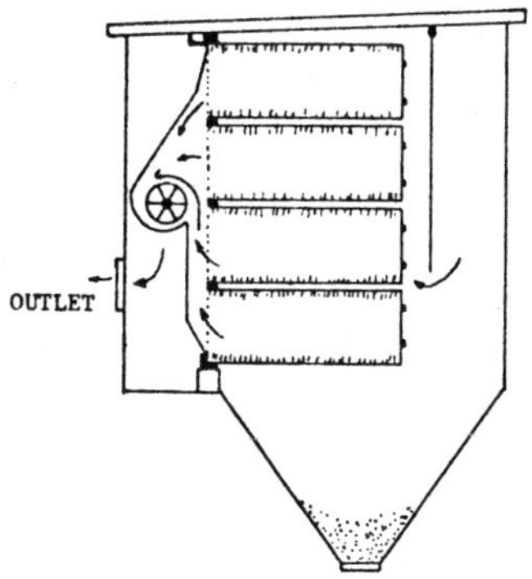

ENVELOPE OR BAG TYPE FABRIC FILTER

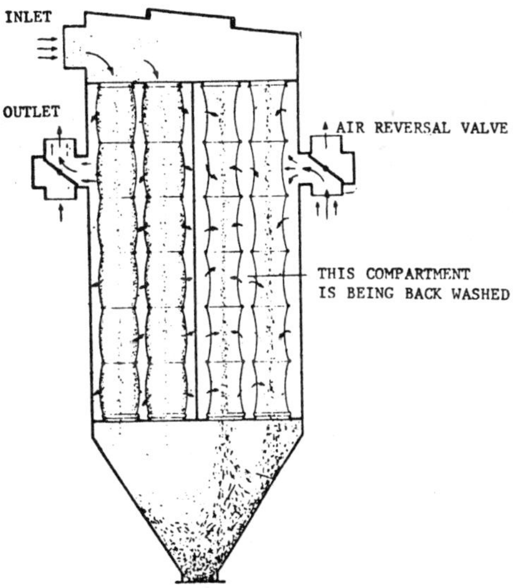

REVERSE JET

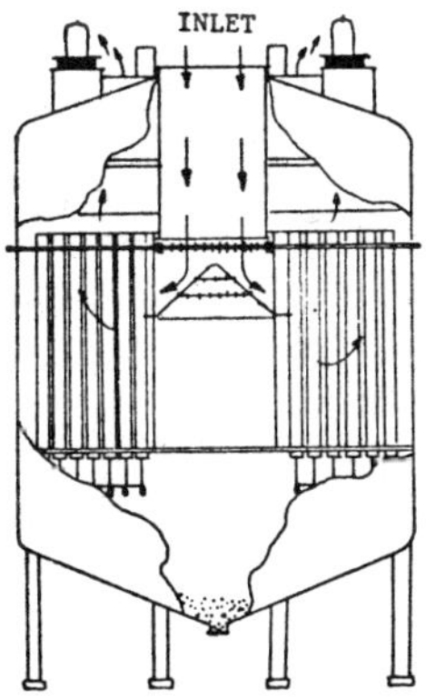

ELECTROSTATIC PRECIPITATOR

Fig. 19-3 (continued)

Table 19-1 *Dust Collector Characteristics*

Type of Collector	*Particle Size Range Readily Collected (Microns)*	*Pressure Drop (Inches Water)*	*Characteristics*
A. Settling Chambers			Particles settle out by gravity. Collection is a function of particle size, specific gravity, and terminal velocity. Used as pre-cleaners but require large space. Not sensitive to flow change unless re-entrainment velocity reached (>600 fpm). Efficiency proportional to horizontal projected area, approaching 100% for 100μ particles and 1% for 10μ particles.
1. Simple	40–100	0.1–0.5	
2. Multiple Troy	10–50	0.1–0.5	
B. Centrifugal (Dry)			Inertia of particles results in migration toward a collecting surface usually the result of a directional change of air stream. Frequently used as pre-filters. Vary in size from small (dynamic, baffle chamber) to large low p.d. cyclone. Sensitive to flow rate changes. Advantages include simple construction, low initial cost, high temp. capabilities, easy maintenance, and dry and continuous disposal of solids. Subject to abrasion and exhibit low collection efficiencies. Dynamic type not suited to sticky or fibrous materials and displays solids build up on rotating elements. All can handle relatively high grain loadings.
1. Baffle Chamber	20–50	1–3	
2. Louver	15–50	1–3	
3. Single Cyclone	15–50	1–2	
4. Multiple Cyclone (High Eff.).	5–20	3–6	
5. Dynamic (Mechanical)	5–20		
C. Scrubbers			Used for gases, mists, and particulates, mainly latter. Variety of operating principles employed including droplet—contaminant contact in spray tower, surface contact in packed tower, centrifugal force—wetted surfaces, centrifugal force—baffles—wetted surfaces—spray nozzles, venturi orifice intermixing of contaminant and water particles, water sprays within fan housing (special blades), contaminant flow through orifices or restricted passages where water contact made by sprays, falling droplets, wetted trays. Water commonly used, sometimes wetting agents and live steam. Chemical absorption of gases and mists a function of solubility and difference in concentration of contaminant in absorbent and solute. Mist and dust collection achieved by: (1) impingement on liquid surface or target droplets of different size and velocity; (2) direct moisture condensation on particulates which serve as condensation nuclei increasing effective particle size and ease of inertial collection; (3) diffusion where dust particles interdispersed among liquid drops by Brownian motion; and (4) gas partitioning and diffusion within bubbles. Effect of wetting agents not impressive. Efficiency function of particle size and power output. Water recirculation common. Some designs sensitive to flow rate change.
1. Spray Tower	5–10	0.5–1.5	
2. Wet Centrifugal	2–5	2–6	
3. Wet Dynamic	2–5		
4. Wet Orifice	2–5	3–10	
5. Venturi	1–5	10–30	
6. Packed Tower (Trays of solid packing).	3–10	1.5–4	
7. Jet	1–5		
D. Fabric Filter			Fabric (glass, cotton, orlon, nylon, dacron, wool felt) serves primarily as a filter support with collection accomplished by gas passages through dust cake. Exhibit high efficiencies down to 0.1μ, cleaned by shaking, reverse jet, collapse cleaning, air pulsation or sonic, or combinations of these. Filter velocity (air to cloth ratio)
1. Tubular and Envelope	1–5	2.5–6	
2. Reverse Jet	1–5	2.5–8	

Table 19-1(continued)

Type			Remarks
			continuous air cleaning, relatively constant pressure drop. Although highly efficient even with variable flow rates they have high maintenance costs, temperature and humidity limitations, and are generally large in size.
E. Electrostatic			Can handle gases which are hot (up to 1100°F), cold, wet, or dry, and tend to work better on smaller particles. Gas distribution and retention time in unit important. Usually require precleaner and single stage industrial type utilizes 25,000 volt potential or more. Particles pass through ionized field, are charged and migrate to collecting electrode. They are deposited, lose charge and removed by vibration or wash-down. Two-stage units consist of an ionization stage and second collection stage and are not intended to handle heavy dust loadings. Resistivity of particulate matter affects collection efficiency. High resistivity particles (silica) can be conditioned by humidification, temperature control or addition of chemical agents such as NH_3, H_2SO_4, or SO_3. High initial cost, generally large in size, sensitive to flow rate changes, and cannot be used for combustible contaminants.
1. Single Stage	1-3	0.25-0.5	
2. Two-Stage	1-3	0.25-0.5	
F. Unit (Portable)			Generally not recommended for production operations where loadings are significant. Suited to intermittent operation, light loads, and easily collectable materials. Shows tendency of being overrated in capacity. Some designs very noisy.

CHAPTER 20

TALL STACKS

Under any given set of meteorological conditions, the ground-level concentrations of a gaseous pollutant emitted at a constant rate into the atmosphere will become smaller as the effective height of emission of the pollutant into the air is increased. This is a truism. A fixed rate of emission at a fixed height does not uniquely determine the resultant ground-level concentrations, nor is it a certainty that sensibly significant concentrations at the ground will always result from even large emission rates from elevated sources. The magnitude of ground-level concentrations and their temporal and spatial variation depend upon the (non-linear) interactions of an impressive number of factors, among which the major ones are:

1. Strength of the source (e.g., emission rate in grams per second).
2. Physical height of the stack or chimney.
3. Topography of the region surrounding the stack.
4. Geometric relationships between the stack and the building complex upon (or near) which it is located.
5. The rate of transformation or decay (if any) of the pollutant.
6. The temperature and efflux velocity of the material at the time of release.
7. The temperature and stability structure of the atmosphere through the depth in which the "plume" is dispersing, and their variations in time and space.
8. The wind velocity and its variation in time and three-dimensional space through the dispersing layer.
9. The intensity and spectral distribution of atmospheric turbulence in space and time through the dispersing layer.
10. The "effective" stack height, i.e., the sum of the physical stack height and the plume rise due to inertial and/or buoyancy forces. (Note: there is no universally accepted definition of conditions that mark when or where the effective height of a plume has been reached.)

Not all of these factors are independent; e.g., number 10 can be a function of all of the others, and 9 strongly depends on 7 and 8, and possibly 3 as well. For existing installations, only factors 1 and 6 may be controlled or modified by the operators. Factors 2, 3, and 4, which bear on aerodynamic effects that might bring elements of the plume prematurely to the ground, are fixed. Factor 5 is a property of each pollutant and varies with environmental factors, such as air temperature, humidity, and sunshine. The meteorological phenomena, 7, 8, and 9, may be highly variable in all space dimensions and in time, although some degree of diurnal and seasonal regularity is usually discernable.

Given this heterogeneous array of fixed, quasi-controllable, and quasi-random functions, the central problem for air pollution meteorologists is to derive and validate the quantitative relationships of these functions to resultant ground-level concentrations over various time periods of interest. That is, as a basis for evaluating a given stack, the meteorologist must derive a diffusion climatology that will establish the occurrence frequency of a range of meteorological conditions and provide estimates of the ground-level concentrations that will occur within a given atmospheric condition.

GROUND-LEVEL CONCENTRATIONS

No general complete theory has been formulated to express the physical relationships between ambient concentrations of air pollutants and the relevant causative factors and processes. Partial or semitheoretical expressions have been derived for limited classes of simplified conditions and situations. These have been based on analogy to heat flow in fluid materials, or on the statistical theory of turbulence, which deals with averaged properties of fluid motions. Although recent extension of these theories to motion in stratified media appears promising, the numerical evaluation of critical parameters, such as exchange or diffusion coefficients for the complete range of atmospheric stability conditions, must be based on experiment. As a result, a large body of empirical data has been accumulated over the past several decades from activities connected with Chemical-Biological-Radiological warfare, atomic energy, and air pollution research. Working equations, including appropriate values of the diffusion parameters, have been derived from these field studies and have proved useful in the solution of practical air pollution problems arising from both ground-level and elevated sources. However, because most of the studies of continuous emissions involved stacks shorter than 300 feet, and because observations at upper levels are technically difficult, few of the data obtained relate directly to plume dispersion from taller stacks.

This dificiency has not precluded the estimation of ground-level concentrations to be expected from tall stacks. In such estimates, however, the values of height-dependent atmospheric variables (diffusion coefficients) used in the working formulae, are largely speculative extrapolations, subject to validation by relevant field observations and experiments. Further, recent ground-level sampling at distances 5 to 20 miles from the base of tall stacks demonstrates that earlier conclusions, based on close-in sampling (up to 5 miles), can significantly underestimate the magnitude of ground-level concentrations possible at greater distances. The importance of the actual relevancy of data collected to validate aspects of tall stack effluent dispersion cannot, therefore, be overemphasized.

Figure 1-4 shows a convenient and often referenced scheme for classifying the dispersion behavior of stack gases on the basis of ambient atmospheric stability conditions in the lower atmosphere.[2] Theoretical calculations of pollutant concentrations in time and space are normally set, explicitly or implicitly, within the framework of such a classification. Although that practice is also followed in the calculations of ground-level concentrations from tall stacks presented in Tables 20-1 and 2, the analogies are only approximately correct. The illustrations in Figure 1-4 refer to conditions in the lower 1000 feet of the atmosphere, but effluents from tall stacks often have effective heights in the range of 1000 to 3000 feet above ground because of plume rise due to buoyancy and inertial effects. The diurnal variation of stability conditions is a function of height and season of the year, and is generally not so marked above 1000 feet. Turbulence is not as intense and variable at such heights as it is nearer the ground.[5] The over-all consequences, in qualitative terms, are that the magnitude and frequency of processes that bring plume effluents to ground level are reduced.

Table 20-1 presents two sets of hypothetical calculations of ground-level concentrations of SO_2, for various effective stack heights as a function of distance, under typical daytime conditions. One set was presented by M. E. Smith in a recent paper,[6] and the other was computed by the Meteorology Program, National Air Pollution Control Administration.

The meteorological situation represented in this table is most nearly characterized as Condition A of Figure 1-4. On an annual basis, this is one of the most frequent situations experienced in the United States. It has a frequency of about 30 percent. Note that the NAPCA estimates of maximum concentrations are lower but at greater distances from the source. This is probably attributable to NAPCA investigators assuming somewhat different dispersion rates at the higher elevations from those used by Smith. This issue can be resolved only by properly designed experiments. From a practical point of view, however, for effective stack heights

Table 20-1 HYPOTHETICAL GROUND-LEVEL CONCENTRATIONS OF SO_2 FROM VARIOUS EFFECTIVE STACK HEIGHTS UNDER TYPICAL DAYTIME CONDITIONS[a]

Effective stack height, m	Author	Distance of maximum, km	Downwind concentrations, ppm				
			Maximum	1 km	5 km	10 km	50 km
1000	Smith	7.6	0.03	0	0.03	0.03	0.004
	NAPCA	14.7	0.02	0	0.001	0.002	0.005
500	Smith	3.4	0.14	0.01	0.12	0.05	0.004
	NAPCA	6.9	0.08	0.00	0.02	0.05	0.005
100	Smith	0.6	3.36	2.20	0.19	0.06	0.004
	NAPCA	1.2	1.85	1.15	0.44	0.14	0.005

[a]Following Smith, assumed SO_2 emission rate of $8 \times 10^5 cm^3 sec^{-1}$ "typical of a 600 Mw - 2 percent S plant", and mean wind speed of 6 m/sec.

Table 20-2 HYPOTHETICAL MAXIMUM GROUND-LEVEL CONCENTRATION OF SO_2 FROM VARIOUS EFFECTIVE STACK HEIGHTS UNDER WEAK LAPSE CONDITIONS

Effective stack height, m	Distance of maximum, km	Maximum concentration, ppm
1000	42.7	0.01
500	18.2	0.05
100	2.5	1.36

above 100 meters, differences in the computed maximum concentrations are only slightly significant. More important, perhaps, is the distance from the stack one must look to find the highest concentrations.

Table 20-2 presents calculations made by the Meteorology Program, National Air Pollution Control Administration of ground-level SO_2 concentrations under conditions similar to those depicted in Part B of Figure 1-4. Data in Table 20-2 is calculated on the basis of the same conditions of SO_2 emission as those assumed for Table 20-1. Although Condition B (coning) occurs much less frequently than Condition A (prevailing 5 to 10 percent of the time), it is another situation in which pollutants emitted aloft are brought to the ground. Although at somewhat lower concentrations than those given in Table 20-1, the pollution from a coning plume will affect a given point for a longer period of time. Plume coning is usually associated with moderate to high winds and overcast skies.

Condition E, inversion break, is a potentially critical situation. As noted by Moore,[7] "when one is concerned with the rise and dispersion of the effluent from a tall stack, the occurrence of stable conditions above the top of the stack may have a very different effect from the presence of a stable layer below the stack."

If the potential temperature increases with height above the source emission point, the air is stable and vertical turbulence and mixing are strongly suppressed. Horizontal mixing processes are still active but somewhat less so than those that produce the coning effect. Inhibited vertical motion together with limited horizontal motion leads to the formation of a relatively thin plume, called "fanning" (Condition C, Figure1-4). During the day, when solar heating establishes the mixing layer, thermal turbulence can reach the plume environment, permitting dispersion of plume effluent to the ground along its length. For a brief period, usually 30 to 60 minutes, high concentrations are brought to the ground. This fumigation was fully studied by Hewson in 1944, during the Trail, British Columbia, investigation.[8]

Fumigation can occur almost daily from a plume emitted from a short stack, as exemplified in Fig. 20-1. In this case the effective plume height was approximately 800 feet above ground level. It is clearly evident that the plume effluent was brought to the ground for a considerable distance downwind. While the plume effluent is contained by the stable layer, it will remain aloft; however, over most of the nation during all seasons of the year the daily maximum mixing depth is deep enough to envelop the layer in which the plume from even the tallest stacks is embedded.[9,10] The resultant ground-level concentrations are a function of source strength, plume height, horizontal spread, and wind speed at plume height.

The experience at Trail, British Columbia, demonstrated that plumes that were high in SO_2 content and trapped in a stable layer retained their physical identity over great distances and damage to vegetation was noticed many miles from the source.[8] Recently Professor Brohult attributed forest damage in Sweden to SO_2 transported from Western Europe.[11] Of course, the damage in Sweden may not have been due to fumigation alone but, if Dr. Brohult's thesis is correct, it may have been due to frequent, low-level insult by transported SO_2. Those concerned with air quality management should note that, regardless of the height of release, the exposure to distant receptors is similar. As Smith[6] has pointed out,

> . . stacks possess no magic power to eliminate a pollutant. They do not reduce by one gram the total amount of pollution released to the atmosphere. They distribute it in a different way than would be true of a low-level source, but a receptor at a great distance from a stack will receive substantially the same concentrations no matter what the source height.

Measured Ground-Level Concentrations

Ultimately, the controversy about whether the tall stack is an interim or complete solution in air quality management must be resolved by meaningful studies of dispersion from tall stacks. According to Gartrell[12]:

> Our experience at large modern plants to date suggests two models that need more intensive evaluation. The first involves high short-term ground-level concentrations which may occur with a high fre-

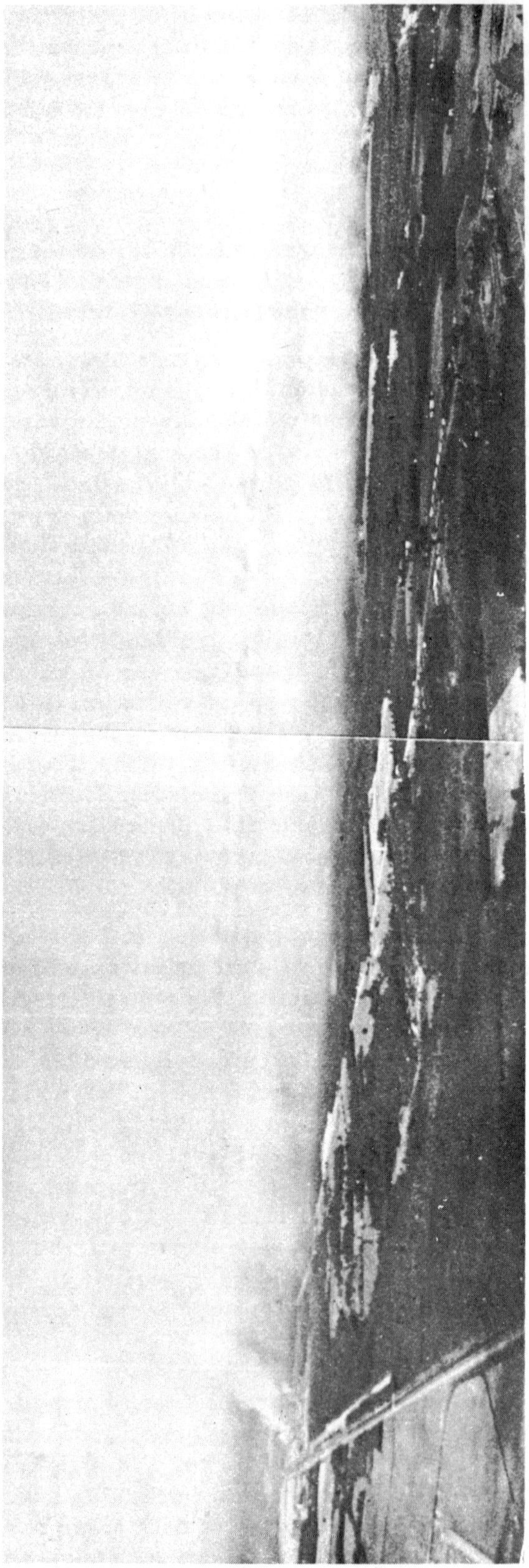

Figure 20-1. An illustration oı a fumigating plume.

quency. While flue gases are emitted from high stacks and have appreciable plume rise, the total plume volume is large and segments may be brought to ground level by thermals typical of moderate instability. Although long-term, say 30 minute average, concentrations are consistently low, 2- to 5-minute concentrations may be relatively high. The second model we believe might involve infrequent, but possible moderately high ground-level concentrations at considerable distances from the source over an extended period of time. This might occur where upward diffusion is limited by a subsidence inversion and strong mixing from ground-level to plume top occurs below the subsidence inversion. This would be characterized by the infrequent stable anticyclonic conditions occurring in the Eastern United States. More time, records, and observations will be necessary for confident evaluation of the significance and validity of this model for large power plants.

Most measurements to date have been made within 3 to 5 miles of a tall stack. In 1966, a rather detailed analysis of ground-level pollution within 3. 5 miles of the High Marnham power station in Great Britain, the contribution to surrounding measuring sites from two 450-foot chimneys were said to be nil.[13] Unusually high measured concentrations (in one instance concentrations of more than 50 parts per hundred million were measured for 8 hours) were attributed to distant urban sources. Other authors, however, in discussing the High Marnham data, report that concentrations at one sampling site average 5. 4 parts per hundred million everytime the wind blows from the station to the site, which it does about 10 percent of the time.[14] These authors state that in one winter the measuring site experienced, on seven occasions, concentrations (attributed to High Marnham) of more than 25 parts per hundred million SO_2 for periods of 28 to 60 minutes. On one occasion, a concentration of this magnitude lasted for a period of 180 minutes.

Table 20-3 presents ground-level observations taken between 2 and 5 miles (Station A in the valley 2. 5 miles, Station B on plateau 2. 25 miles, and Station C on plateau 4. 9 miles), from the Clifty Creek Power Plant in Indiana.[15] The Clifty Creek chimneys extend 683 feet above plant grade and about 300 feet above the plateau northeast of the plant. An analysis of data taken for this study over a 4-year period showed no hourly mean concentrations above 1 part per million at the three SO_2 monitoring stations, although the authors stated, that it was agreed that concentrations slightly above that level may occur infrequently on the plateau north of Clifty Creek Plant, with an occasional peak just reaching the odor threshold (2 parts per million).[15] However, because of the limited sampling network it is difficult to determine how frequently concentrations of a given value occurred around the plant.

Table 20-3 TYPICAL SO_2 CONCENTRATIONS MEASURED NEAR CLIFTY CREEK POWER PLANT, ppm

Gustiness class[a]	Station A		Station B		Station C	
	Average	Peak	Average	Peak	Average	Peak
B_2	0.10	0.40	0.10	0.25	0.05	0.20
B_1	0.10	0.40	0.07	0.30	0.10	0.25
C	0.05	0.20	0.10	0.40	0.10	0.40

[a]Brookhaven gustiness classification.[16]

Table 20-4 presents ground-level observations of 30-minute average SO_2 concentrations taken with four autometers by the Tennessee Valley Authority at the Bull Run Plant near Knoxville, Tennessee. The instruments are sited at the given distances along a line approximately northeast of the 950-megawatt plant, which has a single 800-foot stack. The data presented are selected maximum 30-minute average SO_2 registrations, which occurred at the four sites. Winds along the line of samplers occurred 13.6 percent of the time during the 18-month period.

Table 20-4 NUMBER OF REGISTRATIONS OF 30-MINUTE AVERAGE SO_2 CONCENTRATIONS NEAR BULL RUN PLANT, ppm

Distance, miles	Concentration, ppm								
	0.34	0.26	0.24	0.22	0.20	0.18	0.16	0.14	Total
2.6	1		2	2	1		5		12
4.9		1	1		3	7			12
6.9	1		1	3	5				10
9.7						1		5	6

A study by TVA revealed that 85 percent of the data in Table 20-4 were recor on 16 separate days. The study showed that the 16 days were associated with a given set of meterological conditions, summarized by TVA as follows:

> Meteorological conditions during maximum 30-minute average SO_2 concentrations at the Bull Run Steam Plant during January 1966-June 1967 are remarkably similar and may be identified with one basic or atmospheric model. The lower atmosphere was under the influence of weak anticyclonic circulation with stable air throughout the first few thousand feet from late afternoon through late morning. Average wind speed generally varied between 8 and 15 miles per hour and was adequate to contain the plume within this stable layer. Skies were mostly clear, under anticyclonic circulation, and solar insolation was sufficient to develop thermally induced atmospheric mixing from surface through the plume environment by late morning or early afternoon. This temperature transition (or stability change) resulted in initial SO_2 registrations appearing at most autometers about 11:30 a.m. to 12:30 p.m. reaching maximum concentrations 1 to 2 hours later when the mixing had penetrated through most of the plume environment, and then decreasing to zero by 4:30 to 6:30 p.m. with the onset of stable conditions in the near-surface layers.[17]

On the basis of TVA experience, one may conclude that significant surface concentrations are attributable to emissions from tall stacks in the presence of light winds and limited mixing layers. Futhermore, preliminary data from another current investigation by the Tennessee Valley Authority, with the support of the National Air Pollution Control Administration, show that inversion-breakup fumigation also occurs with tall stacks; however, these fumigations are infrequent because surface-based radiation inversions are rarely deep enough to entrap the plume.

OTHER CONSIDERATIONS

Plume Rise

The height to which a buoyant plume rises is one of the primary factors con trolling ground-level concentrations of pollution downwind of an industrial source. Obviously, the higher a plume rises in the atmosphere, the greater the volume of air available for dilution of the atmospheric contaminants before they are dispersed to ground level.

The rise of a plume after it leaves a stack is due to its buoyancy and momentum. These elements depend on the physical characteristics and emission factors of the particular stack and also on the prevailing meteorological conditions. Stack factors affecting plume rise are exit velocity and temperature of the stack gas, and inside stack diameter. Meteorological elements determining plume rise are wind speed, air temperature, atmospheric stability, turbulence, and wind shear.

Although a number of plume rise equations have been developed over the years, experts in atmospheric dispersion have not been able to agree on a given equation to calculate plume rise under all circumstances. In general the semi-empirical plume rise formulae have been developed from data on relatively small sources and consequently may lead to significant errors in stack height estimates when they are applied to the larger industrial sources of today.

Stack Height Versus Sulfur Dioxide Emission Rate

It was pointed out earlier that the ground-level concentration is dependent upon the effective stack height and the source emission rate. In today's literature, considerable attention is directed toward the effect of increasing stack heights, but there is little or no attention to the fact that emission rates per unit have also increased relative to those of the lower stacks.[15, 19] Figure 20-2 shows a comparison of physical stack height with approximate emission rates and rated capa-

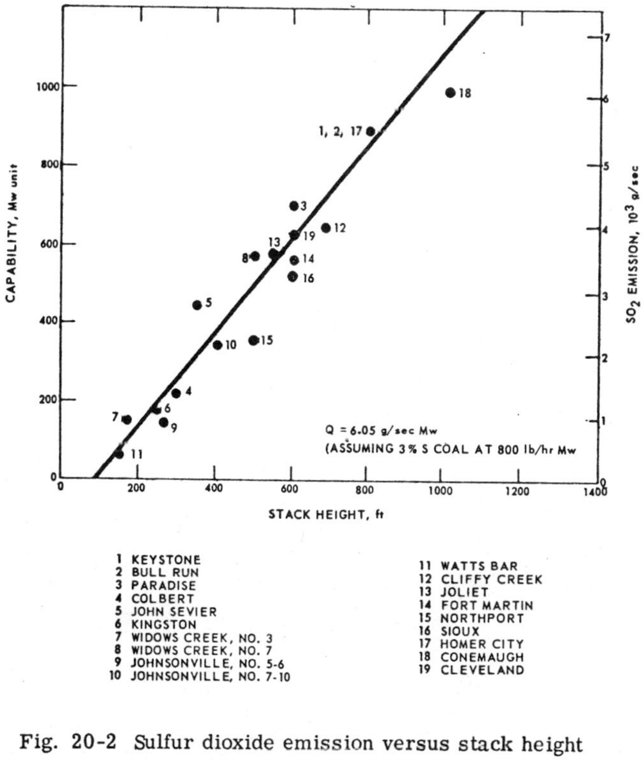

Fig. 20-2 Sulfur dioxide emission versus stack height

bility. The figure shows that, whereas stack heights have increased by a factor of about 4, emission rates have increased by a factor of approximately 6. Thus, a good share of the benefits of increased stack heights are offset by the release of more contaminants from the stack.

Washout By Precipitation

Proponents of the idea that a stack is the most economical solution for control of large discrete sources of pollution should not overlook the problem of atmospheric scavenging, in particular washout by precipitation. Both field and laboratory investigations show that sulfur dioxide can be oxidized to sulfuric acid or an acid salt in the atmosphere, and that precipitation scavenging is the primary cleansing mechanism for airborne gases and fine particles. Since SO_2 is highly soluble in water, the washout process for this gas is absorption of the gas by drops of rain as they fall through the plume.

REFERENCES

The publications referred to in Chapter 20 may be found in "Interim Guide for Incineration at Federal Facilities", National Air Pollution Control Administration, Nov. 1969, pp. 16-1, 16-2

CHAPTER 21

INCINERATOR DESIGN

A. GENERAL - REFUSE INCINERATORS

Tests on incinerators designed according to these recommended standards have shown that when properly operated, the incinerators can meet, without the use of scrubbers, the applicable Federal emission standards for incinerators with rated burning capacity of 200 pounds per hour or less. At capacities of more than 200 pounds per hour, incinerators should be equipped with scrubbers to ensure that applicable Federal emission limits are met.

TYPES OF GENERAL-REFUSE INCINERATORS

Multiple-chamber incinerators are of two general types. Figure **21-1** illustrates the retort type, named for the return flow of gases through the "U" arrangement of the component chambers; and Figure **21-2** shows the in-line type, so called because the three chambers follow one another in a line.

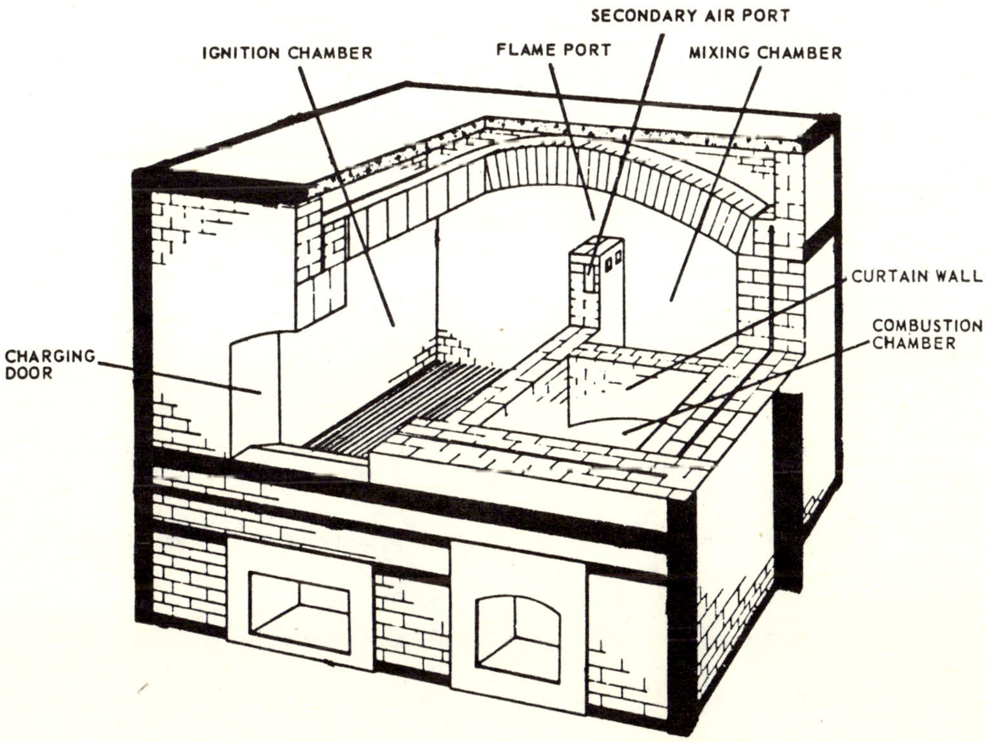

Fig. 21-1 Cutaway drawing of multiple-chamber retort incinerator.

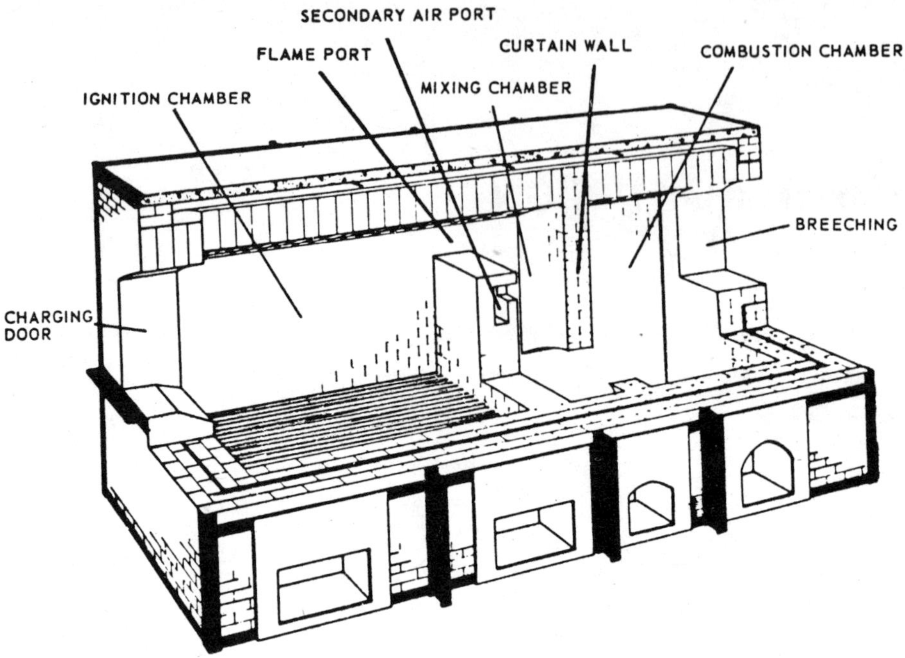

Fig. 21-2 Cutaway drawing of multiple-chamber in-line incinerator.

Multiple-Chamber Retort Incinerators

The following guidelines are recommended:

1. That general-refuse retort incinerators have the configuration shown in Figure **21-3.**
2. That retort incinerators with rated capacities of over 1,000 pounds per hour not be built.
3. That incinerators of over 200 pounds per hour rated capacity be equipped with gas washers.

Multiple-Chamber In-Line Incinerators

The following guidelines are recommended:

1. That all in-line incinerators installed in Federal facilities for the purpose of burning wastes of Types 0, 1, or 2 have the configuration shown in Figure **21-4**.
2. That in-line incinerators with a rated burning capacity of less than 750 pounds per hour not be built.
3. That all in-line incinerators be equipped with gas washers

Types of Natural Gas Burners Recommended

Incinerators having a capacity of less than 200 pounds per hour that use burners rated at less than 400, 000 Btu per hour may be of either the atmospheric or power-burner type. In either case, a continuously or intermittently burning stable pilot adequate to ensure safe, reliable ignition should be installed. A flame safeguard should be used so that no gas can flow to the main burner unless satisfactory ignition is assured. The response time of this flame safeguard to de-energize the gas shutoff device on flame failure should not exceed 180 seconds.

Auxiliary burners on incinerators with ratings of 200 pounds per hour or more, i.e., those equipped with a fan and scrubber, should be of the power-burner type, because this type of burner usually retains its flame better when a fan is used to induce draft. For burners with ratings of more than 400,000 Btu per hour input, the burner equipment shall be of the power type that utilizes a forced-draft blower to supply air needed for combustion under controlled conditions. A continuously or intermittently burning pilot should be used to ensure safe and reliable ignition. Automatic spark ignition should be used on pilots for burners with input of more than 1,000,000 Btu per hour. A suitable flame safeguard should be used so that no gas can flow to the main burner unless satisfactory ignition is assured. On burners with inputs of from 400,000 to 1,000,000 Btu per hour, the response time of the flame safeguard to de-energize the gas shutoff device on flame failure should not exceed 180 seconds. In capacities of more than 1,000,000 Btu per hour, the response time of the aforementioned flame safeguard should not exceed 4 seconds.

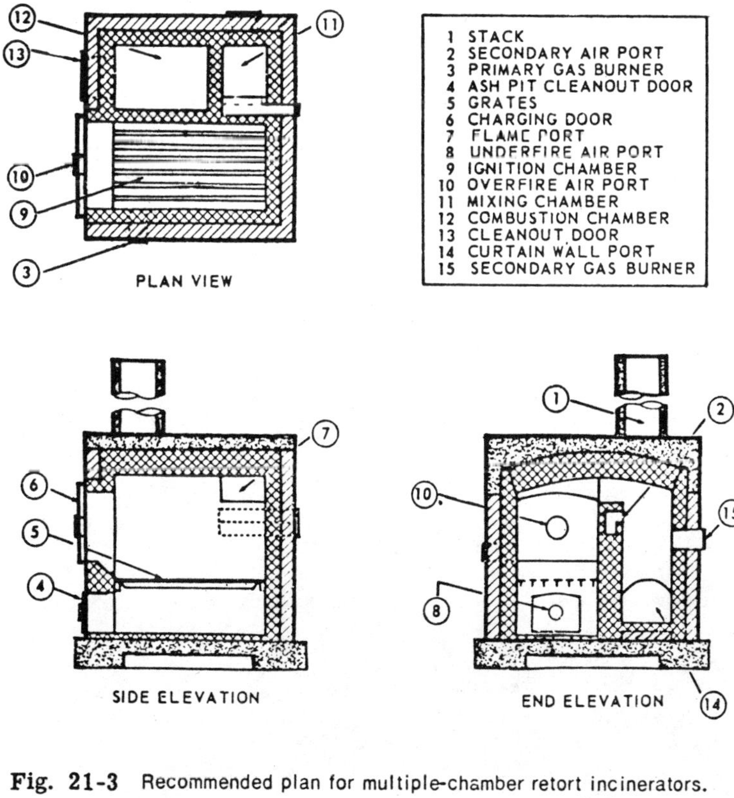

Fig. 21-3 Recommended plan for multiple-chamber retort incinerators.

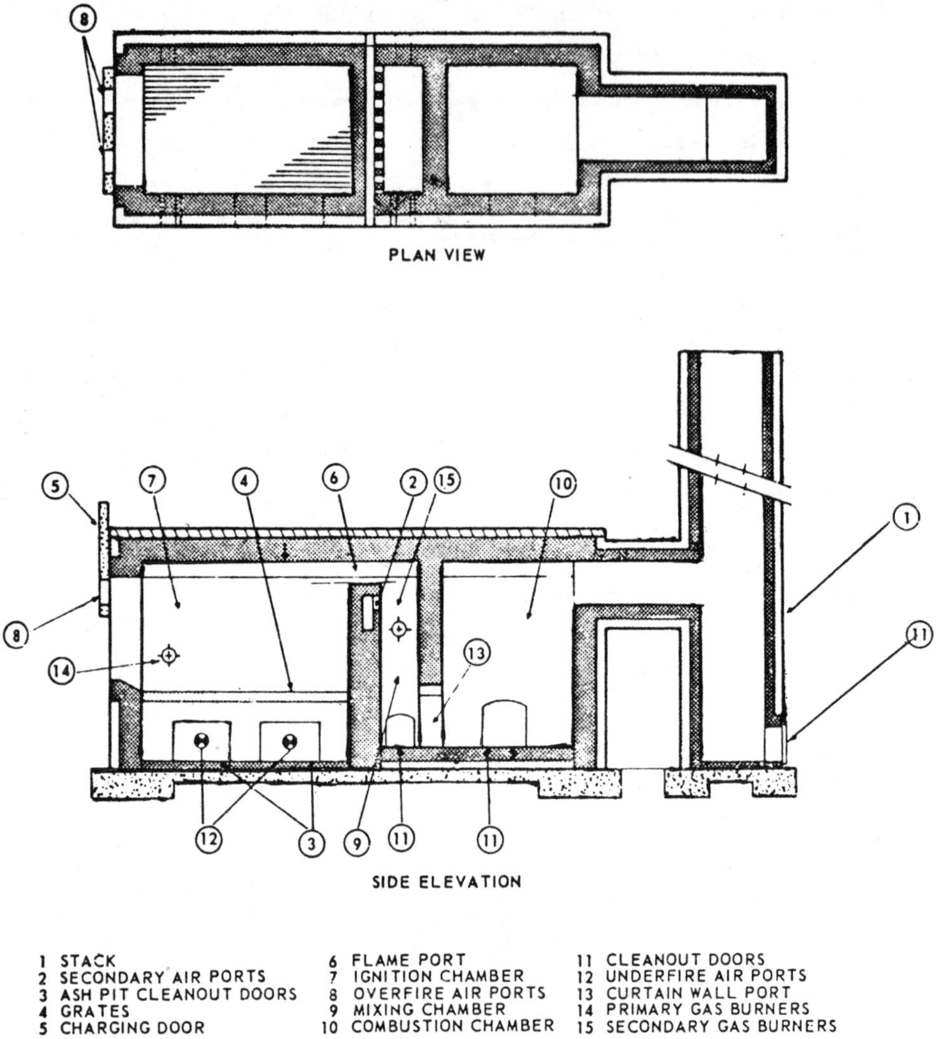

Fig. 21-4 Recommended plan for multiple-chamber in-line incinerators.

The burner assembly should consist of the main burner, pilot burner, automatic valve, the necessary manual valves, and accessory equipment, plus interconnecting pipes and fittings with provision for rigid mounting. The burner should be constructed so that parts cannot be incorrectly located or incorrectly fitted together. Power burners sealed to the walls of incinerators with capacities of more than 100, 000 Btu per hour must be supplied with a means of proving air supply before the main gas valve can be energized.

Electrical motors of more than 1/12 horsepower on power burner equipment should be designed for continuous duty and should be provided with thermal overload protection or current-sensitive devices.

When a complete automatic pilot shutoff system is utilized, the controls should be readily accessible and arranged so that the main burner gas can be manually shut off during lighting of the pilot. When a complete automatic pilot system is not utilized, a readily accessible, manually operated, quarter-turn, lever-handle, plug-type valve should be provided to shut off or turn on the gas supply to the main burner manifold. This valve should be upstream from all controls except the pilot control valve.

Clearly defined and complete instructions for lighting and shutting down the burner should be provided in durable, weatherproof material for posting in a position where they can be read easily.

Sizes of Burners Recommended

Where auxiliary burners are used, their capacity range should include the values shown in Table 21-1.

Table 21-1 GAS BURNER RECOMMENDATIONS FOR GENERAL-REFUSE INCINERATORS

Capacity of incinerator, lb/hr	Size of burners, 10^3 Btu/hr		
	Primary burners		Secondary burners
	Type 1 refuse	Type 2 refuse	All refuse
50	150	250	200
100	200	550	300
150	250	650	400
250	300	750	650
500	550	1100	1000
750	750	1500	1300
1000	900	1700	1700
1500	1100	2200	2100
2000	1600	3300	2700

Other Fuels

If natural gas is not available, equivalent amounts of liquid fuels may be used. Fuel oils of grades higher than Number 2, however, should not be used. The National Fire Protection Association Standard No. 31, Installation of Oil Burning Equipment (1965), should be adhered to where oil burners are used.

If liquified petroleum gas is used, burners should be equipped with a device that will automatically shut off the main gas supply in the event the means of ignition becomes inoperative. The arrangement should be such as to shut off the fuel supply to the pilot burner also.

B. PATHOLOGICAL INCINERATORS

Multiple-Chamber Pathological Incinerators

The pathological incinerators specified here are of one configuration, that of a retort incinerator, as shown in Figure 21-5. To be heated before entering the expansion chamber, gases from the mixing chamber pass under the hearth.

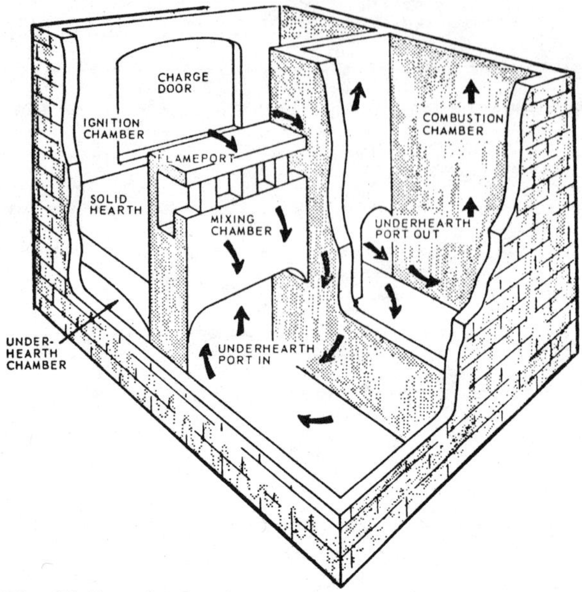

Fig. 21-5 Multiple-chamber pathological incinerator

This flow pattern holds for sizes of 100 pounds per hour and greater. For smaller sizes, a hot hearth and underhearth ports are not required, and the gases pass, as in a conventional retort incinerator, directly from the mixing chamber into the expansion chamber.

Small amounts of Type 4 waste may be burned in a chamber built on the side of a general-refuse incinerator. An incinerator with such a side chamber is shown in Figure 21-6.

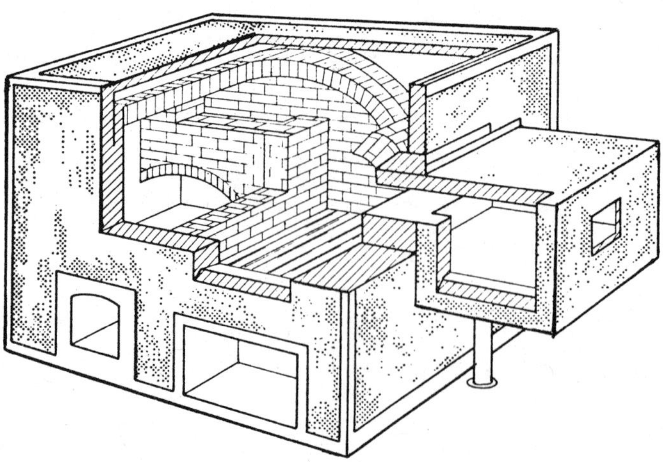

Fig. 21-6 Multiple chamber incinerator with pathological retort

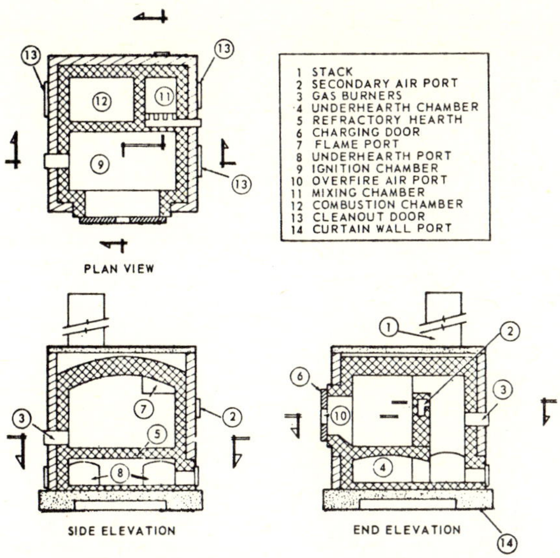

Fig. 21-7 Recommended plan for pathological incinerators

C. INCINERATOR SCRUBBERS

Effluents from general-refuse incinerators burning more than 200 pounds per hour should be cleaned in scrubbers to meet the particulate limit requirement. Since this will generally mean that scrubbers will be widely employed, the disadvantages associated with their use should be recognized.

When scrubbers are used, power lines may have to be installed to furnish energy to operate the induced-draft fan and water pump. Provisions must also be made to supply scrubbing water and a means of disposing of contaminated water from the scrubber. In some areas it will also be necessary to adjust the pH and process the contaminated water through a clarifier to remove fly ash and other collected solids before the water is sewered.

A scrubber may require considerable maintenance as a result of corrosion caused by the acidic water continuously flowing from it. Scrubber water is seldom recirculated because this increases its acidity and, therefore, the rate of corrosion. Even when scrubbers are lined with dense refractory material, corrosion of the steel casing may ultimately occur. In addition, there may be noticeable corrosion and erosion of the fan impeller and, to some lesser degree, of the fan housing. The continuous contact of the acidic water in the sump of the scrubber may gradually attack this surface.

SCRUBBER DESIGN PARAMETERS

Several basic factors are considered in designing scrubbers. To satisfactorily collect the fly ash, the water-gas mixture must be retained within the scrubber for 1 to 1-1/2 seconds at gas velocities not exceeding 15 feet per second. The residence time in the scrubber should also be sufficient to vaporize all the water droplets within the effluent gas stream. Complete vaporization is important

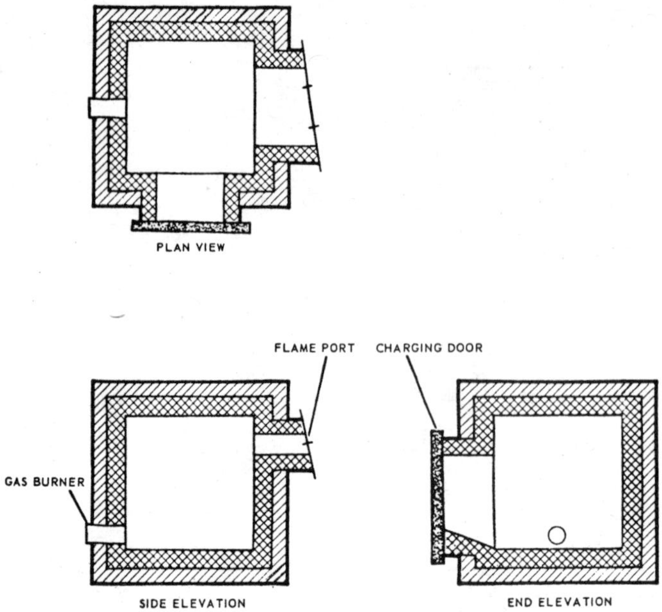

Fig. 21-8 Side chamber for pathological refuse

since nuisance complaints may result from the carryover of water droplets deposited on the surrounding area. From an appearance standpoint, the scrubber should not be longer or higher than the incinerator. The scrubber width should be limited to allow the scrubber to be easily located either adjacent to or at the rear of any incinerator of the retort type. The usual location for scrubbers serving the in-line type incinerator is at the rear of the combustion chamber. Placing the scrubber adjacent to the final combustion chamber is also feasible.

Air dilution of the gases from the incinerator prior to entering the scrubber is unnecessary. Water is introduced into the effluent as it enters the scrubber and flows concurrently down its first pass. By immediately introducing the water into the gas stream, the water has a longer period to mix and evaporate, which accomplish the desired cooling. The average velocity of the gas-water mixture in the first pass ranges from 9 to 10 feet per second. The velocity of the gases in the upward pass is determined by calculating the remaining time requirement so that the gases are within the scrubber for a total time of approximately 1-1/4 seconds. The curtain wall port is sized to permit an air velocity range of 18 to 20 feet per second to prevent excessive pressure drop from occurring and to prevent water from the sump from being re-entrained in the effluent. The gases exit from the extreme top of the uppass so that its full length can be used for the evaporation of any remaining water in the gas stream. This location of the exit at the top also prevents water traveling up the back side of the scrubber from becoming re-entrained in the gas stream. Another feature that reduces re-entrainment of water droplets is a 4-inch channel at the bottom of the curtain wall. The channel collects the larger droplets and carries the water across the width of the scrubber, down its side walls, and into the sump below. Additional structural support for the refractory of the dividing wall is also provided by this channel.

Water in the base of the scrubber collects fly ash and other materials removed from the gas stream may be easily deposited and retained. The water depth is maintained at approximately 3 inches by extending the end of the overflow pipe 3 inches above the floor of the scrubber. Another drain pipe should be installed at floor level so that fly ash and other solids can be washed down the sloping floor of the scrubber.

Design parameters recommended are as follows:

1. The water rate to the scrubber should be 1 gallon per minute for every 100 pounds per hour of rated capacity of the incinerator. This gives a water-to-gas ratio of 1 gallon per minute for every 400 standard cubic feet of effluent stack gas.

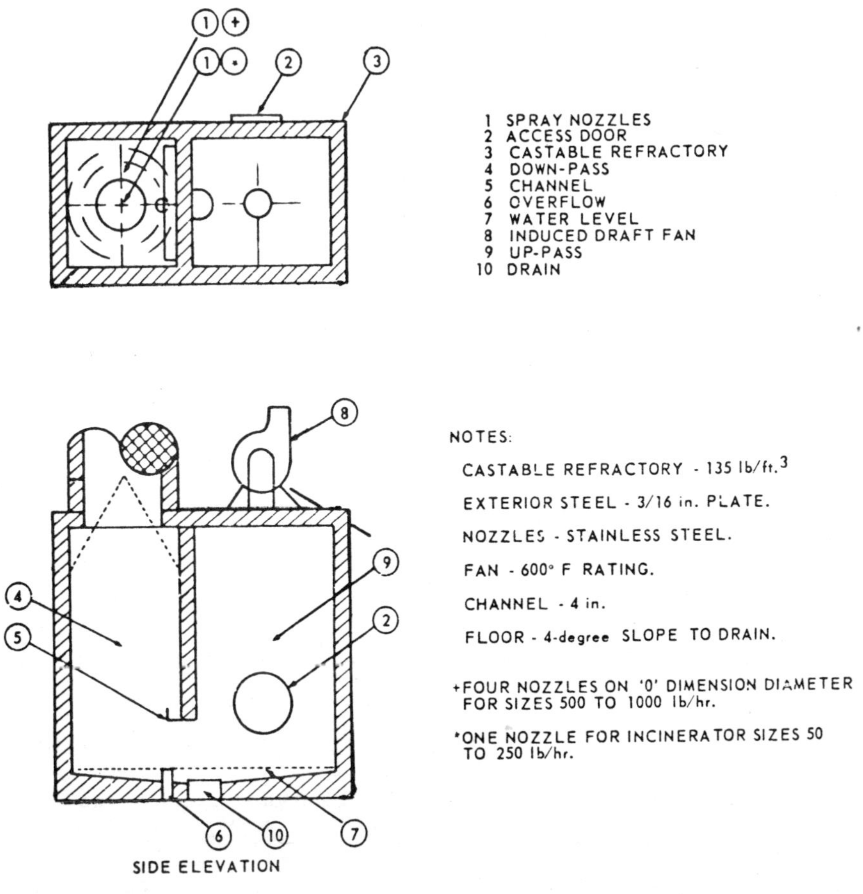

Fig. 21-9 Design recommendations for retort incinerator scrubbers.

2. Configuration of scrubbers for retort and in-line incinerators are given in Figures 21-9 and 21-10, respectively. A graph showing internal areas of the various ports in scrubbers versus incinerator size is given in Figure 21-11.

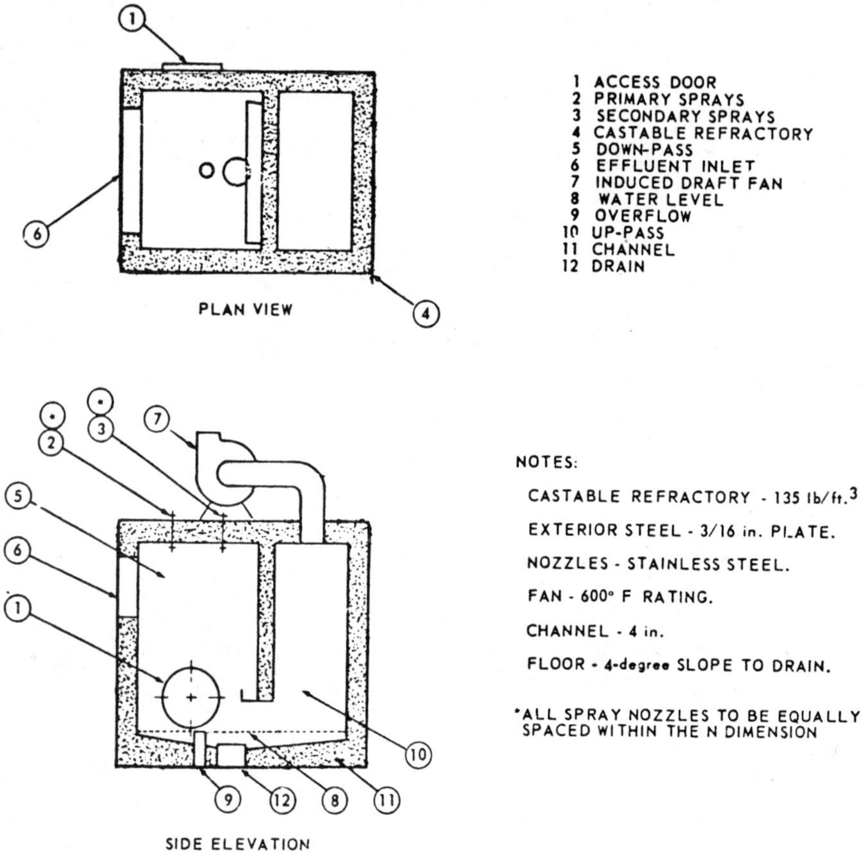

Fig. 21-10 Design recommendations for in-line incinerator scrubbers.

SCRUBBER CONTROLS

While it is recommended that the scrubber controls described herein be installed, it should be realized that a special maintenance and testing program must be established to keep the control systems in good operating condition.

Many types of automatic controls are used to regulate the temperature of the gases leaving the scrubber. Satisfactory controls, which have proved to be both simple and economical, consist of a hand-operated control valve and two automatic solenoid valves. The hand-operated valve is installed in parallel with the solenoid valves between the water supply and the nozzles.

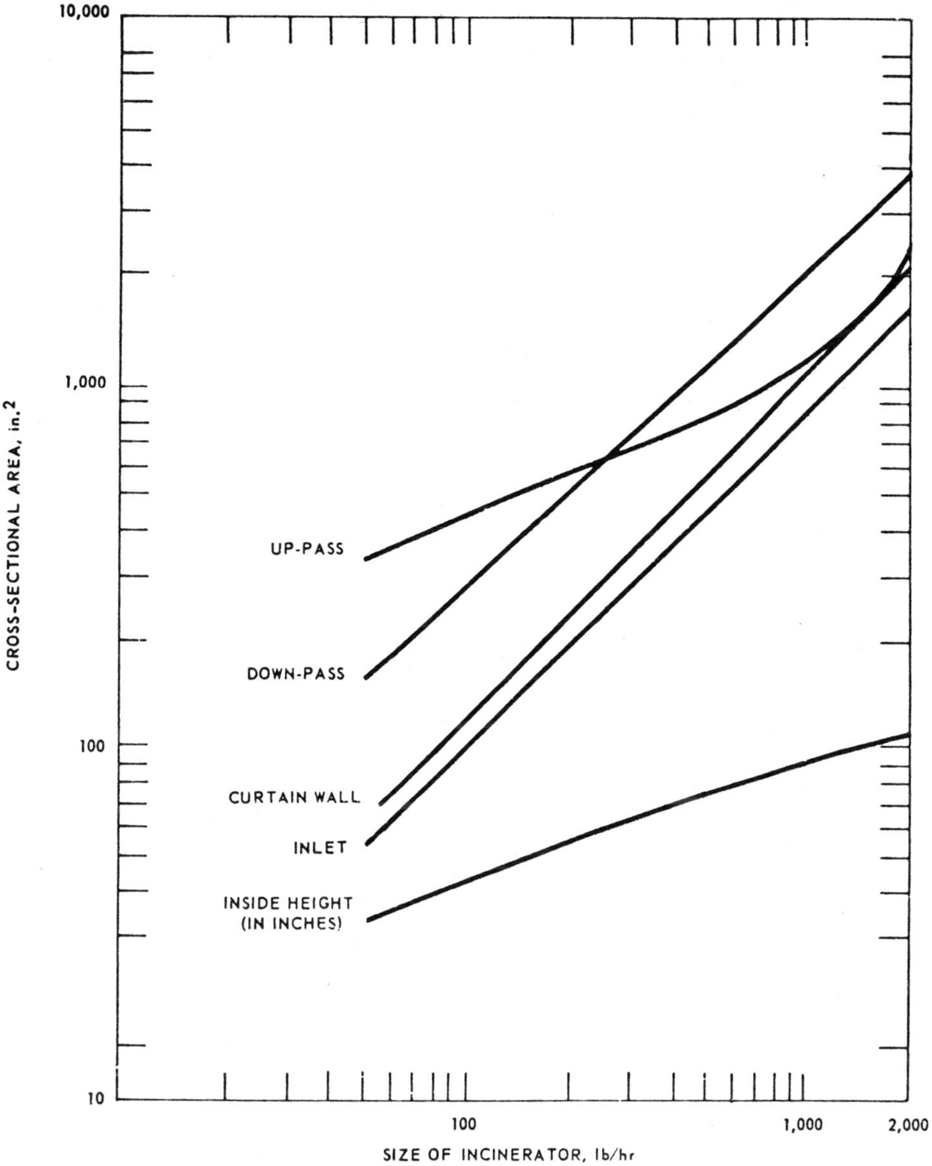

Fig. 21-11 Internal sizing of scrubbers.

The solenoid valves are electrically connected so that one opens when the fan is placed in operation. The flow of water through this valve is adjusted to approximately 40 percent of the scrubber needs. The other valve is controlled by a thermocouple located at the fan inlet. When the temperature at the fan inlet reaches 220°F, the second solenoid valve opens and the remainder of the water is delivered to the nozzles. This arrangement is used to keep the temperature of the gases from exceeding 350°F. Should the automatic control system fail, the operator may open the hand valve and furnish sufficient water to the scrubber.

A back-up system also may be installed to prevent heat damage to the fan in case the automatic system just described fails. One such system frequently used consists of a thermocouple located at the fan inlets and additional solenoid-valve-controlled nozzles located in either the downpass or the uppass of the scrubber. The nozzles should be capable of supplying at least the same quantity of water as the combined volume of the primary and secondary sprays. Should the temperature at the fan inlet exceed 500°F, the back-up solenoid valve opens, and the full volume of water flows to the nozzles to cool the effluent to an acceptable level. As a final precaution, a warning alarm, actuated at 550°F, by a thermocouple at the fan inlet, may be installed to alert the operator to excessive temperature increases.

Back-up systems must be tested frequently so that they are operable when the need arises. Consequently, a safety system of this type would be of doubtful value unless a regular maintenance and testing program were established.

SCRUBBER CONSTRUCTION

The steel exterior of the scrubber should be constructed of 3/16-inch-thick steel plate. Hangers should be mounted on the walls and top of the scrubber on 9-inch centers to hold the lining firmly to the walls. Linings of 135-pound-per-cubic-foot castable refractory should be 3 inches thick for incinerators with capacities of 750 pounds per hour or less. Units with capacities in excess of 750 pounds per hour should utilize 4-inch linings. The castable refractory floor should be sloped upward from the center of the scrubber at a 4-degree angle to facilitate the removal of collected fly ash and solids. The primary spray nozzles should be of the flat-spray type so that water droplets do not enter the connecting breeching and damage the refractory in the final combustion chamber of the incinerator. The secondary nozzles should be of the full cone type with a discharge angle of approximately 60 degrees. Nozzles mounted within the inlet duct should be provided with an access opening for cleaning or replacement. Nozzles mounted in the top of the unit should be installed out of the hot gas stream and should be removable from the exterior of the scrubber. Nozzles should be constructed of brass or stainless steel.

INDUCED-DRAFT FAN

The induced-draft fan should be constructed of mild steel and be capable of withstanding 600°F. The fan should be capable of at least two-speed operation or have a variable speed drive that is adjustable from maximum delivery volume to one-third of maximum delivery volume. Controls of this type permit the operator to reduce the volume handled by the fan when the incinerator is operating at less than the rated capacity. The resultant reduction in cooling in turn will increase the operating temperature within the incinerator and reduce the possibility of water carry-over from the scrubber. The controls for such a fan should be readily accessible to the operator so that he can reduce the fan speed and thus increase the overall efficiency of the incinerator. The fan housing should have a cleanout door and a water drain.

Design Parameters

Volume Requirements - The fan should be sized to deliver 700 cubic feet per minute of gas at 350°F for every 100 pounds per hour refuse capacity of the incinerator.

Static Pressure Requirements - The fan should provide 1/2-inch of water static pressure, at 350°F, for a 50-pound-per-hour incinerator. Its ability to develop static pressure should increase uniformly so that it will develop 1-1/2 inches of water, at 350°F, for a 2000-pound-per-hour incinerator. Fans operating at 350°F develop approximately two-thirds of the static pressure for which they are rated at ambient temperatures. Consequently, the induced-draft fan selected should be able to develop static pressures 50 percent higher than those desired at 350°F. For example, a fan selected for a 50-pound-per-hour incinerator should develop a static pressure of 3/4 inch of water at ambient temperatures, and a fan for a 2000-pound-per-hour incinerator should develop a static pressure of 2-1/4 inches water at ambient temperatures.

Horsepower Requirements - The horsepower requirements of the fan should be based upon the full capacity of the fan at ambient temperature, not at 350°F.

By-Pass Arrangements - For inside installations, a by-pass arrangement of breechings, or flue connections with dampers, to by-pass the scrubber and induced-draft fan is recommended.

MIST ELIMINATORS

Installation of mist eliminators is not usually necessary. There are, however, occasions when water droplets may be discharged from the exhaust fan. Should this be a serious problem, the inclusion of an eliminator section near the top of the uppass is desirable. In general, eliminators need be installed only when the performance of a unit has proved to be unsatisfactory.

ALTERNATE SCRUBBER DESIGNS

The following criteria should be used in the design of alternate scrubbers or gas washers.

1. The scrubber or gas washer should contain sprays, wetted baffles, or orifices arranged singly or in combination so as not to permit the discharge of particulate matter in violation of the Code of Federal Regulations.
2. Unlined gas washers or scrubbers should have welded or gasketed seams

and be corrosion resistant. Lined gas washers or scrubber casings should be made of at least 12-gauge steel and be welded or gasketed. The density of refractory lining should be no less than 120 pounds per cubic foot. The refractory should never be less than 2 inches thick and must be adequately anchored to the casing.

3. Scrubbers requiring an induced-draft fan should have a motor capable of cold startup (70°F). When the impeller of an induced-draft fan is in the gas stream, the fan must be equipped with a cleanout door and drain.

4. Where spray nozzles are employed, an optimum spray pattern must be provided to cover all the area of the gases as they pass through the gas washer or scrubber. Nozzles and valves should be arranged for independent removal by means of unions or flanges. When water is recirculated, a pressure regulator and a strainer should be provided.

5. An access door for cleanout should be provided on all scrubbers.

6. Interlocks should be provided when induced-draft fans and sprays are used.

7. When the outside skin temperature of a gas washer or scrubber exceeds 260°F, protection should be provided.

8. For inside installations, a by-pass arrangement of breeching, or flue connections with dampers, to by-pass the scrubber and induced-draft fan is recommended.

PART II

WATER POLLUTION CONTROL

CHAPTER 22

WATER POLLUTION BIOLOGY

AQUATIC environments are as numerous as the very waters themselves. Rising in snowcapped mountains small streams collect the snow melt and transport it to the plains. As these streams meander through the countryside they take from the lands that which is released to them. Small streams soon form larger ones that eventually join to form the great rivers and these in turn terminate in coastal estuaries. Each change in size and shape forms a habitat that becomes unique and supports an assemblage of organisms that is adapted to life in that particular environment. Reservoirs, built by man on rivers, in turn form a particular habitat that is influenced greatly by the reservoir's morphometric features. The reservoir in turn may influence the downstream environment because of the depth of the penstock that releases water of lower temperature, or of less dissolved oxygen, or of higher mineral quality, than the waters that receive it. The landscape is dotted with ponds and with many larger lakes of varying sizes and shapes.

Organisms that may be found in great numbers in the stream environments are often not adapted to life within the lake or reservoir environments and vice versa. There are many features that tend to make a particular aquatic environcment suitable or unsuitable, completely or to some degree, to a particular organism or group of closely associated organism.

Common to all aquatic environments would be the changes brought about by differences among the water habitats. For the pond and lake group these may include:

Altitude
Latitude
Maximum depth
Area of different depth zones
Volume of different depth strata
Length of shoreline
Littoral slope
Drainage area
Runoff rate
Average inflow
Area
Mean depth
Average outflow
Detention time
Water level fluctuation
Number of islands
Island areas
Island shoreline length
Penstock depth (reservoirs).

Features that create particular aquatic environments in flowing water may include:

Altitude	Drainage area to collection site
Latitude	Runoff rate
Relative extent of pools and riffles	Physical composition of stream bed
Depth at collection site	Physical nature of surrounding terrain
Width at collection site	Area geology.
Velocity of flow	

The estuarine environment is influenced by morphometric features that are common to both the flowing water and static environments. In addition it is influenced by tidal cycles and their fluctuations.

Life in waters is influenced also by water temperatures, dissolved oxygen, pH, color, turbidities, total dissolved solids, total alkalinity, nutrients and mineral composition. Maximum values, and in some cases minimal values also, of these constituents often create an environment that becomes intolerable to particular organisms and will limit their production or interfere subtly with physiological processes that in turn reduce their ability to compete with others within the environment.

Pollutional Effects

Effects of pollution assume many characteristics and an infinitesimal variaton in degree when pollution enters the aquatic environment. The specific environmental and ecological responses to a pollutant will depend largely on the volume and strength of the waste and the volume of water receiving it. As a basic introduction, five types of responses will be described in subsequent paragraphs, and within each of these response types there can be many changes in magnitude and degree.

The classic response that has often been described in the literature is the effects of organic wastes that may be discharged from sewage treatment plants and certain industries. As these wastes enter the water they create turbidity, decrease light penetration, and may settle to the bottom in substantial quantity to form sludge beds. The wastes are attacked immediately by bacteria and this process of decomposition consumes oxygen from the water and liberates essential nutrients that in turn stimulate the production of some forms of aquatic life.

Pollutional Zones

Upstream from the introduction of organic wastes, classic description details a clean water zone or one that is not affected by pollutants. At the point of waste discharge and for a short distance downstream there is formed a zone of degradation where wastes become mixed with the receiving waters, and where the initial attack is made on the waste by bacteria and other organisms in the process of decomposition.

Following the zone of degradation there is a zone of active decomposition that may extend for miles, or days of stream flow, depending in large measure on the volume of dilution that is afforded the waste by the stream, and the temperature of the water. The biological processes that occur within this zone are similar in many respects to those that occur in a "typical" sewage treatment plant. Within this zone, waste products are decomposed and those products that are not settled as sludge are assimilated by organisms in life processes.

A zone of recovery follows the zone of active decomposition. The recovery zone is essentially a stream reach in which water quality is gradually returned to that which existed prior to the entrance of pollutants. Water quality recovery is accomplished through physical, chemical, and biological interactions within the aquatic environment. The zone of recovery may extend also for many miles an its extent will depend principally upon morphometric features of the waterway.

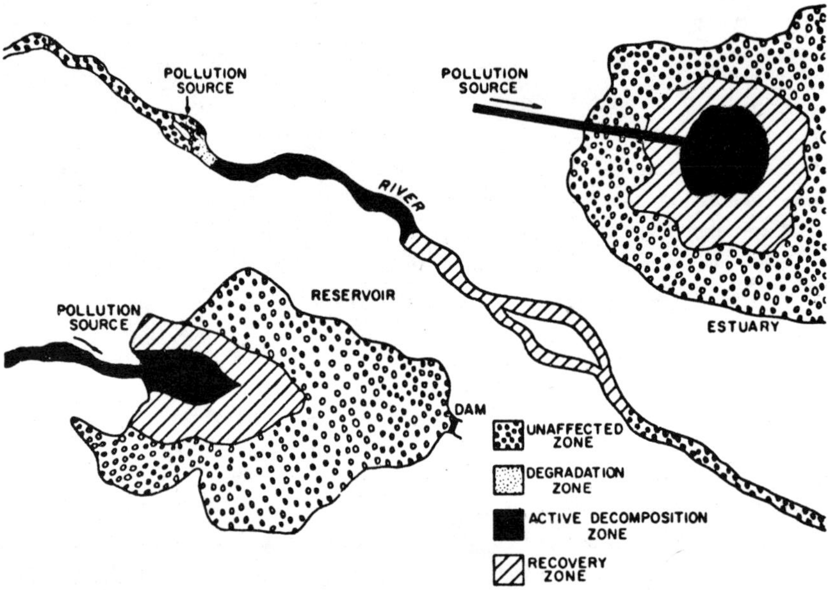

Fig. 22-1 **Pollutional Zones**

Finally the zone of recovery will terminate in another zone of clean water or area unaffected by pollution that is similar in physical, chemical, and biological features to that which existed upstream from the pollution source.

Organic Wastes

The classic description of the effects of organic wastes on the receiving stream often becomes confused in a specific stream investigation, because additional sources of pollution may enter the environment before the receiving water has been able to assimilate the entire effects of an initial source. When this occurs the effects of subsequent introductions become superimposed on those of the initial source and the total effect may confine large reaches of stream to a particular zonal classification.

Effects of organic wastes in the static water environment, as opposed to the flowing water environment, are modified principally by the morphometric features of the receiving water. Zonal changes that have been described for flowing water do exist but may be compressed in great measure either laterally or vertically when the discharge is to a lake or estuary. Such compression may tend to decrease the severity of pollution that is often observed in the flowing water environment and, on the other hand, may increase substantially the development of biotic nuisances such as algae or rooted aquatic plants that may develop from the nutrients released with and decomposed from the introduced organic materials.

Organism communities that may be related to pollution principally are those that are by nature associated with the bed or bottom of the waterway; those that attach themselves to objects such as rocks, aquatic plants, brush or debris submerged within the water; those that are essentially free floating and are transported by currents and wind, such as plankton and other microscopic forms; and those motile free swimming organisms such as fish. Considering each of these common organism groups, a number of observations can be made on their reaction to the introduction of organic wastes to a flowing stream.

Upstream from the waste source such limiting factors as food and intense competition among organisms and among organism groups, predation, and available habitat for a particular species will limit organism populations to those that can be sustained by the particular environment. Most often the limiting factor will be available food. Within this population, however, there will exist a great number of organism species. Thus, the old biological axiom for an environment unaffected by pollution is one that supports a great number of species with the total population delimited largely by food supply.

Following the introduction of organic wastes, conditions of existence for many organisms become substantially degraded. Increased turbidity in the water will reduce light penetration that in turn will reduce the volume of water capable of supporting photosynthesizing plants. Particulate matter in settling will flocculate small floating animals and plants from the water. As the material settles, sludge beds are formed on the stream bed and many of the areas that formerly could have been inhabited by bottom associated organisms become covered and uninhabitable.

The zone of degradation is a transition area between the clean water unaffected reach and a zone of decomposition of organic wastes. In such, the dissolved oxygen may be diminished but not completely removed. Sludge deposits may be initiated but are not formed in maximum magnitude or extent. Conditions of existence become impaired and typically there is a reduction in both the organism population and the number of species that can tolerate this environment.

Within the zone of active decomposition conditions of existence for aquatic life are at their worst. The breakdown of organic products by bacteria may have consumed available dissolved oxygen. Sludge deposits may have covered the stream bed thus eliminating dwelling areas for the majority of bottom associated organisms that could be found on an unaffected area. Fish spawning areas have been eliminated, but perhaps fish are no longer present because of diminished dissolved oxygen and substantially reduced available food. Here, aquatic plants will not be found in large numbers because they cannot survive on the soft shifting blanket of sludge. Turbidity may be high and floating plants and animals destroyed. Water color may be substantially affected. When organic materials are decomposed a seemingly inexhaustible food supply is liberated for those particular organisms that are adapted to use this food source. Thus, bacterial and certain protozoan populations may increase to extremely high levels. Those bottom associated organisms such as sludgeworms, bloodworms, and other worm-like animals may also increase to tremendous numbers because they are adapted to burrowing within the sludge, deriving their food therefrom, and existing on sources and amounts of oxygen that may be essentially nondetectable by conventional field investigative methods. Within the zone of active decomposition the organism species that can tolerate the environment are reduced to extremely low levels. Under some conditions those bottom associated animals that are visible to the unaided eye may be completely eliminated. Because of the tremendous quantity of food that is available to those organisms that are adapted to use it, the numbers of individuals of the surviving species may, indeed, be great. For example, it may be possible to find 50,000 sludgeworms or more living within each square foot of bottom area with the above-described conditions.

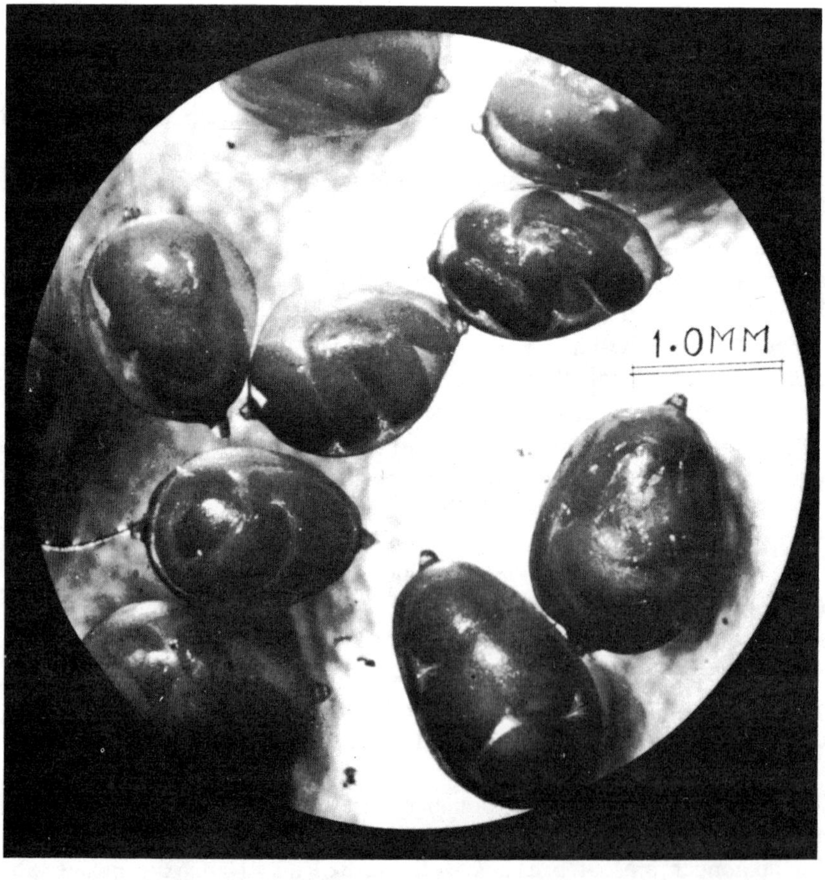

Fig. 22-2 Sludgeworm eggs with embryos.

The zone of recovery is essentially the downstream transition zone between the zone of active decomposition and an environment that is unaffected by pollution. This zone features a gradual cleaning up of the environment, a reduction in those features that form adverse conditions for aquatic life, an increase in organism species, and a gradual decrease in organism population because of decreased food supply and the presence of some of the predators that are less sensitive individually to pollutional affects.

Because of some variation in response among species to conditions of existance within the environment, and because of inherent difficulties in aquatic invertebrate taxonomy, the ecological evaluation of the total organism community is the acceptable approach in water pollution control investigations. At the present time, investigators tend to place organism in broad groups according to the general group response to pollutants in the environment. As we are able to advance our knowledge and determine more specifically the water quality requirements of identifiable species, the use of specific organism indicators may become more prevalent in biological interpretation. The general group known as "sludge-worms," for example, is found in both the unpolluted, as well as the organically polluted environment. Its value as a group lies in the fact that the numbers of individuals within the group is exceedingly low in unpolluted water, whereas in the organically polluted environment its numbers may be very high. Examples of organisms that may inhabit both the unpolluted and polluted environments are presented in table 22-1.

The converse of the effects of pollution on organisms is the effects of organisms on pollutants. Organic wastes, especially, supply food which in turn produces an abundance of a few types of organisms greater than that

Table 22-1 **Organism Associations**

Clean water association		Polluted water association	
Algae	*Cladophora* (green) *Ulothrix* (green) *Navicula* (diatom)	Iron Bacteria	*Sphaerotilus*
Protozoa	*Trachelomonas*	Fungi	*Leptomitus*
Insects	Plecoptera (stoneflies in general) Negaloptera (hellgrammites, alderflies, and fishflies in general) Trichoptera (caddisflies in general) Ephemeroptera (mayflies in general) Elmidae (riffle beetles in general)	Algae	*Chlorella* (green) *Chlamydomonas* (green) *Oscillatoria* (blue- green) *Phormidium* (bluegreen) *Stigeoclonium* (green)
Clams	Unionidae (pearl button)	Protozoa	*Carchesium* (stalked colonial ciliate) *Colpidium* (non-colonial ciliate)
Fish*	*Etheostoma* (darter) *Notropis* (shiner) *Chrosomus* (dace)	Segmented Worms	*Tubifex* (sludgeworms) *Limnodrilus* (sludge-worms)
		Leeches	*Helobdella stagnalis*
		Insects	*Culex pipiens* (mosquito) *Chironomus* (*-Tendipes*) *plumosus* (bloodworms) *Tubifera* (*Eristalis tenax*) (rat-tailed maggot)
		Snail	*Physa integra*
		Clam	*Sphaerium* (fingernail)
		Fish*	*Cyprinus carpio* (carp)

* Names from: American Fisheries Society Special Publication No. 2, "A List of Common and Scientific Names of Fishes from the United States and Canada" (Second Edition) Ann Arbor, Mich. (1960), 102 pp.

produced in an unpolluted environment. In consuming organic wastes, the organisms stabilize the waste in a given number of feet or miles of horizontal stream in a manner similar to that in a vertical trickling filter that is designed especially for maximum stabilizing efficiency by the organisms.

Purdy (1930) found long ago that sludgeworms eat continuously. Observations during 21 out of 24 hours showed no perceptible decrease in the foraging activity. Evacuation of a string of fecal pellets about 68 inches in length in a 24-hour period was recorded for each worm. An incubation of 24 hours showed an oxygen demand of 2.8 mg./l. by these pellets, whereas the original mud beneath the surface showed a demand of 6.7 mg./l. Purdy's conclusion was that the large surface area of fecal pellets exposed to the flowing water possessed a far greater purification potential than did the same mass of material an inch or more beneath the sludge-water interface.

As organic wastes become more stabilized, other organism types predominate within the aquatic animal community. Midge larvae have been

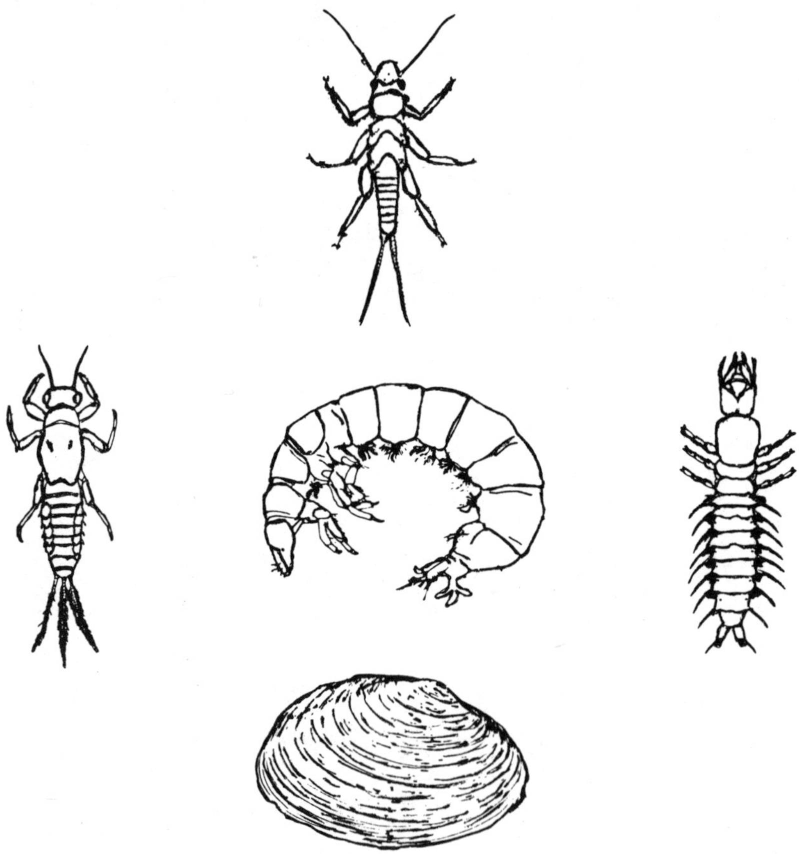

Fig. 22-3 **Representatives of stream bed associated animals (The clean water (sensitive) group).**

From left:

Stonefly nymph
Mayfly naiad; Caddisfly larvae; Hellgrammite
Unionid Clam

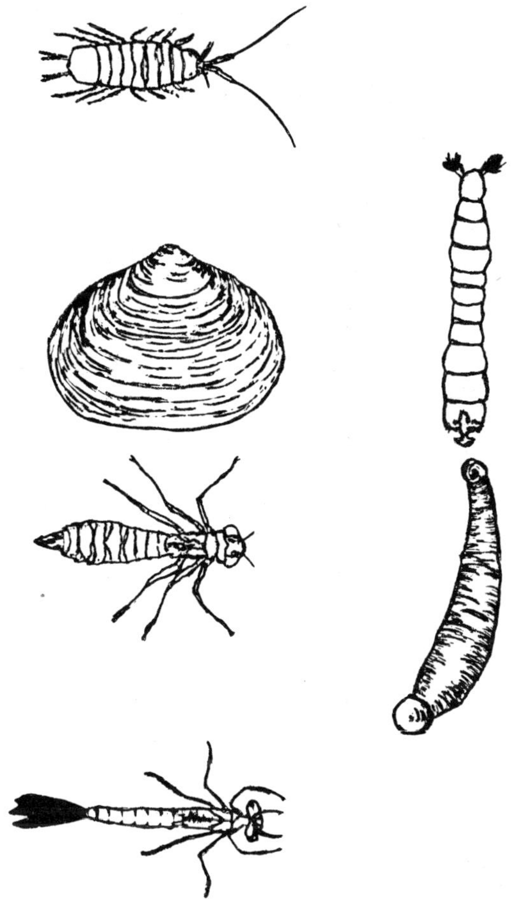

Fig. 22-4 **Representatives of stream bed associated animals (The intermediately tolerant group).**

From left:

Scud; Sowbug; Blackfly larvae
Fingernail Clam
Snail; Dragonfly nymph; Leech
Damselfly nymph

found to "paint" the stream bed a brilliant red with their undulating bodies. Caddisfly larval populations greater than 1,000 per square foot of stream bed or mayfly nymphs numbering more than 300 per square foot have been found on several occasions.

Inorganic Silts

The general effect on the aquatic environment of inorganic silts is to reduce severely both the kinds of organisms present and their populations. As particulate matter settles to the bottom it can blanket the substrate and form undesirable physical environments for organisms that would normally occupy such a habitat. Erosion silts alter aquatic environments chiefly by screening out light, by changing heat radiation, by blanketing the stream bottom and destroying living spaces, and by retaining organic materials and other substances that can create unfavorable conditions.

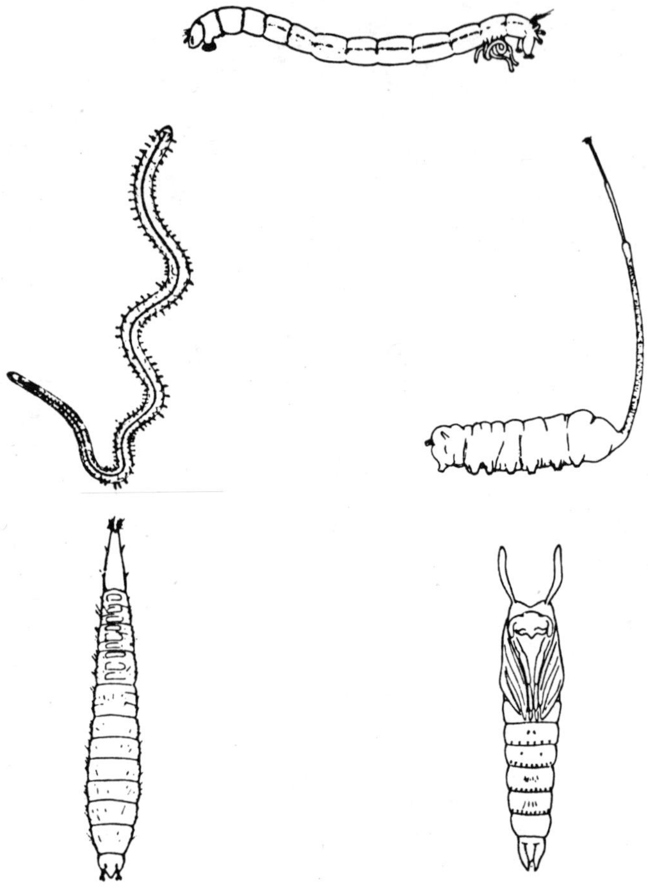

Fig. 22-5 **Representatives of stream bed associated animals (The very tolerant group).**

From left:

Bloodworm or midge larvae
Sludgeworm; Rat-tailed maggot
Sewage fly larvae; Sewage fly pupae

Developing eggs of fish and other organisms may be smothered by deposits of silt; fish feeding may be hampered. Direct injury to fully developed fish, however, by nontoxic suspended matter occurs only when concentrations are higher than those commonly found in natural water or associated with pollution.

Toxic Metals

Wastes containing concentrations of heavy metals, either individually or in combination, may be toxic to aquatic organisms and, thus, have a severe impact on the water community. A severely toxic substance will eliminate aquatic biota until dilution, dissipation, or volatilization reduces the concentration below the toxic threshold. Less generally toxic materials will reduce the aquatic biota, except those species that are able to tolerate

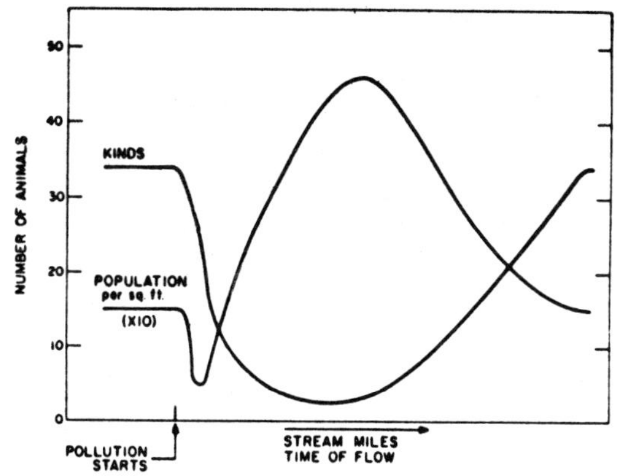

ORGANIC (NON-TOXIC) POLLUTION

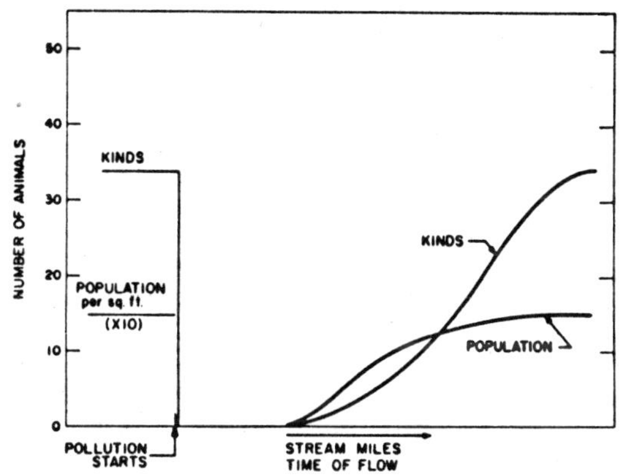

TOXIC POLLUTION

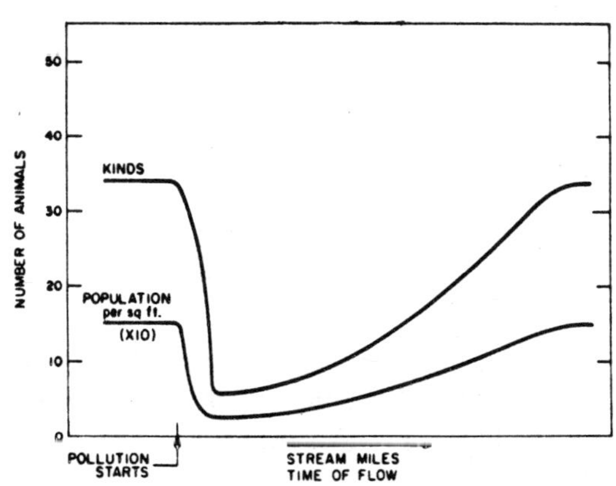

SILT (INERT) POLLUTION

Fig. 22-6 **Pollutional Effects on Animals**

the observed concentration of the toxicant. Because toxic materials offer no increased food supply, such as has been discussed for organic wastes, there is no sharp increase in the population of those organisms that may tolerate a specific concentration. The bioassay is an important tool in the investigation of these wastes, because the results from such a study indicate the degree of hazard to aquatic life of particular discharges; interpretations and recommendations can be made from these studies concerning the level of discharge that can be tolerated by the receiving aquatic community.

Temperature

TEMPERATURE is a prime regulator of natural processes within the water environment. It governs physiological functions in organisms and, acting directly or indirectly in combination with other water quality constituents, it affects aquatic life with each change. These effects include chemical reaction rates, enzymatic functions, molecular movements, molecular exchanges between membranes, etc., within and between the physiological systems and organs of an animal. Because of the complex interactions involved, and often because of the lack of specific knowledge or facts, temperature effects as they pertain to an animal or plant are most efficiently assessed on the basis of net influence on the organism. Depending on the extent of environmental temperature change, organisms can be activated, depressed, restricted, or killed.

Temperature determines those aquatic species that may be present; it controls spawning and the hatching of young, regulates their activity and stimulates or suppresses their growth and development; it can attract and kill when the water becomes heated or chilled too suddenly. Colder water generally suppresses development; warmer water generally accelerates activity.

Temperature regulates molecular movement and thus largely determines the rate of metabolism and activity of all organisms, both those with a relatively constant body temperature and those whose body temperature is identical to, or follows closely, the environmental temperature. Because of its capacity to determine metabolic rate, temperature may be the most important single environmental entity to life and life processes.

Variations in temperature of streams, lakes, estuaries, and oceans are normal results of climatic and geologic phenomena. Waters that support some form of aquatic life other than bacteria or viruses range in temperature from 26.6° F. in polar sea waters to 185° F. in thermal springs. Most aquatic organisms tolerate only those temperature changes that occur within a narrow range to which they are adapted, whether it be high, intermediate, or low on this temperature scale.

Within the same species, the effects of a given temperature may differ in separate populations, in various life cycle stages, or between the sexes, and such effects may depend on the temperature history of the individual tested, as well as on present or past effects of other environmental factors.

Freshwater has the greatest density at 38° F.; higher and lower temperatures result in waters with lower density. Seasonally induced temperature changes are greatest in the midlatitudes.

In lakes, insolation warms the surface waters in spring, reducing their densities compared to the deeper waters until eventually the density differences are sufficient to prevent the wind from mixing the body of water; thermal stratification then occurs. The warm upper layer (epilimnion) is well mixed to a depth determined by wave and other wind induced currents. The cool bottom waters (hypolimnion) become stagnant except for minor currents confined to this strata. A strata of sudden temperature changes (thermocline) separates these regions. In autumn, the lake radiates heat, surface temperatures decrease, surface water density increases, and water viscosity increases. Soon the wind, aided by reduced density differences between water layers, mixes the surface with the bottom waters resulting in a homogenous water mass. Depending on altitude and local climatic conditions, the lake continues to mix until the following spring in latitudes of less than about 40°. In latitudes north of about 40°, winter surface water temperatures are less than 38° F. and these are superimposed over the water mass until they are cooled to freezing. An ice cover eliminates wind induced mixing and stagnation occurs.

Thermal stratification in reservoirs may assume many patterns depending on geographical location, climatological conditions, depth, surface area, and type of dam structure, penstock locations, and hydropower use. In general, large, deep impoundments will cool downstream waters in the summer and warm them in winter when withdrawal ports are deep; shallow, unstratified impoundments with large surface areas will warm downstream waters in the summer; water drawn from the surface of a reservoir will warm downstream waters; a reduction in normal flow downstream from an impoundment will cause marked warming in summer; and "run-of-river" impoundments, where the surface area has not been increased markedly over the normal river area, will produce only small changes in downstream water temperatures.

In the deep, stagnant, summer bottom waters, as well as in ice covered waters, atmospheric reaeration is absent and oxygen from photosynthesis by plants is limited. Decomposing organisms (especially those settling to the bottom waters in summer) remove oxygen from the water and the gaseous byproducts of decomposition are trapped. Undesirable soluble phosphorus, carbon dioxide, iron, and manganese concentrations increase in these stagnant waters. Designed thermal discharges can reduce some of these problems. Ice cover can be limited, thus allowing wind and thermally induced currents to reduce winter stagnation. A deepwater summer discharge could warm hypolimnetic waters to decrease density and permit total water mass mixing where a cold water fishery would not be damaged by such action.

Stratification may occur in streams receiving heated effluents. There are three recognized forms of stream stratification: overflow, interflow, and underflow; the forms are determined by the relationship between the density of the influent and the density of the stream water.

Surface freshwaters in the United States vary from 32° to over 100° F. according to the latitude, altitude, season, time of day, duration of flow, depth, and many other variables. Agents affecting natural water temperature are so numerous that no two water bodies, even in the same latitude, are likely to have the same thermal characteristics. Fish and other aquatic life occurring naturally in each body of water are those that have become adapted to the temperature conditions existing there. The interrelationships of species, length of daylight and water temperature are so intimate that even a small change in temperature may have far-reaching effects. An insect nymph in an artifically warmed stream, for example, might emerge for its mating flight too early in the spring and be immobilized by the cold air temperature, or a fish might hatch too early in the spring to find its natural food organisms because the food chain depends ultimately on plants, and these in turn, upon length of daylight, as well as temperature. The inhabitants of a water body that seldom becomes warmer than 70° F. are placed under stress, if not killed outright, by 90° F. water. Even at 75° to 80° F., they may be unable to compete successfully with organisms for which 75° to 80° F. is favorable. Similarly, the inhabitants of warmer waters are at a competitive disadvantage in cool water.

An animal's occurrence in a given habitat does not mean that it can tolerate the seasonal temperature extremes of that habitat at one time. The habitat must be cooled gradually in the fall if the animal is to become acclimatized to the cold water of winter, and warmed gradually in the spring if it is to withstand summer heat.

Some organisms might endure a temperature of 92° to 95° F. for a few hours, but not for days. Gradual change of water temperature with the season is important for other reasons: an increasing or decreasing temperature often "triggers" spawning, metamorphosis, and migration. The eggs of some freshwater organisms must be chilled before they will hatch properly.

The temperature range tolerated by many species is narrow during very early development; it increases somewhat during maturity, and decreases again in the old adult. Similarly, the tolerable temperature range is often more restrictive during the reproductive period that at other times during maturity. Upper lethal temperatures may be lower for animals from cold water than for closely related species from warm water. Many motile organisms such as fish, some zooplankton, certain algae, and some associated animals can avoid critical temperatures by vertical and horizontal migration into more suitable areas. However, some organisms may be attracted to areas with critical temperatures, and, upon arrival, succumb.

Changes in fish populations can result from many types of artificial cooling and heating of natural waters. These changes result from the discharge of condenser cooling water from thermal electric generating plants, industrial waste cooling waters, and other heated effluents, and irrigation waters. Streams are warmed also by the sun when the shade from stream bank trees and other vegetation is eliminated. The discharge of cold water from stratified impoundments may provide an ideal habitat for trout and other cold water fish, when sufficient dissolved oxygen is present, but not for the warm water fish that inhabited the stream before impoundment.

For every 18° F. increase in temperature, the chemical reaction rate is approximately doubled in an organism or in an environment. Life processes in the water are accelerated with temperature increases and slowed as the water cools.

The solubility of gases, including oxygen, in water varies inversely with temperature. In fresh water, the solubility of atmospheric oxygen is decreased by about 55 percent as the temperature rises from 32° to 104° F. under 1 atmosphere of pressure (760 mm. Hg.). Because all desirable living things are dependent on oxygen in one form or another to maintain the life processes that produce energy for growth and reproduction, dissolved oxygen is of imposing significance in the aquatic environment.

When organism metabolism increases because of higher temperatures, organism development is speeded, and more dissolved oxygen is required to maintain existence. But, bacterial action in the natural purification process to break down organic materials is also accelerated with increased temperatures, thus reducing the oxygen that could be available in the warmer water. When organisms use larger amounts of oxygen, and when oxygen has been reduced by temperature action and interaction, organisms may perish. Life stages that are especially vulnerable are the eggs and larvae. At higher temperatures, phytoplankton have been found to need greater amounts of certain growth factors such as vitamin B_{12}. Between 96.8° and 98.2° F., for example, the vitamin requirement has been found to increase over 300 times for some species.

Fish and other motile organisms seek a preferred temperature at which they can best survive, which is several degrees below a temperature that is lethal. Larger individuals tend to move out of areas that are too hot, but larvae and juveniles cannot often move fast enough to avoid a sudden temperature increase. Large fish and fish in schools avoid heated areas in summer but may be attracted to such areas in winter. This phenomenon may result in good fishing during the cooler months, but an absence of this sport at other times.

Reproduction cycles may be changed significantly by increased temperature because this function takes place under restricted temperature ranges. Spawning may not occur at all because temperatures are too high. Thus, a fish population may exist in a heated area only by continued immigration. Disregarding the decreased reproductive potential, water temperatures need not reach lethal levels to wipe out a species. Temperatures that favor competitors, predators, parasites, and disease can destroy a species at levels far below those that are lethal.

Fish food organisms are altered severely when temperatures approach or exceed 90° F. Predominant algal species change, primary production is decreased, and bottom associated organisms may be depleted or altered drastically in numbers and distribution. Increased water temperatures may cause aquatic plant nuisances when other environmental factors are favorable.

Synergistic actions of pollutants are more severe at higher water temperatures. Given amounts of domestic sewage, refinery wastes, oils, tars, insecticides, detergents, and fertilizers more rapidly deplete oxygen in water at higher temperatures, and the respective toxicities are likewise increased.

The National Technical Advisory Committee on Water Quality Criteria (Anon., 1968), composed in part of the nation's leading fishery experts, recommended that to maintain a well-rounded population of warm water fishes, heat added to a freshwater stream not exceed that which would raise the water temperature more then 5° F. at the expected minimum daily flow for the month involved. In lakes, the temperature of the upper waters should not be raised more than 3° F. above that which existed before heat was added. The increase should be based on the monthly average of the maximum daily temperatures. Temperature should be measured in those areas where important organisms are most likely to be affected adversely.

The Committee recommended provisional maximum temperatures as compatible with the well-being of various fish species and their associated biota as follows:

93° F.: Growth of catfish, gar, white or yellow bass, spotted bass, buffalo, carpsucker, threadfin shad, and gizzard shad.

90° F.: Growth of largemouth bass, drum, bluegill, and crappie.

84° F.: Growth of pike, perch, walleye, smallmouth bass, and sauger.

80° F.: Spawning and egg development of catfish, buffalo, threadfin shad, and gizzard shad.

75° F.: Spawning and egg development of largemouth bass, white and yellow bass and spotted bass.

68° F.: Growth or migration routes of salmonids and for egg development of perch and smallmouth bass.

55° F.: Spawning and egg development of salmon and trout (other than lake trout).

48° F.: Spawning and egg development of lake trout, walleye, northern pike, and sauger.

Because of the large number of trout and salmon waters that have been destroyed, made marginal or nonproductive, remaining trout and salmon waters must be protected if these resources are to be preserved. The Committee further recommended that inland trout streams, headwaters of salmon streams, trout and salmon lakes and the deeper waters of lakes that contain salmonids not be warmed. No heated effluents should be discharged in the vicinity of spawning areas.

Little work has been done regarding thermal addition effects in subtropical estuarine ecosystems. In the subtropical environment, optimum temperatures for many forms are only a few degrees lower than maximum lethal temperatures. Organisms may be existing under stress with naturally occurring summer temperatures. Great care should be exercised to prevent harmful temperature increases.

In general, marine water temperatures do not change as rapidly or range as widely as those of freshwaters. Marine and estuarine fishes, therefore, are less tolerant of temperature variation. Although this limited tolerance is greater in the estuarine than in the open water marine species, temperature changes are more important to those fishes in estuaries and bays than to those in open marine areas.

Marine surf-zone discharge from large-scale coastal power plants may be expected to significantly alter the shore environment for species of invertebrates and fish that are commonly found there.

Some investigators have becomed alarmed over the loss in organisms contained in the water pumped across condensers and through a generating plant. These are subject to thermal shock, physical damage, and perhaps commercial additives. These organisms include phytoplankton, crustaceans, zooplankton, and shellfish larvae, such as clams and oysters that have stages of drift in the water column for a few weeks before they settle to the bottom. Studies have shown a 95 percent mortality of these organisms when they are subjected to the rise in temperature in crossing the condenser.

Available data indicate that commercial and key food-chain estuarine animals cannot tolerate temperatures greater than approximately 90° F. regardless of the temperature to which they have been acclimated. Thus, natural peak summer water temperatures in a subtropical or tropical estuary may be near the tolerance threshold for a number of desirable marine organisms.

In subtropical waters, organisms that find the environment undesirable are not replaced by organisms of greater temperature tolerance, as so often happens in northern latitudes.

The National Technical Advisory Committee on Water Quality Criteria, in reporting to the Secretary of the Interior, recommended that the discharge of any heated materials into coastal waters be closely managed. This Committee stipulated that any rise owing to such discharges should be restricted to 1.5° F. during the critical summer months, outside of established mixing zones. To make water quality standards more meaningful, mixing zones must have definition. The National Technical Advisory Committee suggested only that adequate passageways be provided at all times for the movement or drift of organisms, and that mixing areas must not be used for, or considered as, a substitute for waste treatment, or as an extension of, or substitute for, a waste treatment facility.

Dissolved Oxygen

Dissolved oxygen (D.O.) is a water quality constituent that, in appropriate concentrations, is essential not only to keep organisms living but also to sustain species reproduction, vigor, and the development of populations. Organisms undergo stress at reduced D.O. concentrations that make them less competitive to sustain their species within the aquatic environment. For example, D.O. concentrations around 3 milligrams per liter (mg/l) or less have been shown to interfere with fish populations through delayed hatching of eggs (Silver et al., 1963), reduced size and vigor of embryos (Silver et al., 1963; Van Horn and Balch, 1957), production of monstrosities in young (Alderdice et al., 1958), interference with food digestion and acceleration of blood clotting (Bouck & Ball, 1965), decreased tolerance to certain toxicants (Cairns and Scheier, 1957), reduced food efficiency and growth rate (Chiba, 1966; Herrman et al., 1962), and reduced maximum sustained swimming speed (Davis et al., 1963).

Oxygen enters the water by absorption directly from the atmosphere or by plant photosynthesis, and is removed by respiration of organisms and by decomposition. That derived from the atmosphere may be by direct diffusion or by surface water agitation by wind and waves, which may also release dissolved oxygen under conditions of supersaturation.

In photosynthesis, aquatic plants utilize carbon dioxide and liberate dissolved and free-gaseous oxygen at times of supersaturation. Since energy is required in the form of light, photosynthesis is limited to the photic zone were light is sufficient to facilitate this process. According to Dice (1952), "... the ultimate limit of productivity of a given ecosystem is governed by the total effective solar energy falling annually on the area, by the efficiency with which the plants in the ecosystem are able to transform this energy into organic compounds, and by those physical factors of the environment which affect the rate of photosynthesis." Verduin (1956) summarized the literature on primary production in lakes; based on computations of photosynthetic oxygen production, he found that the yields of several lakes were mostly between 42 and 57 pounds of dissolved oxygen per acre per day. A year-round study under completely natural conditions in western Lake Erie showed winter yields of about 11 pounds of dissolved oxygen production per acre per day, and summer maxima of about 85 pounds per acre per day. The annual oxygen production curve closely followed the solar radiation curve. The net oxygen production rate for East Okoboji Lake in Iowa, a producer of large plankton populations, was 79 pounds per acre per day, with production largely confined to the first 2 meters (Weber, 1958). Whipple et al. (1948) noted that supersaturation in the upper waters is not cumulative to a great extent because circulation is maintained by wind action and convection currents both of which promote contact of the water and the air with a consequent loss of oxygen. Higher saturation is frequently found in the upper region of the thermocline in infertile oligotrophic lakes. Wind action seldom disturbs the waters of this zone, convection currents are absent, and diffusion is a slow process. Plants find an abundant supply of carbon dioxide and sufficient light in this area to stimulate photosynthesis, resulting in supersaturation values that may exceed 300 percent.

During respiration and decomposition, animals and plants consume dissolved oxygen and liberate carbon dioxide at all depths where they occur. Because excreted and secreted products and dead animals and plants sink, most of the decomposition takes place in the hypolimnion; thus, during lake stratification there is a gradual decrease of dissolved oxygen in this zone. After the dissolved oxygen is depleted, anaerobic decomposition continues with evolution of methane and hydrogen sulfide.

In the epilimnion, during thermal stratification, dissolved oxygen is usually abundant and is supplied by atmospheric aeration and photosynthesis. Phytoplankton are plentiful in fertile lakes and are responsible for most of the photosynthetic oxygen. The thermocline is a transition zone from the standpoint of dissolved oxygen, as well as temperature. The water rapidly cools in this region, incident light is much reduced, and photosynthesis is usually decreased; if sufficient dissolved oxygen is present, some cold water fish abound. As dead organisms that sink into the hypolimnion decompose, oxygen is utilized; consequently, the hypolimnion in fertile lakes may become devoid of dissolved oxygen following a spring overturn, and this zone may be unavailable to fish and most benthic invertebrates at this time. During the two brief periods in spring and fall when lake water circulates, temperature and dissolved oxygen are the same from top to bottom and fish can use the entire water depth.

The National Technical Advisory Committee (Anon., 1968) recommended that D.O. concentrations be above 5 mg/l assuming normal seasonal and daily variations for a diversified warm water biota. The Committee stated that under extreme conditions concentrations may range between 5 and 4 mg/l for short periods during any 24-hour period, providing that the water quality is favorable in all other respects. For cold water biota, it is desirable that D.O. concentrations be at or near saturation especially in spawning areas. D.O. levels in the hypolimnion of lakes should not be lowered below 6 mg/l at any time because of the addition of oxygen demanding wastes. The Committee further specified that D.O. concentrations in surface coastal waters should be greater than 5.0 mg/l, except when upwellings and other phenomena may cause this value to be depressed. D.O. concentrations in estuaries and tidal tributaries should not be less than 4.0 mg/l at any time or place, except in naturally dystrophic waters or where natural conditions cause D.O. to be depressed.

pH

The world's literature on pH published prior to about 1950 has been critically evaluated by Doudoroff and Katz (1950). They concluded that ". . . under otherwise favorable conditions, pH values above 5.0 and ranging upward to pH 9.0, at least, are not lethal for most fully developed freshwater fishes. Much more extreme pH values, perhaps below 4.0 and well above 10.0, also can be tolerated indefinitely by resistant species. However, regardless of the nature of acid or alkaline wastes responsible, such extreme conditions, associated with industrial pollution, are evidently undesirable and hazardous for fish life in waters which are not naturally so acid or alkaline."

Lloyd (1968) summarized the conclusions of the European Inland Fisheries Advisory Commission, Food and Agricultural Organization of the United Nations, with the statement:

> "There is no definite pH range within which a fishery is unharmed and outside which it is damaged, but rather there is a gradual deterioration as the pH values are further removed from the normal range. The pH range which is not directly lethal to fish is 5–9; however, the toxicity of several common pollutants is markedly affected by pH changes within this range; and increasing acidity or alkalinity may make these poisons more toxic. Also, an acid discharge may liberate sufficient carbon dioxide from bicarbonate in the water either to be directly toxic, or to cause the pH range 5–6 to become lethal.
>
> "Below a pH value of 5.0 fish mortalities may be expected, although some species may become acclimated to values as low as 3.7. However, the productivity of the aquatic ecosystem is considerably reduced below a pH value of 5.0, so that the yield from a fishery would also become less. Some acid waters may contain precipitated ferric hydroxide which may also act as a lethal factor."

The National Technical Advisory Committee (Anon., 1968) recommended for fish and other aquatic life that:

(1) No highly dissociated materials should be added in quantities sufficient to lower the pH below 6.0 or to raise the pH above 9.0.

(2) To protect the carbonate system and thus the productivity of the water, acid should not be added in sufficient quantity to lower the total alkalinity to less than 20 mg/l.

(3) The addition of weakly dissociated acids and alkalies should be regulated in terms of their own toxicities as established by bioassay procedures.

Light

Rooted, suspended, and floating aquatic plants require light for photosynthesis. Light penetration into waters is exceedingly variable in different lakes. Clarke (1939) pointed out that the diminution of the intensity of light in its passage through water follows a definite mathematical formula. The relationship between the depth of water and the amount of light penetrating to that depth can be plotted as a straight line on semilogarithmic paper. Even the clearest waters impede the passage of light to some extent; light passed through 100 meters of distilled water is reduced to 1 or 2 percent of its incident value.

The principal factors affecting the depth of light penetration in natural waters include suspended microscopic plants and animals, suspended mineral particles such as mineral silt, stains that impart a color, detergent foams, dense mats of floating and suspended debris or a combination of these. The region in which light intensity is adequate for photosynthesis is often referred to as the trophogenic zone, the layer that encompasses 99 percent of the incident light. The depth of the trophogenic zone may vary from less than 5 to greater than 90 feet.

The length of daylight in water varies inversely with the depth of the water. The seasonal variation in the intensity of solar radiation influences the potential rate of photosynthesis. In winter the presence of ice with an over layer of snow further limits the amount of relatively poor incident light energy that reaches the water. The work of Birge, reported by Neess and Bunge (1957), indicates that the absorptive quality of clear ice is very similar to that of water, although the addition of air bubbles or particulate matter reduces the transmission of light. Snow further reduces light penetration through ice. Greenbank (1945) found 84 percent light transmission through 7½ inches of very clear ice, and 22 percent through 7½ inches of very cloudy ice. A 1-inch snow cover permitted only 7 percent light transmission through the ice and snow; 2 inches of snow permitted only 1 percent light transmission. Bartsch and Allum (1957), studying sewage stabilization ponds, found that in the absence of snow 20 to 55 percent of the incident light passed through 10 to 12 inches of ice, whereas, with a 1- to 3- inch snow cover 93 to 99 percent of the incident light was absorbed by ice and snow, when the ice was 1 to 2 feet thick. Mackenthun and McNabb (1961) found less than 1 percent of light passing through 16 inches of ice covered by 2 inches of snow.

Flow

The velocity of water movement is extremely important to aquatic organisms in a number of ways including the transport of nutrients and organic food past those organisms attached to stationary surfaces; the transport of plankton and benthos as drift, which in turn serve as food for higher organisms; and the addition of oxygen to the water through surface aeration. Silts are moved downstream and sediments may be transported as bed load. These in turn are often associated with major nutrients, such as nitrogen and phosphorus, which may be released at some point downstream from their introduction.

The determination of flow is necessary to compute pounds of materials passing a given point. Computations are often made on pounds of nitrogen, phosphorus, or other elements of concern, on amounts of plankton and benthos as drift, on amounts of pollutants, such as wool, fibers, or other microscopical identifiable materials that may be associated with a point source.

Flow determines those species of stream bed organisms that may be present in a particular stream reach. Some of these, such as the black fly larva, require fast water. Others, such as the immature forms of caddisflies and mayflies, will develop to large populations in more sluggish water. Among many invertebrate genera there are those particular species that are adapted for life not only under the two extremes of flow but also under its many variations.

Silt

The deposition of sediments in streams can and often does destroy insect and mussel populations. Ellis (1931), in studying the Mississippi, Tennessee, and Ohio Rivers, reported that erosion silts had destroyed a large portion of the mussel population in various streams by directly smothering the animals in localities where a thick deposit of mud was formed, and by smothering young mussels even where the adults could maintain themselves. Ziebell (1957) reported a marked reduction in organisms 100 yards downstream from the discharge of a gravel washing operation entering the South Fork Chehalis River in Washington. Ziebell and Knox (1957) investigated the effects of yet another gravel washing operation on the Wynooche River in Washington. The results of bottom samples collected downstream from the operation revealed reductions of 75 to 85 percent at distances exceeding 1 mile. Silt from a gravel washing plant located on Cold Creek and the Truckee River, Calif. reduced bottom organisms over 75 percent for a distance of more than 10 miles downstream (Cordone and Pennoyer, 1960). Reports published by the Oregon State Game Commission summarized the results of extensive collections of bottom organisms upstream and downstream from gold dredge operations on the Powder River (Anon., 1955). During siltation, production of fish food organisms decreased to nearly zero in the zone of heaviest pollution and the effect of siltation extended for a distance of 20 miles. In about 1 year after the dredge closed operations, the silt was flushed from the pools and riffles by freshets and bottom organisms increased 8 to 10 fold in weight per unit of bottom area.

Few data are available regarding the direct harm of sediments to fish. In most cases indirect damage to the fish population through destruction of the food supply, eggs, or changes in the habitat probably occur long before adult fish are harmed directly. Ellis (1944) stated that particulate matter of a hardness greater than one, if held in suspension by current action or otherwise, will injure the gills and other delicate exposed structures of fishes, mollusks, and insects when the particles are large enough. Kemp (1949) stated that mud or silt in suspension will clog or cut the gills of many fish and mollusks, and he considered 3,000 p.p.m. dangerous when maintained for a period of 10 days. Wallen (1951) conducted controlled aquarium investigations on the direct effects of turbidity on warm water fishes; he found that observable behavioral reactions that appeared as a turbidity effect did not develop until concentrations of turbidity neared 20,000 p.p.m. and in one species reactions did not appear until turbidities reached 100,000 parts per million (p.p.m.).

Sediment is believed to destroy algae by molar action, by simply covering the bottom of the stream with a blanket of silt, or by shutting off the light needed for photosynthesis. Tarzwell and Gaufin (1953) found that turbid waters may transport the byproducts of bacterial action on organic wastes and the effluent of sewage treatment plants considerable distances before they are utilized. When water clears as a result of impoundment so that phytoplankton can grow, fertilizing materials are utilized and may produce troublesome blooms far from the source of pollution.

Attached algae and vegetation are affected by silt principally by: (1) Covering bottom materials with a layer of sediment, (2) reducing light transparency and preventing light penetration, and (3) grinding algae by action of abrasive particles. A reduction of plant food is accompanied by a reduction in the poundage of plant feeding animals that can be supported, and this in turn limits the production of carnivorous animals including fish.

The European Inland Fisheries Advisory Commission, Food and Agricultural Organization of the United Nations, prepared water quality criteria on finely divided solids (Anon., 1965). With respect to chemically inert solids and to waters that are otherwise satisfactory for the maintenance of freshwater fishes they made the following conclusions:

> "(a) There is no evidence that concentrations of suspended solids less than 25 p.p.m. have any harmful effects on fisheries.
>
> (b) It should usually be possible to maintain good or moderate fisheries in waters which normally contain 25 to 80 p.p.m. suspended solids. Other factors being equal, however, the yield of fish from such waters might be somewhat lower than from those in category (a).
>
> (c) Waters normally containing from 80 to 400 p.p.m. suspended solids are unlikely to support good freshwater fisheries, although fisheries may sometimes be found at the lower concentrations within this range.
>
> (d) At the best, only poor fisheries are likely to be found in waters which normally contain more than 400 p.p.m. suspended solids."

"In addition although several thousand p.p.m. solids may not kill fish during several hours or days exposure, such temporary high concentrations should be prevented in rivers where good fisheries are to be maintained. The spawning grounds of salmon and trout require special consideration and should be kept as free as possible from finely divided solids."

Oil

McKee (1956) in summarizing the effects of oil substances on aquatic life in freshwater, stated that:

"(1) free oil and emulsions may coat and destroy algae and other plankton;

(2) heavy coatings of free oil on the surface may interfere with the natural processes of reaeration and photosynthesis, while light coatings would be less detrimental because wave action and other turbulence would maintain adequate reaeration; and

(3) water soluble principles may exert a direct toxic action."

The deleterious effect of crude oil and lubricating oils on fish is due to a film formed over the gill filaments of fish, which prevents the exchange of gasses and results in suffocation (Klinke, 1962).

The effects of oils on marine animals may include the tainting of fish and shellfish flesh, poisoning by ingestion of oil or soluble fractions, such as phenol, ammonia, and sulfides, disturbances of marine food chains, physical fouling of animals with heavy coats of oil, and repellent effects (Hawkes, 1961).

Many thousands of waterfowl have been destroyed by the effects of oil pollution. This wasteful loss has deprived nature lovers, waterfowl hunters, and bird watchers of immeasurable enjoyment. The destruction of many duck species, such as the canvasback, redhead, and scaups, comes at a critical period for these species that are fighting for survival against the forces of nature. In future years additional waterfowl will be destroyed if oil dumping is continued, especially in late winter. In this modern age of technical development, the discharge of oil into a river system indicates man's lack of responsibility for the preservation of our natural resources.

Erickson (1965) pointed out that the effects of oil on birds depend upon a variety of factors including the type of oil, extent of contamination of plumage, temperature of the air and water, and the quantity of oil ingested. He found that migratory birds are affected indirectly by deposits of oil on the bottom, in shallow water, or along the shore that reduces the available food supply of both plants and animals. Elements within the food chain are eliminated by chemical or physical properties of the oils and food for waterfowl may become unavailable by being overlayed or embedded in the oily materials. Accumulation of petroleum sludge may also prevent germination and growth of plants and the production of invertebrates important as food, either by smothering or by toxic effects.

Oil causes matting of the duck feathers so that ducks become waterlogged, lose their ability to fly and drown if they cannot get out of the water soon enough. It breaks down the insulating power of the feathers; body heat and stored reserves of energy are rapidly lost. Diving ducks may starve, and, following the preening of oil from contaminated feathers, bleeding ulcers may be produced in the digestive tract causing mortality.

Major Nutrients

Eutrophication is a term meaning enrichment of waters by nutrients through either man-created or natural means. Present knowledge indicates that the fertilizing elements most responsible for lake eutrophication are phosphorus and nitrogen. Iron and certain "trace" elements are also important. Sewage and sewage effluents contain a generous amount of those nutrients necessary for algal development.

Lake eutrophication results in an increase in algal and weed nuisances and an increase in midge larvae, whose adult stage has plagued man in Clear Lake, Calif., Lake Winnebago, Wis., and several lakes in Florida. Dense algal growths form surface water scums and algal-littered beaches. Water may become foul-smelling. Filter-clogging problems at municipal water installations can result from abundant suspended algae. When algal cells die, oxygen is used in decomposition, and fish kills have resulted. Rapid decomposition of dense algal scums, with associated organisms and debris, gives rise to odors and hydrogen sulfide gas that creates strong citizen disapproval; the gas often stains the white lead paint on residences adjacent to the shore.

Nitrogen and phosphorus are necessary components of an environment in which excessive aquatic growths arise. Algal growth is influenced by many varied factors: vitamins, trace metals, hormones, auxins, extracellular metabolites, autointoxicants, viruses and predation and grazing by aquatic animals. Several vitamins in small quantities are requisite to growth in certain species of algae. In a freshwater environment, algal requirements are met by vitamins supplied in soil runoff, lake and stream bed sediments, solutes in the water, and metabolites produced by actinomycetes, fungi, bacteria, and several algae.

Evidence indicates that: (1) High phosphorus concentrations are associated with accelerated eutrophication of waters, when other growth promoting factors are present; (2) aquatic plant problems develop in reservoirs or other standing waters at phosphorus values lower than those critical in flowing streams; (3) reservoirs and other standing waters collect phosphates from influent streams and store a portion of these within consolidated sediments; and (4) phosphorus concentrations critical to noxious plant growths vary, and they produce such growths in one geographical area, but not in another. Potential contributions of phosphorus to the aquatic environment have been indicated in the literature(table 22-2).

Keup (1968) in flowing water studies found that phosphorus is temporarily stored in bottom sediments or transported as a portion of the stream's bed-load after its removal from the flowing water. Long-term storage is affected when the phosphorus is pooled in deltas or deposited on flood plains. Keup reviewed the literature on phosphorus discharges by specific streams (table 22-3).

Once nutrients are combined within the ecosystem of the receiving waters, their removal is tedious and expensive; removal must be compared to inflowing quantities to evaluate accomplishment. In a lake, reservoir, or pond, phosphorus is removed naturally only by outflow, by insects that hatch and fly out of the drainage basin, by harvesting a crop, such as fish, and by combination with consolidated bottom sediments. Even should adequate harvesting methods be available, the expected standing crop of algae per acre exceeds 2 tons and contains only about 1.5 lbs of phosphorus. Similarly, submerged aquatic plants could approach at least 7 tons/acre (wet weight) and contain 3.2 lbs/acre of phosphorus. Probably only half of the standing crop of submerged aquatic plants can be considered harvestable. The harvestable fish population (500 lbs.) from 3 acres of water would contain only 1 lb. of phosphorus.

Dredging has often been suggested as a means of removing the storehouse of nutrients contained within the lake bed sediments. These sediments are usually rich in nitrogen and phosphorus, for they represent the accumulation of years of settled organic materials. Some of these nutrients are recirculated within the water mass and furnish food for a new crop of organic growth.

Hasler (1957) found that, in an undisturbed mud-water system, the percentage of nutrients, as well as the amount of phosphorus that is released to the superimposed water, is very small. In laboratory experiments, when P^{32} is placed at various depths in the mud, the diffusion into the overlying noncirculating water is negligible, if the phosphorus is placed more than 1 c.m. in the mud. Application of lime to the water or mud reduces the amount of soluble phosphorus released. Acidification of previously alkalized mud will, upon agitation, increase the amount of phosphorus entering solution. In an aquarium experiment, circulation of the water above phosphorus-rich mud, with the aid of air bubbles, increased the phosphorus in solution.

Zicker et al. (1956) found in laboratory experiments that the percentage of phosphorus released to water from radioactive superphosphate fertilizer placed in an undisturbed mud-water system was very small, with virtually no release of phosphorus from fertilizer placed at depths greater than ¼ inch below the mud surface. Radiophosphorus placed ½ inch below the mud surface showed only a very slight tendency to diffuse into the water, while the radiophosphorus placed at a 1-inch depth did not diffuse into the water at all.

Dredging deepens an area within a lake and can be beneficial if the increased depth is sufficient to prevent growth of larger nuisance plants. Dredging uncovers yet another soil strata that will contain phosphorus in some quantity, subject to solution in water. The newly dredged area immediately begins to receive organic fallout from waters above, and forms

Table 22-2 **Pounds of Phosphorus Contributed to Aquatic Ecosystems**

Major Contributors:

- Sewage and Sewage Effluents: 3 lbs. per capita per year.*
- Some industries, e.g., potato processing: 1.7 lb. per ton processed.
- Phosphate rock from 23 States (Mackenthun and Ingram, 1967).
- Cultivated agricultural drainage: 0.35–0.39 lb. per acre drained per year (Engelbrecht and Morgan, 1961) (Sawyer ,1947) (Weibel, 1965).
- Surface irrigation returns, Yakima River Basin: 0.9–3.9 lbs. per acre per year (Sylvester, 1961).
- Benthic Sediment Releases.

Minor Contributors:

- Domestic duck: 0.9 lb. per year (Sanderson, 1953).
- Sawdust: 0.9 lb. per ton (Donahue, 1961).
- Rainwater.**
- Groundwater, Wis.: 1 lb. per 9 million gals. (Juday and Birge, 1931).
- Wild duck: 0.45 lb. per year (Paloumpis and Starrett, 1960).
- Tree leaves: 1.8–3.3 lb. per acre of trees per year (Chandler, 1943).
- Dead Organisms; animal excretions.

*Various researchers have recorded the annual per capita contribution of phosphorus in pounds from domestic sewage as 2 to 4 (Bush & Mulford, 1954), 2, 3 (Mentzler et al., 1958), 1.9 (Owen, 1953), and 3.5 (Sawyer, 1965).

**Influenced by pollution present in atmosphere "washed out" by rainfall.

Table 22-3 **Phosphorus Discharged by Selected North American Streams (from Keup, 1968)**

Principal land use	River	Number of analyses	Season of sampling	Drainage area (mile²)	P—P (lb/annum/mile²)	Population density mile²	Reference*
Forested	West Branch Sturgeon R, Mich	27+	July	14	37	Sparse	A.
	Pigeon, Minn	4	Aug. and Sept	600	28	Sparse	B, C, D.
	Poplar, Minn	4	Aug. and Sept	114	21	Sparse	B, C, D.
	Baptism, Minn	4	Aug. and Sept	140	42	Sparse	B, C, D.
	St. Louis, Minn	4	Aug. and Sept	3430	58	Sparse	B, C, D.
	Bois Brule, Wis	4	Aug. and Sept	113	97	Sparse	B, C, D.
	Bad, Wis	4	Aug. and Sept	611	78	Sparse	B, C, D.
	Montreal, Wis	4	Aug. and Sept	281	98	Sparse	B, C, D.
	Black, Mich	4	Aug. and Sept	202	65	Sparse	B, C, D.
	Presque Isle, Mich	4	Aug. and Sept	260	39	Sparse	B, C, D.
	Ontonagon, Mich	4	Aug. and Sept	1290	44	Sparce	B, C, D.
	Yakima, Wash	?	Annual	182	473	Sparse	E.
	Tieton, Wash	?	7 months	237	492	Sparse	E.
	Cedar, Wash	?	Annual	125	204	Sparse	E.
	Mulligan, Maine	12	4 seasons	21	4	Sparse	F.
	Stetson, Maine	19	4 seasons	29	20	Sparse	F.
	East Branch Sebasticook, Maine	56	4 seasons	56	128†	> 63‡	F.
	Ellerslie, Prince Edward Island	≈44	April-Dec	10	113	Sparse	G.
	Pigeon, N. C	18	July	133	97	Light	This article.
	Johnathans, N. C	5	July	65	201	Light	This article.
	Kankakee, Ind. and Ill	6	June-Sept	5280	139	28	H, I.
	Vermillion, Ill	8	June-Sept	1230	179	36	H, I.
	Fox, Ill. and Wis	7	June-Sept	2570	489	145	H, I.
Agricultural	Kaskaskia, Ill	100	April-Dec	5220	225	>174‡	J.
Urban	Streams near Madison, Wis	?	?	?	235-262	?	K.
	Du Page, Ill	5	June-Sept	325	18	380	H, I.
	Des Plaines, Ill. and Wis						
	Above confluence with Chicago River	5	June-Sept	635	570	1270	H, I.
	Total basin (includes Chicago River)	19	June-Sept	2180	4020	2570	H, I.
	Chicago, Ill	16	June-Sept	810	6540	5650	H, I.

*Data given, or computed from data in the refernces. A. Ball and Hooper (1963), B. Putnam and Olson (1959), C. Putnam and Olson (1960), D. Anon. (1964), E. Sylvester (1961), F. Anon. (1966), G. Smith (1959), H. Hurwitz, et al. (1965), I. Anon. (1963), J. Engelbrecht and Morgan (1961). K. Sawyer (1947).

†One seasonal (9 months) industry contributes approximately 75 per cent. ‡Only sewered population known.

a new interface at which nutrient exchange is substantial. Sediments disturbed during a dredging operation liberate nutrients at a rate more rapid than sediments left undisturbed and all of these factors must be considered when recommending dredging for nutrient removal. Based entirely on nutrient considerations, dredging can be advantageous only when it removes sediments that contain a higher concentration of nutrients than the interface likely to be formed by fallout.

A RECENT U.S. Department of the Interior report indicated that 4,800 miles of streams and 29,000 surface acres of impoundments and reservoirs are affected seriously by surface coal mining operations alone (Anon. 1967). Inventories of stream pollution magnitude and source have highlighted the seriousness of the problem. There are about 66,500 sources of coal mine drainage pollution in Appalachia in active and inactive mines. Pollutants in this mine drainage reduce about 10,500 miles of streams below desirable levels of quality* (fig. 22-7).

The pollution of streams by coal mine drainage can be extremely damaging to aquatic life. Streams so polluted generally support only a few species of particularly tolerant plants and animals.

Damages to aquatic life from acid mine drainage are attributed usually to high concentrations of mineral acids, the ions of iron, sulfates, and the deposition of a smothering blanket of precipitated iron salts on the stream bed.** In addition, zinc, copper, and aluminum have occurred at lethal concentrations to aquatic life in acid mine drainage; arsenic and cadmium have been found at near lethal levels. The toxicities are increased by synergism among several of the elements including zinc with copper, zinc with cadmium, and copper with cadmium. Toxicities of iron, copper, and zinc solutions are much greater in acid waters polluted by coal mine drainage than in neutral or alkaline waters. Because of the complex chemical nature of coal mine drainage, it is impossible to assign its toxicity towards aquatic life to any single chemical constituent.

Toxic chemicals in acid mine drainage eliminate sensitive aquatic life; tolerant organisms flourish occasionally to great numbers apparently unaffected by the pollutants. Fish are usually not found when the pH of the stream is lower than 4.5. Conversely, populations of midge larvae, *Chironomus* spp., may develop to nuisance proportions.

With a stream pH of 4.0 or lower, cattails, *Typha* spp., and some mosses will flourish; other vascular plants are not found generally. Moore and Clarkson (1967) studied the Monongahela River, W. Va., and concluded that acid mine drainage does not affect the occurrence and distribution of vascular aquatic plants. They found that nitrogen, calcium, total acidity, iron, pH, and rate of flow had no significant effect on plant distribution, but, that type of substratum was the most important factor affecting plant growth and phosphate content and water level fluctuations were significant.

* Stream Pollution by Coal Mine Drainage in Appalachia. U.S. Department of the Interior, Federal Water Pollution Control Administration, 1967.

** Richard W. Warner, Biologist, Technical Advisory and Investigations Branch, personal communication.

Fig. 22-7 **Acid mine discharges kill natural stream bed organisms. Note polluted stream bed on viewers left compared to relatively unpolluted stream bed on viewers right.**

With a stream pH of 4.0 or lower, dense mats of the green alga, *Ulothrix tenerrima* Kuetzing, are usually common enough to attract the attention of casual observers. Gelatinous mats of chlorophyll-containing flagellates, *Euglena* spp., often color stream beds dark green. Microscopic examination often reveals other species of green algae including *Microspora* spp., *Microthamnion* sp., the flagellate *Chlamydomonas* spp., great numbers of diatoms *Eunotia* spp., *Pinnularia* spp., and *Navicula* spp., and lesser numbers of *Surirella* spp.

In severely polluted stream reaches, especially near the mine adits from which polluted water flows, no stream bed animals will be found. In less severely polluted reaches, common inhabitants include midges (*Chironomus* spp.), alderflies (*Sialis* sp.), fishflies (*Chauloides* spp.), craneflies (*Antocha* sp.), dytiscid beetles, and caddisflies (*Ptilostomis* sp.). Conspicuous by their absence are crayfish, blackflies, mayflies, stoneflies, and most species of caddisflies.

Upstream reaches, not polluted by acid mine drainage, may support several species of rooted and floating vascular plants, 20 or 30 species of algae, 15 or 20 or more species of stream bed animals, and a mixed community of fishes. Severely polluted stream reaches may support only three or four species of algae. Less severely polluted reaches may support one or two species of vascular plants, three or four species of algae, three or four species of stream bed animals, and no fish.

Dr. Max Katz* summarized much literature on low pH effects and concluded that: pH 6.5 to 7.0 delayed spawning in some fish species but otherwise was harmless except when concentrations of heavy metals or cyanides may be made toxic to fish; pH 6.0 to 6.5 interfered with spawning and hatching of eggs of some fish species; pH 5.0 to 6.0 may be lethal to eggs and larvae of sensitive fish species and will favor dominance of some blue-green algae; pH 4.5 to 5.0 will prohibit reproduction among salmonids and is the threshold range for blackflies, mayflies, and stoneflies in numbers; pH 4.0 to 4.5 permits pike to survive but perch, bream, and roach can only become temporarily acclimated. Observations on pH ranges <4.0 were similar to those described earlier.

Acid pollution of a stream may have additional effects not generally recognized. When water is too acid, it cannot be used for household purposes, livestock watering, or industrial use, without expensive treatment. Wildlife, unwilling to drink acid water, is usually absent near badly polluted areas. Iron hydroxide deposits in stream channels and lack of desirable fish and wildlife in or near acid streams result in decreased land values (Parsons, 1952).

KRUMHOLZ (1960) reports that radioactive materials may be accumulated by aquatic organisms in three different ways: (1) By ingestion along with water or food materials; (2) by absorption from the surrounding medium through the body surfaces; and/or (3) by adsorption to the body surfaces. The first two methods are largely biological in nature and the radionuclides enter into the metabolic processes of the organism. The third method is primarily physical in nature and the radio-materials adhere to outer body surfaces. In the latter instance, the size of the orga-

* Dr. Max Katz, University of Washington, personal communication.

nism, because of the tremendous differences in surface-volume ratios of the different ones, is of great importance. In addition, the diversity of structure and surface configuration increase the surface area. Foster and Davis (1955) pointed out that the high radioactivities in plankton and sponges in the Columbia River were associated, in part, with their extensive surface areas.

Radioactive and stable isotopes of the same element have the same chemical behavior, and living organisms are not capable of distinguishing between the two. It is well known that all organisms do not concentrate all chemical elements, but that they select certain elements for their metabolic processes and discard others. Because of this selective power, it is to be expected that they will accumulate any radioactive isotopes of those elements that may be present along with the stable ones.

In an aquarium experiment, Whittaker (1953) showed how biological processes effectively removed phosphorus from the water with algae concentrating the tracer material to levels 300,000 times those in the water. During the first few hours after introduction, the phosphorus was absorbed rapidly by the phytoplankton, but after 15 hours the amounts of radioactivity in the phytoplankton decreased, and more and more of the radiophosphorus was accumulated by the algae on the sides and bottom of the aquaria. Maximum accumulations of radiophosphorus by the algae were reached in 18 days.

Davis, et al (1952, 1953) found concentration factors for radiophosphorus by phytoplankton in the Columbia River to be 100,000 to 200,000 times that of the surrounding water, and Krumholz (1954) reported that the attached freshwater alga (Spirogyra) concentrated radiophosphorus by a factor of 850,000 times that of the water in White Oak Lake.

From the culture of *Escherichia coli* cells in a phosphorus-poor medium containing 25 millicuries (mc) of phosphorus-32, and subsequent release into a small unpolluted stream, the release of the radioactive phosphorus (P^{-32}) downstream was investigated (Hooper and Ball, 1966). In moving downstream, approximately 90 percent of the radioactivity remained inside the bacterial cells while 10 percent diffused from the cells and appeared in the water in filterable form. There was a gradual loss of radioactivity from the water because of fallout of labelled cells and uptake by filter-feeding invertebrates. A small amount of the phosphorus-32 released in soluble form by the bacteria was taken up by the periphyton and by aquatic macrophytes. Failure to find important concentrations of radioactivity in any part of the food chain or in the environment suggested that *E. coli* cells effectively dispersed radioactive phosphorus and minimized the transfer activity through segments of the food web leading to man.

There is a general decrease in the level of radioactivity with the decrease in food chain level. Phytoplankton, which absorb nutrients directly from the water, concentrate the largest amounts. Next in order are the herbivorous insect larvae; and then juvenile fish, which feed on bottom organisms. Bass and other game fish, at the end of the food chain, are least radioactive. In a study of smallmouth bass in the Hanford, Washington, vicinity in the Columbia River, fish caught in September had the highest concentration of gross beta activity and fish caught in April the least (Foster and Henderson, 1957). The range in gross beta activity in units of 10^{-6} microcuries per gram of tissue in September fish was 63 to 250 in skin and muscle, 310 to 3,520 in scales and bone, 98 to 840 in internal organs and 700 to 8,240 in stomach contents. Davis (1962) found the insects in the Columbia River within the Hanford Reservation to be many more times radioactive than the waters they inhabit. The most abundant nuclides found in insects were phosphorus-32, copper-64, chromiun-51, zinc-65, and sodium-24.

Echo and Hawkins (1966), in studying radionuclides in settling ponds, found that algae not only concentrated the radionuclides in their structure, but that the release of the nuclides might be retarded for considerable lengths of time, especially in waters of low pH. This might form a zone of high activity of undesirable magnitude at the mud-water interface.

IT IS well documented that many lakes throughout the country, and the world, have been fertile reservoirs for algal development for many years and have been called eutrophic. Notable for their algal growths in the United States are Lake Zoar in Connecticut, Lake Sebasticook in Maine, the Madison Lakes in Wisconsin, Lake Erie, the Detroit Lakes in Minnesota, Green Lake and Lake Washington in Washington, and Klamath Lake in Oregon.

As nutrient concentrations increase the numbers of algal cells increase. Nuisance conditions occur such as surface scums and algal-littered beaches. The water may become foul smelling. Filter-clogging problems may occur at municipal water treatment installations. Filamentous algae, especially *Cladophora,* grow profusely on any suitable subsurface; these can cause nuisances when they break loose and wash ashore to form windrows of stinking vegetation. The abnormal acceleration of a process that is considered as normal often is not in the best interests of man.

To properly assess a nutrient problem, consideration should be given to all of those sources that may contribute nutrients to the watercourse. These sources could include sewage, sewage effluents, industrial wastes, land drainage, applied fertilizers, precipitation, urban runoff, soils, and nutrients released from bottom sediments and from decomposing plankton. Transient waterfowl, falling tree leaves, and ground water may contribute important additions to the nutrient budget. Flow measurements are paramount in a study to quantitatively assess the respective amounts contributed by these various sources during different seasons and at different flow characteristics. In the receiving lake or stream the quantity of nutrient contained by the standing crops of algae, aquatic vascular plants, fish, and other aquatic organisms are important considerations. A knowledge of those nutrients that are harvested annually through the fish catch, or that may be removed from the system through the emergence of insects, will contribute to an understanding of the nutrient budget.

The interaction of specific chemical components in water, prescribed fertilizer application rates to land and to water, minimal nutrient values required for algal blooms, vitamins required, other limiting factors, and the intercellular nitrogen and phosphorus concentrations are likewise important. Usually, it is necessary to determine that portion of the nutritive input attributable to man-made or man-induced pollution that may be corrected as opposed to that input that is natural in origin, and therefore, usually not correctable. A nutrient budget is used to determine the annual input to a system, the annual outflow, and that which is retained within the water mass to recycle with the biomass or become combined with the solidified bottom sediments.

Reservoirs or lakes are the settling basins of drainage areas. The potential productivity of a body of water is determined to a great extent by the natural fertility of the land over which the runoff drains and by the contributions of civilization. Biological activity within the lake influences such chemical characteristics as dissolved oxygen, pH, carbon dioxide, hardness, alkalinity, iron, manganese, phosphorus, and nitrogen; it is varied through temperature fluctuations and stimulated by nutrient variations (e.g., phosphorus and nitrogen). A lake's basin gives dimension to biological activity and may, because of unique physical characteristics, concentrate the nutrients it receives as well as the developing biomass.

The examination of waterway sediments and other solids fits well in a discussion devoted to eutrophication. Sediments are best collected with a device that permits a core of the material to be extracted from which segments may be selected for examination. These identifiable segments may be examined for pollen, diatom skeletons or chitinized remains of cladocerans or midges, as well as selected chemical constituents. Generally carbon and nitrogen, and often phosphorus, are determined. The carbon, nitrogen, phosphorus, and their respective ratios are important values to aid in the identification of a material, to calculate the amount of major nutrients contained within a segment of the biomass or a stratum of sediment, and from which to judge the relative input of nutrients to the water mass when the ecosystem component undergoes decomposition, or natural chemical change (table 22-4).

Gerloff has been a proponent of a tissue analysis technique to evaluate nitrogen and phosphorus supplies in waters for growth of algae and angiosperm aquatic plants and to determine availability of the elements in lakes from which plants were collected (Gerloff and Krombholz, 1966). They found the critical levels (minimum tissue content associated with maximum growth) to be approximately 1.3 percent nitrogen and 0.13 percent phosphorus for several species of vascular plants. In all but 1 of 9 lakes studied, phosphorus was found more likely to be limiting to higher aquatic plant growth than was nitrogen. In Lake Mendota, Wisconsin, a highly fertile lake, plant samples were collected in June, July, August, and September. The percentage of nitrogen in plant tissues of *Ceratophyllum demersum* varied from 2.11 to 4.43 among these monthly samples while percentage phosphorus varied between 0.51 and 0.75. Among 6 plant species, the percentage of nitrogen varied from 1.98 to 4.43 and phosphorus from 0.23 to 0.75. In Lake Nebish, Wis., a relatively infertile lake, the variance among 4 plant species among the same months was 1.48 to 3.19 percent nitrogen and 0.10 to 0.33 percent phosphorus.

The carbon-nitrogen and nitrogen-phosphorus ratios are of greater value in data interpretation than are total to soluble phosphorus ratios. Total to soluble phosphorus ratios may vary from 2 to 17 or even 90 percent dependent upon the particular water, season, aquatic plant populations, and probably other factors (Table 22-5). These ratios are of value

Table 22-4 **Carbon, Nitrogen, and Phosphorus in Freshwater Environmental Constituents**

Constituent	Standing Crop, lbs/ac		%C[1]	%N[1]	%P[1]	Ratio		Reference
	Wet	Dry				C:N	N:P	
Phytoplankton	1,000 to 3,600	100 to 360						Birge and Juday, 1922.
				6.8	0.69		10	Gerloff and Skoog, 1954.
			39	6.1	0.64	6.5	10	Mackenthun et al., 1968.
				9.0	0.52		17	Birge and Juday, 1922.
Attached Algae	2,000	200						Neil, 1958.
				2.8	0.14		2	Birge and Juday, 1922.
Vascular Plants	14,000	1,800						Rickett, 1922, 1924.
				1.8	0.18		10	Harper and Daniel, 1939.
Myriophyllum				3.2	0.52		6	Birge and Juday, 1922.
Vallisneria				1.8	0.23		8	Schuette and Alder, 1928, 1929.
Potamogeton				1.3	0.13		10	Schuette and Alder, 1928, 1929.
Castalia				2.8	0.27		10	Schuette and Alder, 1928, 1929.
Najas				1.9	0.30		6	Schuette and Alder, 1928, 1929.
Myriophyllum				3.0	0.5		6	Anderson et al., 1965.
Bottom Organisms								
Midges	200 to 400	40 to 80						Dineen, 1953; Moyle, 1940.
Chironomus				7.4	0.9		8	Borutsky, 1939.
Hyalella				7.4	1.2		6	Birge and Juday, 1922.
Hirudinea				11.1	0.8		14	Birge and Juday, 1922.
Sialis				8.1	0.6		14	Birge and Juday, 1922.

Table 22-4 **Carbon, Nitrogen, and Phosphorus in Freshwater Environmental Constituents** (continued)

Constituent	Standing Crop, lbs/ac		%C[1]	%N[1]	%P[1]	Ratio		Reference
	Wet	Dry				C:N	N:P	
Fish	150 to 600							Swingle, 1950.
				[2]2.5	[2]0.2		10	Beard, 1926.
				2.8	0.18 to 0.49			Borgstrom, 1961.
					0.19			Love et al., 1959.
					0.20			McGauhey et al., 1963.
					0.29			Sylvester and Anderson, 1964.
				2.6 to 3.3	0.18 to 0.24			Ingalls et al., 1950.
Domestic Wastes[3]					5.1 to [4]10.6			Engelbrecht and Morgan, 1959.
				[4]45	[4]8		6	McGauhey et al., 1963.
				20 to [4]40	5.3 to [4]10.6		4	Bush and Mulford, 1954.
				[4]61.3	[4]10.7		6	Oswald, 1960.
				18 to [4]28	3.5 to [4]9.0			Anon, 1967.
Sediments								
Lake Tahoe			0.6 to 19.8	0.6 to 1.6		4 to 25		McGauhey et al., 1963.
Wisconsin Lakes			4.4 to 40.5	0.6 to 3.6	0.12 to 0.6	8 to 14	5 to 6	Black, 1929; Juday et al., 1941.
Madison, Wis. Lakes				0.7 to 0.9	0.1 to 0.12		6 to 9	Sawyer et al., 1945.
Green Lake				0.6	0.17		4	Sylvester and Anderson, 1964.
Lake Sebasticook			10 to 34	0.3 to 1.8	.06 to .16	8 to 44	5 to 16	Mackenthun et al., 1968.
Klamath Lake			8.6	1.2		7		Thomas, N. A., Unpublished.[6]
Boston Harbor			2.3 to 5.0	.06 to .41				Stewart, R. K., 1968.[5]
Organic River Sediments			0.03	0.0027		12		Finger and Wastler, 1969.

[1] As the total element in percentage of the dry weight, unless specified otherwise

Table 22-4 **Carbon, Nitrogen, and Phosphorus in Freshwater Environmental Constituents** (continued)

Constituent	Standing Crop, lbs/ac		%C[1]	%N[1]	%P[1]	Ratio		Reference
	Wet	Dry				C:N	N:P	
Pulp & Paper Wastes in River			5.3	0.23		22		Finger and Wastler, 1969.
Untreated Domestic Wastes			3.54	0.3		12		Finger and Wastler, 1969.
Untreated Chemical and fertilizers and domestic wastes			3.15	0.12		26		Finger and Wastler, 1969.
No tributary wastes			0.55	0.05		11		Finger and Wastler, 1969.
Sand; silt; clay; loam			0.4 to 2.1	.02 to .10		20		Ballinger and McKee, Unpublished.[6]
Stable sludge; peat; organic debris			2.0 to 5.0	.10 to .20		20 to 25		Ballinger and McKee, Unpublished.[6]
Paper mill wastes			6 to 15	.10 to .30		50 to 60		Ballinger and McKee, Unpublished.[6]
Packinghouse Wastes			2.8 to 4.3	.30 to .50		8 to 10		Ballinger and McKee, Unpublished.[6]

Table 22-4 **Carbon, Nitrogen, and Phosphorus in Freshwater Environmental Constituents** (continued)

Constituent	Standing Crop, lbs/ac		%C[1]	%N[1]	%P[1]	Ratio		Reference
	Wet	Dry				C:N	N:P	
Fresh sludge; decaying algae; sewage solids			5 to 40	.70 to 5.0		7 to 8		Ballinger and McKee, Unpublished.[6]
Log Pond Bark			50.6	.5	.02	100	25	Thomas, N. A. Unpublished.[6]
Sewage sludge in river			5.8	0.28	.18	21	2	Thomas, N. A. Unpublished.[6]
Algae; sawdust; sewage			14.6	0.93	.11	16	9	Thomas, N. A. Unpublished.[6]
Leaf litter			28.3	1.63	0.11	17	15	Warner, R. W. et al., 1969.[7]
Sand			0.2	.02	.005	10	4	Warner, R. W. et al., 1969.[7]
Loam			2.7	.19	.02	14	10	Warner, R. W. et al., 1969.[7]
Muck			7.3	.52	.04	14	13	Warner, R. W. et al., 1969.[7]
Floating Waste Wool			37 to 43	3.4 to 4.7	.08 to .09	9 to 11	38 to 58	[8].

1 As the total element in percentage of the dry weight, unless specified otherwise.
2 Calculated on wet weight.
3 Average sewage flow can be calculated at 100 gallons per capita per day.
4 mg/1
5 Biological Aspects of Water Quality, Charles River and Boston Harbor, Massachusetts by R. K. Stewart. Technical Advisory and Investigations Branch, Cincinnati, Ohio (1968).
6 Technical Advisory and Investigations Branch, Cincinnati, Ohio.
7 Analyses of soil types from "Black-Water Impoundment Investigations," by R. W. Warner, R. K. Ballentine and L. E. Keup, Technical Advisory and Investigations Branch, U.S. Department of the Interior, Cincinnati, Ohio (1969).
8 Fertilization and Algae in Lake Sebasticook, Maine. Department of Health, Education, and Welfare, Technical Advisory and Investigations Activities, Cincinnati, Ohio (1966).

when they can be determined periodically within the same water body and changes in them correlated with volumetric response changes within the algal mass.

Table 22-5 **Total to Soluble Phosphorus Ratios in Water**

Water	Total P:Sol.P	Reference
Western Lake Erie	3.5	Chandler and Weeks, 1945.
Detroit River mouth	5 to 7	PHS Detroit Project.
Linsley Pond, Conn.	10.0	Hutchinson, 1957.
Northern Wisconsin Lakes	7.0	Juday and Birge, 1931.
Northeast Wisconsin Lakes	2 to 10	Juday et al., 1927.
Ontario Lakes (8)	17	Rigler, 1964.
Southeast Wisconsin Lakes (17)	9	Mackenthun, unpublished.
Rock River, Wis.	2 to 15	Mackenthun, unpublished.
Sebasticook Lake, Maine	2.8 winter	Mackenthun et al. 1968.
Do	12.7 spring	Do.
Do	7.0 summer	Do.
Do	4.1 fall	Do.

The nutrient loading to the lake on a unit basis gives some measure of comparability among various water bodies. Likewise, a lake or reservoir usually retains a portion of those nutrients that it receives from its various sources. The amount or percentage of the nutrients that may be retained by a lake or reservoir is variable and will depend upon:

1. the nutrient loading to the lake or reservoir;
2. the volume of the euphotic zone;
3. the extent of biological activity;
4. the detention time within the basin or time allotted for biological activity; and
5. the level of the penstock or discharge from the basin.

ORGANISMS that may cause, or have been known to cause, trouble in water supplies include several species of algae, protozoa, and diatoms that produce tastes and odors and clog filters, iron bacteria that produce nauseous tastes and odors and clog pipes, copepods whose eggs pass through filters, very small nematode worms, sowbugs in the distribution system, midge larvae or bloodworms, and snails and molluscs.

There may be present in surface waters, various types of organisms both plant and animal, which vary in complexity and size. They are uncommon or absent in groundwater supplies unless stored in uncovered reservoirs, but are common and widespread in surface waters. Perhaps none of the organisms found in surface waters that may be used for domestic purposes is injurious to health. The chief complaints against them are interferences with filtration or other water treatment, and their effects upon the palatability and aesthetics of the water. Organisms may accumulate in such numbers that waters become unsightly and turbid, and many organisms impart disagreeable odors and colors to the water.

The need for aquatic biologists to participate in water works engineering was well stated by Lackey (1950) when he concluded that before a source for a water supply is finally selected it should be surveyed by a competent aquatic biologist to evaluate present and probable future biological conditions that might affect water quality. Reservoirs and storage facility plans should be reviewed by him to determine possible biotic changes that may occur. The elimination of undesirable shallow areas can often be accomplished more cheaply during construction than maintaining these areas after reservoir filling. In areas where growths of algae and other aquatic plants present an annual problem, a study by a qualified aquatic biologist may indicate causes for such growths and methods of control.

Reservoir Bed Preparation

Impounding water may result in the leaching of undesirable materials from flooded soils. When present in excessive concentrations, several materials can interfere with desired water uses:

(1) Color is important in water supply sources because it is objectionable aesthetically. The maximum recommended limit in *Water Quality Criteria* (Anon. 1968) is 75 units in raw water, and less than 10 units in finished water are desirable. Complaints from consumers occur at 15 units and increase with additional color. The measurement of color is based on an empirical scale that is affected by the true color of the water, as well as the material suspended in the water.

(2) Iron in water supplies should not exceed 0.3 mg./l, and its absence in finished water is desirable. In excess, it stains laundry, produces a disagreeable taste, interferes with filtration, and supports iron bacterial growths within a water system. The amount of iron present in water is dependent upon other chemical characteristics of the water:

   (a) It is more soluble when oxygen is absent.
   (b) It is more soluble in acidic waters.
   (c) Organic materials increase its solubility.

(3) Manganese should not exceed 0.05 mg./l in raw water and should be absent from finished water because of stains imparted to laundry, and objectionable tastes. Factors that affect its quantity in water are similar to those for iron.

(4) Phosphorus concentrations in excess of 0.1 mg/l may interfere with coagulation in water treatment plants and, in excess of 0.05 mg/l in water supplies, may stimulate the excessive growth of algae and other aquatic plants. Excessive growths impart undesirable tastes and odors to water, interfere with water treatment, become aesthetically unpleasing, interfere with recreation and aquatic life, and alter the chemistry of waterways.

(5) The chemical state of nitrogen is dependent on the overall limnology of the waterway. The decomposition of organic materials produces ammonia that, in an environment without abundant oxygen, may not be oxidized to nitrate; ammonia reacts with chlorine to form chloramines, which interfere with disinfection. In addition, nitrogen is a nutrient that supports excessive aquatic vegetation that can interfere with water uses.

Tastes and Odors

From a nationwide survey reported in 1957 (Sigworth, 1957), it was indicated that algae were considered by 241 waterworks officials to be the most frequent causes of tastes and odors in water supplies, with other types of decaying vegetation second in importance. Decay or decomposition is brought about by fungi and bacteria, including the actinomycetes. Often a considerable proportion of the decaying vegetation is composed of dead algal cells. The odors that are produced through the activities of fungi and bacteria may be either from the intermediate products formed during the decomposition or from special substances that are synthesized within the cells of the microorganisms. The latter appears to be true in the case of actinomycetales. Algae were listed as the causative agent in 82 percent of the reports, decaying vegetation in 67 percent, industrial wastes in 38 percent, and other causes in 23 percent.

A few algae are well know for the production of specific distinctive tastes and odors, while a larger number of others are associated with tastes and odors that vary in type according to local conditions. Often certain diatoms, blue-green algae, and pigmented flagellates are the principal offenders but green algae may also be involved. Thirty-nine species have been selected by Palmer (1959) as representative of the more important taste and odor algae. They are listed alphabetically under their respective groups in table 22-6. Other genera and species, in addition to the ones selected, must be considered also as potential offenders.

Silvey (1963) noted a relationship between summer blooms of blue-green algae and rapid growths of actinomycetes three to four weeks later Most concentrated tastes and odors appear as associated with the latter. Silvey and Roach (1964) state that ". . . it has been observed for many years in the literature that the degradation of algal remains may result in a multitude of types and intensities of odors and tastes in the water supply. Some odors are contributed directly by the algal degradation products dissolved in the water, whereas others are merely described as decaying vegetation." Disintegrating algae are associated with increases in gram-negative heterotrophic bacterial populations, following which there is generally a new growth of algae, either diatoms or blue-green algae. When blue-green algae predominate and algal mats form on the surface or in protected water areas, actinomycetes attack the remains of the algae thus reducing the gram-negative heterotrophic bacteria. Various odors, depending upon the predominant organism, its degradation products or metabolites, are released. Silvey and Roach (1964) point out that when a stream or reservoir is highly polluted, one cycle may impinge upon another to the extent that total confusion results if attempts are made to determine what stage of the microbiotic cycle is operating in a particular water supply at a specific time.

Menasha, Wisconsin draws its water from Lake Winnebago, an algal-laden surface supply. Taste and odor problems were severe in 1939 based upon Marx's (1951) statement, ". . . as an example of how bad the water can get, during August of 1939 the water temperature was 85, the algae count reached 50,000 units per ml., turbidity was 200 p.p.m. and the odor reached 300. A treatment of over 100 p.p.m. of carbon and 9 grains of alum, cut the odor down to about 15 and the turbidity down to about 10, in the finished water. This water was obviously not satisfactory for use, but nothing could be done about it." In 1946, a 25-million gallon pretreatment basin, giving a detention period of 7 or 8 days, with concrete baffles to ". . . force the water from one shore of the basin to the other five times as it travels from the inlet to the outlet" was placed in op-

Table 22-6 **Odors, tastes, and tongue sensations associated with algae in water**

Algal genus	Odor when algae are— Moderate	Odor when algae are— Abundant	Taste	Tongue sensation
Actinastrum		Grassy, musty		
Anabaena	Grassy, nasturtium, musty.	Septic.		
Anabaenopsis		Grassy		
Anacystis	Grassy	Septic.	Sweet	
Aphanizomenon	Grassy, nasturtium, musty.	Septic.	Sweet	Dry.
Asterionella	Geranium, spicy	Fishy.		
Ceratium	Fishy.	Septic.	Bitter	
Chara	Skunk, garlic.	Spoiled, garlic.		
Chlamydomonas	Musty, grassy.	Fishy, septic	Sweet	Slick.
Chlorella		Musty.		
Chrysosphaerella		Fishy.		
Cladophora		Septic.		
(Clathrocystis)				
Closterium		Grassy		
(Coelosphaerium)				
Cosmarium		Grassy		
Cryptomonas	Violet	Violet	Sweet	
Cyclotella	Geranium	Fishy.		
Cylindrospermum	Grassy	Septic.		
Diatoma		Aromatic.		
Dictyosphaerium	Grassy, nasturtium	Fishy.		
Dinobryon	Violet	Fishy.		Slick.
Eudorina		Fishy.		
Euglena		Fishy.	Sweet	
Fragilaria	Geranium	Musty.		
Glenodinium		Fishy.		Slick.
(Gloeocapsa)				
Gloeocystis		Septic.		
Gloetrichia		Grassy		
Gomphosphaeria	Grassy	Grassy	Sweet	
Gonium		Fishy.		
Hydrodictyon		Septic.		
Mallomonas	Violet	Fishy.		
Melosira	Geranium	Musty.		Slick.
Meridion		Spicy		
(Microcystis)				
Nitella	Grassy	Grassy, septic.	Bitter	
Nostoc	Musty	Septic.		
Oscillatoria	Grassy	Musty, spicy		
Pandorina		Fishy.		
Pediastrum		Grassy		
Peridinium	Cucumber	Fishy.		
Pleurosigma		Fishy.		
Rivularia	Grassy	Musty.		
Scenedesmus		Grassy		
Spirogyra		Grassy		
Staurastrum		Grassy		
Stephanodiscus	Geranium	Fishy.		Slick.
Synedra	Grassy	Musty.		Slick.
Synura	Cucumber, muskmelon, spicy.	Fishy.	Bitter	Dry, metallic, slick.
Tabellaria	Geranium	Fishy.		
Tribonema		Fishy.		
(Uroglena)				
Uroglenopsis	Cucumber	Fishy.		Slick.
Ulothrix		Grassy		
Volvox	Fishy	Fishy.		

eration. As the water flows into the basin it is given a dosage of 2 mg/l of copper sulfate throughout the algal season regardless of type of algae, pumpage, or any other variable (Marx, 1951). By the time the water reaches the water treatment plant inlet, the algae have been killed, they have decomposed, the odors have been reduced an average of 67 percent and just a very fine turbidity is left. In five years of use, less than an inch of material settled on the bottom of the basin, according to Marx.

Oshkosh, Wis., gets its water from Lake Winnebago. Conditions in the lake are ideal for prolific algal growths, reducing filter runs to as low as 60 minutes, which results in a great reduction in plant output and causes water shortages during the summertime (Hartman, 1961). In view of the fact that the city of Menasha, Wis., which also uses Lake Winnebago as its source of water supply, had operated a pretreatment basin for several years, it was concluded that pretreatment of this water with copper sulfate should be considered. Treatment of the water with 2 parts per million of copper sulfate followed by a 7 day retention period under maximum conditions was determined to be the proper pretreatment. A presettling basin was constructed in a bay by enclosing an area of approximately 30 acres with an earthen dyke. This basin was baffled to prevent short circuiting. The pretreatment provided has reduced the amount of water required for backwashing from as much as 9.8 percent of the plant capacity in July and August to 2.5 percent for the same 2 months. The number of filter backwashes per day was reduced from 18.6 in 1959 to 3.8 in 1960, and the water used for backwashing was reduced from an average of 8 percent in 1959 to 2.6 percent in 1960. Pretreatment has reduced the cost of chemicals by almost 31 percent and the backwash water by 67 percent while at the same time a better quality water was produced.

Dickson (1968) discussed the control of actinomycetic odors. He noted that actinomycetes growing on the remains of blue-green algae produce the most intense odors and luxuriant growths. Listing 3 "fairly successful" control techniques, he suggests algicides to control growths of blue-green algae, activated carbon to absorb taste and odor compounds and other organic substances, and the use of *Bacillus cereus* in concentrations of 1.2×10^5 to reduce rapidly the tastes and odors produced by actinomycetes. The latter technique was successful at the Lake Hefner treatment plant, Oklahoma City, when the bacteria were cultured in 5,000-gallon tanks for 24 to 36 hours and introduced into the reservoir or pretreatment basin as either a preventative or control measure.

Filter Clogging Problems

As water passes through a sand filter in the water treatment plant the spaces between the grains of sand become filled with colloidal and solid particles that have been dispersed in the water. When the raw water comes from a surface supply such as a reservoir, lake, or stream, the algae that are invariably present will be well represented in the material collected by the sand filter. They are frequently the primary causes of filter clogging.

Efficient coagulation and sedimentation may remove 90 to 95 percent ot the algae from the water. The algae remaining in the water may still be sufficient to cause gradual or even rapid loss of head in the sand filter. The clogged filter must then be backwashed. Filter runs may often extend for 30 to 100 hours before backwashing is required, or may be shortened to less than 10 hours when algae are present (Tarlton, 1949). In Chi cago, when the water to be filtered contained approximately 700 microorganisms per ml., principally two diatoms *Tabellaria* and *Fragilaria,* the filter runs were 4.5 hours. Three days later, when the count was down to approximately 100 per ml., the filter run increased to 41 hours (Baylis, 1955). In Washington, D.C., filter runs were reduced from an average of 50 hours to less than 1 hour because of a sudden influx of the diatom *Synedra,* which occurred in the raw water and reached 4,800 cells per ml (Lauter, 1937).

Most of the microscopic organisms present in water that is passing through a rapid sand filter generally will be caught in the top one-half inch of the sand. This is particularly the case when the organisms are abundant in the water. A few of the organisms will penetrate deeper into the filter bed while others disintegrate quickly as they come in contact with the sand. The longer the filter run the greater the percentage of organisms that will penetrate below the top one-half inch.

In Chicago, a study was made of the number of certain diatoms that were caught on the surface of rapid sand filters. The samples were collected immediately before the filters were backwashed. During 1 year of this study, *Tabellaria* ranged in numbers from 496,000 to 936,000 per square inch of filter surface in April, and from 1,824,000 to 8,016,000 in November. In contrast, the range for *Melosira* was from 784,000 to 2,624,000 in April, and only 16,000 to 416,000 in November (Baylis, 1957).

Gamet and Rademacher (1960) found in a study of filter runs with Lake Michigan water that 37 percent of the runs for all test filters in 12 filtration plants were less than 20 hours. A filter run of 20 hours was taken as the economic breakpoint time. Loss in revenue for the 12 plants was estimated at $226,000 for the 9-month period. *Tabellaria* was the most important of the Lake Michigan plankton that exhibited filter-clogging tendencies. In a followup study, Poston and Gamet (1964) reported that excessive filter clogging at Cleveland was prevented by use of slightly higher chlorine and alum doses; at Toledo, Ohio, short filter runs were remedied by treatment with excess lime, with the lime floc carrying most of the diatoms down in the sedimentation basin and by lengthening the filter runs from 3 to 24 hours; at Detroit, Mich., shortened filter runs during winter and early spring have lasted as long as a month and alum costs in excess of that required for turbidity were as high as $15,000 per month. They reported that *Tabellaria* was beyond doubt the plankton genera causing the most serious problem with filter runs with Lake Michigan waters although *Asterionella, Melosira, Synedra* in combination also reduced filter runs, and *Cyclotella, Stephanodiscus, Oscillatoria, Fragilaria* and *Tribonema* were reported to cause filter clogging.

Microstraining

Microstraining has been succesfully used to alleviate organism problems in water supplies. At Kenosha, Wis., microstraining equipment was installed at the water plant at an initial cost of $330,000 primarily to reduce microorganisms prior to filtration (Nelson, 1965). The output of finished water from this plant has been increased by up to 25 percent, thereby providing 5 m.g.d. of extra water for sale to the public during peak demand periods which otherwise would have been used to backwash filters. Algal removals by microstraining have ranged from 46 to 97 percent.

On 29 sampling dates extending from July 22 through September 6, 1961, water pumping rates at Kenosha varied from 10 to 27 m.g.d. and from 1½ to 10 hours of sustained pumping.* The efficiency of the microstrainer depends on the number, size, and shape of the algal cell,and whether or not the cells occur singly or in chains. Rate of pumpage did not appear to affect efficiency in this study. The microstrainer was very efficient in removing those algal cells occurring in chains, since the water following the strainers contained only single cells or chains of 2 or 3 cells in length. The algal mass was reduced from 141±10 to 11±3 pounds per day (wet weight) in passing through the microstrainers for an overall volumetric efficiency of 91.9 percent. The number of individual algal cells was reduced from 47±11 trillion per day to 5±1 trillion in passing through the microstrainers for an average reduction of 88.8 percent. The efficiency of the microstrainer varies from one species of alga to another. For example, there was little reduction in *Stephanodiscus* spp. with the passage of water through the microstrainer, because the species occurring in these waters were small enough to pass with ease through the 35-micron mesh of the microstrainer.

Berry (1961) stated that during the 15 years that microstrainers have been used at Belleville, Ontario, they have proven their value under prescribed conditions. They are effective in the removal of algae. Care is needed in the evaluation of conditions where they can be expected to prove of service, and intelligent operation is required. Average filter runs were increased from 14 to 52 hours at Belleville, and the number of washes were cut to ¼ of their former frequency (Scriven, 1960). About 1¼ percent of production was required to wash the microstrainers.

Turre (1959) reported that a microstrainer had been installed at the Marston Lake South Side Filter Plant, Denver, Colo., and that microorganism removals exceeded 91 to 95 percent.

Corrosion Problems

Algae sometimes contribute to corrosion either directly in localized places where they may be growing or through their modification of the water by physical or chemical changes. Green and blue-green algae, growing on the surface of submerged concrete, have caused the concrete to become pitted and friable. The effect has been most pronounced when the percentage of SO_3 in the mortar was 0.6 or higher. It has been assumed that the gelatin present in the living plants, together with the carbonic ox-

* Nelson, O. F., K. M. Mackenthun, and L. A. Lueschow, 1961. Microstraining at Kenosha. Paper Presented at Wisconsin Section, American Water Works Association, Milwaukee, Wis. (Sept. 28) 15 pp. (mimeo.).

alic, and silicic acids produced by them, are adequate to corrode the cement (Oborn and Higginson, 1954).

Algae have been reported to cause corrosion in metal tanks and basins open to sunlight. *Oscillatoria* growing in abundance in water in an open steel tank has caused serious pitting of the metal. The pits were bright and clean, the iron apparently going into solution and not producing any covering compound such as an oxide or sulfide. The algal growth permitted the pitting to take place by releasing oxygen which combined with the protecting film of oxide over the steel. When the steel tank was covered to prevent entrance of light, the algae disappeared and the corrosion stopped (Myers, 1947).

Indirectly, algae may affect the rate of corrosion in a number of ways. Increases in organic deposits in the pipe, increases in the dissolved oxygen in the water through photosynthesis in the raw water supply, and changes in the pH, CO_2 content and calcium carbonate content of the water can all be stimulated by algae. Such changes can, in turn, affect the rate of corrosion.

Algae on Reservoir Walls

Drastic measures often have to be taken to control the attached algae, especially if the growths are neglected until large quantities threaten to cause trouble. One city adapted a floor-cleaning machine, fitted with a cylindrical wire brush, for use in scraping the algal growth from the concrete floor of a 13 million gallon reservoir. The machine proved to be much more effective than hand scrapers in removing the attached portions of the algae from the more than 100,000 square feet of concrete. The detached algae were then flushed from the reservoir floor by streams of water from a fire hose (Kuran and Angell, 1953). The use of chemicals to kill attached algae when they are present in quantity may not solve the problem because dead algae, as well as living ones can clog filters and screens, release slimes into the water, and cause tastes and odors.

MUCH of sewage treatment is a biological process. Plants in waste treatment are restricted mostly to bacteria and the higher fungi. When the operation proceeds in light, green filamentous algae, green non-filamentous algae, or flagellated organisms thrive. Blue-green algae may also be present, although generally not in large numbers. Perhaps the least helpful of the common groups of algae in sewage treament are the diatoms. Many bacteria and fungi will grow in light, but their preferred habitat is a dark one and they, therefore, work in the depths of filters, Imhoff tanks, digesters, or the interior of sand filters. Sunlight is necessary for the growth of algae, and most algal forms require about 6 to 7 percent of full sunlight to carry on active growth.

Most of the lower plants and animals found in sewage treatment are cosmopolitan in occurrence. They are well known and seem to be substantially alike in similar treatment plants. Some organisms, however tend to occur very abundantly in certain types of wastes so much so that blooms of them can be called indicative for the waste and for the condition it causes. The dominance of a particular species is in response of the organism to some particular component or condition of the waste. Only a few chlorophyll-containing types can utilize the dissolved organic substances in sewage, raw or partly digested, and these then become of value as indicators of particular conditions. The great majority of algae prefer to thrive in waters low in organic matter, and utilize phosphorus, carbon dioxide, nitrates, oxygen, and other inorganic substances.

Bacteria and other colorless plants initiate the breakdown of organic substances, and while particular organic substances are attacked by particular species, generally there is a large number of bacterial species in any waste disposal unit. At the present time there is far too little knowledge of either the work done by individual species or on the use of particular types for purification of particular wastes.

Visible active inhabitants of anaerobic habitats, Imhoff tanks and digesters, are generally bacteria, together with about 20 species of protozoa divided equally among flagellates, ciliates, and amoeba. None of these generally are present in large numbers, and often the protozoa are extremely scarce. It has been stated that it is probable that these organisms afford a fair criteria to the proper working conditions in an Imhoff tank. When the tanks foam or seem to digest poorly, the number of protozoa is high, and when there is but little solid matter in the liquid, the number of protozoa tends to be small.

Conditions in filters and aeration chambers are less severe, and the animal populations are usually larger. Here it seems well proven that animals exert a decided influence by consuming bacteria to the extent that the bacterial population curve never levels off. The bacteria are always growing and reproducing, and this is their most effective stage in purification. Also certain insect larvae burrow in the film of filters, thereby keeping it loose and porous and capable of unloading.

A complete list of the animals in trickling filters would comprise some hundreds of species that would include psychoda flies, several genera of worms, roaches, snails, other insect larvae, and many species of protozoa. Activated sludge contains much less diversity of fauna, but does frequently contain branching fungi and filamentous bacteria.

Protozoan fauna are abundant and are usually dominated by solitary or colonial Vorticellidae. Other ciliates may occur in large numbers. There are a few flagellated species and amoeboid protozoa. Rotifers, nematodes, and annelid worms may become very important.

Trickling Filters

The biotic populations of the trickling filter are relatively uniform in composition. Bacteria, fungi, protozoans, algae, nematodes, rotifers, snails, sludgeworms, and larvae of certain insects may vary from filter to filter at the species level, but the biota remains relatively constant. Identification of the members of this population has been of little concern to the sanitary engineer as long as the population became established and functional relatively quickly after placing a filter in operation. A good operating condition is one in which little clogging occurs in the spaces between stones and one which carries on its function of removing solid, colloidal, or dissolved materials from the sewage, thus its biochemical oxygen demand (BOD) is reduced.

In certain areas, populations of snails can develop in trickling filters.

Ingram et al. (1958) have reported that at least five species of snails have been found in the United States on trickling filters. *Physa anatina* was found at Urbana-Champaign, Illinois, *Physa halei* at Fort Worth, Texas, *P. cubensis* at Gainsville, Flordia, and *P. integra* at Dayton and Hillsboro, Ohio. In addition, *Lymnaea humilis modicella* was found once in low numbers at Urbana-Champaign. Serious damage to equipment was reported as a result of the activities of *P. integra* at Dayton, and of *P. cubensis* at Gainsville.

Twelve tubular perspex experimental filters, each 91 cm. high and 15 cm. in diameter, were used by Williams and Taylor (1968) to determine the effect of *Psychoda alternata* (Say) and *Lumbricillus rivalis* (Levinsen) on the efficiency of sewage treatment in percolating filters. The volume of each filter was 16.7 liters and all were filled with sterilized 1.3 to 1.9 cm. clinker. Ventilation was provided by two ports, covered with fine gauze, one at the top and one at the base of each filter. A constant temperature of 15° C. was maintained during the experiments.

Artificial sewage, with BOD and organic carbon values of about 350 mg./l and 180 mg./l respectively, was introduced with an initial loading equivalent to 80 gal./yd.^{3} per day and 0.16 lb. BOD/yd.^{3} per day, and was increased to 100 gal./yd.^{3} per day and 0.35 lb BOD/yd.^{3} per day. Three species of fungi and 6 species of protozoa were introduced at the start of the experiment. Each of 4 groups of triplicate filters received fly larvae at a density of 17 larvae/l, worms at a density of 24 worms/l, a mixture of both fly larvae and worms, and no macrofauna respectively.

The film in the control group grew rapidly, blocked the filter, caused ponding and produced anaerobic conditions. The efficiency of treatment, measured in terms of BOD and organic carbon removal, decreased from 90 percent to 40 percent. The two groups containing fly populations had treatment efficiencies greater than 90 percent within 2 months with the larvae successfully controlling the film accumulations. The filter group containing worms achieved similar high treatment efficiencies, but not until 5 months after inoculation. The worm populations increased more slowly because of longer developmental periods.

Many wastewater treatment plants are troubled seriously by *Psychoda* flies. When circumstances are favorable, large numbers of these, especially of the species *P. alternata Say* ("the trickling filter fly") may leave the filter beds.

The *Psychodidae* are not restricted to sewage filter beds. These flies frequent decaying organic material, in which they lay their eggs. The fact that they are able to become a real pest at many treatment plants results from the favorable environment commonly prevailing in a filter bed. The relative humidity is high, sufficient oxygen is present, food for the larvae is abundant, and the pupae can find many places where they can nestle and where the imagoes can hatch.

The flies do not sting or suck blood, but they are a real annoyance because they settle on the eyes, ears, nose, and clothes, and, because of their minute size they easily enter surrounding houses. They are not known to transmit disease, but there are indications which make them suspect.

Scott (1961) presented an excellent concise bibliographic review of filter fly control. Although drying, flooding, and treatment of the slime layer with insecticide give good control, they are all unpopular with the sewage plant operator. The efficiency of the biofilter is dependent upon the activity of its slime layer. Yet, drying destroys it, flooding disrupts it, and both involve cessation of operation for considerable time. While insecticides may be applied without stopping the filter, they throw the slimelayer out of balance, and may poison aquatic life downstream from the sewage outfall. The oil present in solutions and emulsions does serious damage to the slime layer, so suspensions are used in treating the filter bed. However, even suspensions kill off large numbers of key organisms (including filter fly larvae which are in themselves important), and therefore lower the efficiency of the filter.

Den Otter (1966) suggested that *Psychoda* pests could be prevented when fly larvae were hampered in their efforts to reach the surface of filter beds by use of a top layer of stones that can be rinsed easily and continuously. A 15-cm. thick top layer of pebbles, each about 5 cm. in diameter, was recommended. Continuous and complete sprinkling was obtained by recirculation and by distribution of the wastewater in jets or in sheets. Clogging was not found to be a problem.

Activated Sludge

McKinney and Gram (1956) stressed that the type of protozoan present in activated sludge was of major concern to the treatment plant operator. They stated that there is a definite succession of protozoa as activated sludge is formed and becomes more efficient. Knowledge of the various protozoa and their significance can be used as a valuable guide to better activated sludge operation. Observations of the protozoa can be made quickly with a microscope of only 100 power. These observations reflect the immediate condition of the sludge. Industrial waste systems fed unusual or toxic organic compounds require careful control to insure continuous operation at maximum efficiency. Since the protozoa are more sensitive to toxic compounds than bacteria, observations on the change in protozoa from normal levels give an immediate indication of trouble before it has had time to affect the bacteria.

Each operator must determine the protozoa characteristics of his own system and relate their activities to efficiency. A general guide as to relative predomination of protozoa and efficiency in an activated sludge system, as presented by McKinney and Gram follows:

1. *Sarcodina* (amoeba) predominate very rarely and only in systems just starting or just recovering from complete toxicity.
2. The holophytic, soluble food eating, *Mastigophora* (flagellates) predominate at low efficiency when the organic concentration is high. In some industrial wastes systems where the organic load is high, the effciency may be mathematically high but the indication is that the effluent still contains a high concentration of organic matter.
3. The holozoic, solid food eating, *Mastigophora* (flageliates) arise as the holophytic flagellates decrease. They indicate an only slightly more efficient system and efforts to distinguish between the two types of flagellates is not warranted by the results.
4. The free-swimming ciliates are found when there are large num-

bers of free-swimming bacteria. The efficency of the system when the numbers are very high is approximately 50 percent. They can indicate a failry efficient system or a poor one. The key is in the other types present. Flagellates and free-swimming ciliates are at the low side of the efficiency scale while the presence of stalked ciliates and higher animal forms such as rotifers indicates the high side of the scale.

5. The precense of stalked ciliates indicates an activated sludge with a low BOD effluent. The stalked ciliates arise as a result of the number of available bacteria being reduced below the demands of the free-swimming ciliates. Having lower energy requirements than the free-swimming ciliates, the stalked ciliates survive. A very stable activated sludge system will have very few stalked ciliates and, usually, no other protozoan forms. Actually, the ciliates are measuring the bacterial activity; but since the bacterial activity reflects the biochemical condition of the system, the ciliates reflect the efficiency with considerable accuracy.

Stabilization Ponds

In 1957 in the United States, there were 631 waste stabilization ponds in use in 27 States serving 2,362,842 persons, of which 430 ponds, serving a population of 759,941, were used as the sole means of waste treatment (Thomas and Jenkins, 1958). The greatest percentage increase of any treatment category during the period 1945-57 was in oxidation ponds; the number of plants increased from 45 to 631, and the population served increased from 0.2 to 2.4 million.

The use of this type of treatment continues to grow. Porges and Mackenthun (1963) reported that a total of 1,304 stabilization ponds in 39 States serving a population of 2,138,085 were used as the sole method of sewage treatment or in combination with primary treatment. It was noted that ponds were used by the larger communities to serve small populations probably until the extension of municipal sewerage to include the smaller communities. Stabilization ponds were used by 343 communities for tertiary treatment serving a population of 2,146,951 people in 27 States. Thus, a total of 1,647 ponds in 41 States were used to treat sewage from a population of 4,285,036. Ponds are especially adaptable to the smaller communities since the majority of the ponds were used by those communities having populations less than 25,000. The growth of pond use is phenomenal and this treatment will probably continue.

Data indicate that 847 ponds were used by 31 industrial groups in the United States in 1962. The canning industry used the greatest number of ponds with 28 percent of the reported installations. Second was meat and poultry with 21 percent. The chemical industry was next with 7 percent.

Treatment efficiencies varied with indications that well-designed and well-operated ponds produced effluents equal to or better than those expected from effective, conventional, secondary treatment plants.

Data indicate meat and poultry wastes were treated in the majority of cases by aerobic ponds. Most dairy wastes were reportedly treated by the aerobic process at relatively low loadings; petroleum wastes were also reported to be treated primarily by aerobic ponds generally at low loadings; emphasis was on phenol removal, and treatment was effective in this respect. Aerobic treatment was employed by sizable numbers of the following industries: chemical, paper, wine, textile, sugar, rendering, hog feeding, laundry, and potato processing.

Anaerobic ponds as compared to aerobic ponds were used to a greater extent by the canning, paper, textile, sugar, and leather industries. Also employing significant numbers of anaerobic ponds were the meat and poultry, chemical, wine, rendering, and potato-processing industries. Greater loadings were imposed upon anaerobic ponds with some loss in efficiency and increase in odors.

Analysis of reported data indicated that the median loading for aerobic ponds treating meat and poultry wastes was 72 pounds BOD per acre per day. Dairy wastes were more difficult to treat from an odor standpoint; the median loading was 22 pounds BOD per acre per day. Median value for petroleum wastes was 28 pounds per acre per day, although the goal for many petroleum waste treatment ponds was phenol removal rather than BOD; median phenol removal was 97 percent. Median loadings for other wastes were: canning, 104; paper, 105; chemicals, 157; and textiles, 165. The last three were based on limited data and may not be reliable.

Loadings for anaerobic treatment were generally much greater; there was indication that sewage loadings exceeding 250 pounds BOD per acre per day would bring about anaerobic conditions. The review of industrial practices indicates that loadings generally exceeded 160 pounds per acre per day for anaerobic operations. Reported median anaerobic loadings were: canning, 392 pounds BOD per acre per day; meat and poultry, 1,260; paper, 347; textile, 1,433; and sugar, 240.

The fungi and yeasts were examined during a 1-year study of the Lebanon, Ohio, stabilization ponds (Cooke, 1962; Cooke and Matsuura, 1963). More than 100 species of filamentous fungi, including 9 species of *Fusarium* and nearly 50 species of yeasts, were isolated from the ponds. It appeared that several species of fungi potentially pathogenic to plants might occur. Six species of yeasts isolated from the ponds appeared indicative of organic enrichment of water and soil. Cooke stated that effluents from a pond system can readily carry spores from place to place. In the drier areas, effluents from such ponds may be developed as irrigation supplements; thus, heavier than normal spore load is applied to the soil, the effects of which can only be anticipated.

Photosynthesis has been termed the most important single factor in determining the course and effectiveness of stabilization pond treatment. Algae are an integral part of photosynthesis. A striking similarity exists among algal speciation in stabilization ponds. Principal algae (fig. 22-8) cited as inhbitants of stabilization ponds, include *Chlorella, Scenedesmus, Chlamydomonas* and *Euglena.* Allen (1955) found that algae can be used as a means of following pond operation in California since *Chlorella* and *Scenedesmus* predominate when sewage is undergoing active decomposition, and *Chlorella* is succeeded by *Scenedesmus* and *Chlamydomonas* when oxidation is well advanced. Some of the highest algal densities in California were observed during December and January. The coming of heavy rains with dilution and rise of water levels created a sharp drop in the algal population. Parker (1962) found high algal populations in unicellular aerobic ponds, but :nuch more variety and a lower population level in multicell ponds. Th.ough the use of multicell ponds it was possible to obtain relatively clear effluents free from objectionable algal turbidity.

Maximum total counts of phytoplankton per milliliter have been recorded at 2,800,000 (Neel et al., 1961), 4,700,000 (Mackenthun and McNabb, 1959), and 1,300,000 (Merz et al., 1957). Algal mass, in p.p.m. per volume, has averaged 2,000 to 7,000 with a maximum of 34,700 in Missouri (Neel et al., 1961) and 50 to 250 in Wisconsin (Mackenthun and McNabb, 1959). As related by Fitzgerald and Rohlich (1958), Meffert (1955) reported algal yields of 11 to 16 tons per acre per year during warm weather and 7 tons per acre per year during cool, cloudy weather; average yields of 30 to 35 tons of dry algae per acre per year were obtained from ponds at Richmond, Calif., with rate yields of 60 to 75 tons per acre per year in July and August (Gotaas and Oswald, 1955). Algal yields of 600 to 1,600 pounds per million gallons of sewage have also been reported (Gotaas et al., 1954).

Bottom organisms in a northern Wisconsin stabilization pond (Mackenthun and McNabb, 1959) were confined principally to three species of midges whose combined population ranged from 36 per square foot in the early spring to 956 per square foot in late fall. There was a 94-percent population reduction caused by anaerobic benthic conditions during the first winter of operation. This benthic population undoubtedly had an impact on the nutrient cycle. Larvae that assimilate nutrients indirectly

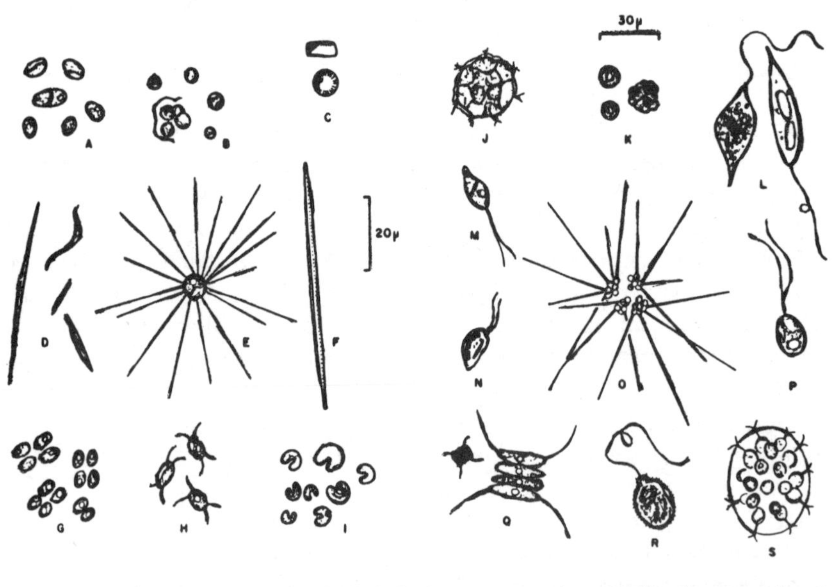

Fig. 22-8 **Algae commonly found in sewage waste stabilization ponds. A, *Chlorella ellipsoidea*; B, *Chlorella vulgaris*; C, *Cyclotella operculata*; D, *Ankistrodesmus* spp.; E, *Golenkinia radiata*; F, *Nitzschia palea*; G, *Crucigenia rectanularis*; H, *Chodatella guadriseta*; I, *Selanastrum minutum*; J, *Pandorina morum*; K, *Coelastrum microporum*; L, *Euglena* spp.; M, *Chlorogonium euchlorum*; N, *Cryptomonas erosa*; O, *Micractinium pusillum*; P, *Chlamydomonas* sp; Q, *Scenedesmus* spp.; R, *Trachelomonas* sp.; S, *Eudorina elegans*.**

through feeding on phytoplankton, and subsequently pupate and develop into adults, may completely remove their share of nutrients from the pond. Those killed by winter anaerobic conditions delay the entrance of their share of nutrients into the superimposed water until late winter or early spring. Final effluent polishing might be accomplished by additional ponds in series, the last one containing fish that could be harvested at periodic intervals.

Mosquitoes in stabilization ponds have usually been associated with growths of aquatic vegetation (Beadle and Harmston, 1958; Myklebust and Harmston, 1962; Beadle and Rowe, 1960; Mallack et al., 1962). Shallow ponds with abundant vegetation produce mosquito problems; ponds free from vegetation have presented no problem. Suggestions for minimizing mosquito problems include specifications on bank slope, depth, the clearing of the pond bottom prior to filling, a system of water level control to permit the use of soil sterilants to control marginal vegetation, and the use of larvicides such as diesel oil, DDT, and malathion to control significant mosquito production in the event of failure of other remedial measures. The use of specific herbicides or larvicides is dependent on local pond conditions and the ultimate fate of effluent water.

Kimerle and Enns (1968) reported that 60 species of aquatic insects were found in 18 central Missouri stabilization ponds. Three species of midges predominated. *Culex pipiens* and *C. tarsalis* bred in pond outlets where overlfow water was allowed to drain across a flat area before reaching a drainage ditch.

In the Wisconsin stabilization pond studies, phytoplankton counts were recorded as the number of organisms per ml., as well as the volume of cell mass in p.p.m. (fig. 22-9). The dissimilar trends between the two approaches of arriving at the plankton standing crop, and the similarity of phytoplankton volumetric data between the two stabilization ponds illustrate the necessity of "going the extra mile" to obtain volumetric data. Both stabilization ponds were receiving biochemical oxygen demand (BOD) loadings of 20 to 23 pounds per acre per day.

Local climatological conditions influence to a great extent the functioning of a stabilization pond. Under the severe winters of northern Wisconsin, where ice and snow cover the pond's surface, unsatisfactory waste stabilization occurred at loadings of 20 pounds and above of BOD (fig. 22-10). Satisfactory pond operation occurred during summer and fall.

The operation of a stabilization pond requires frequent inspection and periodic maintenance to be reasonably free from odors. Under certain operational and climatic conditions blue-green algae may appear as a substantial floating mass within a period of 2 or 3 days that can be wind congregated along the leeward shore. Under a hot sun, the surface of the algal mass will bake, decompose, and emit pungent odors. Specific causes of algal mat formation cannot be stated specifically, and operational and climatic factors may indeed interact to create the problem.

Three avenues of approach are available to combat odors arising from floating blue-green algal and other solids on stabilization ponds: (1) Floating solids may be skimmed off and buried whenever they appear, or as necessary; (2) the solids mass may be broken up by the application of a hard spray of water from a pump and firehose type spray nozzle, or by

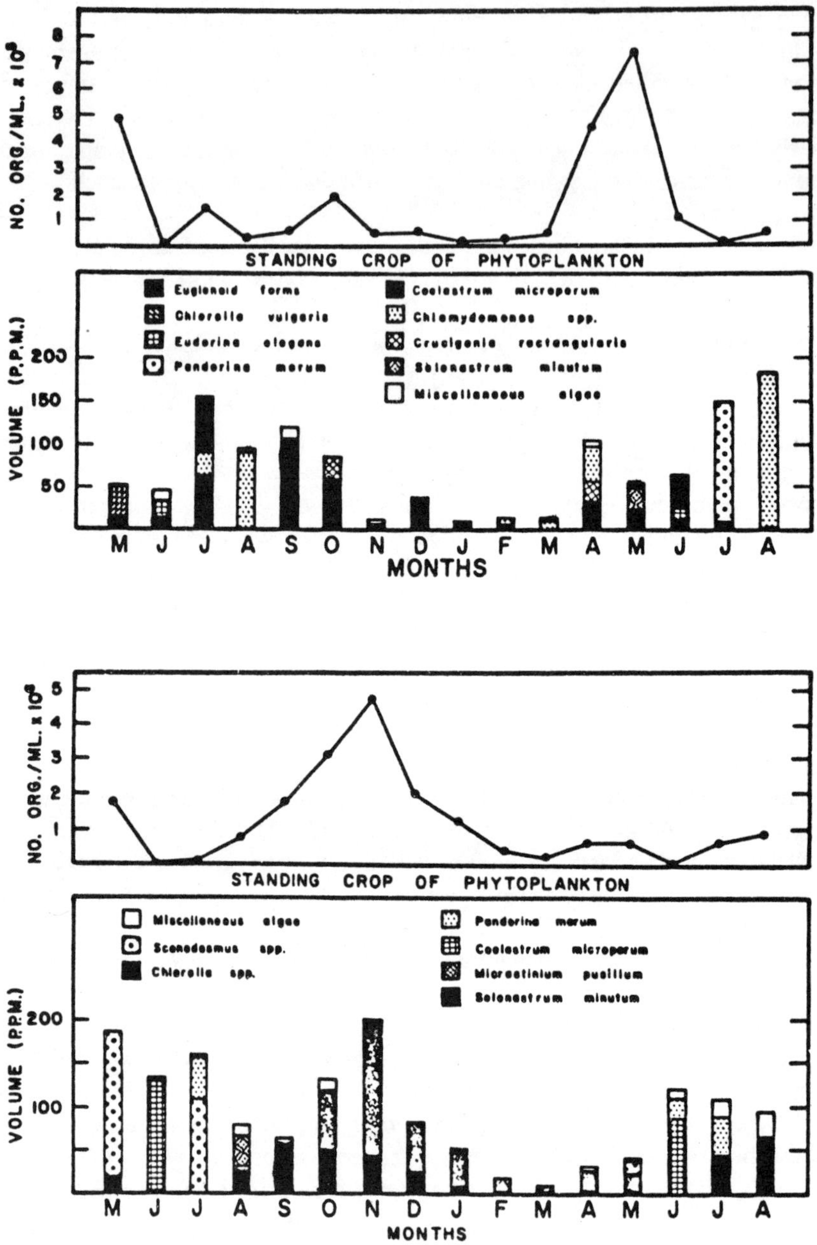

Fig. 22-9 **Phytoplankton standing crops in Wisconsin stabilization ponds May 1957, to August 1958, reported as No./ml. and as p.p.m. by volume.**

running a boat with outboard motor through the nuisance area; (3) the formation of the floating algal mass may be controlled through the periodic use of an algal-killing chemical such as copper sulfate. Users of chemical controls have indicated that blue-green algae may be killed by copper sulfate, but that green algae are not injured sufficiently to handicap the stabilization or organic wastes. In treating a stabilization pond with copper sulfate, the commercial grade crystals are placed in a burlap bag, and this, in turn, is towed in the water behind a boat as it is

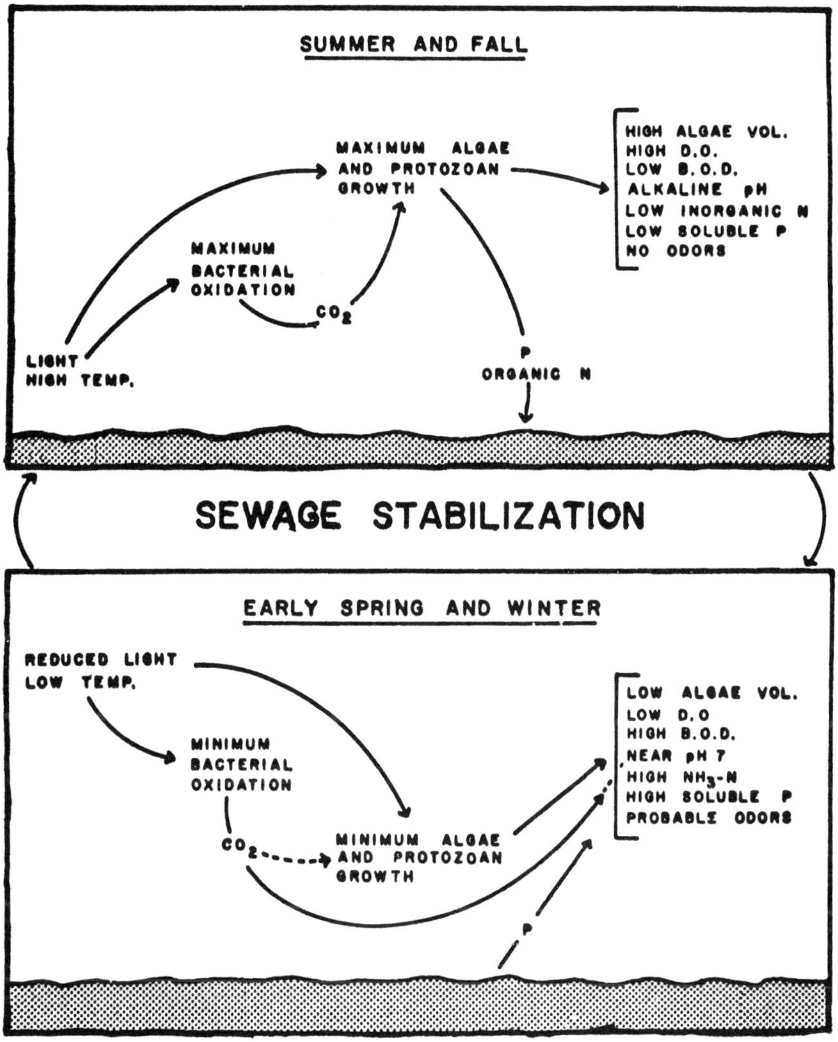

Fig. 22-10 **Diagram of sewage stabilization in ponds influenced by climate in northern Wisconsin.**

propelled back and forth across the pond until the crystals dissolve, and the pond is treated completely.

The copper sulfate dosage rate is based on the total methyl-orange alkalinity of the water. With a methyl orange alkalinity of less than 50 mg./l, a calculated dosage of 0.5 p.p.m. commercial copper sulfate is used; with higher alkalinities, 1 p.p.m. copper sulfate is used. Thus, for a high alkalinity pond that is 4 feet deep, 10 pounds of commercial copper sulfate per acre is required. Application might be necessary as often as every two weeks depending on pond conditions. The pond should be observed closely following such a treatment to ascertain that the purpose for which the pond was initially designed is not seriously impaired. Chemical application will have no advantage when applied to an area that already has substantial floating solids. These solids must be removed and buried, or dispersed prior to treatment, because they will continue to emit odors.

UPON the introduction of waste nutritive materials into water courses, biological slimes may develop to the extent that visible masses appear (fig. 22-11). These are woolly coatings on submerged objects or tufts and strands, sometimes 15 inches or more long, streaming in the current from point of attachment. They vary in color from milky white in fresh new growth to dull grey-white, brown or rusty-red, depending on age, nutrition, and type and amount of solids they entrap from the passing water. In many cases, a rather heterogenous population makes up what is commonly called the slime floc. For example, Tiegs (1938) described the "sewage fungus" * community as including the following organisms: *Sphaerotilus natans, Beggiotoa alba, Thiothrix nivea, Fusarium aquaeductum, Leptomitus lacteus* and *Mueor* sp. The two organisms that have been described repeatedly as ecologically dominant are *Sphaerotilus natans* and *Leptomitus lacteus*. Butcher's (1932) description of his "sewage fungus" community also included the fungus *Shenospora* and the protozoan *Carchesium*.

Numerous problems arise from the presence of slime growths in streams. Where commercial or game fishing exists, drifting *Sphaerotilus* may foul gill nets rendering them ineffective (Lincoln and Foster, 1943**; Ingram and Towne, 1960; Wilson et al., 1960), interfere with fish hatching by coating fish eggs, and smother aquatic fauna that serve as food for fish. Biological slimes and the materials they entrap, such as plant fibers, wood chips, and debris, blanket the streambed and destroy the homes of clean water associated organisms. Because organic food is usually abundant where slime growths occur, pollution tolerant organisms such as sludgeworms may become abundant in association with slime growths. These organisms do not offer the fishfood potential that clean water associated species do. Conditions may become so severe that all benthos are eliminated.

Kramer and Smith (1965) found that suspended conifer groundwood fibers had no effect upon egg survival, respiration rate of embryos, or growth rates of alevins and juveniles from eggs incubated in fiber but hatched and grown in clean water. When alevins were held in woodfiber suspensions, survival was reduced from 98 to 100 percent in controls to 0 to 72 percent in 250-p.p.m. fiber. Walleye (*Stizostedion vitreum vitreum*) eggs were incubated in a temporary jar hatchery using Rainy River water taken downstream from paper mills discharging sulfite, kraft, insulating board, and groundwood wastes (Smith and Kramer, 1963). Eggs were also incubated in trays held on and off the bottom in the river downstream from the mills, upstream from the mills, and in tributary streams. In two years, survival of eggs in jar controls was 31.2 to 73.5 percent. Survival to hatching of eggs jar-incubated in polluted water was 0.02 to 6.0 percent. At river stations downstream from the mills, survival of tray

* *Sphaerotilus* is not a genus of a fungus, but is a filamentous bacterium; for years in the literature of sanitary science it has been referred to incorrectly as a fungus.

** Lincoln, J. H. and R. F. Foster, 1943. Report on the Investigation of Pollution in the Lower Columbia River. Washington State Pollution Commission and the Oregon State Sanitary Authority.

Fig. 22-11 Massive Slime Accumulation on Commercial Fisherman's Net (Photo by; Horn Photo Clatskanie, Oregon, May 6, 1964)

eggs on the bottom did not exceed 1.2 percent; and offbottom, 3.5 percent. Eggs held on the bottom upstream from the mills had maximum survivals of 37.6 percent; and off the bottom, 49.1 percent. In most experiments in polluted Rainy River water the principal cause of mortality was *Sphaerotilus* growths on the eggs, which prevented successful emergence of fry. Prior to hatching, egg survival rates in jars were similar in controls and in experiments. Chemical treatment to remove slime increased the hatch of eggs. *Sphaerotilus*-covered eggs removed to fresh water lost the bacterium and hatched at a high rate.

Biological slimes bring about an aesthetically unpleasant stream. To the public, they are an obvious sign of stream pollution (fig. 22-12). Prolific slime growths destroy the recreational potential of the water, thus interfering with one of the major public-associated uses of water.

Because most bacteria respond directly to the introduction of organic materials to the water course, they may be considered the major factor in the self-purification processes occurring in most natural waters. Slime growths in natural waters function similarly to those in treatment plants in the stabilization of organic materials. In simulated stream studies, biochemical oxygen demand (BOD) reductions as high as 87 percent were obtained at waste retentions as low as 33 minutes after extensive slime growths had developed. In one experiment, 11.5 mg./l BOD were removed during this short time interval. Similar high purification rates have also been observed in many shallow, turbulent streams in which slime growths were the major force in the self-purification process (Amberg and Cormack, 1960).

APART from the investigation of water pollution problems, waterworks infestations, or waste treatment characterizations, the aquatic biologist is often called upon to consult on nuisance aquatic organisms, and to proffer recommendations for their control. Nuisance organisms may be either plant or animal, and there are many varieties of each.

Plant nuisances may curtail or eliminate bathing, boating, water skiing, and sometimes fishing; perpetrate psychosomatic illness in man by emitting vile stenches; impart tastes and odors to water supplies; shorten filter runs or otherwise hamper industrial and municipal water treatment; impair areas of picturesque beauty; reduce or restrict resort trade; lower water front property values; interfere with the manufacture of a product in industry, such as paper; on occasion become toxic to certain warm-blooded animals that ingest the water; foul irrigation siphon tubes and trashracks; and cause skin rashes and hay-fever-like symptoms in man. These plant nuisances may be grouped into the algae and the higher aquatic plants. Algae appear as floating scums; suspended matter giving rise to murky, turbid water or water having a "pea soup" appearance; attached filaments; and bottom dwelling types that may be confused with the rooted higher aquatic plants. The higher aquatic plants grow submersed, floating, or emersed.

When aquatic animals become excessively abundant, they become pests and nuisances to man because of their biting habits and their sheer mass of numbers. Some aquatic animals serve as intermediate hosts for parasites that may attack man directly and some serve as vectors of diseases that affect the health of man. Animal congregations often cause hydraulic

Fig. 22-12 **Slimes from waving masses in polluted streams that destroy the habitat for animals, as well as the aesthetics of the waterway.**

problems in water transport systems.

Algae

Most algal problems occur when growth conditions permit the formation of a "bloom." A bloom is an unusually large number of cells (usually one or a few species) per unit of surface water, which often can be discerned visually by the green, blue-green, brown or even brilliant red discoloration of the water. Lackey (1949) arbitrarily defined a bloom as 500 individuals per ml. of raw water. He found bloom conditions 509 times during a 2-year survey of 16 southeastern Wisconsin lakes and of 3 rivers in 1942–43. Of this number, only 13 percent were blue-green algae, generally the most troublesome nuisance-producing group. Diatoms are rarely obnoxious except in water supplies; they predominated in 40 percent of the blooms. The lake with the highest concentration of inorganic nitrogen (N) and inorganic phosphorus (P), 0.79 and 0.38 mg./l respectively, had the most blooms—112 during the 2-year period.

Algae are found in every non-toxic aquatic habitat. Constituting a primary source of food for fish and other aquatic animals, they may be free-floating and free-swimming, or attached to the bottom substratum.

There are some 1,500 genera and 17,400 species of algae, according to Fuller and Tippo (1954). Fresh-water algae fall into major groups including (a) blue-green algae; (b) green algae; (c) yellow-green algae; (d) golden-brown algae and diatoms; (e) red algae; (f) euglenoids and (g) dinoflagellates.

The blue-green algae (Cyanophyta) are extremely primitive in several respects. The plant body is a single cell; colonies may be formed that are loose aggregations of similar cells among which there is little or no differentiation. Their nuclear material is not organized into a definite nucleus, but is scattered throughout the center of the cell. Green chlorophyll is not localized in definitely formed bodies, but is diffused throughout the peripheral portion of the cell. In addition to the green chlorophyll there is a blue pigment and sometimes a red pigment. Reproduction is by simple division (fission). Blue-green algae produce "water blooms," "pea soup" appearance, septic "pig-pen" odors; impart a "fish taste"; and cover rocks with slimy gelatinous masses.

Green algae (Chlorophyta) have pigments that are principally chlorophyll confined to chloroplasts or definite bodies. There is an organized nucleus, and the motile cells, either vegetative or reproductive, have flagella.

The yellow-green algae, golden-brown algae, and diatoms (Chrysophta) have the pigment confined to definite bodies. There is a greater proportion of yellow or brown pigment than chlorophyll. The cell wall of the diatoms is composed of silica.

The euglenoids (Euglenophyta) are unicellular, motile, and bear one to four flagella. They have a definite nucleus with grass-green chlorophyll localized in definite chlorophyll bearing bodies (Plastids).

The group of dinoflagellates (Pyrrophyta) includes a great diversity of mostly pigmented and mobile unicellular organisms. Two flagella are present. Brown pigments predominate, although chlorophyll is present.

Excessive algal growths are troublesome in many ways. When present in great numbers, algae may: give an unsightly blue-green or green appearance to the water, form a massive floating scum that obstructs navigation or impairs water use, in decay produce noxious odors that drive people from the area, remove dissolved oxygen from the water when decomposing and kill aquatic life as a result, foul fish harvesting equipment and water intake devices, reduce the carrying capacity of water distrubution systems, and destroy bathing and other water use areas. *Stigeoclonium, Oedogonium, Ulothrix,* and *Cladophora* have been problems in irrigation systems where they have restricted flow in the canals, and fouled pumps and tubes. *Cladophora* has been a problem of great concern in Lakes Ontario, Erie, and Michigan where abundant growths have become detached through wave and wind action to be washed upon a beach where they decompose to make the area uninhabitable.

Blue-green algae have been severe nuisance problems, especially to those engaged in water oriented recreation in most eutrophic lakes. *Chara,* a branched erect alga that becomes encrusted with calcareous deposits giving the plant a rough surface, has become a problem in many high alkalinity lakes and ponds. Often *Chara* becomes abundant following the destruction of aquatic vascular plants. Should the lake or pond bed be composed of a peat-type material, *Chara* can become detached from its mooring and bring substantial quantities of bottom with it when it rolls to the surface in 9-inch thick rafts that obstruct navigation, destroy aquatic aesthetics, and emit vile odors. The author has witnessed dead tree stumps, with root systems that measure 20-feet in diameter, embedded in these floating *Chara* rafts.

In highly enriched streams, green algal streamers that may exceed 50-feet in length provide an excellent substrate for blackfly larvae, midge larvae, and other nuisance animals.

More and more interest is being aroused in the toxic effects of freshwater algae on animals. This interest stems partly from a recent upswing in the number of reported cases of animal poisonings related to algal blooms, and from a growing appreciation of the strictly scientific and biological problems involving the physiology of algae, especially those that produce toxic substances (possibly toxins), antibiotics, and growth-stimulating excretions (Ingram and Prescott, 1954).

Wheeler et al. (1942), stated that no human outbreaks of gastroenteritis have ever been traced to algal contamination of drinking water, and that it is probable that the tastes and odors almost invariably associated with severe algal pollution would cause humans to seek other sources for drinking water before a harmful amount of the polluted water would be consumed. These authors state: "It should be noted that the presence of algae in drinking water, in addition to causing tastes and odors, may have some importance from the standpoint of the allergist as algae may, on occasion, liberate considerable amounts of protein in water. If allergic reactions to algae proteins do occur, they would be as uncommon as they would be obscure and hardly likely to occur as a public health problem."

Schwimmer and Schwimmer (1955) chronologically summated 38 incidents of animal intoxications by phytoplankton from 1878 to 1951. In most cases the attacks occurred after the animals had drunk from lakes or ponds containing heavy algal growths, usually during successive days of hot weather. The reported symptoms of algal intoxications vary but the most striking clinically were the involvements of neuromuscular and respiratory systems in cattle. As aptly described by Francis (1878), "the animals developed stupor and unconsciousness, falling and remaining quiet, as if asleep unless touched, when convulsions came on, with the head and neck drawn back by rigid spasm which subsided before death."

Fitch et al. (1934), observed that when samples of toxic algae were fed in the fresh state to guinea pigs, rabbits, and chickens, death occurred in a very short time. When toxic algae were stored for a brief period at ice box temperature, or dried, it was further observed that it was almost impossible to produce death by feeding, but with a similar dose, animals could be killed quickly by inoculating intraperitoneally. The specific alga or algae used are not listed.

The syndrome of symptoms, described by Fitch et al., exhibited by guinea pigs from the feeding or from the inoculation intraperitoneally of a fatal dose of toxic algae was: (1) restlessness; (2) urination; (3) defecation; (4) deep breathing; (5) weakness in the hind quarters; (6) sneezing; (7) coughing; (8) salivation; (9) lachrymation; (10) clonic spasms and death. In studying experimental deaths of guinea pigs, it was noted that they occasionally show symptoms of intoxication and then recover.

Olson (1952) pointed out that the toxicity of unpurified algal material from field collections varies a great deal. For example, 0.02 ml. injected into mice intraperitoneally may kill a 20-gram individual in an hour, whereas from another sample it may require 2.0 to 3.0 ml to produce death in 18 hours. Certain samples of toxic algae (0.1 ml fatal dose) taken from a lake 1 day kill mice in 20 minutes, whereas 9 days later, samples from the same lake with a greatly increased dosage (2.0 ml) may require 5 hours to produce death.

The poisoning of wild and domestic animals by a toxic waterbloom of *Nostoc rivulare* Kuetz from a pond in central Texas has been described by Davidson (1959). Cattle, fish, frogs, and fowl that drank the water containing a heavy growth of the alga became acutely ill, and many died within a short period of time. Most of the accompanying symptoms were similar to those described for a number of other toxic waterblooms. Intraperitoneal injections of albino mice with fresh algal material or with extracts prepared from preserved material produced symptoms that were uniform. The characteristic symptoms were restlessness, increased respiratory rate, renal and intestinal disorders, decrease in blood sugar, decrease in blood coagulation time, decrease in red blood cell count, increase in the rate of heart beat, and paralysis of the hindquarters. Subcutaneous injections resulted in complete necrosis of the abdominal wall, including the skin. The mice that survived developed tumors on the shoulders; two mice experienced evisceration, and the eye of one completely atrophied.

The minimal lethal dosage of the fresh alga was found to be 0.0933 mg./g. of body weight in mice. The toxic factor was soluble in ethanol or water and was rendered inactive when autoclaved at 15 psi pressure for 20 minutes; it affects the neuromuscular and respiratory systems of albino mice.

Gorham (1964) reviewed the literature on toxic algae as a public health hazard. He concluded that the fish and livestock poisons produced by waterblooms were nuisances and economic hazards rather than public health hazards. It was estimated that the oral minimum lethal dose of decomposing toxic *Microcystis* bloom for a 150-pound man would be 1 to 2 quarts of thick, paint-like suspension. Gorham stated that this amount would not be ingested voluntarily; however, in the case of an accident, such a quantity might be ingested involuntarily.

Mackenthun and Ingram (1967) chronologically listed 20 outbreaks of human gastrointestinal disorders, 14 recorded human respiratory disorders, and 10 human skin disorders, all associated with algae.

ULTIMATE control of nuisance aquatic organisms can be accomplished only by drastic alteration of the basic cause(s) of the problem. Generally, the introduction of organic materials and inorganic nutrients into the Nation's waterways are chiefly responsible for the explosion of an aquatic plant or animal population into the nuisance category. Drainage from fertilized lands, inadequately treated domestic or industrial waste discharges, street and storm drainages, and municipal and private refuse dumps each contributes its abundant share to the depredation of the receiving water; each must be adequately treated or controlled if this Nation's waters are to be used to the extent to which they have been projected.

The maintenance concept must be considered by all users of streams, lakes, or reservoirs. The waterfront is an aquatic extension of the surrounding land. To achieve the most lasting beauty, it must be maintained periodically in a fashion similar to that of the adjoining lawn or the abutting parkway; otherwise nuisances and unsightliness will prevail. Controls developed to cure water ills are not singular operations. The mechanism triggering nuisance development is usually such that it will reestablish itself another year. Thus, continuous surveillance as well as maintenance are necessary in water management.

Methods have been developed and perfected that effect an adequate temporary reduction and control of plant nuisances under a number of circumstances. Controls may be either mechanical or chemical, and their selection and use depend upon the nature and scope of the problem, the type and extent of the control desired and comparative costs. Mechanical controls are limited principally to rooted aquatic vegetation, whereas chemical controls have been developed for algae, rooted aquatic vegetation, and other nuisance organisms. Every control has its limitations, based upon the dimensions of the area to be treated; these limitations are broad and do not exclude overall use of the water. Controls that are recommended most generally have not been shown to seriously disrupt general lake ecology.

Chemical Usage

As reported in Weeds Trees and Turf,* the use of chemicals to control excessive growths of algae and aquatic weeds in lakes has become more widespread in Wisconsin in recent years. Dr. M. Starr Nichols, researcher with the University of Wisconsin's departments of agricultural and civil engineering is quoted as stating that from 1950 to 1958, 81 Wisconsin lakes received treatments of about 78,000 pounds of sodium arsenite per year, while 41 lakes were treated with 671,964 pounds of copper sulfate.

During the past several years, the number of lakes chemically treated has continued to grow. In 1966, 150 lakes received the following treatments: 48 with copper sulfate for algal control; 16 with copper sulfate and copper carbonate for swimmer's itch control; 38 with sodium arsenite for submersed weed control; 37 with organic chemicals such as diquat and 2,4-D for weed control. Sixteen of those lakes needed more than one treatment; one lake required 13 treatments.

The British Columbia Research Council (1967) did an extensive investigation on the market for algicides. The present maximum market, predicated on probable needs rather than actual usage, and the probable growth rates were estimated as follows:

irrigation waters—600 tons per year, increasing by 2.5 percent per year

farm ponds—5,000 tons per year, increasing by 3 percent per year

recreational waters—5,000 tons per year, increasing by 5 percent per year

Red Tide—several hundreds of tons, unpredictable

swimming pools—up to 300 tons per year, increasing by 10 percent per year

municipal water supplies—15,000 tons per year, increasing by 3 percent per year

industrial waters—probably small.

The study further concluded that copper compounds, principally copper sulfate, would remain dominant in the large-scale treaments of irrigational and recreational waters, but would likely lose out to the more expensive chemical toxicants in the intermediate-scale market, unless a longer lasting copper algicide can be found that would allow a once-a-season treatment. Copper compounds were not considered to be serious competitors in the small-scale markets of swimming pools and industrial cooling ponds where fast acting bactericides are necessary and the cost per pound of the applied chemical is of less importance.

Harvesting

From the standpoint of nutrient removal, harvesting the aquatic crops annually would be advantageous. The economics of present methods of harvesting and the scope of the problem, however, necessitate a critical appraisal of benefits versus costs. The expected standing crop of algae approaches 2 tons per acre (wet weight) and would be expected to contain 15 pounds of nitrogen and 1.5 pounds of phosphorus. Submerged aquatic plants would be expected to approach at least 7 tons per acre (wet weight) and contain 32 pounds of nitrogen and 3.2 pounds of phosphorus. Values may be higher under severe nuisance conditions.

Some methods and equipment used for the physical and mechanical removal of water weeds are reservoir drawdown and drying, burning, hand-pulling, hand-cutting, hand-raking, chain-dragging, hand-operated underwater weed saws, and power-driven underwater weed cutting and

* Vol. 8, No. 3, p. 3 (March 1969)

weed removal units. The type of control employed depends on the field situation, the extent of control desired, and the labor and cost involved. The mechanical harvesting of aquatic weeds involves equipment capable of handling large tonnage of materials(fig. 22-13). Mechanical controls are especially valuable to reclaim shallow nuisance areas.

Algae

Chemical control measures to be employed depend upon the type of nuisance and local conditions. A good algicide or herbicide must: (1) Be reasonably safe to use; (2) kill the specific nuisance plant or plants; (3) be nontoxic to fish, fishfood organisms and terrestrial animals at the plant-killing concentration; (4) not prove seriously harmful to the ecology of the general aquatic area; (5) be safe for water contact by humans or animals, or provide suitable safeguards during the unsafe period; and (6) be of reasonable cost. Some of these factors assume added significance when based on the physical dimensions of a particular control operation. What might be suitable from both a cost and toxicity standpoint in a control program for a pond might not be feasible in a control program for specific areas on a large body of water.

For the chemical control of algae it may be necessary to know only the acreage of water requiring treatment and the methyl orange alkalinity of the water. For the chemical treatment of many aquatic pests, it is necessary to know the volume of water receiving treatment. A formula is used to ascertain the amount of chemical to apply to a given area:

$$\frac{\text{Length (ft.)} \times \text{Width (ft.)} \times \text{Average Depth (ft.)} \times 62.4 \text{ (wgt. of a cu. ft. of water)}}{1{,}000{,}000} =$$

pounds of chemical (active ingredient) needed to give a concentration of 1 mg./l. This, multiplied by the required chemical concentration in milligrams per liter for treatment equals the pounds of chemical needed for the measured area. Various formulations may be purchased. For example, a formulation containing 2 pounds of active ingredient per gallon would necessitate dividing the pounds of chemical by 2 to arrive at the gallons of commercial formulation required to control the nuisance. A dosage chart is often convenient (fig. 22-14).

From 1904 (Moore and Kellerman, 1904) to the present, the chemical that has most nearly met specifications for the control of algae has been copper sulfate (blue vitriol). Despite its extensive usage, copper sulphate has shortcomings in that it may, in excessive concentrations poison fish and other aquatic life, it may accumulate in bottom muds as an insoluble compound following extensive usage, and it is corrosive to paint and equipment.

Definite dosages of copper sulfate for the control of various types of algae were first prescribed by Moore and Kellerman (1905); these have been reprinted extensively in tabular form with specific dosages for some 70 organisms (Hale, 1954). The practical application of such a table is limited, however, because of the many variables encountered in nature. Because the solubility of copper in water is influenced by pH and alkal-

Fig. 22-13 Mechanical weed cutting and removal.

inity, as well as temperature, the dosage required for control depends upon the chemistry of the water itself, as well as on the susceptibility of particular organisms to the copper. Thus, rather arbitrary dosage rates have been succesfully used, especially in the midwestern States (Bartsch, 1954; Mackenthun, 1958). Since a total alkalinity of 40.0 mg./l seems to be a natural separation point between soft and hard waters (Moyle, 1949), those lakes that have a total methyl orange alkalinity of 40 mg./l or greater are treated with blue vitriol (commercial copper sulfate) at a rate of 1 mg./l ($CuSO_4 \cdot 5H_2O$) for the upper 2 feet of water regardless of actual depth. On an acreage basis, the concentration would amount to 5.4 pounds of commercial copper sulfate per surface acre. The 2-foot depth has been determined to be about the maximum effective range of a surface application of copper sulfate in such water, since algae will be killed with increasing depth only if the rate of downward diffusion exceeds the rate of copper precipitation. The algae killed by such a treatment are those that are suspended near the water surface and occur commonly as blooms in calm weather. For lakes with a total methyl orange alkalinity below 40 mg./l, a concentration of 0.3 mg./l commercial copper sulfate for the total volume of water has been recommended. This is comparable to 0.9 pounds of copper sulfate per acre-foot of water. It is obvious that when a low-alkalinity lake has an average depth of about 6 feet, the dosage would be about the same as in a high-alkalinity lake having the same area. When lesser average depths are involved, a greater concentration of chemical results in the high-alkalinity lakes. This apparent paradox would be even more striking were it not for the fact that in low-alkalinity lakes the algae frequently are of the filamentous types that may lie at the bottom and, therefore, the entire volume of water must be calculated to insure that a sufficient concentration of the chemical reaches the algae to effect a kill. In high-alkalinity lakes, algae frequently are planktonic and tend to con-

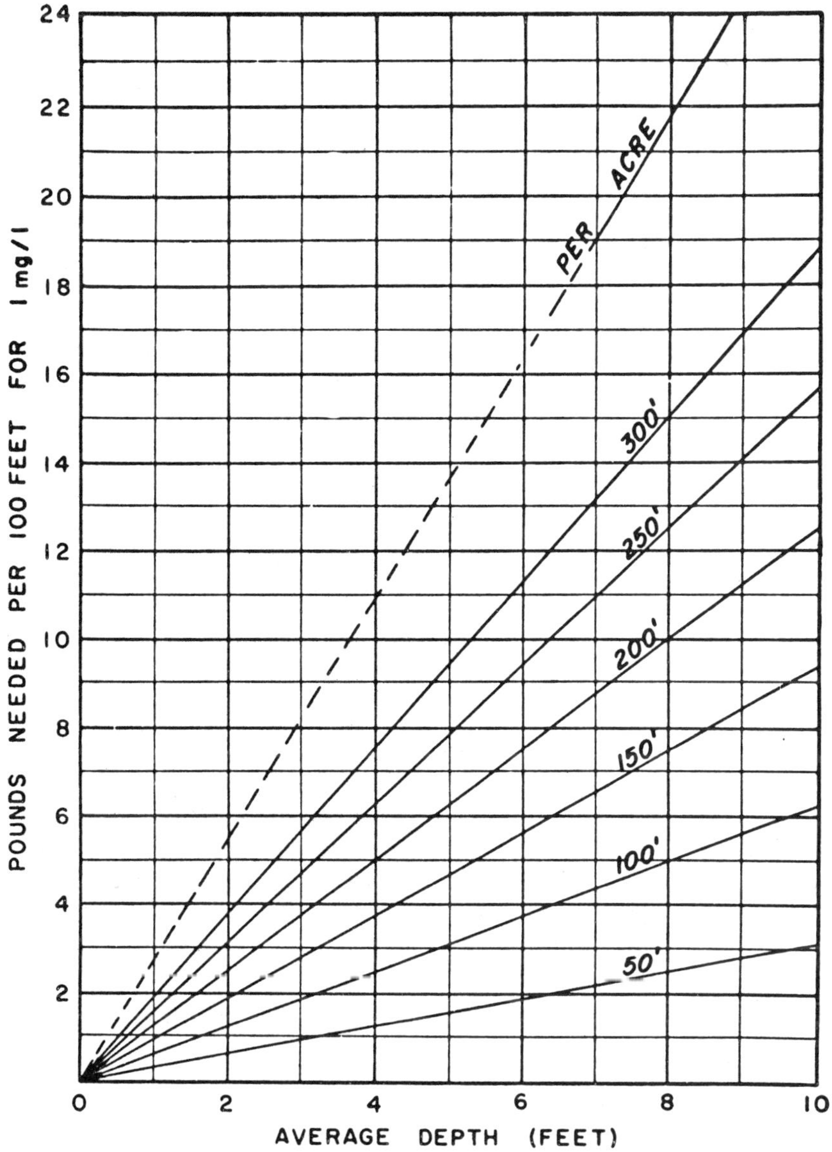

Fig. 22-14 **Chemical Dosage Chart. To achieve a chemical concentration of 1 mg./l in water having an average depth of 8 feet requires 10 pounds of the active chemical for an area 200 feet by 100 feet, or 21.8 pounds per acre.**

centrate near the surface, which is the only stratum in which appreciable concentrations of soluble copper can be produced. Certainly, when copper sulfate is used as the algicide, the best and most lasting control will result when the lake water has a total alkalinity around 50 mg./l or less.

Algal control treatments can be marginal or complete, the type applied to a given body of water must be determined by the size, shape, and relative fertility of the water, and the estimated cost of the project. Complete treatment in which the calculated amount of copper sulfate is systematically applied over the entire surface area is the most satisfactory. It insures that a major portion of the total algal population is eliminated at one time, and that a longer recovery period is required before an algal bloom condition recurs. The interval between necessary treatments will be directly correlated with climatological conditions and the available nutrients utilized by the remaining algal cells that are not killed as a result of chemical application. One to three complete treatments per season may be sufficient to give reasonable control.

Marginal treatment, on the other hand, is a method designed to obtain temporary relief in a restricted area where more extensive activity is not feasible or financially possible. In this procedure a strip, 200 to 400 feet wide, lying parallel with the shore, and all protected bays are sprayed in the same manner as in complete treatment. No other part of the area is treated even though many algae may be present. As a result of treatment, the algal population and the intensity of odors along the periphery of the lake are reduced. The duration of freedom from the algal nuisance following marginal treatment is dependent upon the density of the algal population in the center of the lake and its ability to infiltrate the treated area through the action of wind, waves, and currents. Any marginal control operation should definitely be considered on a periodic repeat basis. If fertility is not excessive, large bodies of water might gain enough relief from marginal treatment to warrant this type of control; however, it probably would be a waste of money for a large, fertile lake that is long and narrow and subject to considerable wind and wave action. In the latter case, complete treatment might be the only present answer, the cost of which might be prohibitive.

Copper sulfate may be applied in a variety of ways: bag-dragging, dry feeding (Monie, 1956), liquid spray (Mackenthun, 1958), and airplane application of either dry or wet material. Because rapid and uniform distribution of the algicide is essential, the size and scope of the problem determine to some extent the method employed. In general, liquid spraying

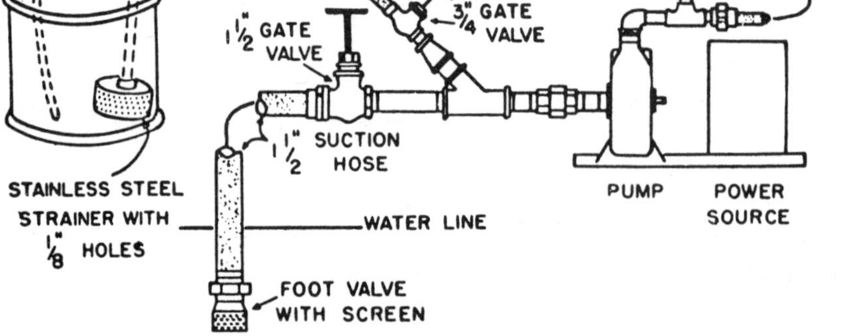

Fig. 22-15 **Equipment design for algal control. Small blue vitriol crystals are placed over perforated drum in chemical solution tank.**

Fig. 22-16 Liquid spray distribution of chemical by small boat.

systems operated from a boat or barge have been most widely used (figs. 22-15 and 16). Because copper sulfate is a highly corrosive chemical, materials that are used in the construction of spraying equipment should be resistant to its corrosive nature.

Other methods of distributing copper sulfate have been devised; one of these is to blow a chemical dust rather than use a slurry or solution (Anon., 1965). The principal advantage of the blower-type machine is the ability to treat large surface areas rapidly with a light dosing of material.

The blower, operating at 3,000 to 3,500 r.p.m., has the tendency to grind the commercial-grade $CuSO_4$ snow into smaller particles. These small particles are blown into the air, and wind currents assist in spreading them over the surface of the water.

Certain disadvantages are found in the blower-type machines. For example, the larger machines are heavy enough to reduce the permissive load of chemical in the boat; and two or more men are required to transport the units in and out of the watercraft. The machines also need continual adjustment by a trainer operator, such as a contract applicator, to maintain a constant feed and to obtain an even distribution of copper. An excessive rate of feed may clog the discharge spout. Use of these blower-type machines is dependent upon the wind for distribution of chemical, and with shifting winds the boat crew, as well as the reservoir may be dusted with the material. There is always the loss of varying amounts of copper sulfate dust that is carried away by the wind and then settles upon the above-water shoreline of the reservoir.

Helicopters have also been used in chemical distribution (Rosenberg, 1964). The East Bay Water Co., Oakland, Calif., found that a more efficient treatment could be obtained with a helicopter and that, as a result, fewer treatments were required per season for a particular reservoir.

The cost of chemical distribution was slightly more by helicopter than by boat. Helicopter rental averaged $0.80 per acre. DeCosta and Laverty (1964), in commenting on this treatment, stated that it was the district's experience that an ounce of copper sulfate properly applied at the reservoir was worth a pound of activated carbon added at the filter plant to correct a problem after it had arisen.

Algal control measures should be undertaken before the maximum development of the algal bloom. If, for some reason, a given area is not treated until the algal population has become dense, or a floating mat has formed, judgment must be used to determine the area that should receive treatment at any given time, lest sufficient organic matter is killed to result in oxygen removal when decomposing. It is good practice to subdivide the total area into sections and control the nuisance in one section at a time. Other sections may be treated after an interval of 7 to 10 days to ensure that sufficient dissolved oxygen is present to satisfy the demands of the decomposing algae.

Particular algal species often require particular chemical application measures and sometimes particular algicides. For example, the Ontario Water Resources Commission (Anon., 1959), in its examination of control measures for *Cladophora* in Lake Ontario, found that copper sulfate merely retarded growth but Aqualin at 5 p.p.m. cleared the test plots of *Cladophora* and they remained clear for 4 weeks during which observations were taken. In inland lakes and ponds, Diquat has been recommended for *Cladophora* control along with copper sulfate (De Vaney. 1968).

A problem often encountered in farm ponds is that of thick yellowish-green or green scums that form on the pond's surface. Scum-forming green algae are *Hydrodictyon, Pithophora, Cladophora, Oedogonium, Zygnema, Mougeotia,* and *Spirogyra.* The optimum time to control these growths is in early developmental stages when they first appear on the water surface. Using copper sulfate dosages that were discussed earlier, it may be necessary to spray-treat the area on successive occasions to achieve control. Care should be exercised to prevent killing too much algae at one time or fish may be killed by a lack of dissolved oxygen when the algae decompose. Mechanical removal of much of the scum, when feasible, will do much to enhance the success of the chemical application.

Another algal problem, discussed earlier, is *Chara. Chara* often is troublesome in high alkalinity (>40 mg./l) ponds and lakes, and in ponds may render them relatively unproductive of fish. Like the control of pond scums, *Chara* control chemicals should be applied early in the season before portions of the plant begin to decompose, and before the growth has "rolled" from the bottom. *Chara* is readily susceptible to copper sulfate applied at 5.4 pounds per acre, but the chemical must be applied directly above the growth. The algicide may be introduced through a perforated "⊥" pipe that is attached to the discharge side of the chemical distribution equipment and then lowered to the desired depth for chemical introduction.

Vascular Plants

In the control of submersed aquatic plants, it is often desirable to chemically treat localized areas along the shoreline, such as bathing beaches and around piers, and to develop channels through weed beds so that boats will have access to deeper water. Sometimes it is advantageous to treat an extensive area in an effort to curtail an advancing population of a weed species, such as Eurasian watermilfoil (*Myriophyllum spicatum* Linnaeus). Best results are obtained when the shoreline areas are treated because currents and wave action are usually minimized, and the diffusion of the chemical can take place only in one direction—into the lake. There are limitations below which it is usually not feasible to attempt liquid chemical weed control. The recommended minimum area to be treated is 200 by 200 feet. The treatment of very small areas permits the diffusion of the chemical on 3 sides, thus reducing the concentration of the chemical within the area to a point below the toxic level for rooted plants. An exception to this recommendation might be a small slough, bay, or stagnant channel with an area of less than 40,000 square feet. Granular aquatic herbicides, applied early in the growing season, are often applied successfully on smaller areas.

For many years, arsenic trioxide as a sodium arsenite solution was used effectively to control submersed aquatic vegetation (Mackenthun, 1958). In recent years the trend has been away from those chemicals that are essentially nondegradable to organic herbicides that do not accumulate in the environment. These herbicidal formulations include the phenoxy compounds such as 2,4-D and silvex, Dalapon, Diquat, Endothall, and others. Dosage rates and application procedures have been detailed by Surber (1961), Lawrence (1962), and De Vaney (1968). De Vaney has listed those herbicides that have been registered in accordance with the Federal Insecticide, Fungicide, and Rodenticide Act for use in aquatic sites (table 22-7). Many of the compounds are available in liquid and granular formulations, the latter being suitable for broadcasting on the water surface (fig. 22-17).

A few details should be kept in mind in the application of a chemical. It has been found advantageous, for example, to divide a large area to be treated into a convenient number of small subareas and to accurately determine the volume of water each contains. Since the quantity of the chemical used is proportional to the water volume to be treated, it is a simple matter to properly adjust the chemical application and evenly distribute the correct amount of chemical into the subarea. This procedure is then repeated in successive subareas. Spraying should be initiated at the shoreline so that fish will not be trapped in shallow water.

Chemical control of aquatic vegetation must currently be regarded as a temporary remedy although it should last for the season in which it is applied. Under certain conditions, the removal of a vascular plant population may promote the growth of a bottom-dwelling alga such as *Chara*. The alga must then be attacked with a suitable algicide.

Biological controls are currently being tested in the control of aquatic weeds. Preliminary studies on use of grass carp, *Ctenopharyngodon idellus.* (Cuvier and Vallenciennes), for aquatic weed control have recently been completed at Auburn University Agricultural Experiment Station. When stocked at a rate of 685 per acre, the fish eliminated 12 species of

Table 22-7 **Herbicides That Have Been Registered In Accordance With the Federal Insecticide, Fungicide, and Rodenticide Act for Use In Aquatic Sites.**
(From DeVaney, 1968)

Acrolein
- Irrigation canals
- Ponds

Amitrole (Not approved for Bureau of Sport Fisheries and Wildlife use)
- Non crop areas
- Drainage ditches

Amitrole T (Not approved for Bureau of Sport Fisheries and Wildlife use)
- Non crop areas
- Drainage ditches
- Marshes

Ammonium Sulfamate
- Around lakes, ponds
- Around potable water reservoirs
- Along inflow streams

Bromacil
- Ditch lands (drainage)

Copper Sulfate
- Farm ponds
- Lakes

Dalapon
- Non crop areas
- Drainage ditches

Dichlobenil
- Lakes
- Ponds

Dichlone
- Lakes
- Ponds
- Irrigation canals

Diquat
- Lakes
- Ponds
- Irrigation canals
- Drainage ditches

Endothall
- Lakes
- Ponds
- Irrigation canals
- Drainage ditches

Fenac
- Lakes
- Ponds
- Reservoirs
- Drainage ditches and banks

Petroleum Distillate (aromatic)
- Irrigation ditches
- Drainage ditches
- Irrigation, drainage ditches and banks

Silvex
- Lakes
- Ponds

Sodium Arsenite (Not approved for Bureau of Sport Fisheries and Wildlife use)
- Lakes
- Ponds

Sodium TCA
- Non crop areas
- Ditch banks (drainage)

2, 4-D
- Lakes
- Ponds

2, 4, 5-T
- Lakes
- Ponds

weeds growing in plastic-lined pools within 6 weeks. In ponds, 3 species of rooted weeds were significantly reduced or eliminated in 1 month after being stocked with 20 to 40 grass carp per acre (Avault, 1965).

Stevenson (1965) reported that early growth of the grass carp at the Fish Farming Experimental Station, Stuttgart, Ark., compared favorably with that reported in semi-tropical countries. The average weight at 18 months was 1,816 grams; the length, 50 centimeters. The fish were given a supplemental ration of commercial fish pellets and cut grass. Observations of the feeding habits indicate that this carp may not be a strict herbivore; it is recommended that a thorough study be made before the fish is released in natural waters. Stevenson calls it the most efficient aquatic plant-eating fish, but cautions that it might become another carp problem if introduced.

Rapid advances are also being made in research on the control of

Fig. 22-17 Helicopter application of a granular herbicide.

marsh weeds and other aquatic forms that project above the water surface. Two factors explain the growing interest in the control of these weeds: one is the advent of new and better herbicides for the purpose; the other is the increasingly critical situation facing the Nation's waterfowl hunting resource—a sport on which 2 million Americans spend millions annually. Already more than half of the country's original 125 million acres of wetlands have been spoiled for waterfowl use. As the national population approaches the 200-million mark, more and more of the places where ducks feed, breed, and are hunted will have to be converted to the needs of advancing civilization. In other words, prospects are for fewer hunting places and fewer ducks for a large number of hunters.

To help offset this trend, it is important to make the best use of available waterfowl habitat. Thousands of poor or fair areas in the United States can be made more attractive for ducks and duck hunters by replacing marsh weeds with plants that furnish food or cover for waterfowl. Commonly, the costs involved are not prohibitive (Martin et al., 1957).

In Florida recently, water hyacinths and other pest plants were cleared from 20,000 acres of ponds and lakes by the Game and Fresh Water Fish Commission. Numerous other programs, both large and small, are being waged against phragmites, cattails, waterchestnut, and other marsh weeds in various States. In the Federal refuge system, more than 2,000 acres of marshlands are treated with herbicides annually to make them more productive for waterfowl.

Martin et al., discussed vegetation control by waterlevel management, mechanical control by bulldozing, burning, and the use of herbicides such as 2,4-D, 2,4,5-T, ammonium sulphamate, Dalapon, Kuron, and others. Specific recommendations for chemical control of particular marsh plants were given.

Animals

Problems have arisen in the chemical control of certain animal pests. Clear Lake, Calif., was treated in September 1949, with a chlorinated hydrocarbon insecticide DDD (dichlorodiphenyl dichloroethane) at a concentration of 1 part of active insecticide to 70 million parts of water to control the Clear Lake gnat. It was determined that a 99-percent kill of larvae resulted. A second chemical application was necessary in 1954 and a third in 1957. These latter applications were calculated to produce a concentration of one part in 50 million parts of water. Following both the 1954 and 1957 applications, 75 to 100 western grebs were found dead along the lake shore. Results of chemical analysis of the fatty tissue from the grebes in 1958 indicated DDD was present at the unusally high concentration of 1,600 p.p.m. (Hunt and Bischoff, 1960). The amount of DDD found during Mar. 1958, in the visceral fat of the brown bullhead ranged from 40 to 2,500 p.p.m. To prevent the possibility of increasing the present hazard of DDD poisoning of wildlife, Clear Lake will receive no further treatment with DDD.

In laboratory tests to control chemically other midge larvae, Hilsenhoff (1959) screened 16 commercially available organophosphate insecticides to determine their relative toxicity. Dipterex and malathion incorporated into granules produced an 80-percent mortality of the larve at concentrations of 0.1 pound per acre of the technical material. These insecticides have a low toxicity to fishes, waterfowl and mammals. Field tests on Lake Winnebago, Wis., with malathion granules, however, did not prove conclusive at low concentrations; the feasibility of chemical control on a large body of water with a nonaccumulative insecticide is questionable.

Various species of midges have become established as significant nuisance problems in San Mateo County and many other suburban areas throughout California (Whitsel et al., 1963). Midge production at all stations in a study of the problem area, was preceded by a pronounced increase in accumulated organic matter. Water known to be polluted with sewage wastes was found to have over 1,100 larve per square foot of substrate. Man-made drainageways were found to produce greater midge infestations than natural drainage creeks. Localized chemical applications were reported to be successful in control practices. "Chemical control by granular Baytex 0.1 and 0.2 pound per acre provided fairly effective reduction of *Chironomus* larvae in impounded and slowly moving waters, with no harm to mosquito fish, *Gambusia affinis.* Granular dieldrin showed promising control in slowly moving waters, with little harm to Gambusia and aquatic invertebrates at 0.5 pound per acre." Patterson and von Windeguth (1964) described Baytex as one of the most promising chemicals for use as a larvicide against chironomid midges. In laboratory experiments, they found that a granular formulation at the recommended larvicidal rates exhibited no overt toxic effects on fresh water copepods, ostracods, hydra, annelid worms, snails, clams or fish over a 4-month period. There was indication that Baytex was very toxic to midge larvae and to *Cladocera.*

In a recent study of the biology of midges in Lake Winnebago, Wis. (Hilsenhoff 1966) found that eggs of *Chironomus plumosus* (Linnaeus) are laid on the lake surface in masses averaging 1,676 eggs. These masses absorb water, swell, and sink to the lake bottom, where the eggs hatch in 3 to 14 days. Larval stages range from 1.40-mm., colorless, first instars to 30-mm, dark red, fourth instars. The newly hatched first instars are free-swimming and positively phototrophic. All instars, except perhaps the first, construct U-shaped tubes in the bottom mud and, at temperatures above 5° C., feed indiscriminately by "filter feeding." Female larvae must weigh 60 mg. or more to pupate, but male larvae will pupate when they weigh 48 to 56 mg. Emergence may occur at any hour, day or night, but is most prevalent just before sunset. Newly emerged flies fly weakly with the wind, eventually accumulating on the lee shore. Swarms of males form about 1 hour before sunrise and disperse about 1 hour after sunset in response to light. The swarms, typically containing several thousand flies, orient over objects that contrast with the background. Females fly into the swarms for mating, which occurs most commonly in the early morning. Mated females, after a preoviposition period of 1½ to 5 days, fly out over the lake to oviposit, extruding their egg masses onto their hind legs while in flight; oviposition occurs commonly at all hours of day or night. The larvae are eaten by many fish and by 1 species of leech, are parasitized by a microsporidian, and are occasionally killed by unknown diseases. There are normally 2 generations a year in Lake Winnebago. Overwintering fourth-instar larvae pupate and emerge during mid-May; their progeny emerge from mid-July to early September. Rate of development and time of emergence are dependent on mud temperatures. If unknown stimuli necessary for pupation are absent, a generation may be delayed or entirely missed.

Moore (1923) describes methods of leech control through freezing in their winter quarters. As water temperatures fall with the onset of winter, leeches become more and more sluggish and finally bury themselves in the mud or beneath stones on the bottom of shallow water. Leeches are readily killed by exposure to a temperature of 20° F. for a few hours. When the first thin ice starts to form and it is certain that the water temperature has attained its minimum, Moore recommends that the water be drawn off as rapidly as possible until the water level is lowered 4 feet, and that this level be maintained for at least 5 or 6 weeks during the coldest part of the winter. He reasons that, under these circumstances, the exposed flats would be frozen hard to a considerable depth, and temperatures would be well below the fatal minimum for the imprisoned leeches. This method of control was effective in Carr Pond, Palisades Interstate Park near New York City. Pennak (1953) states that leeches may be temporarily controlled in localized bathing beach areas by applying 100 pounds of powdered lime per acre per day in the shallows.

In general, chemical controls have not proved successful, although chemicals such as copper sulfate will kill leeches that are exposed to it. The measures employed in snail control have been used with some success in leech control. Weekly distribution of a slurry composed of 10 pounds of copper sulfate and 5 pounds of copper carbonate or lime per acre of bathing area has shown some success. Chelated copper compounds applied marginally as a spray at concentrations sublethal to fish have also been of value.

Regulations

Most states have laws or regulations to regulate specifically the use of pesticides to control nuisances in aquatic habitats. Some States regulate only the commercial applicators; others regulate all chemical users. Before chemicals are applied to control any aquatic pest, it is well to check with the appropriate state authority to ascertain the legality of the proposed chemical application.

The application of a chemical to water involves certain hazards that must be understood and against which public rights must be protected. Factors that must be considered are the short- and long-range toxicity to all aquatic life; the deposition and possible accumulation of the chemical upon the lake bottom, and the subsequent reaction upon the community of bottom organisms; the impact resulting from the destruction of too much biological growth at one time; and the possible disturbance or impairment of the general aquatic environment.

De Vaney (1968) cites the basic policy of the Department of the Interior with respect to the use of pesticides as follows:

> "It is essential that all pesticides, herbicides, and related chemicals be applied in a manner fully consistent with the protection of the entire environment. Any contemplated use of these chemicals must take into account both known and possible environmental effect. The guiding rule for the Department shall be that where there is a reasonable doubt regarding the environmental effects of the use of a given pesticide, herbicide, or other chemical, no use should be made."

REFERENCES

The publications referred to in Chapter 22 may be found in "The Practice of Water Pollution", by K. M. Mackenthun, Federal Water Pollution Control Administration, 1969, pp. 253-270

CHAPTER 23

HOUSEHOLD WASTE WATER

An investigation was carried out under the direction of the Swedish National Institute for Building Research, to obtain information about the quantity and nature of waste water from dwelling houses. The results provide a basis for estimating the loading capacity of recipients and the necessary sewage treatment procedures as well as the scaling of drainage networks and treatment plants. The AB Stockholmshem Building Society and Stockholm City Real Estate Department granted permission of a block of 25 flats in Bromsten (a suburb of Stockholm) to be used as the subject of the investigation. A Liljendahl vacuum drainage system is installed in the building. This system made it possible to divide up the principal components of household waste water into separate drainage stacks.

A further division was made so that waste water could be obtained from 10 kitchens, 10 bathrooms, 15 kitchens + bathrooms, a laundry, and 28 vacuum toilets. These waste waters have been measured and analysed separately, and the sum total of the pollution in the waste water from bathrooms, kitchens and laundry (grey water) has been compared with the pollution in the waste water from toilets (black water).

However, the principal interest was devoted to a careful analysis of grey water, since information about such water is imperfect. The components of black water have been dealt with in other connections, principally in medical literature. In addition, the National Road Board has carried out an investigation of faeces and urine in black water. The Institute, therefore, considered that spot-check studies of black water would suffice. Samples taken at random have also formed the basis of the analyses of waste water from the laundry. The flow from kitchens and bathrooms has been carefully measured throughout between 06.00 and 22.00 hours as it arrived in special collection tanks. The flow from the vacuum toilets has been measured by registering of the number of flushings, since the type of toilet installed in the subject building utilises a constant quantity of water for each flushing.

The flow from the laundry has been measured by means of a water meter installed just before the taps (specific supply, see Table 23-1).

Some 3500 analyses form the basis of an estimate of the quantities of pollution. Samples were collected during a trial period of 12 weeks (from 11 January to 4 April 1965). The analytical values have been statistically processed for the whole period of measurement for months, weeks, days of the week and parts of a day. A careful proportioning of all the samples has been made relative to the flow.

Table 23-1 shows the different analyses which have been carried out. These analyses principally refer to biochemical oxygen demand, BOD_5, which was determined every day. For economical reasons, one week (No. 3) was chosen for an intensified investigation. During this week, the daily variations of all analyses were studied. During the other weeks, $KMnO_4$, phosphorus and nitrogen determinations were obtained on proportioned weekly samples.

In addition to the BOD_5 determinations, a number of BOD curves have been compiled in order to make it possible to assess the rate of reaction and the various steps of decomposition (L_a' and L_a'') as well as total BOD, ΣL_a (see Table 23-1).

Bacteriological analyses were performed on waste water from kitchens and bathrooms during week 3, while the combined waste water from kitchens and bathrooms, as well as water from the toilets (black water), was analysed bacteriologically during week 12.

Temperature, acidity (pH) and specific conductivity (μS) were determined continuously at the sampling site.

During the trial period, a number of investigations were carried out in order to check the reliability of the results obtained. Thus BOD_5, for one thing, was checked by different laboratories which were commissioned to analyse samples taken from the same waste water at the same time. The results of these parallel analyses gave widely varying results (10 to 30 %). In making comparisons with other investigations, however, an underrepresentation of the BOD value by ten per cent should be taken into consideration. For most practical purposes this uncertainty should not be regarded as excessive. During the measuring period it

Table 23-1 Summary of the specific arithmetic average values (M) per person per day for the waste water from households and its various components from kitchens, bathrooms, combination group kitchens and bathrooms, laundry and vacuum toilets.

		(1)	(2)	(3)	(4)	(5)	(6)	(7)	(8)	(9)	(10)
		Kitchens		Bathrooms		Kitchens & bathrooms	Kitchens & bathrooms combined	Laundry	Col. 5+7	Black water	Col. 8-9 Grey and black
Analysis	Unit	Week 1-12	Week 3	Week 1-12	Week 3	Week 1-12	Week 3	Week 1-12	Week 1-12	Week 1-12	Week 1-12
L_a	p/c.d	≈ 24	-	≈ 9	-	≈ 32	-	-	-	-	
L_a'		15	-	5	-	19	-	-	-	22	
L_a''		≈ 9	-	≈ 4	-	≈ 13	-	-	-	-	
BOD_5		17		5	4	22	20	3	25	20	45
$KMnO_4$		34	41	7	6.0	41	47	7	48	72	120
Total P		0.3	0.3	0.6	0.6	0.9	0.9	1.3	2.2	1.6	3.8
Phosphate P		0.01	0.01	0.01	0.01	0.02	0.02	-	-	-	-
NH_4-N		0.04	0.04	0.03	0.011	0.07	0.051	-	-	-	-
NO_2-N		0.001	0.001	0.001	0.001	0.002	0.002	-	-	-	-
NO_3-N		-	-	-	-	-	-	-	-	-	-
Kjeldahl N		0.6	0.6	0.3	0.3	0.9	0.9	0.2	1.1	11	12.1
Total residue		36	31	22	19	58	50	19	77	53	130
Fixed total residue		9	8	12	10	21	18	12	33	14	47
Volatile total residue		27	23	10	9	37	32	7	44	39	83

Table 23-1 (continued) Summary of the specific arithmetic average values (M) per person per day for the waste water from households and its various components from kitchens, bathrooms, combination group kitchens and bathrooms, laundry and vacuum toilets.

Analysis	Unit	(1) Kitchens Week 1-12	(2) Kitchens Week 3	(3) Bathrooms Week 1-12	(4) Bathrooms Week 3	(5) Kitchens & bathrooms Week 1-12	(6) Kitchens & bathrooms combined Week 3	(7) Laundry Week 1-12	(8) Col. 5+7 Week 1-12	(9) Black water Week 1-12	(10) Col. 8-9 Grey and black Week 1-12
Non-filterable residue		13	11	3	3	16	14	2	18	30	48
Fixed non-filterable residue		1	1	1	1	2	2	1	3	5	8
Volatile non-filterable residue		12	10	2	2	14	12	1	15	25	40
pH		7.3	7.2	8.1	7.8	-	-	9.8	-	8.9	-
μS		346	387	287	292	-	-	1521	-	-	-
Temp. °C		28	28	25	25	26	26	-	-	-	-
Plate c 35°	number/c.d	-	$35.5 \cdot 10^9$	-	$40.6 \cdot 10^9$	$76.1 \cdot 10^9$	-	-	-	$62.2 \cdot 10^9$	-
Coli 35°		-	$11.9 \cdot 10^9$	-	$1.0 \cdot 10^9$	$12.9 \cdot 10^9$	-	-	-	$4.8 \cdot 10^9$	-
Coli 44°		-	$2.3 \cdot 10^9$	-	$0.2 \cdot 10^9$	$2.5 \cdot 10^9$	-	-	-	$3.8 \cdot 10^9$	-
Specific flow	l/c.d	51	44	62	55	113	99	8.5	121.5	8.5	130

was also established that the way in which the extracted samples were cooled (stored) before analysis had a considerable effect on the result. Two methods of cooling were used - one where the sample was put directly into the refrigerator (Method I) and one where the sample was pre-chilled in a water bath at 3°C before being put into the refrigerator (Method II). It was thus found that samples of kitchen waste water treated according to Method I gave a 14.5 % *higher* value than those treated according to Method II. The reverse was the case for waste water from bathrooms, where cooling Method I gave a 10 % *lower* value. It was considered that cooling Method II, generally speaking, is to be preferred, since this method for one thing gives higher reproducibility, and the results are thus more amenable to comparison in various contexts.

Control analyses were carried out on the stored weekly samples as well, and it was found that it was possible to aggregate samples taken every day and from these samples build up a representative mean sample for a week without affecting the results.

The results of the determinations of phosphorus in the waste water from the laundry were checked, too. In this case it was noted that there was a good measure of agreement between the quantity of phosphorus introduced into the washing machine and the amount of phosphorus established by analysis. To facilitate study of the consumption of synthetic laundry and dish-washing detergents, the Institute supplied all the preparations used throughout the period of the investigation. The consumption proved to be: synthetic laundry detergent 0.51 dl per person per day and dish-washing detergents 0.11 dl per person per day.

In order to be able to express the quantity of pollution per capita and day, an investigation based on interviews with the occupants was carried out, and they were characterized as to total number, occupation, sex and age. Two series of interviews were conducted, one immediately before and the other after the investigation.

A collation of the arithmetic mean values (M) of the specific pollution, expressed in g/c.d, both with reference to the category of waste water and the type of analysis, can be seen in Table 23-1 (cols. 1-7). Column 8 gives the values for grey water and column 9 those for black water. The values for the total household waste water are listed in column 10.

It was established that the flow as well as the pollution, with respect to both volume and nature, was the same from day to day without any noticeable differences between the different days of the week. On the other hand, the variations within the day itself were so great and so unsystematic that the analysis results from separate parts of the day do not supply any usable representative figures of the quantities of pollution.

If the composition of the grey water is first studied, it is found that nearly 70 per cent of the pollution in terms of BOD_5 emanates from the kitchens. Of the quantities of phosphorus, nearly 60 per cent originates from the laundry, and only 14 per cent from the kitchens, while the quantities of nitrogen are split up in such a way that the kitchens supply 51 per cent, the bathrooms 31 per cent and the laundry 18 per cent.

If one compares the percentage contribution of grey water and black water respectively to the quantities of different pollutants present in the mixed waste water, see Table 23-2, one finds that the share of the grey water is higher for BOD_5, phosphorus, dry matter, fixed and volatile total residues, as well as plate count 35° and Coli 35°. Grey water contributes between 53 and 70 per cent in these cases.

On the other hand, black water gives larger proportional contributions of $KMnO_4$, non-filterable residues fixed and volatile non-filterable residues, Coli 44° and nitrogen. Particularly noticeable is the part played by black water as the principal source of nitrogen (91 % of the total amount of nitrogen).

In order to establish the effect of sedimentation and filtration on the different components of grey water, 36 samples were sedimented and 12 were filtered. BOD analyses of these samples showed that the BOD_5 values were 10 per cent lower for sedimented than for non-sedimented samples, and 25 to 35 per cent lower for filtered than for non-filtered samples.
A point of interest was that $KMnO_4$ was not affected by filtration by more than 10 per cent. The proportion of phosphorus was not affected by sedimentation but by a few per cent through filtration, while nitrogen was affected both by sedimentation and by filtration by 10 and 25 % respectively.

The BOD curves which were compiled show the loss of oxygen up to 32 days. In this way, an approximative measure of the ultimate

Table 23-2 Comparison between the quantities of pollution per person per day in grey water and black water.

Analysis	Unit	Grey water	Black water	Total grey + black water	% division Grey water	% division Black water
BOD_5	g/c.d	25	20	45	56	44
$KMnO_4$-O_2		48	72	120	40	60
Total P		2.2	1.6	3.8	58	42
Kjeldahl N		1.1	11.0	12.1	9	91
Total residue		77	53	130	59	41
Fixed total residue		33	14	47	70	30
Volatile total residue		44	39	83	53	47
Non-filterable residue		18	30	48	38	62
Fixed non-filterable residue		3	5	8	38	62
Volatile non-filterable residue		15	25	40	38	62
Plate c 35°	Number of bacteria/c.d	$83.0 \cdot 10^9$ [a]	$62.2 \cdot 10^9$	$145.2 \cdot 10^9$	57	43
Coli 35°		$8.5 \cdot 10^9$ [a]	$4.8 \cdot 10^9$	$13.3 \cdot 10^9$	64	36
Coli 44°		$1.7 \cdot 10^9$ [a]	$3.8 \cdot 10^9$	$5.5 \cdot 10^9$	31	69
Flow	l/c.d	121.5	8.5	130	93	7

[a] The average of the weighted arithmetic average values for the grey water from kitchens, bathrooms and the combination group kitchens and bathrooms. The laundry is not included.

BOD (C and N steps) was obtained. Further attempts, using more refined methods, are being made to supply more exact information about this important parameter.

The biological process of decomposition shows different behaviour patterns for different categories of waste water. However, in general one can distinguish different - most often two - phases of decomposition or oxidation steps. These can be defined through the values L_a' and L_a'', respectively, which apply to the first and the second phase of decomposition, and the corresponding rate-of-reaction constants.

For grey water the first phase is virtually equal to BOD_5.

The rates of decomposition of the waste water from kitchens, bathrooms and both combined during the first phase of decomposition are equally fast (about 65 % per 24 hours as against 20 % for municipal waste water).

On the other hand, the black water from toilets decomposes relatively slowly, i.e. approximately at the same rate as municipal waste water. The oxygen consumption is thus slower for black water than for the other categories of waste water. This suggests that the organic substances in black water, regarded as a substrate, are not easily accessible to micro-organisms.

From the BOD curves which were compiled during the investigation it can be seen that, after the first day's biological decomposition, about 40 per cent of the *total* BOD-type pollutants have been oxidized in the waste water from kitchens, bathrooms and both combined, while only about 10 per cent of the black water pollutants have been oxidized.

This show that the high values obtained from kitchens, bathrooms and the combination of kitchen plus bathroom are not directly comparable with the results of measurements taken at sewage purification plants to which long sewers most often lead and where a certain decomposition of the organic matter can take place. If the BOD_5 for grey water is compared with that for black after one day in the sewers it is found that the respective proportions of black and grey water have changed and that the black water now represents 18 g/c.d while the grey water gives only 15 g/c.d. The values in the fresh mixture were 20 g for black water and 25 g for grey.

These circumstances may possibly - at least in part - supply an explanation of why strong divergences are obtained in the BOD values (g/c.d) when investigations are carried out on the BOD deliveries from different built-up areas.

In this connection it should be pointed out that the consumption of oxygen due to the oxidation of organic carbon differs from the oxygen consumption due to nitrogen oxidation (nitrification) both in its nature and - to a certain extent - in its further consequences to the oxygen economy of the recipient. The former leads to an irreversible and often quickly started loss of oxygen in the water. On the other hand, the latter brings about a relatively slow build-up of what can be called a "chemically bound oxygen reserve" (in NO_2 and NO_3). However, it should be emphasized that the favourable effects of the nitrification process can be utilised only at a level of loading where the waters, regarded from different points of view, have already reached a critical condition. The advantage to be gained from the stored oxygen reserve of the nitrates lies mainly in that it can be used to retard the formation of hydrogen sulphide.

From the results mentioned above it is evident that special attention must be paid to the grey water, since it is the grey water that supplies such a large proportion of primary (= direct) pollution. This means that even if black water was eliminated this would not reduce the pollution to such an extent that the remaining grey water can be discharged into the recipients without special treatment.

However, the composition of waste waters and their properties should not only be looked upon from the point of view of primary pollution. Just as important is the secondary (indirect) aspect of pollution. Interest here centres primarily on the phosphorus content. As is well known, the phosphorus supply to the recipient is, in the majority of cases, the decisive factor of growth for algae and the higher forms of plant life. If, schematically, we assume that the phosphorus content of plants in relation to the carbon content reflects their need of phosphorus for the building up of a certain quantity of organic substance, and if we use stoichiometric rules to derive the quantity of oxygen required for complete oxidation from the carbon content of this substance together with the quantity of nitrogen assimilated in the process, we can obtain a measure of the potential oxygen demand represented by the phosphorus in a given quantity of waste water.

Considering the matter this way, we obtain a weight relationship P:N:C = 1:7.3:41. Since 1 C corresponds to 2.7 O_2 and 1 N corresponds to 4.6 O_2, we find that 1 P corresponds to 143 O_2. It is interesting to note in this connection that a weight unit P can be calculated as corresponding to 1700 weight units of plankton, or 136 units of plankton dry matter.

Based on data material from the investigation, we can calculate that phosphorus via synthesized organic matter gives rise to the theoretical secondary biochemical oxygen demand as shown below - providing that no other limiting factor is present - which can thus be compared with primary oxygen demand (total BOD).

	Biochemical oxygen demand (g/c.d)					
	Bathrooms	Kitchens	Laundry	Grey	Black	Grey & black
Secondary	85	45	180	310	230	540
Primary	9	24	7	40	80	120

From numerical calculations of the above type, which, by the way, have only during the last few years been shown to be applicable to nutrient salt problems, it appears that the release of household waste water - at least in the case of some types of recipients - can have secondary effects which may far and away exceed the effects of primary pollution such as they can be judged or forecast through BOD analyses and BOD calculations.

CHAPTER 24

COMBATING POLLUTION FROM OIL SPILLS

The major operational consideration in any oil spill situation is that, if at all possible, the oil spill should be treated at sea to prevent the contamination of the coastline and attendant damage to the coastal ecology and economy. Operational procedures for combating oil spills at sea must be based on a 24-hour day capability and point to a logical order of priority. First, the oil should be, if possible, contained at the source. Second, the oil inside the containment area should be removed. And, third, the oil that has spread outside the containment area should be removed – the combatant method depending on the kind of oil, its age, the sea-state, and other considerations.

1. Combating the Oil Afloat–State-of-the-Art

Our investigations have shown that although a large amount of information on oil spills has been developed, there is insufficient in-depth technical information on oil spill combatant methods available on which to base a definitive technical evaluation. Current research and development programs in the United States and Europe have not yet provided any meaningful conclusions on which of the various methods will be most successful in terms of technical, ecological, operational, and cost effectiveness. The state-of-the-art for combating oil spills in the open ocean has not progressed significantly since the Torrey Canyon incident in March 1967, as demonstrated by the agonizing and expensive cleanup operations at Santa Barbara. In contrast, the state-of-the-art for combating the more frequent oil spills in harbors is developing rapidly because of public pressures, harbor regulations, and the less severe conditions.

Despite the scarcity of sound technical information on the operating characteristics and performance of oil spill combatant methods under actual conditions, we have drawn some specific conclusions for each generic method on the current state-of-the-art, potential applicability and limitations.

a. Mechanical Containment

Mechanical booms are commercially available and have been successfully demonstrated in protected waters. However, none of these systems has proven effective in containing an oil spill in the open ocean, nor is there any boom system designed specifically for open-ocean use. The boom concept of open ocean containment, however, offers potential for all cleanup operations and reduction of cleanup costs by reducing the area to be treated.

Air-curtains also have been demonstrated to be quite useful as barriers in calm waters where boat traffic must be maintained. This concept holds promise for open ocean use and appears best suited to permanent installations, but its potential and limitations have not been assessed. Limited experimental work continues.

By treating the periphery of a spill with a chemical gelling agent or a polymerizing agent, a chemical "boom" could be created to retard the spreading of the oil. This technique has not been tried even in protected waters and its usefulness is, at best, problematical. It appears that chemical booms will be severely limited by their mechanical properties (strength, flexibility, elasticity, etc.) and the height (above and below the surface) that they present to the spreading oil.

b. Mechanical Removal

The skimmer concept offers a potentially effective way of removing the oil from the ocean environment without introduction of other foreign materials. Mechanical skimmers are commercially available (and are used regularly) for limited application in harbors. The rate at which these devices can collect oil is limited by the thickness of the oil on the water surface, the rate at which the oil-water mixture can be separated, the storage capacity of the vessel, and the area swept. No skimming device has been demonstrated to be effective under open ocean conditions.

c. Physical Sinking Methods

Sinking agents were used with some success in the Torrey Canyon disaster. However, systems for efficiently spreading sinking agents are not available for treating large spills on the open ocean. More importantly, little is known about the mechanism of sinking, and the behavior of sunken oil on the ocean floor and its effect on the benthic ecology. Once sinking agents are better understood and the long-range fate of sunken oil is ascertained, the method could be useful for treating large areas of medium to heavy or weathered oil where there is little danger of damage to the benthic ecology and little possibility of bottom currents washing the oil ashore.

d. Chemical Dispersion

Chemical dispersion has been the most extensively used of any combatant method. Dispersants are most useful on freshly formed slicks of light oil in a moderate-temperature environment. However, the toxicity of chemical dispersants to marine life has limited their wider application. We believe that the use of dispersants has been unduly restricted, particularly in the open ocean where biological activity is low. At present, chemical dispersion offers the most effective method of treating open ocean spills.

Nevertheless, a better understanding of the physical chemistry and dynamics of dispersions is needed as well as that of the ultimate fate of the dispersed oil and the toxicity of oil-chemical mixtures. Equipment for the efficient application of dispersants must be developed before this method can be rated as operationally effective for large spills.

e. Physical Absorption

Inexpensive absorption materials which could be easily distributed are available for the treatment of an oil spill with minimum damage to the ecology. The major limitation of absorption, however, is that the spent, oil-soaked materials must be collected. Equipment now available for the spreading and collecting of these materials either on calm or open waters is ineffective. If effective equipment can be developed, this technique would be best suited for thin to moderate oil slicks where the wind will not unduly impede the distribution of the materials and where the biology is of sufficient importance to preclude the use of sinking agents or chemical dispersants.

f. Combustion

Small-scale experiments on relatively calm waters have shown that oxidents and wicking agents can be used to augment the burning of freshly spilled oil, leaving a smaller amount of residue than the ∿ 1/8 inch which remains after burning the oil without enhancement. However, the feasibility of improving combustion of a large oil spill on the open ocean, has not been demonstrated. Burning would be effective on thick slicks of freshly spilled oil in calm waters if the hazards to ships and shoreline property could be minimized and the resulting air pollution could be tolerated.

g. Biological Degradation

Biological seeding of oil slicks with special bacterial cultures is neither necessary nor especially effective for the treatment of an oil spill. The natural marine environment provides an adequate supply of microbial species which are capable of metabolizing oil. However, the rate of microbial degradation of oil in the ocean is limited – not only by density of organisms but by the availability of the oxygen and nutrients needed to support the metabolic processes; therefore, the addition of phosphorus or nitrogen compounds to floating or dispersed oil might improve the rate of microbial degradation. Acceleration of the natural process is particularly attractive in the open ocean where nutrients are not available in the necessary concentrations for rapid degradation. But, in any case, the rate of bacterial degradation (accelerated or natural) of floating oil is slow and will not be useful if the oil is threatening a coastline. Nutritional enhancement of the degradation of dispersed oil, however, may be desirable if the nutrients can be applied with chemical dispersants.

2. Combating the Oil Ashore–Shoreline Restoration

Shoreline restoration technology is almost nonexistent and is highly labor intensive. The cleaning method to be used will depend upon the economic and recreational value and the nature of the shoreline. Restoration of sandy beaches may involve the loss of large quantities of valuable sand; the cleaning of rocky shores has had only limited success; and all restoration methods may have severe effects on the littoral ecology. For low-value areas, the natural biological degradation of oil and the cleaning action of the tides may be the best alternative.

Currently, the Federal Water Pollution Control Administration (FWPCA) is developing new methods for processing oil-coated sand. The FWPCA's program holds promise for the development of effective methods for treating contaminated beaches without the loss of valuable sand. However, the problem of restoring other features of the coast remains.

3. Costs

Because of the undeveloped state-of-the-art of the various combatant methods, the indicated costs are not so much quantitative measures as they are comparative measures. The materials and operating costs for combating an oil spill are largest for physical absorption because of the need to collect the spent absorption materials. Mechanical removal, presently applicable only for calm waters, chemical dispersion, and physical sinking have the lowest direct costs; combustion is somewhat higher because residue resulting from the combustion must be collected. We estimate that these direct costs range from $1700 to $4100 for a small (1,000-gallon) harbor spill; from $64,000 to $115,000 for a medium (100,000-gallon) off-shore oil spill; and from $4.5 million to $8.5 million for a large (10,000,000-gallon or 35,000 ton) off-shore spill.

The capital costs for equipment and facilities associated with each of the combatant methods are nearly equal for treating small harbor spills. This cost, about $100,000, is mainly attributed to the cost of the containment boom.

The capital costs for combating off-shore spills (based on estimates of equipment not yet developed) are greatest for physical absorption because of the high costs of large collection equipment to collect spent absorbents. The capital costs for chemical dispersion, physical sinking, and combustion are close to the same, with sinking agents becoming relatively more expensive for larger spills because of the sinking agent storage costs and high costs of large sand dredging equipment. The capital costs for treating off-shore oil spills are estimated to range from $50,000 to $600,000 for a 100,000-gallon spill and from $700,000 to $6,500,000 for a 10,000,000-gallon (35,000 ton) off-shore spill.

C. RECOMMENDED R&D PLANS

The following R&D plans cover specific potential combatant methods which will (a) contain the oil at its source; (b) remove oil from within the containment area, and (c) treat oil that has spread beyond the containment area. Specific combatant methods must be investigated for their technical feasibility and their cost effectiveness compared to the other methods. If feasibility is demonstrated, the development and testing of the method can proceed, the size of each combatant method test to be decided on the basis of oil spill size, location, frequency and severity and combatant method deployment and operating costs.

a. *Towable and static containment devices* for open ocean operation have the highest potential for combating oil spills: they could substantially add to the effectiveness of cleanup methods by reducing the size of the area to be treated. A methodical program to consider floating and submerged, static and mobile, mechanical and pneumatic containment systems should be pursued to ascertain their limitations in the severe environment of the air-sea interface and to select the containment concept most promising for the widest application. The initial work must consist of demonstrating feasibility to resist the effects of waves, wind, and anchoring loads.

b. *Skimming equipment* should be developed to remove the oil from the inside of the containment area; however, the feasibility of mechanically removing oil from the sea surface under open ocean conditions must be assessed. The ability of a skimmer to respond rapidly to the changes in elevation in a wave field and to conform to the surface of the water will be the limiting factor in open ocean operation. Furthermore, concentrating the oil by containing the spill or by "funneling" it into the inlet of the skimmer by means of a towable boom will substantially improve the effectiveness of the method. The major need, therefore, is for an understanding of the tradeoff between sea state, oil thickness, and effects of an oil-water ratio on collection and separation equipments for any single skimming concept as applied to stationary and moving systems.

c. *Chemical dispersants*, although extensively used in combating oil spills, are not well understood in terms of the mechanism of dispersion energies, the resulting concentrations, the effects of the crude oil chemistry, the temperature of the environment, the microbial rate of degradation, and the toxicity. Once a better understanding has been reached, development work should proceed on distribution methods which will result in effective use of chemical dispersants in the open ocean. Chemical dispersants specifically developed for treating oil spills might have higher effectiveness than commercially available dispersants; analyses and laboratory tests are recommended.

d. *Sinking agents* are not well understood in terms of behavior and limitations on oil slick thickness and age or in terms of how the oil-agent combination sinks and how the oil behaves once it reaches the ocean floor. When a better understanding is reached, devices for effectively distributing sinking agents should be developed for open ocean conditions.

e. *Absorbing agents* offer the least ecologically damaging combatant technique (next to skimming equipment), but a careful evaluation of their behavior and effectiveness is needed before equipment for absorbent collection is developed. The absorbing agent and its effectiveness as a function of oil properties and thickness should determine the type of distribution and collection system which would be developed for open ocean conditions.

f. *Microbial degradation* could be accelerated with the addition of phosphorus or nitrogen nutrient compounds to floating or dispersed oil; the feasibility of such an accelerating technique should be investigated and compared to the natural process under various conditions. The nutrients, if used in conjunction with dispersants, may provide further improvements in the rate of microbial degradation. The results of laboratory and field tests should be evaluated and cost effectiveness estimates made for comparison with similar estimates on other methods.

g. *Combustion* methods, augmented by wicking or oxidizing agents, are complex technical processes which must be better understood before a realistic assessment of their value can be made. Analyses and tests are recommended to understand the processes; an evaluation of the existing agents should be conducted to assess their limitations and their potential cost effectiveness in comparison to other methods.

The combatant equipment and methods resulting from the R&D projects should, wherever possible, be tested on accidental spills to determine their performance under actual conditions. It also may be feasible to create oil spills in remote areas for the purpose of testing.

To insure the successful development and utilization of combatant technology, a detailed operations research program is required to define the environment in which combatant methods must operate; assess the operating parameters of the vessels that may be used in cleanup operations; develop meaningful criteria against which equipment can be designed and tested; and provide cost information to make cost-effectiveness decisions on various technical approaches. As information is generated from the R&D work on each combatant method it will be necessary to update the operational and systems analyses.

CHAPTER 25

WASTE WATER TREATMENT METHODS

At present, municipal wastewater is usually given varying degrees of treatment before being discharged into a receiving stream. The primary object of the treatment has been to reduce the amount of oxygen-demanding materials in the water along with reduction of suspended solids and bacteria. There is no doubt that the conventional processes can be made to perform better. The investigation of ways that will improve treatment by conventional methods cannot logically be separated from studies of new processes that may be added after or combined with the conventional processes. It should also be pointed out that processes added after conventional treatment may, for optimal results, require better conventional treatment than is usually available today. All treatment stages must be integrated in such a way as to give the desired quality water as economically as possible.

More and more frequently, the question arises of whether conventional treatment methods will soon be obsolete. There are inherent disadvantages to parts of a conventional system for which dependable and economical substitutes would be welcomed. Some possible substitute processes are being studied. None of these have reached the stage of development, however, where use on a plant scale can be expected in the near future. It is safe to assume that the conventional processes will have widespread application for many years. Designers, however, should take into account the future need for chemical additions at various points within plants and the present availability of processes that can be added after conventional treatment.

Principle of Operation

Conventional treatments are classed as primary and secondary. Primary treatment usually consists of coarse screening or grinding, grit removal by settling, and suspended organic removal by further settling. Removal of pollutants as measured by the 5-day biochemical oxygen demand (BOD_5) is about 35 percent.

Conventional secondary treatment consists of some form of biological oxidation. In the trickling filter process, settled effluent is sprayed over a bed of rocks or other media that accumulates microorganisms on its surface. The mixed biological life utilizes the organic materials in the wastewater for food and thus purifies the water. There is a slow sloughing-off of the organisms into the product water. These solids are removed in a final settler. Some of the product water may be recycled back through the biological bed to increase contact time.

Another common secondary treatment method is the activated sludge process. After primary settling air is bubbled through the wastewater for up to six hours. This causes growth of aerobic bacteria and other organisms which use organics in the wastewater for food. A final settler is used for removal of the organisms. A large part of the biological sludge is returned to the aeration chamber. BOD removal, including the removal by primary treatment, can reach 95 percent. This is slightly better treatment than is possible in a trickling filter.

The biological sludges that result from the oxidation processes and the sludge settled in the primary settler are wastes that must be further treated. A common treatment method is anaerobic biological digestion, followed by drying of the residue. This is then used for land fill or as a soil conditioner or, in some cases, it is burned.

Another secondary treatment method, less effective than the others, is the use of stabilization ponds. There is considerable variation in design and operation of these ponds, but in all designs the water is given a long enough residence time in shallow lakes for oxidation of organic matter to take place. Growths of algae occur in most cases. The algae are a good source of oxygen due to photosynthesis, but they contribute secondary pollution. Their removal, which is difficult, would lower the product-water BOD_5. This removal is, in fact, a subject for advanced waste-treatment study. Stabilization pond treatment is a good example of a method that may be improved by proper combination of a conventional oxidation stage with a new process to give an advanced treatment system.

System Optimization and Process Control

The designing of waste treatment plants has not been carried out in the past in such a way that a given quality water would necessarily be produced at the lowest possible cost. True optimization of plants has not been possible because of a lack of adequate mathematical descriptions for the performance of component parts. Similarly, automatic adjustment of controllable variables in response to changes in flow rate and contaminant concentration was not possible because of lack of mathematical models and measurement techniques for the various parts of the treatment system. Whether optimization and automatic control will have a significant effect on treatment cost or treatment efficiency remains to be seen. They are important enough, however, to warrant exploration, and studies are underway at Cincinnati in this area of work.

As the first step in the study of optimization and control, two principal types of mathematical models are being developed for each wastewater treatment process. Initially, primary and secondary treatment are being studied, but work will be extended to other processes later. The first mathematical model is the pseudo-steady-state model which relates the effluent stream vectors from a process to the influent stream vectors and the decision vector which specifies how the process is to be operated. Stream vectors are a list of parameters which completely characterize the stream, such as volume flow, temperature, and the concentration of up to twenty species of contaminants. The principal classification of contaminants is chemical, that is, carbon, nitrogen, phosphorus, and fixed inorganic matter. The carbon is sub-divided into biodegradable carbon and non-biodegradable carbon. Each class is further sub-divided into settleable and dissolved fractions. For processes which remove dissolved salts, the stream vector also contains the concentration of each ionic specie.

The pseudo-steady-state model is based on 24-hour composite values and is expressed in terms of material balances around control volumes. Stoichiometry, pseudo-rate constants, mixing assumptions, and assumptions about separation of phases such as occurs in settlers or filters are included in the model. Each computed effluent stream vector must be suitable for use as an influent stream vector to any other downstream process.

Models have been developed for the conventional system which included preliminary treatment, primary settling, activated sludge, sludge thickening, digestion, elutriation of sludge, vacuum filtration with addition of chemicals, and sludge incineration. A digital computer program has been developed which contains pseudo-steady-state models for these processes arranged in the most commonly used configuration. Many assumptions not fully justified by engineering investigation were made. Thus, development of the models has drawn attention to the need for additional measurements of process performance and has helped to show what additional data are needed.

A serious limitation of the existing computer program is the assumed arrangement of the processes within the system. An executive digital computer routine is being developed which will allow the performance of any set of processes arranged in any configuration to be computed without rewriting the system computer program.

The ultimate goal of the pseudo-steady-state modeling effort is to find mathematical relationships for the performance and cost of all practical wastewater renovation processes. These relationships must have sufficient validity that they can be used with confidence by plant designers and system analysts to make decisions concerning the best way to construct a system to produce water of any specified quality from any wastewater stream for which the stream vector is known.

The second principal kind of mathematical model is the time-dependent model which computes the process performance as a function of time when the influent stream or the operating decisions, or both, vary with time. One digital computer program has been completed which simulates the time-dependent behavior of the activated sludge process. This computation, which is a numerical integration utilizing the Runge-Kutta fourth-order method, integrates the Monod differential equations which express the growth of three classes of bacteria: heterotrophic carbon users, *Nitrosomonas* which converts ammonia to nitrite, and *Nitrobacter* which converts nitrite to nitrate.

The aerator is sub-divided into ten sub-volumes, any number of which can be used to simulate mixing, and a sludge storage tank is provided in the sludge return line. A surge tank, in which no biological activity is assumed, is also provided upstream from the aerator. Sinusoidally varying flow and BOD concentration will be applied to this system, and the concentration of three classes of microorganisms and three substrates, as well as flow, will be computed as a function of time at each of seventeen stations. Various control schemes will be studied including pumping of sludge from the sludge storage tank to minimize the effect of the varying load on the plant, or the use of the surge tank to smooth out the flow. The biological growth equations were taken from the British literature and a need exists to compare the computed results with the performance of a full-size plant under a load varying with time.

The ultimate goal of the time-dependent model is to study methods for control of the activated sludge process to give optimum performance when the flow and BOD concentration entering the plant vary throughout the day.

Experimental Investigations

Two investigations involving conventional treatment have been undertaken. One of the investigations, involving the contact stabilization method of treatment, was tested on a municipal waste with a large fraction of industrial waste. Contact stabilization is a modification of the activated sludge process that has a very short aeration time for the wastewater. The sludge that is to be returned to the aerator is, however, reaerated first in a separate tank. Using a 250-ml/min pilot system, the process was found to give up to 79 percent BOD removal with an aeration time of 25 min and a sludge reaeration time of 115 min. High product quality was not a primary aim of this work. One of the functions of the system was to supply secondary effluent for other studies.

An investigation, was undertaken to learn whether there could be a substantial saving in treatment cost if an activated sludge process with very short aeration time were combined with chemical treatment for improved removal of biological floc. Phosphorus would also be removed by the process. Short aeration under the conditions proposed produces a poorly settling sludge. A final chemical clarification would be necessary to bring the effluent BOD into the range obtained by conventional operation. From experimental results, it was concluded that aeration could be reduced to 1 hour compared to a standard of 6 hours. A lime dose of 200 mg/1 as $Ca(OH)_2$ and an alum dose of 20 mg/1 were shown to be required for proper settling. Although the capital cost of the combined system was slightly less than for the standard activated sludge process, the operating cost, including amortization of equipment, was slightly higher. There does not appear to be a clear-cut advantage to the combined system except for improved phosphate removal that results from chemical addition.

SUSPENDED SOLIDS REMOVAL

Secondary effluent contains materials ranging in particle size from easily-visible, insoluble solids down through colloids to definitely soluble molecules and ions. It is impossible to precisely classify all these materials as either insoluble or soluble. There is, however, a significant amount, often 10 to 20 mg/l, of the organic material and some of the inorganic material in secondary effluent that can be removed by the various processes used to carry out clarification. These materials are usually called suspended solids. The smallest particle to be removed varies with the process. A simple straining of the water might only remove particles greater than 10 microns. The combination of flocculation, sedimentation, and filtration removes most particles down through the colloidal range and even removes some soluble materials by adsorption on the floc.

The suspended solids are detrimental for a number of reasons. They are a major source of the remaining organic pollution in secondary effluent, that is, they constitute a major part of the BOD in the water. Their removal would, therefore, significantly reduce the pollutional load on a receiving stream. The suspended solids interfere with or prevent the operation of a number of other advanced waste-treatment processes. These materials also are undesirable or prohibitive in most envisioned direct reuses of the water.

Comparatively little work was done on removal of suspended solids from secondary effluent before 1965. Since then, however, several investigations have been completed and a number of new ones initiated. Included are processes involving only mechanical separation and processes combining chemical flocculation and mechanical separation.

MICROSCREENING

Principle of Operation

Microscreening is a form of simple filtration. The minimum particle size removed depends upon screen mesh size and pattern of weave. Because of the layer of material that collects on the screen during use, the minimum particle size is actually somewhat less than the size of the screen openings.

Experimental Investigation

An investigation was made using a Microstrainer* to filter Lebanon, Ohio, secondary effluent (1). This particular device, shown in Fig. 25-1, uses a screen in the form of a partly submerged rotating drum or cylinder. Water flows continuously by gravity through the submerged portion from inside the drum to a clear-water storage chamber outside the drum. Because of the rotation of the drum, the filtering portion is constantly renewed to maintain a practical filtration rate and the former filtering portion is raised above the water for cleaning. Cleaning is carried out by backwashing with sprays of product water. Backwash requirement is 5 percent or less of the feed volume. An ultraviolet light is used to control growths on the screen.

The drum of the unit tested was 5 ft in diameter and 1 ft in width. This gave about 15 ft^2 of screen surface of which about 9 ft^2 was submerged. With this unit it was possible to treat up to about 60 gpm when suspended solids were low. Since pressure-drop rather than flow rate was controlled, the flow rate was not constant. Flow rate is greatly affected by the amount of suspended solids.

**Equipment loaned by Glenfield & Kennedy, Inc., King of Prussia, Pa.*

Fig. 25-1 Pilot microscreening equipment at Lebanon, Ohio. Wastewater enters the compartment on the left and flows by gravity radially outward through the rotating drum screen in the center to the compartment on the right. At the top of the drum, streams of product water are pumped downward on the screen to backwash solids off. The backwash stream is taken off as waste. Here a sample of backwash water is being taken for analysis.

Table 25-1 Average Contaminant Removals Obtained with a Microstrainer

Screen Openings (microns)	Percent Removal			
	Suspended Solids	Biochemical Oxygen Demand	Total Organic Carbon	Turbidity
23	89	81	30	76
35	73	61	30	60

Suspended solids, BOD, total organic carbon (TOC) and turbidity-removal data were taken using two types of screens, one having 23 micron openings and one having 35 micron openings. The average removals are shown in Table 25-1. Although the finer screen has the better removal capability, as one would expect, both screens remove a considerable amount of the suspended solids. The high degree of BOD removal shows that most of the biodegradable material was insoluble. TOC measurements are subject to considerable error when suspended material is present, but these results show roughly what gross organic removal might be expected.

Cost estimates were made for treating secondary effluent by microscreening. For a 10-mgd plant the total operating cost, including amortization of equipment, should be about 1.5 /1,000 gal. The type of screen will have only a small effect on cost.

DIATOMACEOUS EARTH FILTRATION

Principle of Operation

Diatomaceous Earth (DE) filtration is a form of mechanical separation that uses a layer of powdered filter aid, diatomaceous earth, built up on a relatively loose septum to screen out the suspended solids. As filtration proceeds, the filter-aid layer will become clogged with deposited solids. To prevent a rapid build-up of pressure, more filter aid must continuously be added to the feed so that a porous cake is produced. This filter aid is called body feed. Eventually the pressure becomes too great to continue operation. At that time, the filter must be backwashed and a new precoat applied.

The minimum size particle that can be removed is related to the size of openings between filter-aid particles which, in turn, is related to the particle size of the filter aid. Since pressure drop increases as filter-aid particle size decreases, there is a practical lower limit. Generally, the minimum size particle that can reasonably be removed is about 0.5 to 1.0μ. By changing the zeta potential of the filter aid through treatment with alum or certain polyelectrolytes, it is possible to improve somewhat the removal capability.

Experimental Investigations

Pomona Pilot Scale Work - A study of diatomaceous earth filtration was made at the Pomona Research Facility. Both a vacuum filter and a pressure filter were tested. The vacuum filter had 8 flat-leaf septa, vertically suspended and covered with polypropylene filter cloth. Filter area was 120 ft.^{2} The pressure filter also had vertical filter leaves. The filter area was 180 ft.^{2}

Data were obtained on both filters while filtering secondary effluent over a range of filter rates, body-feed concentrations, and filter-aid types. Although each filter-aid type used has a wide range of particle sizes, the manufacturer* indicates that 50 percent by weight will be less than 31 microns for Celite 545, 15 microns for Celite 503, 11 microns for Hyflo Super-Cel and 7 microns for Standard Super-Cel.

Results for the vacuum-filter tests are shown in Table 25-2. These results indicate the run-length that might be expected for a given set of conditions. As one would expect, longest runs were obtained with the coarsest grade filter aid. Unusually short runs were obtained with Celite 503 because of the low ratio of body feed to turbidity. Turbidity removal for all runs averaged roughly 85 percent. Data from runs not shown in Table 25-2 indicated for Celite 545 at 0.52 gpm/ft^2 a chemical oxygen demand (COD) decrease of about 21 percent.

Estimation of the minimum cost of diatomaceous earth filtration requires optimization with respect to body feed and flow rate. A rough estimate of 5¢/1000 gal. for a 1-mgd plant has been made, however, for vacuum filtration at 0.5 gpm/ft.^{2} using Celite 545. This figure does not include cost of disposal of sludge nor does it consider recovery of filter aid from the sludge. For large installations, filter-aid recovery might provide a significant saving.

Pressure filter results are shown in Table 25-3. Run-lengths are generally longer than for the vacuum filter, primarily because it is possible to operate to considerably higher pressure drops. Turbidity removal was poor compared to the vacuum filter. The reason for the difference is not definitely known. Organic and biological removals were checked on some runs. COD reduction of about 13 percent was observed using Celite 545. For the same filter aid, average coliform reductions of about 44 percent and total bacteria reductions of about 83 percent were observed. Hyflo Super-Cel gave a coliform reduction of 53 percent and total bacteria reduction of 69 percent. Although the biological results must be considered to show only very approximately the removals that may be expected, they do demonstrate clearly that significant numbers of bacteria will remain after diatomaceous earth filtration.

Table 25-2 Vacuum Diatomaceous Earth Filtration of Secondary Effluent

Flow Rate (gpm/ft^2)	Filter Aid	Body-feed Conc. (mg/l)	No. of Runs	Run-length (hrs)	Turbidities (JTU)* Feed	Turbidities (JTU)* Product
0.53	Celite 545	42	5	19.6	5.5	0.8
0.75		33	3	10.7	5.2	0.8
1.0		19	1	5.4	4.4	0.4
0.51	Celite 503	36	2	6.0	10.5	3.3
0.75		29	1	3.0	9.2	0.7
1.0		21	1	1.2	8.0	0.75
0.54	Hyflo	35	6	12.7	4.6	0.85
0.75	Super-Cel	21	2	4.4	5.8	0.8

Precoat - 0.1 lb/ft^2

Headloss at end of run - 18 in.Hg

*Jackson Turbidity Units

Johns-Manville Products Corp., New York, N.Y.

Table 25-3 Pressure Diatomaceous Earth Filtration of Secondary Effluent

Flow Rate (gpm/ft^2)	Filter Aid	Body-feed Conc. (mg/l)	No. of Runs	Run-length (hr)	Turbidities (JTU)* Feed	Product
0.50	Celite 545	50	1	>50.0	8.2	3.1
0.75		19	2	24.2	5.7	2.5
0.81		42	2	28.4	8.3	3.9
1.0		20	2	7.3	6.4	2.1
1.0		45	3	31	7.5	3.0
0.76	Celite 503	18	1	14.5	7.0	4.0
0.78	Hyflo	21	1	22.3	8.1	4.9
1.2	Super-Cel	29	1	9.7	6.5	3.8

Precoat - 0.1 lb/ft^2

Head loss at end of run - 35 psi

*Jackson Turbidity Units

The cost of pressure filtration is estimated to be very nearly the same as for vacuum filtration. Although the capital cost is greater for pressure filters, the longer runs that can be obtained compensate for the difference.

Some work was done with the vacuum filter on secondary effluent that had first been passed through granular activated carbon. Although the primary purpose of the granular carbon is to remove soluble organic materials, it also acts as a filter for the larger suspended solids in the secondary effluent. As a result, carbon-treated effluent has very low turbidity. Filtration results are shown in Table 25-4. Because of the low feed turbidity, no body feed was used. Run-lengths were considerably greater than for untreated secondary effluents. Flow rates up to 2 gpm/ft.^{2} appear practical. Turbidity removals were, with one exception, less than 50 percent. The carbon treatment apparently removes much of the larger particle size turbidity, leaving a large fraction of material that is not removed by the filter aids commonly used for water. Bacterial removals were also poor. Celite 545 showed almost no removal of bacteria. Data from one run, using Hyflo Super Cel, showed 60-percent removal of coliforms and 43-percent removal of total bacteria. Because of the small consumption of filter aid and the high filtration rates that can be used, the cost of filtering carbon column effluent should only be 1.5 to 2¢/1000 gal including amortization.

Lebanon Pilot Operation - Although a detailed study has not been made, some operating data were obtained on a pressure filter producing water for an electrodialysis study at the Lebanon, Ohio Advanced Waste Treatment Pilot Plant. The filter had 96 ft ^{2} of filtering area and was operated at a rate of about 0.7 gpm/ft ^{2} to a maximum pressure drop of 16 psi. The filter aid was Celite 503 that had been treated with 7.5 percent-by-weight alum and neutralized with soda ash. The secondary effluent suspended solids concentration was usually slightly higher and more variable than at Pomona. As a result, a higher body-feed concentration was required. To allow a margin of safety for upswings in turbidity, a body feed of 80 to 100 mg/l was usually employed. Occasionally, runs of more than 30 hours were observed; runs were frequently between 4 and 8-hrs duration.

The product turbidity was found to vary with water temperature. In the winter,

Table 25-4 Vacuum Diatomaceous Earth Filtration of Carbon Column Effluent

Flow Rate (gpm/ft^2)	Filter-aid	No. of Runs	Run-length (hrs)	Turbidities (JTU)* Feed	Product
1.5	Celite 545	3	33	1.11	0.72
2.0		2	31	1.24	0.77
1.0	Hyflo	3	22	0.80	0.41
1.5	Super-Cel	3	15	0.43	0.21
1.0	Standard Super-Cel	1	10	1.05	0.65

Precoat - 0.1 lb/ft^2

Body feed - none

Headloss at end of run - 18 in.Hg

*Jackson Turbidity Units

with water temperatures as low as 11°C, the product water usually had a turbidity of about 1 JTU. In the summer, at temperatures near 25°C, product turbidities were usually 0.1 JTU or less. The reason for this variation is not clear, but is probably related to differences in the biological treatment.

Because of the greater consumption of filter aid at Lebanon, the cost of treatment would be several cents per 1,000 gal. higher than estimated from Pomona results.

CHEMICAL CLARIFICATION

Principle of Operation

Chemical clarification usually includes a coagulation step, a flocculation step, a sedimentation step, and a filtration step. Coagulation or destablization of colloidal suspensions is achieved by addition of chemicals which are rapidly mixed with the water. The formation of small floc particles results. Further slow stirring allows floc particle size to increase by agglomeration. This is the flocculation step. Usually the floc is at least partly composed of a hydrous oxide that has a slight positive charge. Metals whose oxides are involved are usually aluminum, iron, or in the case of lime treatment, magnesium. The floc particles have ability to aggregate colloidal particles and adsorb or entrap them. The mechanisms involved are not completely known. There is a lowering of the usually negative zeta potential of the colloidal particles that would tend to destablize colloidal dispersion. There is also an attraction between the slightly negatively charged colloid particles and the slightly positive floc particles. The result, when proper chemical doses are used, is a very effective tying-up of suspended solids into floc particles, most of which are large enough to settle well.

Most of the flocculated material is removed by sedimentation. The water is allowed to flow slowly enough through a basin that gravity settling occurs. In some cases, water of good clarity can be obtained by settling alone. Usually, however, further removal of floc particles by filtration is necessary.

Filtration of chemically clarified water is usually carried out in beds of porous media such as sand or coal. A combination of straining, sedimentation, and further flocculation in the filter contributes to removal of most of the floc remaining after the sedimentation step. In the most common type of filter, the sand filter, much of the removal is by straining at the sand surface. As the floc layer builds up, pressure drop over the filter increases. When the pressure drop reaches a predetermined value, feed-water flow is stopped and clear water is passed backward through the filter at a high enough rate to fluidize the sand and remove the collected floc. If a larger-size medium is used so that the floc can penetrate some distance into the filter, the other mechanisms become more important. Because floc retention occurs in the volume of the filter rather than just near the surface, the pressure-drop build-up is slowed and run-lengths increased. By constructing the filter of two or more layers with each layer having a smaller particle size than the one above, and with the top layer of sufficient particle size that little straining occurs, it is possible to take advantage of all three mechanisms. Filters built in this manner are called multi-media filters and are sometimes used to carry out all three steps of the clarification process. Coagulating chemicals are mixed with water, which is immediately applied to the filter. Although floc size is initially too small for much removal to occur at the filter surface, the size increases sufficiently during passage of the water through the filter that the suspened solids are effectively strained out and stored in the interstices of the media.

When lime is used in chemical treatment,softening or removal of calcium and magnesium occurs along with clarification. Lime adds hydroxyl ion to the water that reacts with bicarbonate ion, forming carbonate ion. Calcium is then precipitated as calcium carbonate. The degree of calcium removal depends on pH and is maximized at a pH of about 10. If the pH is further increased to about 11 with additional lime, magnesium precipitates as magnesium hydroxide. At this pH there is not sufficient carbonate to precipitate all the calcium added with the lime. Additional carbonate must be added in one of several ways. When this is done, the pH is again reduced to about 10. Operation for both magnesium and calcium removal requires two separate reaction chambers and is referred to as two-stage treatment. Partial softening obtained by raising the pH high enough for only calcium removal is one-stage treatment.

The sludge from a lime clarifier contains a large fraction of calcium carbonate. This material forms a crystalline precipitate that can be carried out the overflow weir of a settler unless the particle size is sufficiently large to hasten settling. To increase particle size,it is common practice to recirculate sludge back into the part of the clarification system where new floc is being formed. This is referred to as solids-contact operation. It may be carried out by pumping settled sludge back into the clarifier or by designing the equipment so that some of the sludge in internally recirculated. Various designs are available from manufacturers of clarification equipment that make use of solids-contact.

Experimental Investigations

Laboratory Studies - Initial studies of the effectiveness of chemical clarification are usually made using jar tests. These are breaker-scale batch tests. Chemicals are added to a beaker of water, the water is stirred rapidly for several minutes and then more slowly for 15 to 30 minutes to allow flocculation to occur. The flocculated material is then allowed to settle. The clarity and other characteristics of the supernatant liquid can be observed with or without filtration. These tests give an indication of the chemical doses and some of the conditions that will be required to obtain good clarification in larger-scale equipment.

Jar tests were conducted on wastewaters from a number of locations before considering work on larger-scale equipment. Generally these tests show that both poorly and well-treated secondary effluent can be chemically treated with lime or alum to give turbidities of less than 1 JTU after filtration. Alum doses as low as 150 mg/1 and lime doses as low as 50 mg/1 have been used to achieve that turbidity level. A somewhat higher dose for lime would, however, probably be used in most cases to achieve removal of certain pollutants besides suspended materials.

Clarification Pilot Plant Studies - Several pilot-scale investigations have been undertaken. The most extensive of these is being carried out at the Lebanon Pilot Plant (2). The equipment shown in Figs. 25-2 and 3 is used to study a range of operating variables affecting coagulation, flocculation, settling, and filtration. It is also possible to bypass the flocculator and settler for study of the mode of operation in which flocculation occurs in the filter itself. Three types of filters are included: anthracite coal, sand, and dual-media consisting of coal over sand. Filter-media characteristics are shown in Table 25-5. Feed and product turbidities are automatically monitored and recorded. Pressures at various points in the filters are also automatically recorded.

The use of alum as a flocculant has been studied in detail, both alone and with activated silica as a floc strengthener. During this work, the flow rate through the settler was varied from 300 to 900 gpd/ft ^{2} and the alum dose was varied from 0 to 500 mg/1. Filters were usually operated at 3 gpm/ft.^{2}

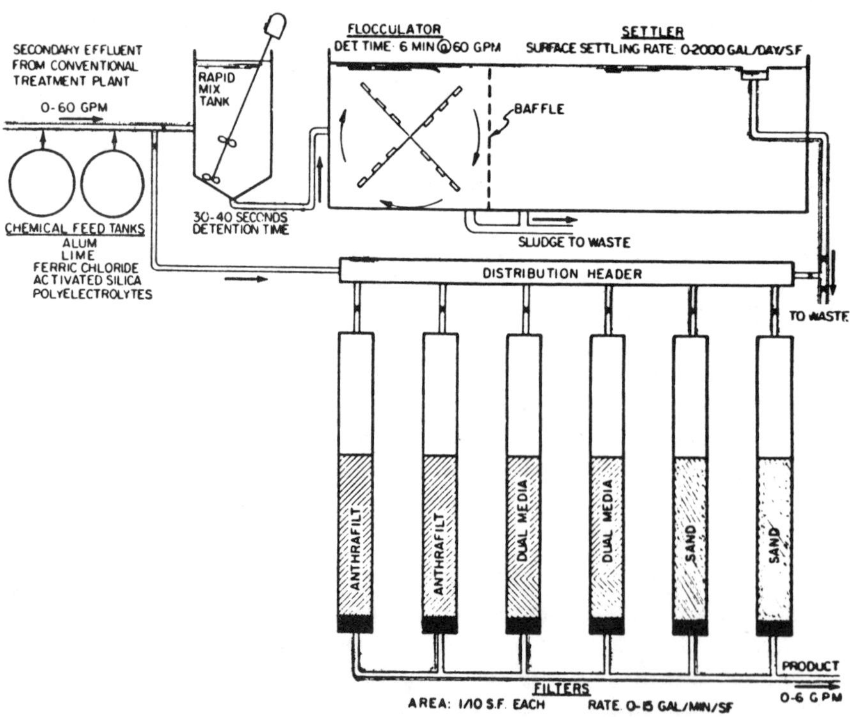

Fig. 25-2 **Suspended solids removal study, Lebanon, Ohio**

Fig. 25-3 Solids removal study at Lebanon, Ohio. At the top of the picture are containers for holding and mixing of coagulating chemicals. Behind the containers are a flocculator and settler. Efficiency of suspended solids removal can be investigated for different chemicals over a range of chemical doses, flocculator speeds, and settler flow rates. At the bottom are six filters that receive water from the settler. Various filter media can be tested for pressure drop characteristics and solids removal capability over a wide range of flow rates. Taps in the walls of the filters are for measuring pressures down through the filter bed.

Some typical run-length results for dual-media filters using alum alone are shown in Fig. 25-4. All runs involved were terminated when product turbidity reached 1.0 JTU. The results give an idea of the maximum run-lengths and the chemical dose at the maximum that might be expected for the stated conditions. Similar relationships were found for the filters using coal alone. Because of the polishing effect of the sand at the bottom of the dual-media filters, these filters

Table 25-5 Filter Media Characterization Data

Filter	Depth (inches)	Effective Size (mm)	Uniformity Coefficient	Porosity (%)
Single Medium				
Sand	24	0.45	1.33	40
Coal	24	0.75	1.50	50
Dual Media				
Sand	6	0.45	1.33	40
Coal	18	0.90	1.75	50

generally gave better product quality than just coal. It is obvious from the two plots in Fig. 25-4 that the conditions in the flocculator-settler have a large effect on filtration results. An increase in settler overflow rate increases the amount of floc in the filter feed and decreases filter run-length. The amount of alum required for maximum run-length was, however, roughly cut in half at the higher settler rate. For the conditions investigated, the net production of the dual-media filters was 34 to 214 gal/ft of headloss. Backwash water ranged from 2 to 20 percent of the feed. Similar results were obtained in the coal filters.

The use of activated silica along with the alum decreased the tendency for turbidity breakthrough and increased the length of filter run for the dual-media filters. Using 2 mg/l of silica, the lowest dose tested, the average length of filter run was almost 3 times that without silica. For this silica dose, some runs reached the maximum operating head loss of 8 ft of water and had to be terminated before the 1-JTU turbidity criterion was exceeded. Above 2 mg/l of silica, the fraction of runs terminated by headloss increased as silica dose increased. The average run length decreased.

For a 1-JTU turbidity criterion, the 0.9-mm coal and the 3-gpm/ft ^{2} filtration rate initially chosen for dual-media-filter tests do not give maximum water production per filter run when silica is used to strengthen the floc. Study of other coal sizes indicates that filter production can be increased by increasing coal size to about 1.1 or 1.2 mm. Investigation of filtration rates of 3, 4, and 5 gpm/ft ^{2} showed that 4 gpm/ft ^{2} gives maximum production per filter run. This was found to hold over a coal particle-size range from 0.9 to 1.8 mm.

Correlation of run-length with alum dose was not possible for the sand filters. There was, however, a more definite relationship between solids load to the filter and run-length than for the dual-media and coal filters. This relation is shown in Fig. 25-5. Even without the use of silica, sand-filter runs were terminated at a head-loss of 8 ft of water rather than by turbidity breakthrough as in the case of the other two types of filters. At 8 ft of head, turbidity was consistently less than 1 JTU. Net production from the sand filters ranged from 4 to 54 gal/ft of head-loss. Backwash requirements were from 3 to 28 percent of the feed.

Pressure readings taken at various points in the filters showed that good distribution of floc throughout the filter volume took place in the coal and dual-media filters. For the sand filters, most of the pressure-drop took place in the first few inches as expected.

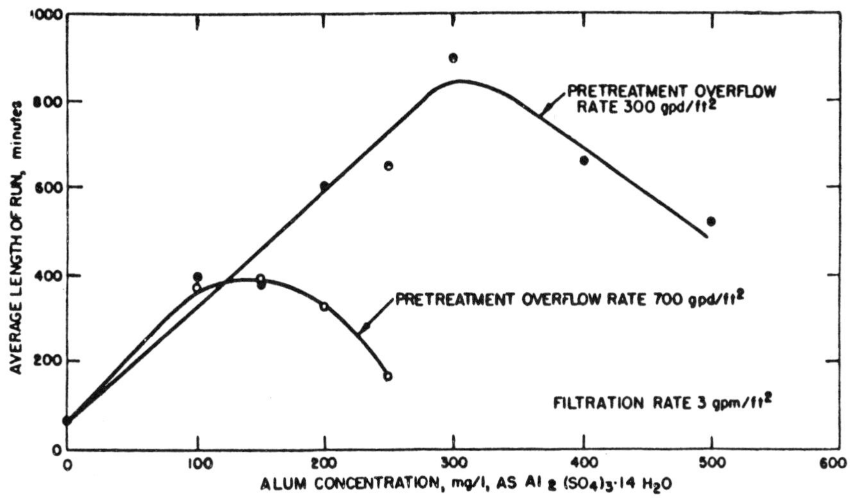

Fig. 25-4 **Effect of alum dosage on length of run for dual media filters.**

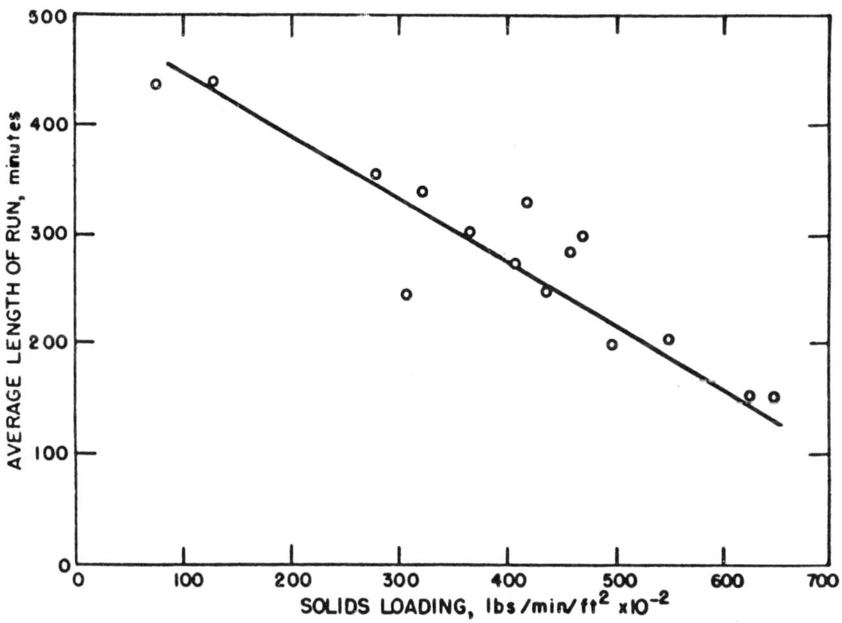

Fig. 25-5 **Effect of solids loading on sand filter performance at 3gpm/ft.^{2}**

An important aspect of clarification processes, especially when dealing with wastewater, is the degree of removal of microorganisms. Table 25-6 shows bacterial-count data for some pilot runs. Removal for all types of filters was excellent.

A tentative cost estimate for alum treatment of secondary effluent in a 10-mgd plant has been made. Using a settler overflow rate of 700 gpd/ft ^{2} and an alum dose of 150 mg/1, the cost should be about 5¢/1,000 gal. If it were desired to add enough alum to obtain 95 percent phosphate removal, about 300 mg/1, the cost would increase to about 8¢/1,000 gal.

Table 25-6 Bacterial Removal from Secondary Effluent by Chemical Coagulation and Filtration

Sample Stream	Secondary Effluent	Settled Product	Filter Product: Coal	Filter Product: Dual-Media	Filter Product: Sand
Average Total Plate - counts per 100 ml	72×10^6	8.0×10^5	2.2×10^5	2.6×10^5	2.3×10^5
Average Fecal Streptococcus - counts per 100 ml	7.5×10^3	200	29	7	2

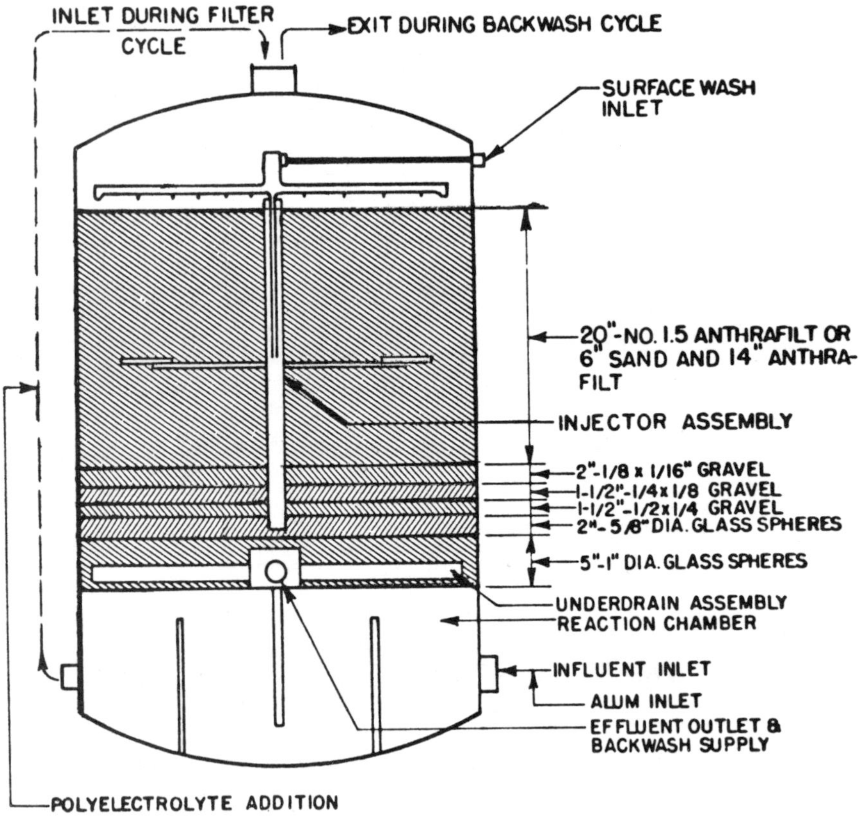

Fig. 25-6 Schematic arrangement of pressure filter unit.

Pressure Coal Filter - A short-term investigation was made of a pressure filter.* The equipment is shown in Fig. 25-6. The filter is designed to use anthracite coal, Number 1.5 Anthrafilt,** as a filter medium and to operate at a very

**Equipment loaned by Zurn Industries, Inc., Erie, Pa.*

***Palmer Filter Equipment Co., Erie, Pa.*

high rate of 10 gpm/ft ^{2} up to a pressure of about 25 in. Hg. Essentially no flocculation time is provided and there is no settler. Instead, floc is retained in the filter bed. During the study, 20 in. of coal and 14 in. coal underlaid by 6 in. of 0.45 mm sand, were used in the filter. Under the filter media there was a bed of graded gravel. Rates of 5 gpm/ft ^{2} and 10 gpm/ft ^{2} were investigated. Most of the work was done with 10 to 15 mg/l alum as the flocculant and up to about 2.5 mg/l of a cationic polyelectroyte Primafloc C-7,*** as a flocculant aid. Another cationic, one anionic, and one nonionic polyelectrolyte were tested, but did not give as good product quality as the C-7.

Some typical data using coal alone in the filter and a rate of 10 gpm/ft ^{2} are shown in Fig. 25-7. It is quite obvious that the polyelectrolyte significantly improves product quality. This is considered to result from a strengthening of the floc. The improvement in product quality is obtained at the expense of a higher rate of pressure build-up. Reducing the flow rate to 5 gpm/ft ^{2} roughly doubled the length of filter run, based on 25 in. Hg pressure drop as the cutoff point. The average run-length at 10 gpm/ft ^{2} was about 2.5 hours.

The filter was run using dual media to see if the amount of polyelectrolyte could be decreased. It was felt that the fine particle size of the sand should prevent turbidity breakthrough with a weaker floc than for the case of coal alone. Surprisingly, it was not possible to reduce the polyelectrolyte dose much below 2.5 mg/l without getting breakthrough. At similar chemical doses, the average run-length for the dual-media filter did not appear to be greatly different from the filter using only coal.

Because of the short filter runs compared to operation with a flocculator and settler, backwash requirements were high. For a 2.5-hr run, 6.6 percent of the feed was required; for a 1.5-hr run, 11 percent was required. Mudball formation was a problem.

Total cost of operating a 1-mgd plant, including amortization of equipment, is estimated at about 7¢/1,000 gal. This is based on a 10-gpm/ft^2 rate and polyelectrolyte and alum dose of 2.5 and 11 mg/l. The cost could be reduced significantly if the dose of polyelectrolyte could be reduced.

Lime Clarification Pilot Plant Studies - A 75-gpm, single-stage, lime-clarification system has been installed at the Lebanon Pilot Plant. The primary purpose of the system is to supply clarified water for other pilot plant studies using secondary effluent for feed. The equipment is shown in Fig. 25-8. It consists of an upflow solids-contact type flocculator-settler combination and two dual-media filters in parallel. Sludge from the settler is gravity thickened to about 10 percent solids by weight and conveyed to drying beds.

Fig. 25-9 shows the interior of the clarifier. Feed water and lime are mixed in the innermost compartment. This compartment contains a dilute slurry of sludge that is continuously being pumped up from the bottom of the clarifier by an agitator. Calcium carbonate and other precipitates begin to form using the sludge particles as "seeds" (solids-contact). The water then flows downward in the space between the inner cylinder and the cone where particle size of the suspended material increases. Most of the solids settle out; others are carried a short distance up with the water in the outer compartment forming a sludge blanket. This blanket helps trap small particles and allows them to agglomerate to settleable size. The water travels upward and out through a ring of holes near

***Rohm & Haas Co., Philadelphia, Pa.*

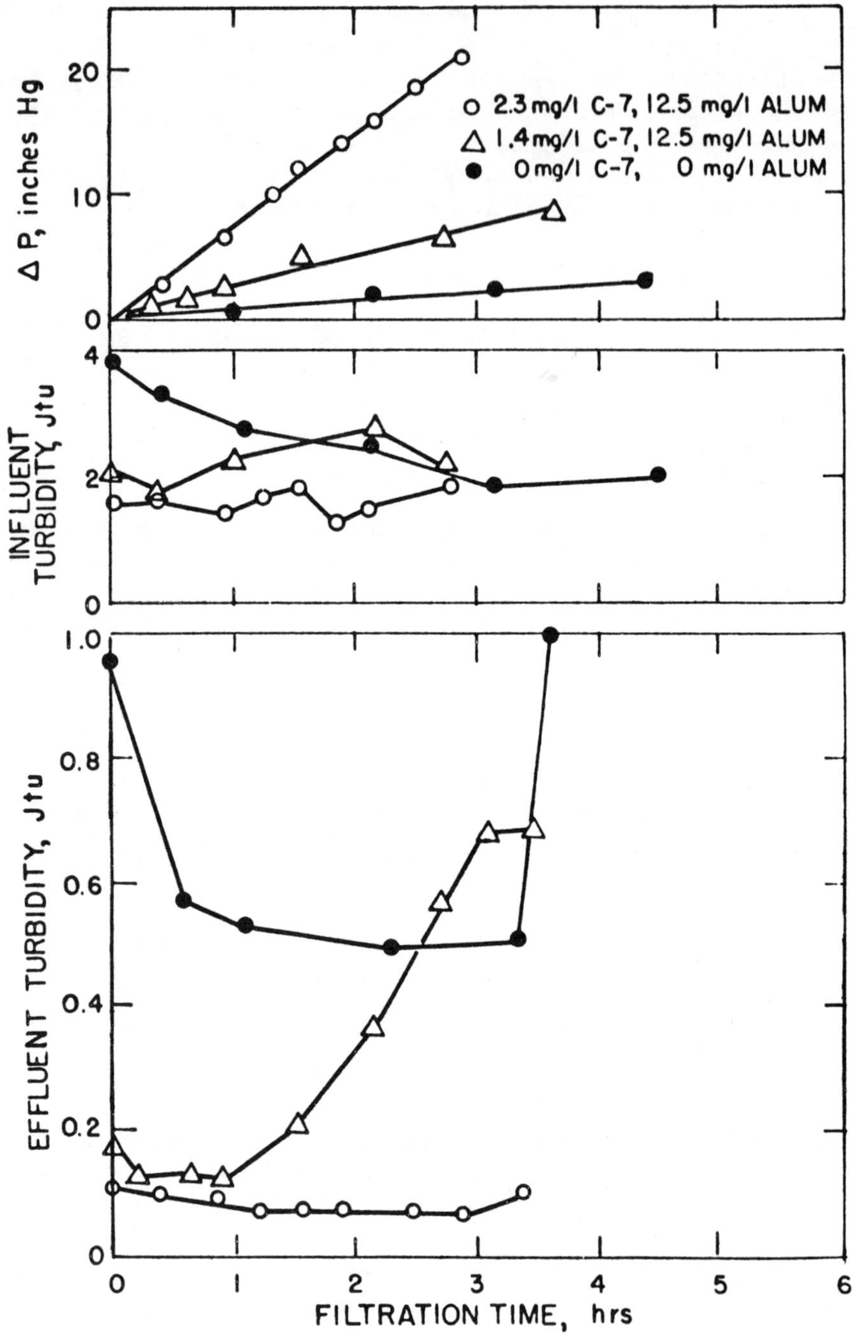

Fig. 25-7 **Pressure coal filter results.**

the top of the clarifier. Upflow rate at the water surface is 1 gpm/ft ^{2} A water-pulse system is used to move sludge out to a sludge collector ring. A scraping mechanism could be used equally well. Sludge is removed from the clarifier as a 1- to 2-percent slurry. The height of the sludge blanket can be controlled to some degree by the rate of sludge removal.

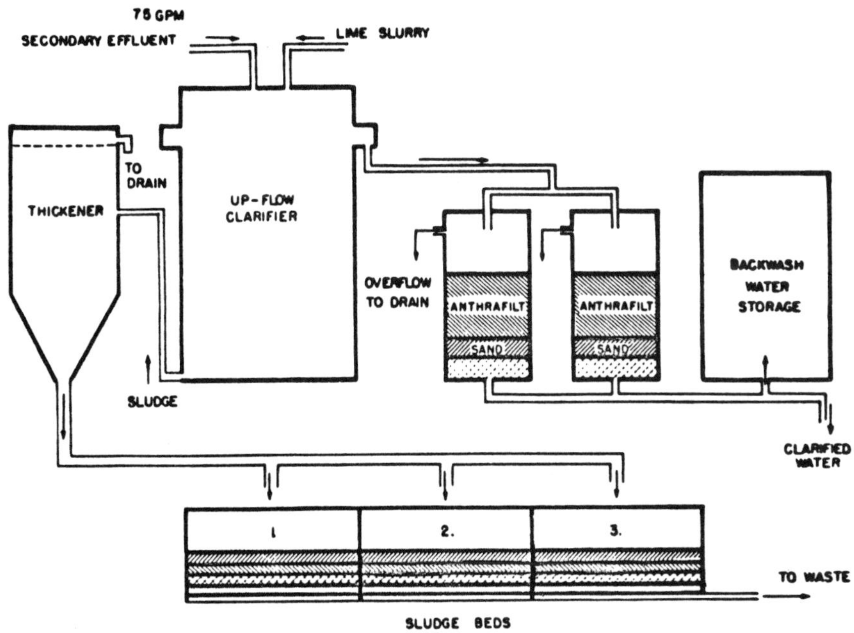

Fig. 25-8 Lime clarification system at Lebanon Pilot Plant.

The dual media filters contain a 6-in. layer of sand overlaid by an 18-in. layer of anthracite coal. The sand has a particle size range of 0.45 to 0.55 mm. The coal is No. 1 Anthrafilt.*

It is intended that the clarification system will be run at the lowest pH that will provide high clarity water. Preliminary results with secondary effluent indicate that a pH of 9.5 will produce water of less than 0.5-JTU turbidity from the filters without use of a flocculant aid. At this pH, phosphate is reduced to about 3 mg/1 or less as PO_4. Calcium is reduced from about 100 mg/1 to about 50 mg/1, but there is no reduction in magnesium concentration. A small amount of data at pH values above 9.5 indicate improved phosphate removal.

ORGANIC REMOVAL

Although the biological sewage treatment methods are used primarily to remove organic materials from wastewater, they are not capable of removing the last traces of these materials. Besides suspended solids, there remain measurable amounts of refractory soluble organics, or more accurately, materials not removed by clarification. It is the removal of these organic materials that we are concerned with here. The composition of these materials is largely unknown. Certain classes of materials have been identified, but a complete analysis has not and may never be made. Although not easily biologically oxidizable, the soluble organics do exercise an oxygen demand in a receiving stream. This in itself is a good reason for removing them, especially when a large fraction of the stream flow is of wastewater origin and has a low dissolved oxygen content. When traces

**Palmer Filter Equipment Co., Erie, Pa.*

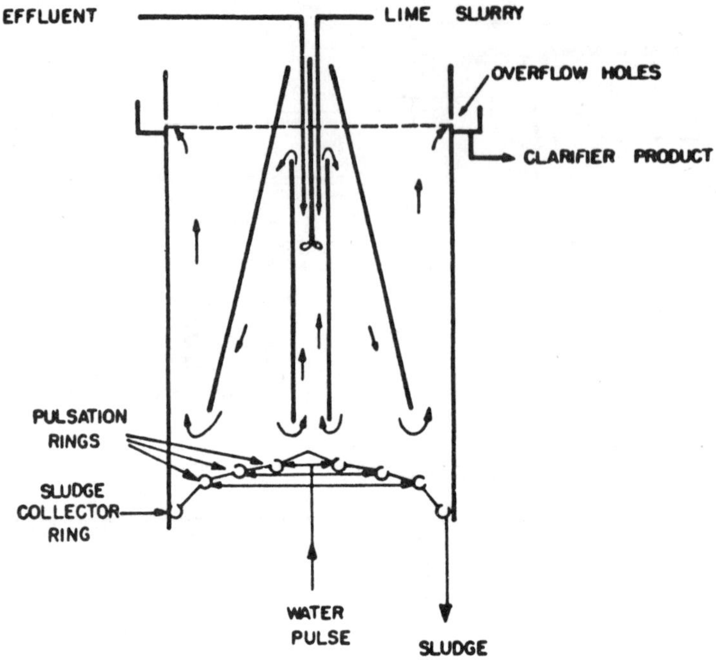

Fig. 25-9 **Details of lime clarifier at Lebanon Pilot Plant.**

of certain synthetic or industrial wastes are present, fish and other animal life can be killed, weakened, or at least made unfit for human consumption. When the wastewater is to be reused for human consumption, either by direct recycle or after dilution in a receiving stream that is largely of wastewater origin, the organics must be removed to prevent taste, odor, and color. Although the toxicological effect on humans has not been established, removal of the organics would be the only safe way to assure against long-term health effects.

While a high degree of removal of soluble organics is desirable for direct reuse of wastewater, lesser removal when there is a very large amount of dilution water or removal of only certain contaminants might be all that is necessary. Processes for organic removal do not, therefore, always have to remove a large fraction of the gross organics. Partial and selective processes also may be of value.

ADSORPTION ON GRANULAR ACTIVATED CARBON

Principle of Operation

Many solids have the ability to attract to their surface soluble materials from solution. The phenomenon is called adsorption. By giving a solid a very porous structure, a very large surface per unit volume results. Certain porous solids are, therefore, excellent adsorbents.

Activated carbon is a porous material having a high adsorption capacity for the soluble organics in wastewater. This material can be contacted with the water either as a fine powder or as granules having a particle size usually in the range of 8 to 40 mesh. The use of granular carbon has been found convenient in treatment of wastewater. Vessels can be filled with the carbon and the water then can be passed through these fixed beds assuring good contact between the water and solid. As the water passes through the beds, the organic content decreases and the carbon becomes loaded with the removed contaminants.

Eventually the carbon nearest the feed point becomes loaded to capacity and must be replaced by fresh or reactivated carbon. The replacement carbon would not actually be added near the water feed point, but would be added to the system where the treated water exits. This countercurrent method of contacting solid and liquid gives the most efficient utilization of carbon. It can be accomplished in a number of ways, the simplest being the use of multiple contractors or columns in series. In this arrangement, the first contactor of carbon is replaced when exhausted and the flow of water is changed to make that contactor last in the series.

To make the cost of operation reasonable, exhausted carbon must be reactivated and reused. Thermal reactivation is presently the only practical method. Using a multiple-hearth furnace or rotary kiln, the exhausted carbon is heated to about 1700°F in a steam-air atmosphere. This procedure burns off the adsorbed organics and restores the carbon to nearly virgin condition.

Experimental Investigations

Pomona Pilot Treatment Plant - Extensive work (3,4,5,6,7,8) has shown the technical feasibility of using granular activated carbon to remove organic material from wastewater. As a result of these laboratory and small pilot-scale studies, a large pilot plant was constructed. The site is the Pomona Research Facility. The pilot plant includes a 200-gpm carbon treatment plant and reactivating furnace.

The treatment plant (9) uses four carbon columns in series with the water flowing downward in each column. Details of column construction are shown in Fig. 25-10. Each column is 6 ft in diameter and about 16 ft high. Construction is of steel with a tar-epoxy lining. The perforated screen on which the carbon is supported is of stainless steel. Maximum pressure is 50 psig. Provision is made for backwash and surface wash. Each column holds 6650 lb of water-treating carbon.* That amount of carbon fills the column to a depth of 9.5 ft. Characteristics of the carbon are shown in Table 25-7.

A 30-in. diameter six-hearth furnace** is used to reactivate the spent carbon. The furnace is equipped with gas burners and points of injection for steam and excess air. Maximum temperature is about 1700°F. The contacting system and the furnace are shown in Fig. 25-11.

Chlorinated secondary effluent with no further pretreatment is fed to the system of carbon contactors. The elimination of a sand filtration step was not attempted in earlier carbon work. The low suspended solids content of the Pomona effluent, however, makes filtration unnecessary. Because there is no provision for suspended solids removal from the feed stream, there is sufficient buildup of solids on the first contactor to make daily backwash necessary. Backwash is carried out with secondary effluent at a maximum rate of 200 gpm using a total flow of about 5000 gal. Operation of the system is continued until there is a significant breakthrough of soluble organics in the product water. The concept of breakthrough is somewhat nebulous here because there is at all times

---

**Pittsburgh Activated Carbon Co., Pittsburgh, Pa.*

***Bartlet-Snow-Pacific Inc., Cleveland, Ohio*

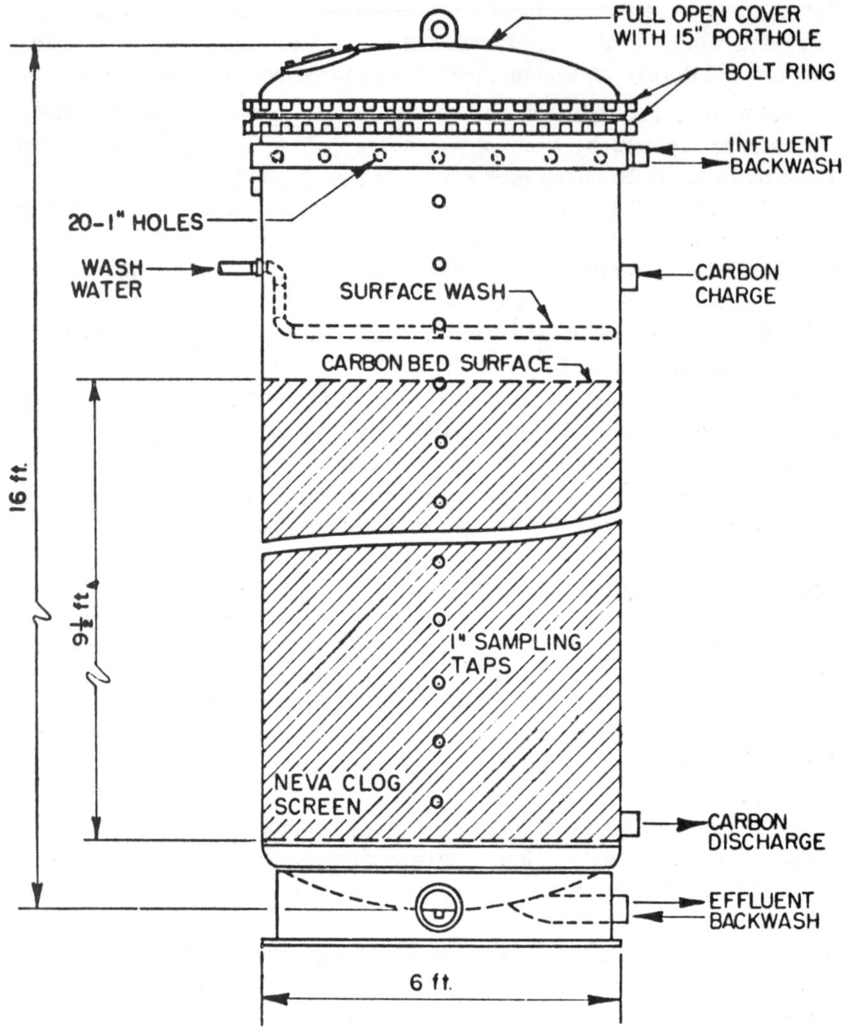

Fig. 25-10 Pilot carbon contactor.

a small organic content in the product water that slowly increases with time. It has been found, however, that approximately 20,000,000 gal. of water can be treated before the product-water COD reaches a level of 12 mg/1. The average organic content will be somewhat less than 12 mg/1. When 20,000,000 gal. of water have been treated, operation is stopped and the carbon in the first contactor is reactivated. This is done by hydraulically conveying the carbon to a dewatering tank, reactivating in the multiple-hearth furnace, quenching, and conveying hydraulically back into the contactor. The flow pattern through the system is then changed so that the former second column becomes the first in line and the reactivated column becomes the fourth in line.

Performance of the carbon adsorption system during more than a year of operation is shown in Table 25-8. During this time, the carbon in all but the last column was reactivated once. Regeneration losses averaged about 10 percent. More than 90 percent suspended solids removal and 80 percent COD removal was achieved. Good removal of color and odor was also achieved. Dissolved COD

Fig. 25-11 200 gpm granular activated carbon pilot plant at Pomona, California. The carbon contractors are on the left. In the center is the six-hearth furnace for reactivating the carbon when it is exhausted. The raised tank to the right of the furnace holds drained carbon to be reactivated. The low tank to the left is used to quench the carbon in water when it leaves the furnace. Carbon is transported hydraulically to and from the carbon columns.

removal, which is a better measure of removal by adsorption than total COD, was 77 percent. A soluble-COD loading on the carbon of 60 lb/100 lb carbon was obtained. Methylene blue detectable detergent removal was excellent, the concentration in the product being 0.1 mg/l or less. Results from continued operation of the system have not shown a significant decline in performance.

Although Table 25-8 indicates a sizable reduction in nitrate concentration, this is not a result of adsorption. It is due to biological action within the carbon column. Biological activity undoubtedly also aids organic removal. If a carbon column is operated well beyond the point of exhaustion of adsorption capacity, it will continue to remove some organics because of biological degradation.

Experience obtained from operation of the 200-gpm Pomona pilot plant was used in making a cost estimate for a 10-mgd plant. Using a four-stage system similar to the pilot plant and assuming a 10-percent loss of carbon during regeneration, the total treatment cost is estimated to be 8.3¢/1,000 gal.

Other Pomona Pilot Studies - Using a single contactor of the design used in the 200-gpm pilot plant, a study was begun to determine the effect of multiple reactivation. With one column, reactivation of the same batch of carbon can be carried out at 4 times the frequency of the 4-contactor system. The carbon in the column, which originally had the characteristics given in Table 25-7, is reactivated after treating about 20 million gal. water. After three reactivations, the iodine number, a measure of adsorbent surface, had declined from 1100 to less than 800. Additional reactivations have shown a further decline in iodine number to about 700. A parallel decline in adsorption capacity for soluble organics has been observed. This is shown in Fig. 25-12 for seven reactivations. Data for

Table 25-7 Carbon Characteristics

Iodine Number	1100
Methylene Blue Number	300
Molasses Number	220
Apparent Density	0.47 g/ml
Ash Content	5%
Mean Particle	0.85 mm
Mesh Size	16 x 40

Table 25-8 Average Water Quality Characteristics of the Carbon Column at the Pomona Research Facility

June 1965 to August 1966

Parameter		Column Influent	Column Effluent
Suspended Solids	mg/l	10	<1
C O D	mg/l	47	9.5
Dissolved C O D	mg/l	31	7
T O C	mg/l	13	2.5
Nitrate, as N	mg/l	6.7	3.7
Turbidity	J T U	10.3	1.6
Color		30	3
Odor		12	1
Carbon Chloroform Extract	mg/l	-	0.014

additional reactivations are required to determine where the curve will finally level out. Although the carbon loadings in Fig. 25-12 are for 9 ft of carbon and cannot be compared exactly with the loadings for the 36 ft of carbon in the 200-gpm carbon pilot plant, it can be expected that some decline in performance of the pilot plant will occur. To maintain the product quality that was obtained for the first year of operation will require more frequent reactivation. Carbon losses experienced during the multiple-reactivation study have been about 10 percent, the same as for the 200-gpm pilot plant. The relatively small size of the reactivating equipment and short, intermittent furnace runs prevent the losses from being reduced to levels found in other carbon applications. It is believed that losses in a larger plant can be reduced to 5 percent or less.

For operation of the large columns, a hydraulic application rate of 7 gpm/ft ^{2} was chosen. This choice was based more upon pressure-drop considerations than mass-transfer considerations since it was not known exactly what the effect of velocity on mass-transfer rate might be. The first of several studies to determine this effect has been completed. Two-inch diameter columns were

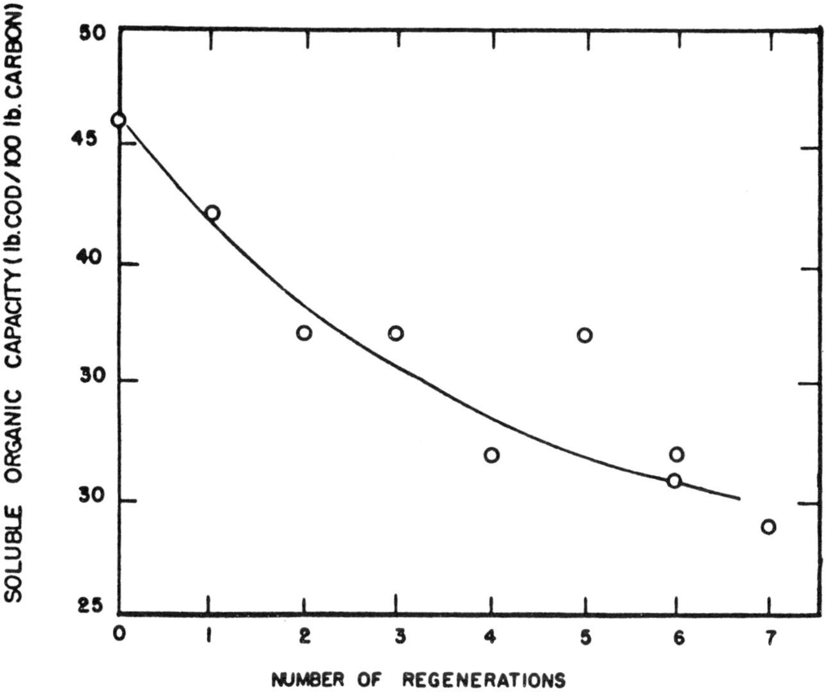

Fig. 25-12 **Effect of reactivation on absorption capacity.**

operated in parallel at different rates using diatomaceous-earth-filtered secondary effluent. Three rates chosen were 4, 7, and 10 gpm/ft ^{2} TOC was used as a measure of column performance. Data for 4 and 10 gpm/ft ^{2} as a function of empty column residence time are shown in Fig. 25-13.To a good approximation, the removal obtained depends only on the contact time between carbon and water over the velocity range investigated. This is a very important point in design of adsorbers since, from a mass-transfer standpoint, only total system volume need be considered, not column geometry. It must be emphasized, however, that the data of Fig. 25-13 were obtained on fresh carbon. Velocity may affect the mass transfer rate when carbon has been in contact with wastewater for long periods of time. Further study is required.

Laboratory Studies - A study was made (10) to determine the effect of feed turbidity upon the quality of the product from a carbon column. Batches of secondary effluent were collected and divided into two parts. One part was filtered through a dual media filter to remove large particles, but retain the fine suspended material contributing to turbidity; the other part was chemically clarified using alum or lime for removal of nearly all turbidity. The filtered water was then passed through a carbon column at 2 gpm/ft.^{2} Length of the column was 163 cm for most runs. Some work was done with a 305-cm column. The filtered water was followed by the chemically clarified water. Results of these runs are shown in Table 25-9. Effluents labeled A were from a trickling filter; those labeled B and C were from activated sludge plants. It is obvious that the total organic carbon in the product water is due largely to the turbidity that remains. Appreciable turbidity removal occurred in the column. Average turbidity

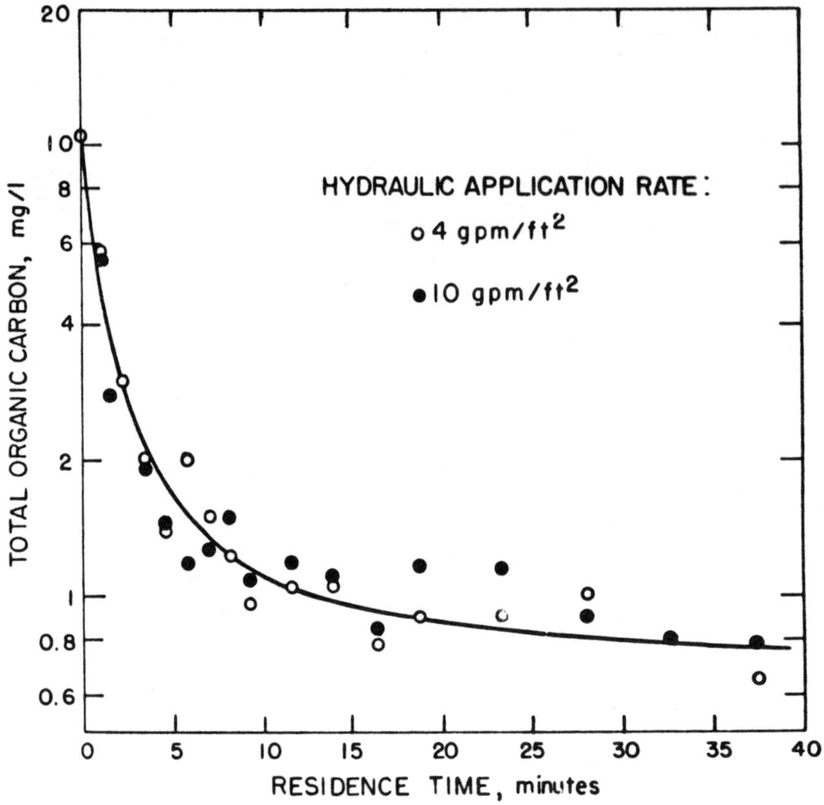

Fig. 25-13 Effect of feed rate on carbon column performance.

Table 25-9 Effect of Feed Turbidity on Carbon Column Product Quality

Pretreatment: Filtered

Effluents	Turb. Units In	Turb. Units Out	TOC (mg/l) In	TOC (mg/l) Out
A-1[a]	--	--	--	--
A-2[b]	--	--	--	--
A-4	10	6.7	17	4
A-5	15	10	19	5
B-3	7.1	5.5	14	4
B-4	15	8.7	18	5
C-2	8.8	6.5	12	2

Chemically Clarified and Filtered

Chem.	Dose (mg/l)	Turb. Units In	Turb. Units Out	TOC (mg/l) In	TOC (mg/l) Out
Al	260	19	11	33	4
Al	400	1.1	0.6	30	1
Al	300	0.2	0.1	9	0
Ca	303	0.5	0.3	11	0
Al	350	0.2	0.2	8	0
Ca	227	1.6	1.3	13	0
Ca	151	0.9	0.7	9	0

[a]Partially clarified with alum.

[b]38-minute empty bed contact time; all others 20-minute bed contact time.

removals of 31 percent, and 44 percent were found in the 163-cm and 305-cm column respectively. The logarithm of the fraction remaining turbidity was proportional to the column length, or residence time, over the range of carbon column lengths studied.

Carbon column treatment of chemically clarified secondary effluent was further studied on a continuous basis. The secondary effluent was batch-clarified with 300 mg/l lime, chlorinated, and fed at a steady rate of 1.85 gpm/ft ^{2} to a 10 ft column of 16 x 40 mesh carbon. The column was constructed of 1-ft sections, each of which could be removed separately. The product was monitored for turbidity, ammonia, TOC, and ultraviolet absorbance at 2537 Å. The ultraviolet absorbance has been found to correlate with TOC and could provide a useful control method for carbon column operation. After operating for 103 days, there was significant organic breakthrough. Two 1-ft sections were removed from the top of the column and two new sections were added at the bottom. Breakthrough then occurred more frequently and carbon changes were required every 31 days. Clarification and chlorination did not prevent biological growths from occurring in the upper part of the column. These growths increased column pressure-drop noticeably. Backwashing with 90°C water for one hour did not prevent the growths, but did slightly extend carbon life. Both hot and cold water backwash effectively controlled pressure-drop increase.

Table 25-9 shows that very little organic material escapes from a column containing unspent carbon when the feed is well-treated clarified secondary effluent. If secondary treatment is poor, however, there usually is breakthrough of soluble organics even when the carbon is new. This result should be expected since certain groups of materials in poorly treated wastewater are known to resist adsorption on carbon. To demonstrate the poor adsorbability of some examples from some of these groups, batch-adsorption tests were run using 1000 mg/l of carbon and solutions of pure materials. Although the experiments were done with powdered carbon, the results apply equally well to granular carbon. Concentrations of compounds tested were 100 mg/l. Table 25-10 shows the results based on TOC measurements. Amino acids and sodium salts of fatty acids with less than 8 carbon atoms were only slightly adsorbed. Glucose is also only slightly adsorbed. Sucrose, surprisingly, is appreciably adsorbed, but probably would be displaced easily by more adsorbable materials.

Adsorption tests on pure materials provide data to explain carbon-column breakthrough, but do not conclusively prove what materials are breaking through. To determine this requires expensive and time-consuming analysis of a very dilute multicomponent solution. Some work along these lines has been carried out. Analysis of several carbon-column effluent samples failed to show significant amounts of non-carbohydrate carbonyl compounds, total carbohydrates and reducing sugars. One sample from an overloaded carbon column showed 1.8 mg/l or 13 percent of the residual TOC as carbohydrate. Analysis of

Table 25-10 Carbon Adsorption of Pure Materials Known to be Poorly Adsorbed

Substance	% Adsorbed	Substance	% Adsorbed
Glucose	8	Sodium Formate (pH 7.5)	2
Sucrose	37	Sodium Acetate "	1
Glycine	1	Sodium Propionate "	4
Alanine	1	Sodium Butyrate "	4
Serine	0	Sodium Valerate "	8
N Phenyl glycine	48	Sodium Caprylate "	54

another carbon-column effluent sample showed removal of straight-chain acids from propionic through caproic. About 90 percent of the acetic acid, but less than 50 percent of the formic acid was removed.

ADSORPTION ON POWDERED ACTIVATED CARBON

Principle of Operation

Powdered activated carbon adsorbs soluble organics from wastewater just as granular carbon does. Because of the much smaller average particle size, about 10 microns, the adsorption rate is much more rapid. The small particle size results in high pressure drops for fixed-bed operation. Some other form of contacting must, therefore, be used. A combination of mixing, flocculation, settling, and filtration if necessary is one possible system. This system has the advantage of clarifying the water at the same time that adsorption is taking place. One mixer-flocculator-settler combination does not, however, allow advantage to be taken of countercurrent contacting and the resulting lower carbon requirements. For countercurrent action, two or more mixer-flocculator-settler combinations must be used in series, with the fresh carbon being added to the last mixer. This carbon, upon settling in the last settler, is then added to the second last mixer. The movement of carbon upstream is continued until the first mixer is reached.

Reactivation of the exhausted carbon is needed to obtain reasonable operating costs. No suitable procedures for reactivation of powdered carbon are yet available. Thermal methods will probably be the only ones feasible.

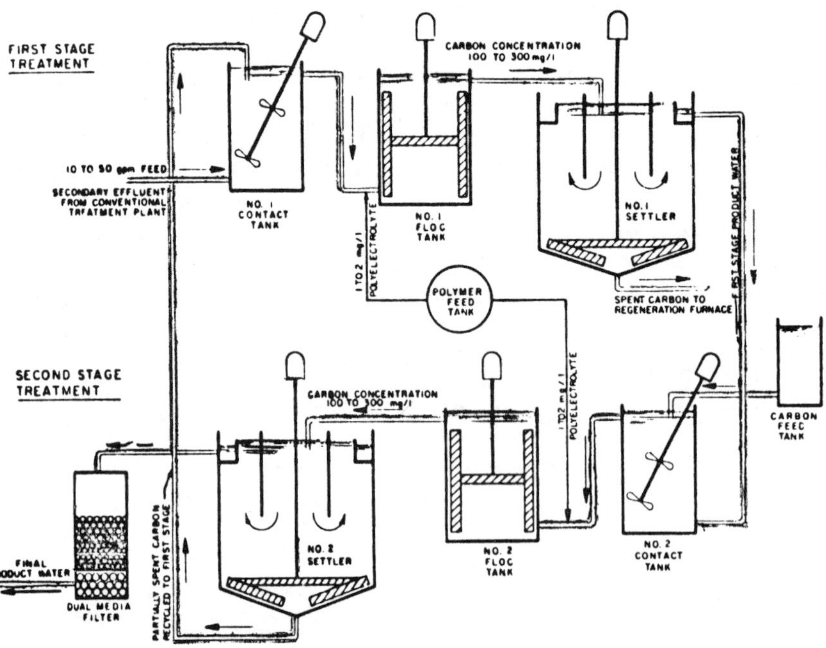

Fig. 25-14 **Powdered carbon adsorption pilot plant, Lebanon, Ohio.**

Experimental Investigations

Dorr-Oliver Inc. conducted the initial laboratory scale study (11) of powdered-carbon treatment of wastewater. They compared contacting methods and demonstrated the practicability of the mixer-flocculator-settler method using a polyelectrolyte to improve flocculation. They also showed the saving that results from using two-stage countercurrent contacting. A preliminary study of thermal reactivation of the carbon suggested that a steam atmosphere and temperatures as low as 750°F might be used. Costs for the process were estimated at less than 10¢/1,000 gal.

As a result of the laboratory work, a 10-gpm pilot system was designed and constructed at the Lebanon, Ohio Pilot Plant. The two-stage countercurrent system is shown in Figs. 25-14 and 15. Each stage consists of a 100-gal. contact tank, a 200-gal. flocculation tank, and a 600-gal. settler. Following the second stage is a dual-media filter for removal of small amounts of carbon escaping the second settler. Aqua Nuchar A,* usually in concentrations of 100 or 200 mg/1,

Fig. 25-15 Powdered carbon pilot plant at Lebanon, Ohio. The row of tanks on each side of the walk makes up one stage of the two-stage contacting system. Wastewater enters the small tank on the right in the rear. Here it is mixed with carbon already used in the stage to the left of the walk. In the second tank on the right, the carbon-water mixture is slowly stirred in the presence of a small amount of polyelectrolyte to cause flocculation. In the third tank the solids are settled from the water. These solids are largely spent carbon. The water then passes to the second stage where it is treated with fresh carbon. Following the second settler is a dual-media filter for removal of small amounts of carbon that do not settle out. Studies to develop methods for reactivating the waste carbon for reuse are being planned.

*West Virginia Pulp & Paper Co., Covington, Va.

Table 25-11 Powdered Carbon Treatment of Secondary Effluent

Feed Rate (gpm)	Carbon Concentration (mg/l)	Product TOC (mg/l)
10	100	2.0, 1.7
	200	0.9, 1.0, 1.7, 1.4
15	200	1.7, 1.4
	250	1.7
20	100	1.5
	200	2.3

Feed TOC varied from 11 to 36 mg/l.

was the carbon used in the pilot plant. The polyelectrolyte used to improve flocculation has been Primafloc C-7. Concentration of polyelectrolyte is 0.5 mg/l in the first stage and 1.0 mg/l in the second stage.

Some early results from the pilot plant are shown in Table 25-11. Results are the average over at least two days of operation. Although tank sizes were designed for a 10-gpm rate, the data show that higher rates do not seriously affect product quality. Higher carbon concentrations do not in all cases give better product quality. At 20 gpm, for example, 200 mg/l of carbon gave a poorer product than 100 mg/l. The TOC of the feed was higher for the 100 mg/l run, therefore, the feed concentration is not responsible for the inconsistency. The nature of the organics appears to be the most important factor. More recent data indicate that generally better results are obtained with higher carbon concentrations, as one would expect. Turbidities of the product water are usually less than 1 JTU.

Fig. 25-16 shows what contaminant loadings on the carbon may be obtained. Included in the loading values are suspended solids which are actually removed by flocculation. The line through the 100-mg/l data has been drawn at twice the loadings of the line through the 200-mg/l data. The fact that the 100 mg/l data lie very near the line indicates that the total organic removal capability of the process is not greatly affected by carbon concentration in the concentration range investigated. On the average, however, the higher carbon concentration gives slightly lower TOC values in the product. To obtain very high-quality water, at least 200 mg/l would be required.

The economics of powdered carbon treatment cannot be accurately predicted until more is known about the feasibility of reactivation. If the carbon can be reactivated with not more than 15 percent loss, the total operating cost, including amortization of equipment, should be about 12¢/1,000 gal. for a 10-mgd plant.

OXIDATION

Principle of Operation

Although the conventional biological processes used to treat wastewater are oxidation processes, they do not completely oxidize the organic material present. Chemical oxidation has the potential for removing the remaining

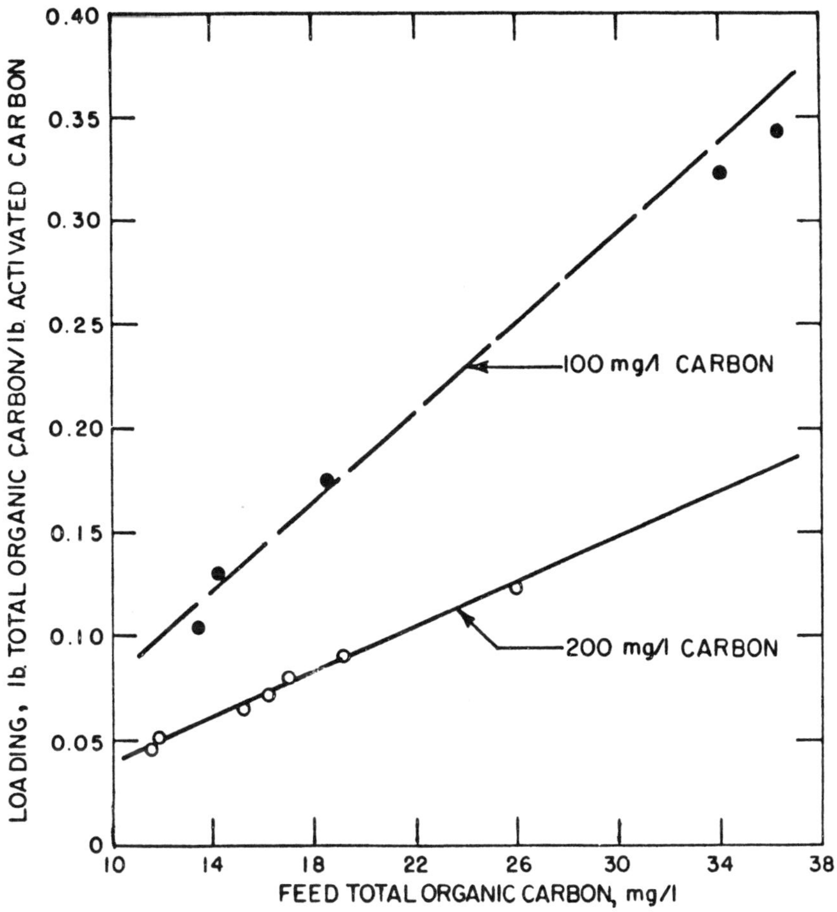

Fig. 25-16 **Powdered carbon loadings.**

organics. These would be converted to carbon dioxide, water, oxides of nitrogen, and oxides of other elements. Another possible application is the polishing of carbon-treated water for destruction of remaining traces of organic materials. Oxidation methods that might be employed include use of active oxygen species such as ozone, hydrogen peroxide, and hydroxyl free radical; oxidation with molecular oxygen; catalytic oxidation with molecular oxygen; use of chlorine and derivitives; use of oxy-acids and their salts such as potassium permanganate; and electrochemical treatment.

The first method, use of active oxygen species, is particularly interesting because of the ways other than direct addition of ozone and hydrogen peroxide in which the active species may be introduced into the water. One method is by corona discharge in moist air in proximity to the water to be treated. Corona discharge is a high-voltage, high-frequency (10,000 cycles per second), and low-temperature electrical discharge. A number of free radicals and other active oxidizing species are formed in the gas phase and are transferred to the water where oxidation occurs. Another way to produce active oxygen species is by radiation from radioactive sources. In this case, the radiation causes free radical formation directly in the water by breaking up water molecules.

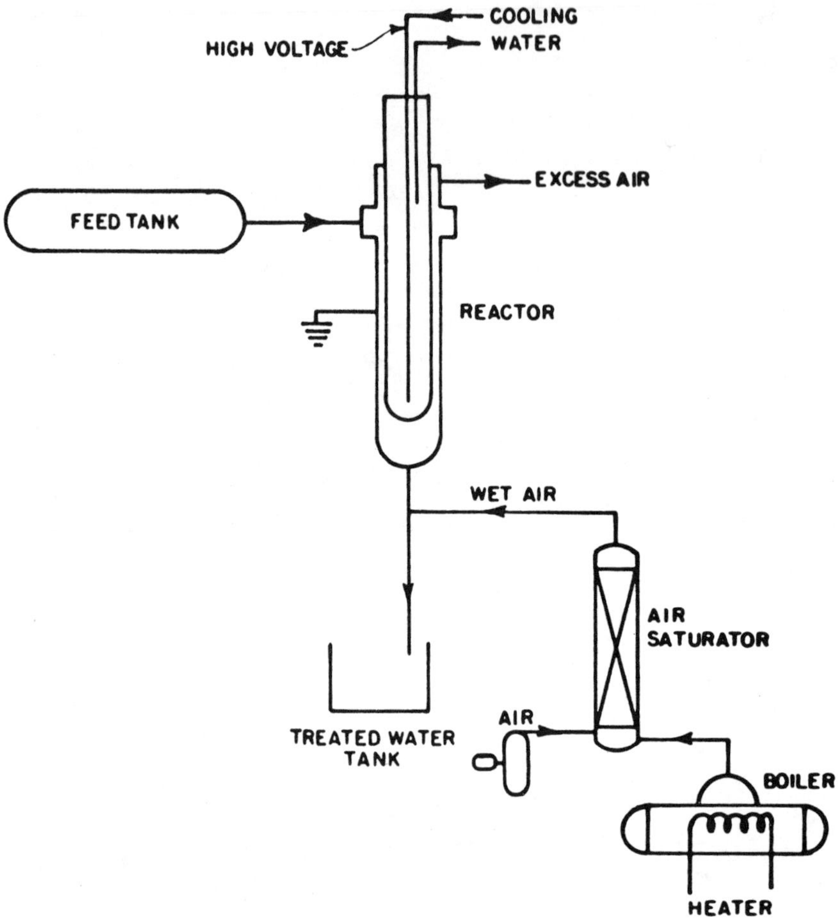

Fig. 25-17 Equipment for studying corona discharge treatment of wastewater.

Experimental Investigations

Although early laboratory-scale (6,12) studies of a number of oxidation methods indicated those methods were not practical, several promising processes remained. One of these was corona discharge, a study of which was undertaken by the General Electric Company. Fig. 25-17 shows a laboratory reactor that was used to obtain data. It was designed so that the water encircled the inside of the outer tube and fell in a thin film through the area of electrical discharge. Table 25-12 shows results for some synthetic and actual wastewater samples. Although there is considerable scatter in the results, it is obvious that the power requirements are impractically high. Of interest is the fact that the oxidation of ferrous ion to ferric ion required about an order-of-magnitude less power. Unless much more efficient use can be made of the power for oxidizing organics, the process is not economically feasible.

A short investigation was made of oxidation by nuclear radiation. Using a cobalt-60 source for 48 hours at an approximate dose of 1.2×10^3 rads/hr, only 20 percent of the COD and 8 percent of the organic carbon was removed from a wastewater sample. The most optimistic cost estimate for oxidation of a normal

Table 25-12 Corona Oxidation

Power Level: 80 watts — Liquid-feed rate: 0.36 liters/min

System	Initial COD	ΔCOD	meq/watt-hr	kwh/1,000 gal/10 COD
Benzoic Acid	20.4	2.2	0.075	63.7
Benzoic Acid	195.0	3.8	0.129	36.9
Benzoic Acid	33.5	1.1	0.037	127.0
ABS	22.1	3.0	0.100	46.7
ABS	21.7	3.5	0.119	40.1
Secondary Eff. I	75.6	0.2	0.014	700.0
Eff. I and H_2O	36.4	1.5	0.050	93.5
Secondary Eff. II	72.2	5.1	0.170	27.5
Eff. II and H_2O	36.1	--	--	--
Carbon Eff.	2.0	1.4	0.047	100.0
	Average power: kwh/1,000 gal/10 COD			58.0

waste effluent by radioactivity is 50¢/1,000 gal. if the source costs 10¢/curie. This does not include equipment costs.

FOAM SEPARATION

Until recently, accumulations of foam on our nation's waterways from discharge of sewage were becoming increasingly frequent and foam at sewage treatment plants was becoming a serious problem. The major cause of this foam was non-biodegradable detergents of the anionic type. A process was developed in which sewage was deliberately aerated to form foam and this was then removed. Because of the tendency of detergents to concentrate in the foam, the result was a relatively low-foaming product water. In addition to removal of detergents, it was also found that removal of other soluble organics and suspended material occurred. In laboratory studies, about 25 percent of the organics were removed from primary and secondary effluents (13, 14). Pilot plant operation at Pomona, California, showed as high as 40-percent removal from secondary effluent. Estimated cost of the process was less than 2¢/1,000 gal. for plants of 10-mgd capacity.

In 1965, more highly degradable or "soft" detergents began to replace the older "hard" types. The result was lower detergent concentrations in treated sewage effluents and streams. The need for and interest in a defoaming process decreased. As a result, study of the foam separation process ended. With increased awareness of the public in the problem of water pollution, it is doubtful whether highly-foaming non-biodegradable materials with as widespread a use as detergents will be marketed in the future. If foaming does again become a problem, however, a proven process will be available for control.

INORGANIC REMOVAL

Secondary effluent usually contains 300 to 400 mg/l more minerals than are found in the water supply to a municipality. The materials making up this increment are, except for ammonia, nitrate, and phosphate, commonly found in surface waters. A list of the predominant ones and approximate average values for their concentrations in once-used water, as well as concentration increases, are shown in Table 25-13. In the usually observed concentrations they are nearly all harmless, although not necessarily without taste. In the case of multiple reuse, however, the concentrations will all build up rapidly to prohibitive levels. The rate of build-up can easily be demonstrated by adding to the secondary effluent concentrations, an incremental increase for each additional reuse. For the purpose of preventing mineral build-up, methods to demineralize wastewater are being studied.

From a cost standpoint it is important to know what fraction of minerals must be removed. Although this fraction will obviously be related to the amount of minerals added, in an actual situation it will also depend upon the amount of unused dilution water available, the selectivity of the removal process for the various ions, and the maximum allowable concentrations. A study (15) of complete recycle for 22 municipalities, assuming no consumptive losses or dilution-water additions and using the non-selective distillation process for demineralization, indicated that between 24 and 64 percent of the wastewater would have to

Table 25-13 Average Mineral Composition of Secondary Effluent and Concentration Increases During Use (15)

Material	Concentration in Secondary Effluent (mg/l)	Increase During Use (mg/l)
Sodium	124	66
Potassium	12	10
Ammonium*	-	Essentially the concentration in secondary effluent
Calcium	66	18
Magnesium	19	6
Chloride	143	74
Nitrate*	-	Slightly less than the concentration in secondary effluent
Bicarbonate	296	100
Sulfate	84	28
Silicate	43	15
Phosphate	25	24
Total Dissolved Solids	703	320

*In most secondary effluents the predominant form of inorganic nitrogen is ammonium ion with only traces of the oxygen-containing forms, nitrite and nitrate ion. If nitrification is carried out at the treatment plant, however, nitrate will be the predominant form with only traces of ammonium and nitrite ion. It is difficult, therefore, to give an average value for the separate nitrogen forms. A very rough average value for total inorganic nitrogen is 20 mg/l. The average tap water contains about 1 mg/l of inorganic nitrogen and this is almost all present as nitrate ion. Nearly all the inorganic nitrogen found in secondary effluent must be considered, therefore, to have been added during use.

be distilled depending on the municipality chosen. Because distillation is assumed to remove all inorganics, except ammonia, these percentages also represent fraction removal. A rough average figure for necessary inorganic removal is 40 percent.

Ammonia would not be tolerable in most waters at a concentration of 1 mg/l or more. Fractional removal would not be adequate. Essentially all the ammonia must be removed either by nitrification in the biological treatment process or by an auxiliary process.

All practical demineralization processes studied thus far have the undesirable property of producing a volume of concentrated waste that is likely to be between 5 and 10 percent of the product. The volume of concentrate cannot be further reduced without the use of specially designed auxiliary equipment because low-solubility materials begin to precipitate out as scale on heat or mass-transfer surfaces. The result is a sizable volume of liquid that must be disposed of in such a way as to prevent further pollution. Depending upon location, the cost of waste disposal can vary widely, but in no case should it be considered negligible.

DISTILLATION

Distillation is well known as a process for producing pure water when the feed contains only materials of low volatility. Recovery of potable water from properly pretreated seawater is a much-publicized large-scale application.

Principle of Operation

To carry out distillation at a reasonable cost, it is necessary to devise special equipment that will allow for multiple reuse of the heat needed to vaporize the water. Three general equipment types are employed: multiple effect evaporators, multistage flash evaporators, and vapor compression evaporators (6). The first two usually are operated with a feed temperature of about 212°F or higher; the third is often operated with a feed temperature less than 212°F. Operation at less than 212°F naturally requires vacuum conditions. Using these special types of distillation equipment, it now appears possible to distill seawater in multimillion-gpd installations for less than $1.00 per 1,000 gal. The cost is still somewhat higher than estimated for most of the advanced waste treatment processes. It should be possible to treat only part of the water by distillation for organic and mineral removal, however, and to treat the remaining fraction by a lower-cost organic-removal process (6). At the present cost of distillation the cost of product water from the split-treatment system would still be too high to consider for pollution control alone. Reuse for a high-quality water purpose would probably be the only justification for such a system.

Experimental Investigations

An initial desk-top study (16) suggested that multiple effect or multistage flash equipment, operating with feed temperatures as high as 350°F, should be most economical for wastewater application. A pilot-scale investigation (6) of the multistage flash evaporator operating on secondary effluent at temperatures above 260°F indicated several serious problems. Rapid organic fouling of heat-transfer surfaces occurred. The product water had a strong disagreeable odor even though the feed was almost odorless. Laboratory batch distillation of secondary effluents (17) at temperatures of 160, 212, and 340°F also gave products with objectionable odors. The odor was less at the lower temperatures and could be removed by aeration or carbon treatment. This work suggests that as low a distillation temperature as possible should be used to minimize organic contamination of the product-water.

In the laboratory investigation, ammonia carryover from all ammonia-containing samples proved to be a serious problem. Decomposition of organic nitrogen compounds increased the amount of available ammonia, especially at the highest temperature. Carryover could only be prevented by lowering the feed pH to about 3. Corrosion of common still-construction materials would be appreciable under these acid conditions.

ELECTRODIALYSIS

Electrodialysis is being used to produce potable water by partially demineralizing brackish ground water in many water-short areas of the world. A plant at Buckeye, Arizona, has a capacity of 650,000 gpd. A plant being operated by the Office of Saline Water at Webster, South Dakota, has a capacity of 250,000 gpd. It is reasonable to assume that, technically, electrodialysis should be applicable to partial demineralization of wastewater. Because wastewater has a lower mineral content than brackish water, cost of the operation should, in fact, be less than for brackish water. Wastewater, however, contains organic contaminants not found in ground water and this factor will affect the economics of operation. It is therefore necessary to study the process in terms of the degree of pretreatment needed for practical operation. Because of the extensive pretreatment required, the product-water should be very high quality, suitable for many uses.

Principle of Operation

An electric potential impressed across a cell containing mineralized water

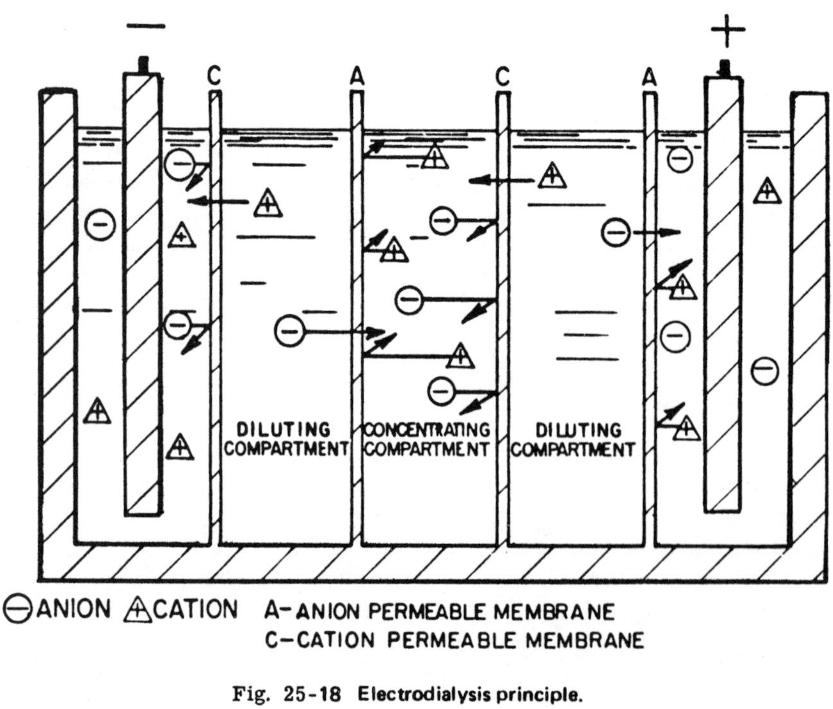

Fig. 25-18 Electrodialysis principle.

will cause positively charged ions or cations to migrate to the negative electrode and negatively charged ions or anions to migrate to the positive electrode. If cation and anion-permeable membranes are placed alternately between the electrodes as shown in Fig. 25-18, alternate compartments become more concentrated, while the intervening compartments become diluted. This separation process is called electrodialysis. Membranes that have the property of being permeable to only cations or anions are composed respectively of cation or anion exchange material similar to ion exchange beads. Many membranes can be placed between the electrodes forming many alternate dilute and concentrate compartments. Manifolds can then be added so that the membrane stack can be fed continuously and product and waste concentrate streams can be removed continuously.

The amount of electric current required to carry out demineralization is proportional to the number of ions removed from the diluting compartments. If the number of ions that must be removed becomes very large the electrical power consumption becomes uneconomical. To demineralize wastewater sufficiently to remove the mineral increment added during use requires only a small amount of electrical power.

Although the electrodialysis membranes freely pass common inorganic ions of either positive or negative charge, they do not necessarily permit passage of larger charged particles such as large organic ions and colloids. These materials are, however, attracted to the membranes, especially the anionic membranes, by electrical forces. The result is a fouling of the membranes. The most obvious effect is an increase in electrical resistance that at constant stack voltage decreases the current and, therefore, the demineralizing capacity of the equipment.

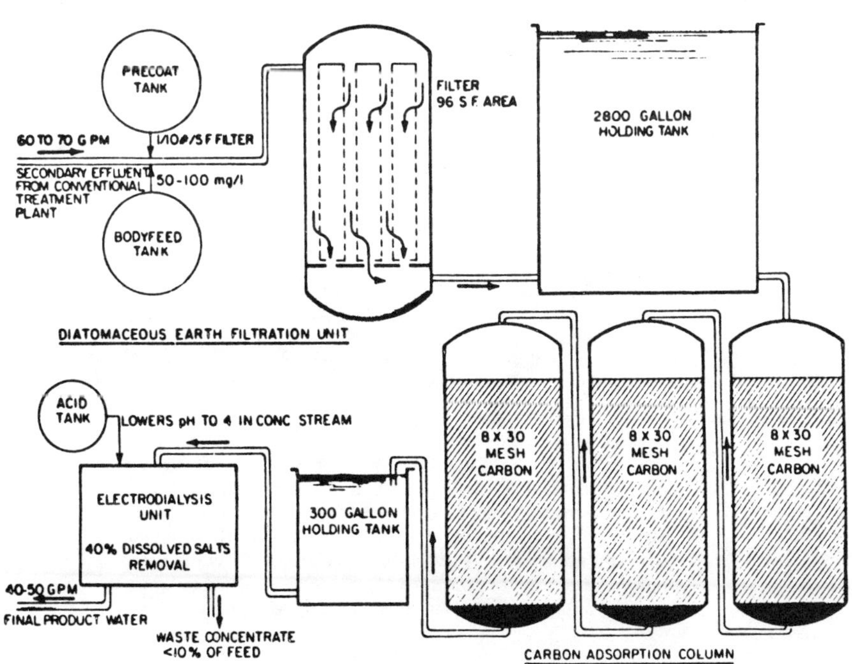

Fig. 25-19 **Electrodialysis pilot plant, Lebanon, Ohio.**

Experimental Investigations

Laboratory-scale work (6, 18) indicated that electrodialysis should be practical for partial demineralization of properly pretreated wastewater. The pretreatment considered necessary was clarification for removal of most colloidal and larger size particulate matter, and carbon adsorption treatment for removal of soluble organics. To determine more realistically the feasibility of electrodialysis, a pilot-scale system (19) was constructed and is being operated on secondary effluent from the Lebanon, Ohio Municipal Wastewater Treatment Plant. The advanced treatment system consists of a diatomaceous earth filter, carbon adsorption columns, and a 150-cell-pair Ionics Mark III* electrodialysis stack, designed for removal of about 40 percent of the total dissolved solids (TDS). The equipment is shown in Figs. 25-19 and 20. The diatomaceous earth filter uses water-filtering-grade diatomaceous earth treated with alum. The columns contain 8 x 30 mesh granular activated carbon.** Figs. 25-21 and 22 show the electrodialysis stack in more detail. Membranes are 18 x 40 in. and contain openings that form internal manifolds for the various liquid streams during operation. Spaces between membranes are of the tortuous path type. Acid feed to the concentrate stream is used to prevent calcium carbonate scale

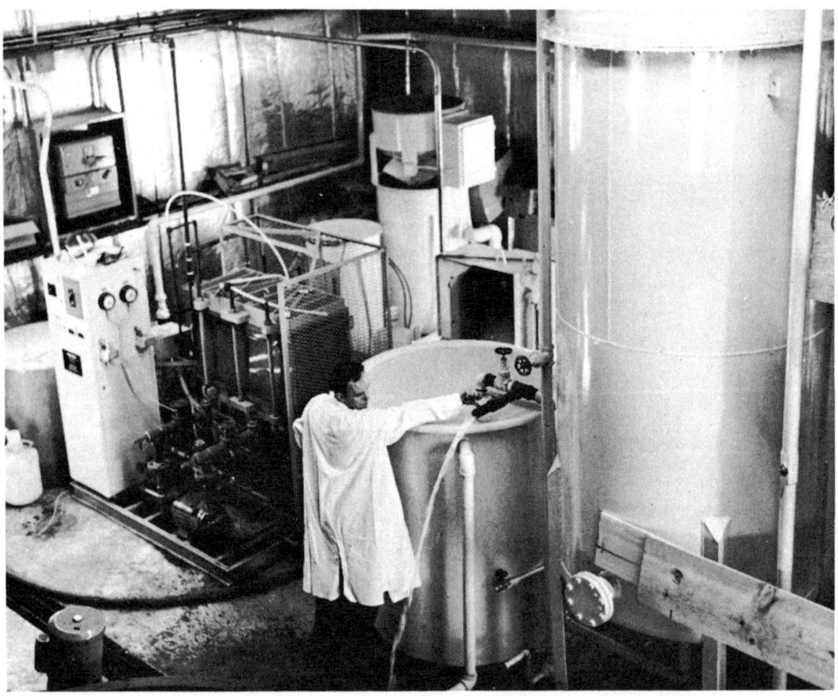

Fig. 25-20 50-gpm electrodialysis pilot plant at Lebanon, Ohio. The electrodialysis equipment is on the left. It is contained on a skid about four feet wide and six feet long. On the right is one of three carbon columns that operate in series to remove soluble organic materials from the electrodialysis feed. In the background are tanks associated with the diatomaceous earth filter used to clarify the wastewater before entering the carbon columns. The carbon columns are shown being started for an electrodialysis run.

Ionics, Inc., Watertown, Mass.

***Pittsburgh Activated Carbon Co., Pittsburgh, Pa.*

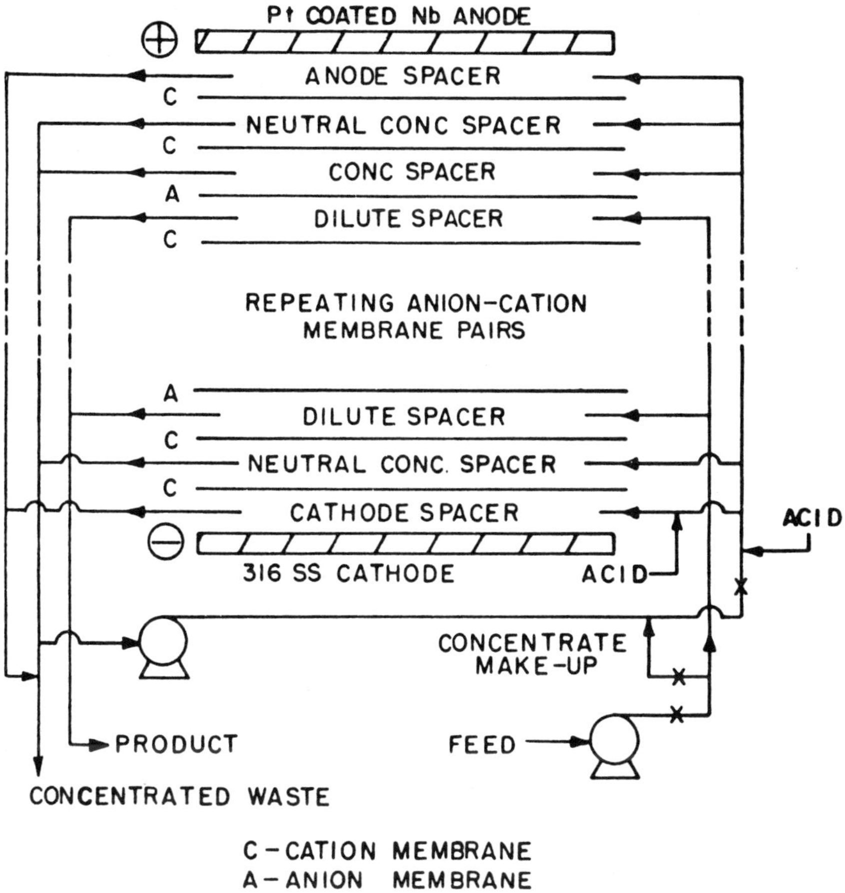

Fig. 25-21 **Electrodialysis stack details.**

formation. Using 150 cell pairs, the product rate was 50 gpm and the amount of concentrated waste was 5 gpm. These rates were reduced to 42 gpm and 4 gpm respectively and the number of cell pairs was reduced to 125 to cut down the rate through the diatomaceous earth filter which was found to be underdesigned. Total running time has been more than 1,500 hrs.

During operation of the pilot plant only one serious problem has arisen, fouling of the anion membranes. Fig. 25-23 shows the effect that fouling has on the demineralizing capability during operation. The top and bottom curves indicate that feed turbidity has a major effect on fouling rate. Obviously a very low feed turbidity, of the order of 0.1 JTU or less, is required to maintain satisfactory demineralization for a significant length of time. Fortunately, the fouling is highly reversible. Most of the fouling materials can be removed merely by shutting down the stack for a day. The middle curve in Fig. 25-23 illustrates the result of stack disinfection by low pH upon fouling. The lower-than-expected fouling rate following acid rinsing of the stack suggests that not only deposition of materials on the membranes but also growth of organisms contributes to fouling. It is believed that fouling can be reduced by using chemical clarification, for example, lime treatment, rather than diatomaceous earth filtration. Chemical clarification equipment has been installed and shakedown begun. Although the resulting turbidity values might not be lower than are obtainable by diatomaceous earth filtration, the composition of the turbidity should be different. Turbidity from an unflocculated secondary effluent is nearly all biological in origin and has a high fouling tendency; turbidity from flocculated water will be a complex mixture of biological material and inorganic floc that hopefully will have lower fouling tendency.

There are methods available that might alleviate stack fouling even with diatomaceous earth filtered water. Two of these are periodic current reversal and the use of turbidity scavenging treatments. These, however, will not be investigated extensively until the effect of chemical clarification is known.

The permanent damaging effect of fouling materials upon the membranes is very small. During 1,500 hours of operation there was only about a 5% increase in membrane pair resistance. Part of the increase is probably due to adsorption of trace soluble organics into the membranes.

Using a waste stream volume that is 9.5 percent of the product stream and keeping concentrate pH less than 5 resulted in occasional calcium carbonate scale formation on the anion membranes. This is not considered a serious problem, however. The particular water studied was high in calcium. Water with less calcium would show less scaling tendency and might even be concentrated significantly more than occurred in this work.

Although fouling has not been completely eliminated as an operating problem, it is felt that electrodialysis is a practical method to partially demineralize wastewater. Periodic stack shutdown, although not desirable, will keep fouling

Fig. 25-22 Pilot electrodialysis stack at Lebanon, Ohio. On the right of the skid-mounted unit is the membrane stack itself. Up to 150 cell pairs can be used. On the left is the rectifying equipment for supplying direct current to the stack. In the foreground are dual pumps operating from a single drive. The pump to the left recirculates the concentrate stream; the pump to the right is for feed.

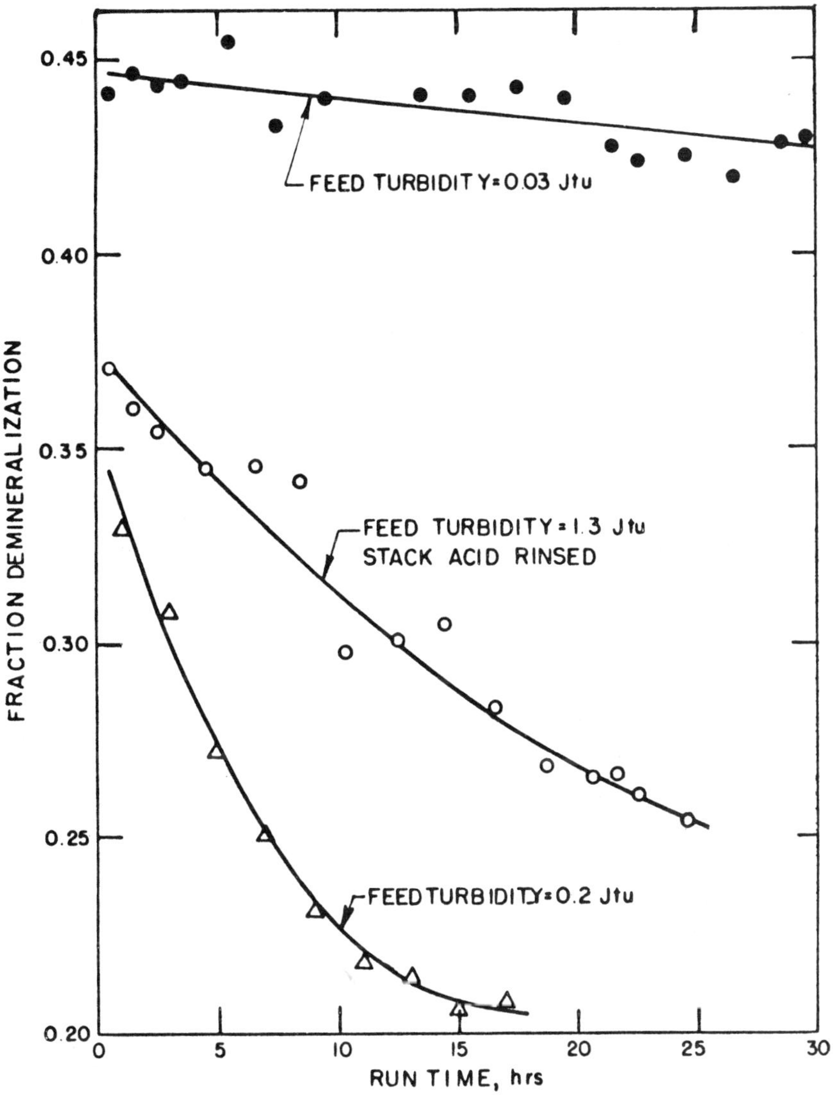

Fig. 25-23 **Effect of turbidity on demineralization.**

under control. A cost estimate has been made for a 10-mgd plant assuming 15 percent down-time. The capital cost of the plant exclusive of pretreatment should be about $3,400,000. The operating cost exclusive of pretreatment and concentrate disposal should be 16¢/1,000 gal or less.

FREEZING

Principal of Operation

When water containing impurities is frozen, the ice formed is of high purity. Freezing of seawater, for example, produces ice of very low mineral content. The same result would be expected from freezing of wastewater.

In the process, feed water is cooled and partly frozen. The ice is separated from the unfrozen brine, washed with clean water, and melted to produce purified water. Whenever possible, energy recovery is practiced to decrease treatment cost. Although the process is usually considered a demineralization process, it also is capable of removing organic materials from water. In the wastewater application, a high degree of organic removal would be expected to occur.

Experimental Investigation

Previous work on freezing of wastewater (6) indicated that problems might arise in obtaining high-purity product water. Washing of impurities from the ice crystals, especially organic materials, was different. To investigate the process under more realistic conditions than were possible in bench-scale equipment, a small pilot plant study was undertaken with Struthers Scientific and International Corporation. This company owned, and had operated on seawater, a 750-gpd freezing pilot plant at Oxnard, California. The plant utilized a secondary refrigerant, butane, to remove heat from the feed water. A schematic diagram of the experimental plant is shown in Fig. 25-24. After minor revisions to the pilot plant and a check-out on seawater, the plant was started up on secondary effluent from the Eastside Bio-filtration Plant of the City of Ventura, California. The effluent was filtered through sand and diatomaceous earth and air-stripped to reduce ABS before freezing.

In the first experimental runs, operations were severely hampered by the formation of an over-abundance of very small ice crystals which plugged up the crystallizers. It was felt that this problem could be surmounted by continuous operation until the brine in the crystallizer was concentrated to a TDS of about 15,000 mg/l, but time did not permit such a procedure. To simulate these conditions, rock salt was added to the crystallizer to build up the TDS content. Several runs were made under these conditions. From these runs it was learned that the inorganic salts are more readily washed from the ice than are the organic materials. Organic carbon in the product was about 4 mg/l.

It was estimated from the work that the capital costs for a 10-mgd wastewater renovation plant would be about 10 percent lower than those for a 10-mgd sea water desalting plant. The main reason for this is that a seawater plant would be fed 18 mgd and a wastewater plant 11-mgd or less to produce

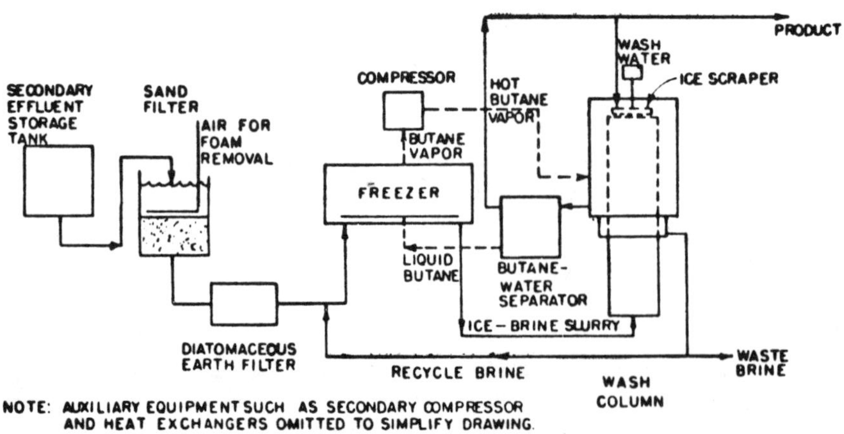

Fig. 25-24 Freezing pilot plant.

10-mgd of product-water. At 10-mgd, the total cost of production of pure water from wastewater would be 48¢/1,000 gallons compared to 53¢/1,000 gallons for the seawater plant.

A wastewater renovation plant utilizing coagulation, sedimentation, filtration, carbon adsorption and electrodialysis can produce water of higher quality than was obtained in the freezing runs for 25-35¢/1,000 gallons. It seems unlikely that the freezing process could be competitive with this process chain.

ION EXCHANGE

Ion exchange is a well-known method for softening or for demineralizing water. Although softening could be useful in some instances, the most likely application for ion exchange in wastewater treatment is for demineralization. The highly demineralized water usually produced by the process is of higher quality than is necessary, therefore, only part of the total stream need be treated.

Many ion exchange materials are subject to fouling by organics. It is possible, that treatment of secondary effluent for suspended solids removal and possibly soluble organic removal will be required before carrying out ion exchange.

Principle of Operation

Many natural materials and, more importantly, certain synthetic materials have the ability to exchange ions from an aqueous solution for ions in the material itself. Cation exchange resins can, for example, replace cations in solution with hydrogen ion. Similarly, anion exchange resins can either replace anion in solution with hydroxyl ion or absorb the acids produced from the cation exchange treatment. A combination of these cation and anion exchange treatments results in a high degree of demineralization.

Since the exchange capacity of ion exchange materials is limited, they eventually become exhausted and must be regenerated. The cation resin is regenerated with an acid; the anion resin is regenerated with a base. Important considerations in the economics of ion exchange are the type and amounts of chemicals needed for regeneration. Often, water to be demineralized is first passed through a cation exchange material requiring a strong acid, usually sulfuric, for regeneration. The exchange material is called a "strong" acid resin. The amount of acid regenerant is somewhat more than the stoichiometric amount, possibly 100 percent excess of more. If sulfuric acid is the regenerating acid, a waste brine is produced consisting of sulfates of the various cations in the water being treated. Because the partially treated water contains mineral acids, it is common to pass it next through an acid absorbing resin or "weak" base resin. This resin can be regenerated with either a weak or strong base. Efficiency of regenerant use is quite high with these resins. If sodium hydroxide is the regenerating base, a waste brine is produced consisting of the sodium salts of the various anions in the water being treated. Certain anionic materials are not removed by the "weak" base resin and must be further treated with "strong" base resin if thorough demineralization is desired. Regenerant usage by the "strong" base resins is poorer than for the "weak" base resins.

The cost of chemicals required for this typical ion exchange system, although within reason for many applications, is probably too high to make the process generally attractive for wastewater treatment. Some advantage can be gained by carrying out regeneration in a more nearly countercurrent manner than is usually done. A more desirable system, however, would be one in which regenerants could be cheaply recovered and reused. A number of these have been devised and are being developed by resin manufacturers and others. Systems designed for regenerant reuse do not always operate with an acid resin bed followed by a base resin bed.

Experimental Investigations

Regenerant Recovery System - A novel scheme for recovering regenerants was studied by Chemical Separations Corporation. To take advantage of the recovery method, it is necessary to use nitric acid to regenerate the cation resin and ammonium hydroxide to regenerate the anion resin. Upon regeneration of the cation resin, a brine is produced with a high concentration of nitrates. Regeneration of the anion resin produces a brine with a high concentration of ammonium salts. The nitrate solution is evaporated and the residue is steam calcined to produce nitric acid and hydroxides of the metal cations removed from the water. The hydroxides can then be reacted with the ammonium salt solution from anion regeneration to recover ammonium hydroxide.

The critical part of the process from an economic standpoint is the calcination of nitrates to recover nitric acid. Good recovery is required or the makeup nitric acid cost becomes prohibitively high. Nitric acid presently costs about 2.5 times as much per pound as the commonly used sulfuric acid. Laboratory-scale and small pilot-scale runs were carried out to determine the degree of recovery of nitric acid from sodium, potassium, calcium, and magnesium nitrates and mixtures of these. Laboratory work indicated that nearly 100 percent recovery could be achieved for calcium and magnesium at about 600°C, one atmospheric steam pressure, and 1.5 hr calcination time. Under the same conditions, sodium and potassium gave yields less than 50 percent. Rotary calciner pilot runs with less than one hour residence time resulted in considerably poorer yields of acid. In the pilot work it became obvious that carryover of undecomposed nitrate salts into the product was a problem. In view of the incomplete recovery of nitric acid, this part of the regeneration scheme does not appear practical. Furthermore, in all of the work it was assumed that no ammonium nitrate would be in the nitrate brine. If there were any ammonium ion, and this is very likely, it would have to be removed prior to calcination. This removal adds to the complexity and cost of the recovery scheme.

It must be emphasized that the economic feasibility of the nitric acid recovery system depends directly upon the relative costs of nitric and sulfuric acid. With rising sulfur costs, the cost of sulfuric acid has been increasing at a rapid rate. As a result, acid users are giving more consideration to substitutes such as nitric acid. Greatly increased consumption of nitric acid could substantially decrease its cost in relation to the cost of sulfuric acid. If this occurs, further study of the nitrate recovery system may be desirable.

Fouling of Anion Resins - Early anion exchange resins were subject to severe fouling by many organic materials. Significant advances have been made in the elimination of fouling tendencies of some of these resins. It is not well known, however, what effects the soluble and suspended organic materials in secondary effluent may have. A small-pilot-scale survey of a number of anion resins has been initiated to determine fouling effects. Preliminary results are available on one weak base resin, IRA-68.* A sample of this resin was contacted in a column with sand-filtered secondary effluent for a total of 23 exhaustion cycles. Regeneration was initially carried out in a cocurrent manner with 4 percent ammonium hydroxide. The resin was then converted to the bicarbonate form with water carbonated under a carbon dioxide pressure of 50 psig. For the first 15 cycles there was a gradual decline in exchange capacity, indicating some fouling. About 50 percent of the organics in the feed were removed on the resin. A reduction of ion exchange capacity from 1.0 meq/ml observed for new resin to 0.6 meq/ml occurred. A hydrochloric acid rinse restored the capacity to that of fresh resin. During the remaining cycles no further acid rinse was required to maintain resin capacity. Regeneration for these latter runs was carried out in a direction counter-current to the feed, however, rather than cocurrent with the feed as in the first 15 cycles. It is believed that cocurrent regeneration caused organics initially held in the resin near the column entrance to be moved through the column, resulting in fouling of much of the resin. Countercurrent regeneration, on the other hand, would keep organics on the resin near the column entrance from being moved through the column. Instead, these organics were removed by the regenerant. Analysis of spent regenerant from organic carbon indicated that organic removal from the resin occurred. Using countercurrent regeneration, organic fouling is not a significant problem for this particular resin.

During the study, breakthrough curves for common anions in wastewater were obtained. A high selectivity was found for phosphate and sulfate ions. This result is in line with the rule of thumb that multivalent ions are usually removed preferentially to monovalent ions. Essentially all the resin capacity could be used for removal of these two ions.

Work being planned for ion exchange includes study of new regenerant recovery schemes, testing of special resins that can be used in the presence of organics or even as scavengers of organics and testing of resins highly selective for certain specific ions. Selective resins are of special interest for removal of ammonium ion and other nutrients from water. Certain natural zeolites are selective for ammonium ion in the presence of sodium. Since sodium is present in municipal wastewater in large amounts relative to other cations, selective removal of ammonia means a substantial savings in regenerant cost. In the planned study, calcium salts will be tested as regenerants. Development of a practical ion exchange process for ammonia removal would represent a significant advance in wastewater treatment.

REVERSE OSMOSIS

Reverse osmosis is one of the few really new water-treating processes. All the important developments in this process have been made in the last 10 years. Its principal application presently is in demineralization of waters too high in minerals for an intended use. Application of the process is not yet widespread, but interest is rapidly increasing. The reason for the increasing interest is the simplicity of the process, the high product quality, and the low treatment cost that could result from further development work. Although reverse osmosis is usually considered a demineralization process, it also has the capability of removing many of the organic materials in wastewater. Suspended material is almost totally removed.

Rohm and Hass Company, Philadelpha

Principle of Operation

Several types of materials, when formed into sheets, have the property of preventing passage of minerals while allowing the passage of water in which the minerals are dissolved. Reverse osmosis makes use of this special property. The name is derived from the fact that, in the process, water is caused to pass from a solution of higher concentration through a membrane to a solution of lower concentration, the opposite of natural osmosis. To cause this flow, a pressure greater than the osmotic pressure must be applied to the more concentrated solution. In a practical situation, pressure considerably in excess of the osmotic pressure is required for useful water flux through the membrane.

The type of membrane that is presently being used in all reverse osmosis equipment is a specially prepared cellulose acetate. The method of preparation gives a dense layer, less than one micron thick, on one side of the otherwise rather porous membrane. This dense layer is believed responsible for rejecting most of the impurities in the water. The mechanism causing rejection is not completely understood.

High concentrations of the soluble materials rejected by the membranes can build up on the concentrate side causing scale. Suspended material can also build up, forming a fouling layer. Equipment design must, therefore, incorporate means for lowering the concentration of materials at the membrane. This is commonly done by moving water past the membrane surface at a high enough rate to cause turbulence.

Experimental Investigations

An initial investigation by the Aerojet-General Corporation of the treatment of municipal secondary effluent with reverse osmosis indicated that a very

Fig. 25-25 Pilot reverse osmosis unit at the Pomona Research Facility.

high degree of inorganic and organic removal was possible (6, 20). This study showed, however, that decline in water flux through the membrane with time might be a problem. The General Atomic Division of General Dynamics then carried out a study of their own at the Pomona, California Water Renovation Plant. Product quality and flux rate were similar to those obtained in the earlier study. A flux decline was not observed in this investigation. As a result of the General Atomic work, a 5,000-gpd pilot unit was purchased from the company and installed at the Pomona Research Facility. The pilot unit is shown in Fig. 25-25. This equipment has spiral-type membrane packages shown in Fig. 25-26. These packages are made up of several sandwiches of two membranes with a porous backing between. Each sandwich is sealed on three edges and the fourth edge is connected to a collection tube around which the sandwiches are wrapped. A thin screen-like spacer is placed between adjacent membranes to permit turbulent flow of water across the membrane surfaces. Water passes through the membranes into the backing and to the collection tube. A large membrane area can be packed into a small pressure vessel. To prevent formation on the membranes of pH-dependent inorganic scales such as calcium carbonate, the feed stream is acidified.

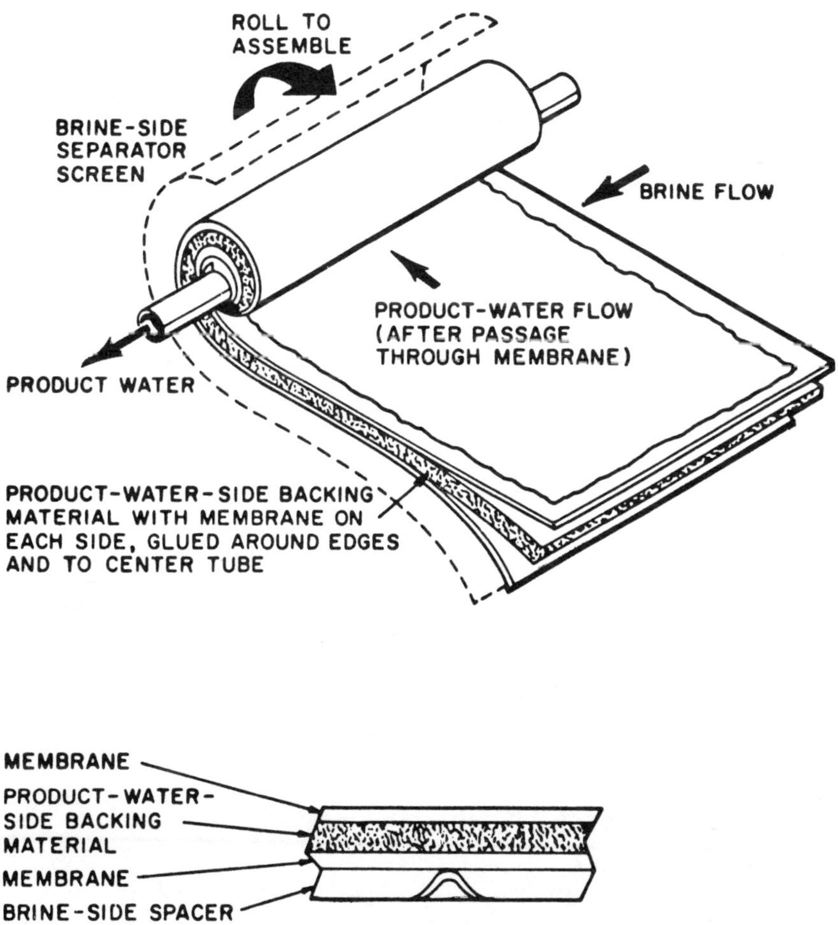

Fig. 25-26 Membrane configuration in reverse osmosis pilot unit.

Mechanical difficulties prevented the taking of meaningful data immediately after completion of installation. After solving the most serious mechanical problems, a run was begun using carbon-treated and chlorinated secondary effluent with pH adjusted to 5. Only a portion of the total membrane area was used in this run. Inlet pressure was 400 psig. The initial flux was about 5 gpd/ft.[2] To prevent an increase in pressure drop from feed point to brine outlet resulting from the presence of suspended solids, a daily air-water cleaning was employed. A flux decline of about 15 percent occurred during the first 250 hrs of operation. Little further decline was noted until after 1,100 hrs of operation. Flux decline then continued until the run was terminated after 1,600 hrs. Difficulties with the chlorination system prevented chlorination after 500 hrs. Examination of the membrane support material when the run was terminated indicated growth of mold that is believed to have caused the flux decline. Chlorination of the feed to the end of the run should have prevented the growth. Product quality during the run was excellent. TDS removal was greater than 95 percent. Removal of ammonia, which is known to be poor for reverse osmosis membranes, was greater than 85 percent. The organic content of the product was negligibly small. This is not surprising since the feed was carbon treated and had an average COD of less than 12 mg/l. Further long term runs at higher operating pressure and with secondary effluent that has not been carbon treated are needed to fully evaluate the effectiveness of this unit.

A study of a tubular reverse osmosis unit* was also carried out at Pomona. In this unit, feed water flows inside the tubes and product water collects in a shell that surrounds groups of seven of the tubes. Both secondary effluent and carbon treated secondary effluent were investigated. The feeds were chlorinated and, for the most part, acidified to a pH of 5. Feed pressure was in the range of 700 to 850 psig. Frequent tube breakage and other operating problems interfered with the experimental program by preventing long runs, but certain information was obtained. It was clearly demonstrated that pH adjustment is necessary to prevent calcium carbonate scale formation. Limited results suggest that membranes treating secondary effluent without carbon treatment undergo more rapid flux decline than when carbon treatment is used. Since extensive studies of effluents with and without carbon treatment are planned, this point will be clarified in the future. Runs with membranes having initial fluxes covering a range from 10.8 to 17.5 gpd/ft.[2] indicated that the rate of flux decline increases with increasing initial flux. Similar results have been obtained by other investigators on waters other than wastewater. Product water from the tubular unit was low in both inorganics and organics. Removal of TDS was about 95 percent, phosphate 99 percent, and ammonia 87 percent. Nitrate, like ammonia, is removed less effectively than gross inorganics measured by TDS. Nitrate removal was about 68 percent. Removal of organics from the secondary effluent was 96 percent measured as COD. Average product COD was only 1.9 mg/l. Whether these excellent removals could have been sustained for long periods was impossible to determine because of operating problems.

NUTRIENT REMOVAL

Certain materials in wastewater act as nutrients to algae and other aquatic plants in receiving waters causing excessive growths of these organisms. The

**Loaned by Havens Industries, San Diego, Calif.*

excessive growths, algae blooms, are an esthetic nuisance. They also cause wide swings in dissolved oxygen content that may kill fish. Decomposition of the plants contribute organic matter which then depletes the dissolved oxygen in the water below levels necessary for other aquatic life. At the same time, bad tastes and odors are produced.

Although many materials act as plant nutrients, the two most important ones in wastewater are phosphorus and nitrogen. In secondary effluent, phosphorus occurs largely as inorganic orthophosphate ion, but also combined in organic materials. Nitrogen also occurs as inorganic and organic forms. In secondary effluent the inorganic forms constitute the greatest nutrient problem. They are predominantly ammonia or nitrate with minor amounts of nitrite. In most secondary effluents ammonia is the important form. Under certain conditions, an important one being an excess of oxygen, it is possible, however, to convert the ammonia to nitrate biologically. This is called nitrification and the water is termed a nitrified effluent.

Besides its problem as a nutrient, nitrate at high concentrations in drinking water can cause methemoglobinemia in infants. A drinking water standard (21) of 10 mg/l NO_3-nitrogen maximum has, therefore, been adopted in this country. Ammonia exhibits oxygen demand and hence adds to the BOD. Although there is no drinking water standard for ammonia, it can be tasted at a concentration of about 1 mg/l.

Removal methods for phosphorus may be biological or physical-chemical. The biological processes ordinarily remove some phosphorus. By regulating the operation of the activated sludge process there is the possibility that the degree of removal can be considerably increased. Physical-chemical processes include ion exchange and precipitation. Presently, precipitation appears most promising.

Removal methods for nitrate may also be biological or physical-chemical. To make use of the biological method, a highly nitrified secondary effluent must be available. A possible physical-chemical removal method is ion exchange.

Ammonia can be removed during secondary treatment by activated sludge if nitrification is maintained. Nitrifying plants are, however, not common at the present. A physical-chemical method of removal is air-stripping at high pH. A study of this method was made at Lake Tahoe and results were sufficiently promising that a plant-scale installation is planned.

PHOSPHATE REMOVAL BY PRECIPITATION

Principle of Operation

A number of chemicals, including aluminum sulfate and lime, when added to wastewater will precipitate orthophosphate. The precipated material can then be removed by either sedimentation alone or by sedimentation combined with filtration. The reactions that occur during precipitation are not completely understood. The precipitates are not simple aluminum and calcium phosphates. When aluminum sulfate is the precipitating chemical, a complex structure made up of aluminum associated with hydroxyl and phosphate groups appears to be formed. When lime is used, hydroxyapatite is one material likely to be formed. Studies of this material indicate the chemical composition to be $Ca_5OH(PO_4)_3$. It is an interesting fact that the concentrations of calcium and phosphate in wastewater before addition of lime can greatly exceed the maximum values estimated from the solubility of hydroxyapatite.

Experimental Investigations

Data on phosphate removal by precipitation has been obtained during a number of studies of clarification. In one of these studies (10), several secondary effluents were flocculated in 165-gal. batches with lime or alum and filtered through a dual-media filter. Phosphate removals obtained from the clarification tests and results from jar tests are shown in Table 25-14. Both lime and alum were effective in reducing the phosphate concentration to low levels, sometimes so low as to be undetected. These results are typical of those found also in other studies.

Table 25-14 Phosphate Removal During Batch Clarification

Chem [a]	Dose (mg/l)	Phos (mg/l) In [b]	Phos (mg/l) Clar
Al	400	--	--
Al [c]	450	58	0.0
Ca [c]	450	58	0.0
Al	300	19	0.8
Ca	303	23	1.0
Ca [c]	133	35	1.0
Ca [c]	151	33	0.4
Al	350	21	0.0
Ca	227	29	0.5
Al [c]	300	24	0.8
Ca [c]	300	24	2.0
Ca	141	20	2.8
Al [c]	150	11	0.0
Ca [c]	150	11	0.4
Ca	378	--	0.2

[a] Ca = lime as CaO; Al = alum as $Al_2(SO_4)_3$ $14H_2O$.

[b] In = input value; Clar = after chemical clarification and filtration.

[c] Indicates Jar tests; not filtered.

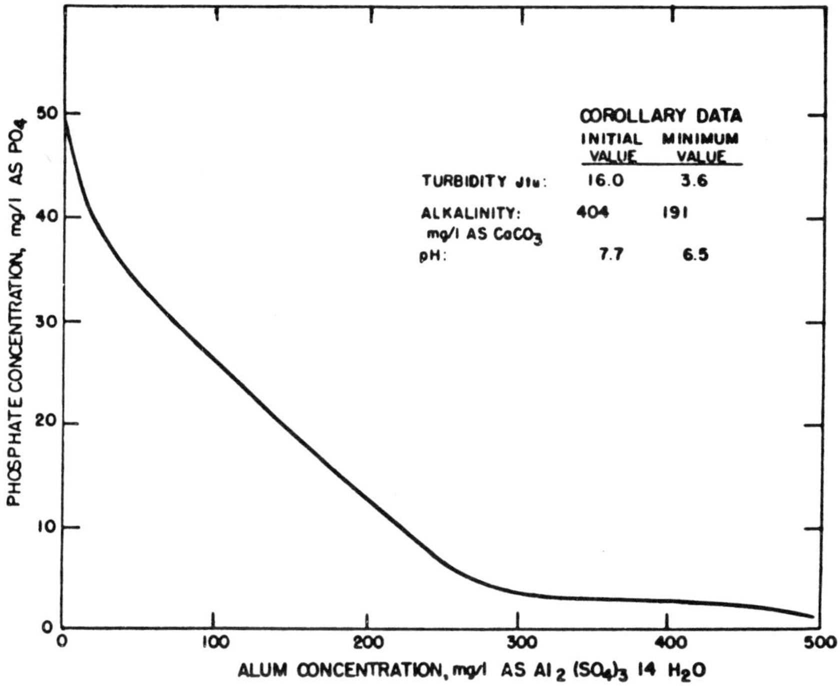

Fig. 25-27 **Results of jar tests showing effects of alum dosage on phosphate removal.**

Fig. 25-27 shows how phosphate concentration decreases with increasing alum addition for a particular secondary effluent sample. In taking these data, samples were flocculated and settled, but not filtered. Filtration would give slightly lower phosphate concentrations. At a dose of 300 mg/l, phosphate was decreased to less than 10 percent of its original value. Considerably more alum is required to significantly reduce the phosphate concentration further. A similar curve would be found for lime. The shapes of these curves vary somewhat among different effluent samples.

NITRATE REMOVAL BY BIOLOGICAL DENITRIFICATION

Principle of Operation

If organic material that can be used for food by denitrifying bacteria is added to a highly nitrified secondary effluent in the absence of oxygen, reduction of nitrate ions to nitrogen occurs. Although the process is a simple one, systems for its application on a large scale remain to be proven effective and reliable.

Experimental Investigations

Two studies of biological denitrification have been undertaken. One of the studies was carried out in conjunction with a small-scale pilot study of the contact stabilization process. The work was carried out at the Mill Creek Treatment Works, Cincinnati, Ohio. A diagram of the system is shown in Fig. 25-28. Nitrification was carried out in a separate aerator following contact stabilization.

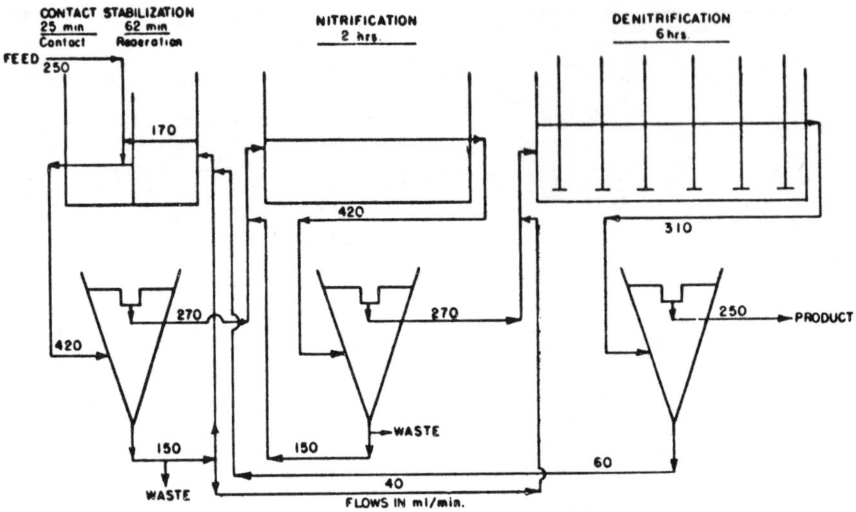

Fig. 25-28 **System for nitrification and denitrification.**

The denitrification chamber was a separate stirred vessel. Sludge from the settler following the contact stabilization aerator was used to supply organic material to the denitrifying organisms. Over a 4-month period, conversion of ammonia nitrogen to nitrate was 95 percent. Reduction of nitrate to elemental nitrogen averaged 86 percent. The cost of nitrification and denitrification was estimated to be 5 to 6¢/1,000 gal.

A second denitrification study was undertaken at Pomona, California, after it was noticed that a small degree of nitrate removal was occurring in the activated-carbon pilot plant located there. An investigation was begun using a 10-ft long carbon column and methanol as a supplemental food for the denitrifying bacteria. Feed rate is 7 gpm/ft.^{2} Available results indicate that removal of nitrate down to 1 mg/l or less may be obtained with the addition of 2 mg/l methanol for each mg/l of NO_3-nitrogen. Cost of the methanol would be about 2¢/1,000 gal. The organic content of the product-water is not raised significantly by the alcohol addition. More recently, sand has been investigated as a substitute for the carbon. Early results indicate sand may be as effective as carbon for carrying out denitrification.

AMMONIA REMOVAL BY AIR-STRIPPING

Principle of operation

Ammonia exists in water either as the un-ionized NH_3 or as the ammonium ion, NH_4^+. While NH_3 is volatile, NH_4^+ is not. The relative amounts of each depend upon pH. At pH of 10 or above, the un-ionized form greatly predominates. Since it is volatile, it can be removed by contacting the water with air. Large quantities of air are required to reach the 1-mg/l concentration that is considered a maximum for some water uses. Although the pH can be raised in several ways, the most economical method is by addition of lime. The air-stripping process, therefore, would probably be used in conjunction with lime clarification.

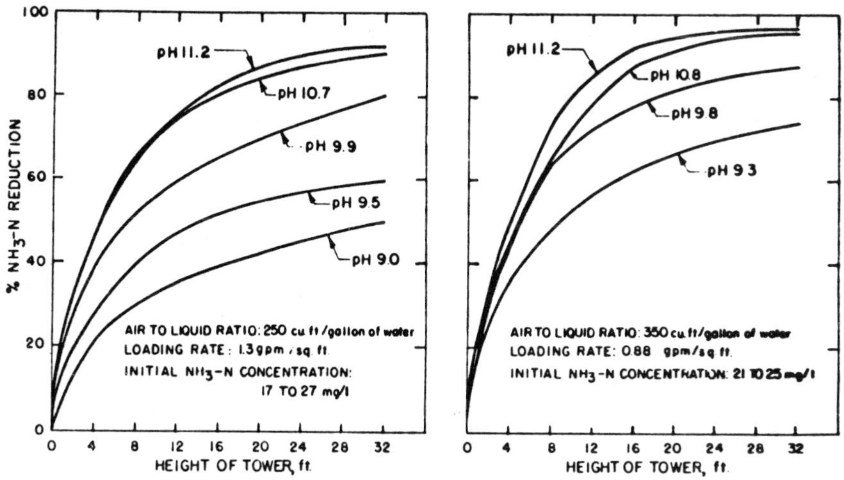

Fig. 25-29 **Ammonia-nitrogen reduction-air stripping with pilot cooling tower.**

Experimental Investigation

Air-stripping of ammonia was studied on secondary effluent at the Pomona Research Facility. Initial work utilized a column packed with plastic Pall Rings. Because of the high pressure drop that develops with this type of equipment, air pumping costs are high. Later work was done using cooling tower equipment. This contactor was about 2 ft high, 3-1/2 ft long, and 3 ft wide and contained 5 ft^2 of wood slats. Secondary effluent with lime-adjusted pH flowed at a steady rate downward over the slats perpendicular to air that was blown horizontally with a fan. Power costs for this equipment are low, but gas-liquid contact is not as efficient as in a packed column. The 2-ft height of the contactor was not sufficient for high ammonia removal. Water that had been treated once could, however, be recycled to give some indication of the effect of an additional 2-ft of tower height for each recycle. This method of multiple contacting was used to obtain the removal results shown for various tower heights in Fig. 25-29. Results are given for two liquid loading rates, 0.88 and 1.3 gpm/ft^2. Air rate was about 1,500 ft^3/min. There is a noticeable effect of pH on removal. For 90-percent removal and for the range of conditions covered, a pH of at least 10 is required. Removal to 1 mg/l, or slightly over 95 percent average, requires a pH of 11 or higher at an air-to-liquid ratio of 350 ft^3/gal. and a tower height of over 20 ft. At 250 ft^3/gal a much taller tower would be required to reach the 1-mg/l level.

Operating at an air-to-liquid ratio of 350 ft^3/gal. and using a 20-ft tower, the total operating cost of the process is about 2¢/1,000 gal. This does not include the cost of lime or solids removal from the product water. For the effluent studied, the lime cost to raise the pH to 9.9 is estimated at about 3¢/1,000 gal. Considerable savings would result if the lime were recovered.

NUTRIENT REMOVAL FROM OXIDATION POND EFFLUENT

A project (22) has been carried out at Lancaster, California, in cooperation with the Los Angeles County Sanitation Districts to improve the quality of water from oxidation ponds to the extent that it could be used for recreational lakes. It was necessary that the water be of high enough quality to allow growth

Table 25-15 Proposed Standards for Recreational Lake Water at Lancaster, California

Turbidity	< 10 JTU
PO_4	2 mg/l
pH	< 8.0
Algae	< 10,000/ml
Coliform & virus	To meet requirements of County Health Officer

The standards were later revised to a PO_4 content of <0.5 mg/l since this was found necessary to prevent excessive algal regrowth.

and propagation of fish. It has been shown that the ponds would not support fish life, probably because of the low dissolved oxygen concentration occurring during the dark hours when the algae were not producing oxygen. The initial water quality objectives set for the study were those shown in Table 25-15.

Principle of Operation

Suspended solids and phosphate removal are types of processes required for production of the recreational water. Chemical clarification processes are appropriate and, with large enough doses of lime and alum, good results can be expected.

Experimental Investigation

Studies of various chemical clarification systems were carried out for a period of more than two years on laboratory and pilot-scale units. Laboratory studies included jar test evaluations of coagulants to determine the most effec-

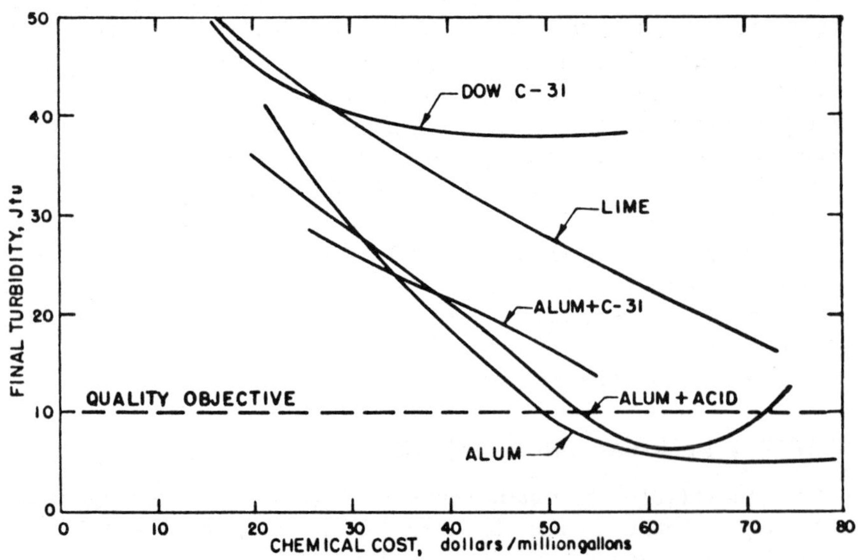

Fig. 25-30 Cost versus final turbidity for different coagulants. Laboratory jar tests.

tive coagulant and coagulant aid, their required dosages and relative costs. The results of the laboratory studies are shown in Figs. 25-30 and 31. Fig. 25-30 shows that alum is the best over-all coagulant for reducing turbidity. Fig. 25-31 compares the efficiency of lime and alum in removing the phosphate ion.

Several pilot-scale pieces of equipment ranging in capacity from one to 25 gpm were evaluated on the site. These included an upflow clarifier, a high density solids clarifier, a dissolved air flotation unit, a sedimentation tank and sand and diatomaceous earth filters. Of all those tried, the best series of processes was found to be similar to that used in water treatment processes; that is, a rapid-mix section for coagulating chemicals followed by flocculation, straight sedimentation, dual-media filtration, and chlorination. Alum was the chemical selected for coagulation. The effect of alum dosage in the pilot equip-

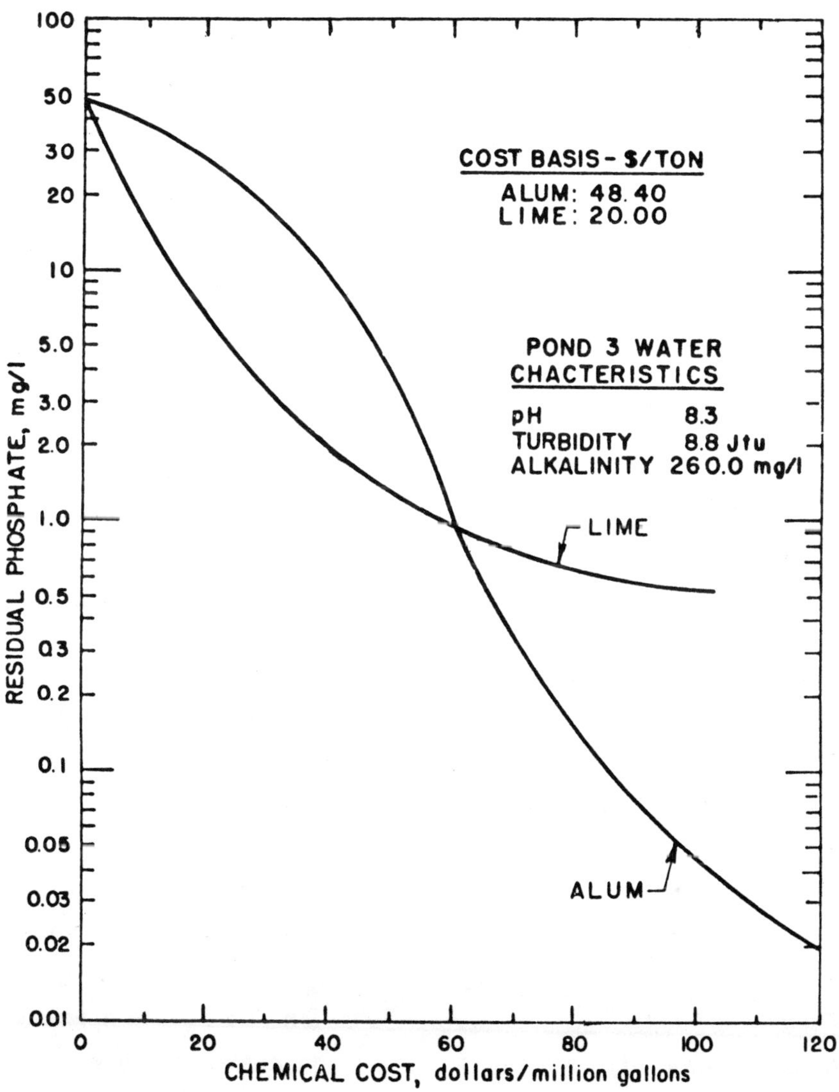

Fig. 25-31 **Residual phosphate for alum and lime coagulation. Laboratory jar tests.**

ment is shown in Figs. 25-32 and 33. Table 25-16 shows the quality of water produced by the process. It has been demonstrated in simulated recreational lakes at the Lancaster site that this water will not produce algal blooms and will retain enough dissolved oxygen to propagate fish.

A 500,000-gpd plant is being designed by Los Angeles County to produce water for 19 acres of recreational lakes that will provide boating, fishing, and picnic areas. At this capacity, the estimated operating cost of the water renovation facility is 18.4¢/1,000 gal. At 3 mgd, the total output of the oxidation ponds, the estimated operating cost is 15¢/1,000 gal.

ULTIMATE DISPOSAL

Many advanced waste-treatment processes produce a waste stream in which the pollutants removed from the treated water are concentrated. The concentrated stream is a minor, but not insignificant, fraction of the total flow. This waste concentrate must be treated, handled, or placed in such a manner that it no longer pollutes the environment. Its disposition has been termed "ultimate disposal".

Table 25-16 Comparison of Oxidation Pond and Chemically Clarified Water Characteristics

Characteristics	Pond Water	Product Water
pH	8.3	6.7
Turbidity (JTU)	90	4
Total Alkalinity (mg/l as $CaCO_3$)	260	95
Hardness (mg/l as $CaCO_3$)	80	100
Suspended Solids (mg/l)	75	6
Dissolved Solids (mg/l)	575	575
Chemical Oxygen Demand (mg/l)	250	50
Biochemical Oxygen Demand (mg/l)	38	<10
Dissolved Oxygen (mg/l)	0.1 to 40	7 - 15
Ammonia-N (mg/l)	0.1 - 20	0.1 - 20
Organic - N (mg/l)	7 - 20	1 - 3
Nitrate - N (mg/l)	1 - 4	1 - 4
Nitrate - N (mg/l)	0.1 - 12	0.1 - 12
Total - Nitrogen (mg/l)	7 - 20	3 - 20
ABS mg/l	3	3
Phosphate (mg/l)	40	0.25
Algae (Counts/ml)	200,000	7,000
Confirmed Coliform (MPN/100 ml)	7,900	<1.8
Chlorine Residual mg/l	-	0.2 - 0.5

Because of the need to prevent further pollution, only a limited number of solutions to the disposal problem are available. These include storage in sealed subsurface cavities or isolated surface ponds, disposal to the ocean in such a way that the seawater will dilute the waste to harmless concentrations, and conversion to innocuous or useful materials. Conversion to innocuous forms is the method often applied to the sludges from conventional treatment processes.

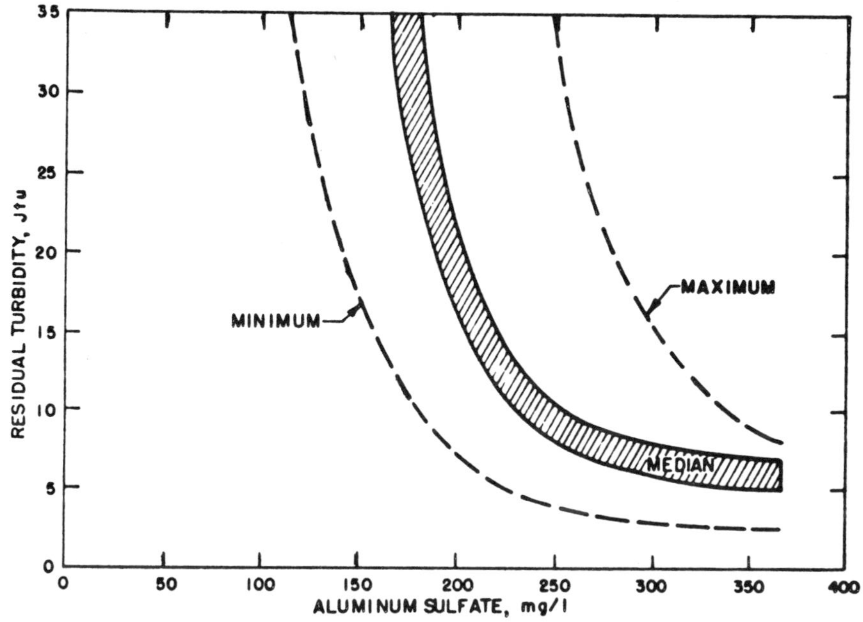

Fig. 25-32 **Residual turbidity versus aluminum sulfate prior to filtration. Pilot plant.**

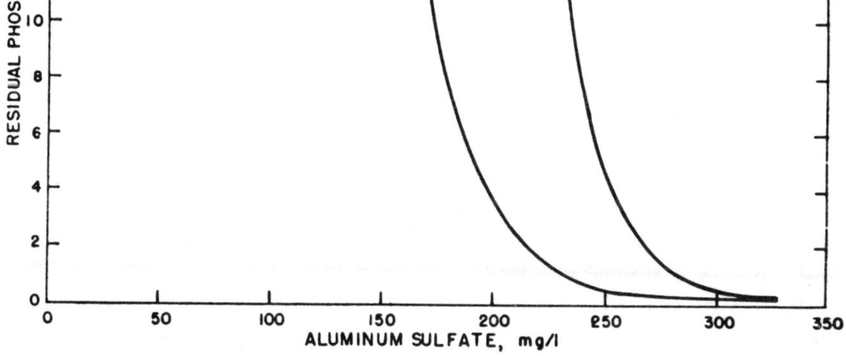

Fig. 25-33 **Residual phosphate versus aluminum sulfate prior to filtration. Pilot plant.**

These sludges are either digested and dried, or incinerated. The dried sludge or ash is then used for land fill.

Study of many advanced waste-treatment processes has now reached the stage where fairly reliable information is available on the volume and nature of the resulting waste. It can be determined, therefore, which processes present the greatest ultimate disposal problem.

Suspended solids removal processes generally should produce small volumes of waste. Mechanical screening processes such as microscreening, when used on secondary effluent, produce essentially no waste for disposal since solids and backwash water can be returned to the treatment plant. There would be a slight increase in the amount of sludge from primary and secondary treatment. Diatomaceous earth filtration wastes could be treated similarly to those from micro-screening. It might be necessary or desirable to dispose of some or most of the diatomaceous earth separately. The result would be a concentrated slurry of small volume. Chemical treatment processes for solids removal ordinarily produce a sizable volume of waste. In an advanced treatment system, this volume would be significantly reduced by recirculating backwash water through primary and secondary treatment, and, hopefully, by recovering chemicals for reuse from the sludge that is formed during treatment. The material remaining for disposal would be the waste from the chemical recovery treatment and should be less than 1 percent of the feed. Insoluble precipitates containing phosphate would be a component of this waste.

Organic removal processes under study produce little or no waste for ultimate disposal. To make adsorption on activated carbon practical, the carbon must be recovered and reused. There is essentially no disposable waste. It might be necessary, however, to purge a small amount of carbon from the system to prevent ash buildup from repeated thermal reactivation. This waste could be recycled through primary and secondary treatment where it would add slightly to the sludge produced. Not enough is known of oxidation processes to accurately assess the volume or nature of wastes produced. The amounts would not be expected to be large enough to cause a significant problem.

Those processes usually thought of as inorganic removal processes, in contrast to the solids removal and strictly organic removal processes, produce significant amounts of waste. Examples are electrodialysis, reverse osmosis, and distillation. Because of scaling problems, it is impossible during treatment to concentrate the wastes in a small volume. Probably 5 percent and in some cases as much as 10 percent of the feed must be discarded. For electrodialysis the waste stream will probably be clarified, have low organic content and probably contain not more than 5,000 mg/l total dissolved solids. Unless clarification and organic removal were required for the feed to other inorganic removal processes, the wastes will contain, in addition to minerals, suspended solids and soluble organics.

Consideration of the wastes from the various advanced treatment methods serves to point out the almost obvious fact that these wastes can be divided into concentrated sludges of relatively small volume and dilute suspensions or solutions of large volume. The sludges are likely to be the digested or undigested waste from conventional treatment mixed with waste from an advanced process. Some differences in the methods of disposal that apply to the two broad categories of waste might be expected.

Desk-top studies have been made of a number of disposal methods by Louis Koenig-Research to determine approximate costs. Included were incineration with disposal of ash (23), wet oxidation (23), injection into underground strata (24), placement in underground cavities (24), and surface spreading (24). More recently, a study was made of the cost of conveying waste with the intention that the waste would be dumped in the ocean. A summary of conveyance costs for rail, trucking, and pipeline are given in Fig. 25-34. Costs from this figure do not include transporting the waste from the shore out into the ocean. Assuming that a 30-mile outfall would be sufficient, an estimate of this cost can be made for pipelines by adding 70 miles to the distance.

Not all the disposal methods studied are practical and some do not completely fulfill the requirements for ultimate disposal. Wet oxidation removes a

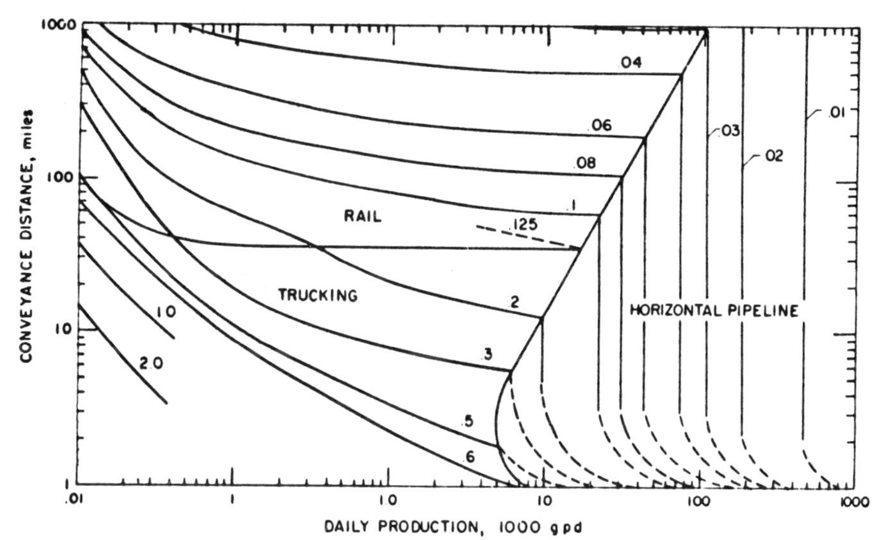

Fig. 25-34 **Cost surface for conveyance: Pipeline, Truck and Rail ($/1,000 gal-mi).**

Fig. 25-35 **Areas of potential injection sites (sedimentary basins).**

large fraction of organics from a waste, but only partially reduces total volume and does little to inorganics. Injection underground, to be useful, requires appropriate geological formations, usually the sedimentary types. Fig. 25-35 shows the distribution of these formations in the United States. Obviously, certain sections of the country could not depend on injection. Placement in underground cavities other than those formed by nuclear explosion does not appear practical. So little information is available on nuclear cavities that these cannot be given strong consideration at this time. Surface spreading could only be considered ultimate disposal if the waste were completely innocuous.

If either of the two general categories of wastes are considered separately, certain disposal methods become impractical. For concentrated wastes of low volume, injection probably would not be feasible because of the presence of a large amount of insoluble material. Remaining complete disposal methods are then, incineration with conveyance of ash to the sea, or direct conveyance of the waste to the sea.

For the case of a 100-mgd plant with 0.5 percent waste by volume, disposal costs per 1,000 gal. of product-water would be about 2.5¢ for incineration without disposal of the ash. Direct disposal of the waste to the sea without incineration would be less than 1¢/1,000 gal. of product for 100-mile conveyance and about 5¢/1,000 gal. for 1,000-mile conveyance. Considering that a conveyance cost for the ash must be added to the incineration cost, there is not a clearcut difference between the costs for the two methods. These figures give some idea of the cost that would be involved in complete disposal of a relatively concentrated low-volume waste. If wet oxidation was all the treatment necessary, it could be done at slightly lower cost than incineration. Spreading would be very cheap, less than 0.1¢/1,000 gal.

Of the complete disposal methods studied, those that might be applied to large-volume, dilute waste are incineration, injection underground, and conveyance to the sea. Incineration must be preceded by extensive preconcentration, assumed to be by distillation. Costs reported for incineration (23) include this preconcentration. In estimating preconcentration cost, however, scaling was assumed to be less of a problem than it probably will be. The reported total incineration costs are, therefore, probably low for dilute wastes. Disposal to the sea without incineration might also involve a preconcentration step for lowest total cost. Again the scaling problem will be an important consideration.

Table 25-17 gives disposal costs by conveyance to the sea for some typical situations in which a dilute waste is produced. Data for the table were taken from Fig. 25-31 and unpublished information from Louis Koenig-Research. For distances of 1,000 miles, preconcentration is necessary for lowest cost. It should be pointed out, however, that preconcentration costs are in doubt and might actually be considerably higher than assumed. It is also possible that if preconcentration were carried to the degree indicated, the product could contain enough precipitated material to be unpumpable. The most important conclusion to be drawn from Table 25-17 is that ultimate disposal to the sea of the large volumes of waste from inorganic removal processes is expensive unless the distance is very short. Dr. Koenig has proposed combining the wastes from a number of cities, "group disposal", to improve the cost picture. More dependable cost figures for preconcentration are necessary, however, before any definite statements can be made of the minimum possible cost of conveyance to the sea.

For one of the cases given in Table 25-17, 100 mgd with 95-percent recovery,

Table 25-17 Effect of Degree of Water Recovery on Cost of Ultimate Disposal to the Sea

Disposal Method: Preconcentration (If Economical), Conveyance, 30 Mile Ocean Outfall

Plant Size, MGD	10				100			
Percent Recovery	95		90		95		90	
Waste Produced, 1000 gpd	500		1000		5000		10000	
Total Solids Concentration Ⓐ mg/l	11000		5500		11000		5500	
Conveyance Distance, miles	100	1000	100	1000	100	1000	100	1000
Preconcentration to Solids fraction = 0.55 Ⓑ	-	Economic	-	Economic	-	Economic	-	Economic
Volume to be Conveyed, 1000 gpd	500	10	1000	10	5000	100	10000	100
Cost for Preconcentration $/day	-	900	-	1400	-	6000	-	7000
Cost for Conveyance $/day	500	310	750	310	2200	3000	2900	3000
Conveyance Mode -	Pipe	Rail	Pipe	Rail	Pipe	Pipe or Rail	Pipe	Pipe or Rail
Cost for Ocean Disposal $/day	420	170	560	170	1400	220	2400	220
Total cost $/day	920	1380	1310	1880	3600	9220	5300	10220
Percent savings in Preconcentration	-	74	-	76	-	61	-	68
Cost per 1000 gal Product water	9.7	15	15	21	3.9	9.7	5.9	11

Ⓐ Assumes 550 mg/l total solids in feed - 50 mg/l organics and 500 mg/l inorganics

Ⓑ Preconcentration to a solid fraction = 0.1 was found to be uneconomical in comparison to 0.55

disposal costs for incineration and injection have been calculated. These are respectively 4.8 and 0.9¢/1,000 gal. product-water. The incineration cost does not include conveyance of ash to the sea. It is also subject to doubt because of possible error in the cost of preconcentration. Incineration appears to be more expensive than direct conveyance of waste to the sea when the distance is about 100 miles, but should be less expensive than direct conveyance for long distances. Injection is significantly less expensive than the other two methods. It should be the method of choice in any locations where the geology is satisfactory.

Future Work

The ultimate disposal studies conducted in the past were undertaken to obtain a broad perspective of possible methods and approximate costs. No attempt was made to demonstrate by experiment the technical feasibility of the methods. Treatments used on the wastes from conventional sewage treatment were not included because their costs were already established. Now that most available disposal methods have been surveyed, it is necessary to conduct detailed studies of those that appear promising.

One point demonstrated by past studies is the high cost of conveying wastes long distances to the sea. Disposal methods such as conversion of waste to useful and innocuous forms deserve extensive investigation since they will reduce or eliminate the need for conveyance. Obvious applications for reuse of the waste include recovery of coagulating chemicals from clarifier sludges and reactivation of exhausted carbon. Presently the dried, digested sludge from conventional waste treatment is sometimes used as a soil conditioner and low-grade fertilizer. Incineration of advanced treatment wastes may in some cases produce a sufficiently harmless residue that disposal as land fill will be satisfactory. This point needs further exploration. In many sewage treatment plants today, sludge is incinerated and the ash is disposed of as land fill.

Unless a waste can be concentrated to a very small volume, the use of conveyance is unrealistic. For the dilute wastes from inorganic-removal processes, where conveyance may in some cases be the only available disposal method, extensive preconcentration is a necessity. Technically, preconcentration is possible. Experimental work is required, however, to determine which types of equipment are practical. Scaling and fouling problems could be severe.

Injection of wastes into the ground is an economically promising disposal method. Little study has been made, however, of the pretreatment requirements for the waste to be injected. Colloidal and soluble organic contaminants could plug the porous rock formations. Inorganic materials in the waste might be incompatible with the strata, also causing plugging. Investigation of these problems is required.

It is likely that conventional sludges will, in the future, contain wastes from advanced processes that have been recycled through the conventional processes. The effect of the added wastes in these sludges requires study. Present disposal methods for wastes from conventional treatment plants are not always satisfactory. Study of ways to get rid of these wastes in a manner that better fulfills the definition of ultimate disposal would be desirable.

REFERENCES

The publications referred to in Chapter 25 may be found in "Summary Report, Advanced Waste Treatment", Publication WP-20-AWTR-19, Federal Water Pollution Control Administration, 1968, pp. 85-87

PHOTOLYSIS MECHANISMS FOR POLLUTION ABATEMENT

Photocatalytic oxidation of dissolved organic matter by irradiation of slurries of zinc titanate (Zn_2TiO_2), zinc oxide (ZnO), titanium dioxide (TiO_2), and beach sand by sunlamps has been demonstrated. The reaction appears to follow first order kinetics in most cases. Zinc oxide appears to be superior for this purpose. At concentrations of 100 to 200 mg/l organic carbon, 80% of phenol, 67% benzoic acid, 44% acetic acid, 40% sodium stearate, and 16% sucrose were oxidized in 24 hrs. with 10gr/l zinc oxide catalyst. Continued illumination reduced organic carbon to a few mg/l in most cases.

The photcatalytic properties of illuminated beach sand (87% of phenol removed in 72 hrs) strongly suggest that photocatalysts are widely distributed in nature and that photocatalytic oxidation is a mechanism whereby dissolved organic matter is oxidized in the natural environment in streams and lakes.

Dissolved organic matter in a sample of domestic sewage was reduced 50% in 24 hrs. and 75% in 70 hrs. The limiting factor appears to be the activity of the photocatalyst since sunlamp intensity dropped approximately 80% without measurable decrease in oxidation rate. Improving catalyst activity by quenching from high temperature and/or by doping is proposed. The effects of ionizing radiation should be studied to determine the economic and technical advantages over UV. A recent publication describing ZnO sensitized polymerization with both UV and gamma radiation reports that zinc oxide caused an enormous enhancement of the polymerizing effect of ionizing radiation in an aqueous system of calcium acrylate and acrylamide which suggests a corresponding increase in free radical generation.

At the present state of knowledge the most pormising area of application appears to be in problems of industrial disposal where high concentrations of organic matter are involved since the amount removed at constant irradiation intensity increases with increased concentration. Small residual concentrations might be economically removed by conventional means.

Certain oxides, notably zinc oxide and titanium dioxide, are known to be photosensitizers or photocatalysts. Goodeve and Kitchener obtained powerful photosensitization with TiO_2, studied by bleaching a blue dye under near-UV irradiation. Jacobsen correlated outdoor chalking behavior of TiO_2 pigmented paints with the loss of reflectance of various TiO_2 pigmented media under sunlamp exposure. Weyl and Forland proposed that atomic oxygen was dislodged by photons from rutile and other oxides; e.g., SiO_2 due to the asymmetric force field acting on the surface oxygen atoms.

Although photo-oxidation by zinc oxide has been much more extensively studied than TiO_2, no method to make it non-photoactive for outdoor pigment use has been found. The use of zinc oxide as a photoconductor in electrophotography has stimulated a great deal of study. Bauer and Neuweiler were first to report the formation of hydrogen peroxide in near-UV- irradiated aqueous suspensions of zinc oxide. Markham et al. studied zinc oxide photocatalysis extensively in aqueous systems and showed that sharply increased hydrogen peroxide concentrations were formed in aqueous media containing organic substances.

Yamamoto and Oster reported that ZnO under UV excitation causes polymerization of vinyl monomers if water and oxygen are present. They concluded, on the basis of end group titration of polymethylmethacrylate polymers formed under these conditions, that hydroxyl radicals initiate the polymerization.

Work at IITRI showed that UV irradiated aqueous slurries of a specially prepared TiO_2, produce an oxidant similar to the hydrogen peroxide formed in UV irradiated zinc oxide slurries. However, this oxidant does not give the amber pertitanic acid coloration specific for hydrogen peroxide. The formation of hydrogen peroxide in irradiated zinc oxide slurries has been confirmed by other investigators by the titanic sulfate reaction. Both oxidants liberate, of course, iodine from potassium iodide. It was first thought the unknown oxidant was ozone; however, the absence of ozone odor at high concentrations and its appreciable solubility in water (concentrations of 500 ppm have been obtained) rule out this possibility.

Either oxidant (produced _in situ_ by UV irradiation of ZnO and/or TiO_2) might provide a means of oxidizing dissolved organic matter in polluted water, since manufactured hydrogen peroxide and ozone have been used for this purpose.

CHAPTER 26

DEEP WELL DISPOSAL METHOD

DEEP-WELL DISPOSAL OF LIQUID WASTES

General

Increasing emphasis on pollution abatement has focused attention in recent years on the concept of discarding liquid waste into deep underground strata through disposal wells. The petroleum industry has shown that deep-well injection, under proper conditions, provides a practical solution to the disposal of oil field waste waters.

A deep-well disposal system that injects liquid waste into underground reservoirs, properly designed with application of sound engineering and geologic principles, will place industrial wastes where they cannot contaminate fresh water or other natural resources.

The concept of subsurface disposal of industrial wastes may be looked upon technically as the inverse of ground water production. Actually, since the same forces and phenomena are involved, the same formulae used for estimating water-well productivity may be used to estimate injectivity. Designed to accomodate various kinds of waste, there are certain differences in well equipment and techniques of completion, but even these are similar in many respects.

Subsurface strata used for liquid waste disposal are, almost without exception, artesian in nature. An artesian formation, as distinguished from one that is unconfined, contains fluids existing at pressures sufficient to cause contained fluids to rise in the bore hole to points above the top of the formation. Until penetration is made, fluids are contained by an impermeable barrier that prevents the native fluids from moving into the formations lying above.

Fluids injected into a disposal reservoir will flow (in an

unfractured formation) from the well bore in a somewhat radial pattern. As they do so, native waters in the formation will be displaced for the most part by injected fluids and will themselves move away from the bore hole. There will be some mixing between injected wastes and natural formation fluids. There are no homogeneous formations in nature (although some may be more uniform than others) but it is practicable to make certain assumptions, including the one that injection strata are homogeneous.

A disposal reservoir must have void space (porosity), the ability to permit fluid movement (permeability), and sufficient lateral extent and thickness (geometry) to accept the type and volume of wastes intended for the system. Porosity and permeability can be a property of the matrix (unfractured) rock or a gross property of the whole rock *in situ*. Gross porosity and permeability are secondary characteristics developed through natural fracturing (jointing) or solution openings or both. Clean sandstones or carbonate rocks having secondarily developed porosity and permeability comprise the better disposal zones.

The disposal formation must be isolated by impermeable formations above and below. These confining beds must be unbroken and capable of preventing the migration of wastes from the selected receptive formation. Evaporite rocks (salt and anhydrite for example), dense limestones, and shales are examples of adequate confining formations.

Whereas some tectonic history of crustal movement may influence positively desirable secondary characteristics in disposal formations, too much structural activity in a specific area may impair the proper functioning of confining beds as a consequence of faulting and other complications.

History of Subsurface Disposal

Subsurface disposal was pioneered in the early 1930's by the petroleum industry. Earlier, salt water produced with petroleum and natural gas had been allowed to flow into and thereby to contaminate surface streams and other fresh water. The petroleum industry, as a solution to the pollution problem, began returning the natural brines to the producing formations or injecting them into shallower, saline aquifers.

Rules and regulations by the various states were soon adopted governing subsurface disposal to protect potable surface and ground waters.

Industrial companies associated with the petroleum industry were not far behind in the use of subsurface disposal. The first industrial disposal wells were also drilled during the 1930's. They were used to dispose of toxic and obnoxious fluids that could not be disposed of safely by dilution with surface water. A large segment of industry, not familiar with drilling, lagged behind the oil companies. This group did not have the understanding of the subsurface; they did not have dry holes nor abandoned producers to convert to disposal service; and the wells used, as a result, were drilled specifically for subsurface disposal. Unfortunately, in too many cases, it was convenient and more economical to dispose of effluents into surface waters. As a result, until 1960, there were fewer than thirty industrial disposal wells in the United States although the petroleum-producing industry numbered them in the thousands.

Geological Requirements for Subsurface Disposal

Past geologic experience has shown that sedimentary rocks are more likely to possess the necessary physical characteristics for a successful disposal system than other rock types. Sedimentary rocks are formed from pre-existing rocks or mineral grains. The matrix or solid components may include rock fragments or particles of various sizes, the remains or products of animals or plants, the products of chemical action or of evaporation, or mixtures of these materials. Sedimentary rocks were derived from pre-existing materials, transported by wind or water as particles or as ions in solution, then deposited and finally indurated or solidified at normal temperatures by one or more of the rock-forming processes which include cementation, recrystallization, compaction, and chemical replacement.

The matrix components of sedimentary rocks possess characteristics important to the physical properties of the rock as a whole. Such characteristics include:

1. Size of individual grain, particle or crystal, and the range of particle sizes;

2. Shape of grains or particles (roundness or angularity);

3. Texture or arrangement of component particles, one to another;

4. Mineral composition of particles, grains or crystals, including the composition of the cementing material.

Properties of the Whole Rock

In addition to specific characteristics of the individual particles, sedimentary rocks possess other properties such as porosity, permeability, and strength. These properties depend greatly on the physical properties of the particles as well as on the degree of cementation and the degree of compaction or structural deformation.

Subsurface rocks may have been fractured by processes of crustal deformation. In such cases, the whole rocks possess two types of void space, one type associated with the fracture network and another associated with the matrix (the unbroken rock).

Porosity

Rocks composed of an aggregate of solid particles or grains may contain voids, fractures, or other openings such as vugs. Porosity is the measure of these void spaces. The ratio of the volume of the void space in a rock to the total or bulk volume of the rock is called porosity and is expressed as a percentage.

Matrix porosity (void space in unbroken rock) of sandstone theoretically can range as high as 47 percent, assuming uniform spherical grains center-packed. In nature, porosity rarely exceeds 35 percent and is commonly less than 10 percent. Gross porosity, a measure of the whole rock including fracture voids and solution cavities (vugs or caverns), may display a very wide range, particularly in rocks such as limestone and dolomite subject to solution phenomena.

Gross porosity is of primary concern because it is a measure of the ability of rocks to contain fluids. Porosity of one percent is equivalent to a total void volume of 3,258 gallons of void space per acre per foot of bulk volume.

The size of individual pores or voids is largely dependent upon the size of the individual solid particles making up the aggregate. A fine-grained rock such as a shale (clay particles size - 2 microns or less) will possess a multitude of small or capillary-size pores between the grains, whereas a coarse clean sandstone (grains - 1-2 mm. in diameter) will contain fewer but much larger voids; yet both rocks can be equal in porosity.

A potential disposal formation must contain sufficient porosity to warrant development and exploitation. Porosity alone, however, is not the unique requirement. The pores or void spaces must be interconnecte and must be of sufficient size to permit the movement of fluids.

Permeability

The permeability of rock is its capacity for transmitting a fluid. The degree of permeability depends upon the size and shape of the pores, and the size, shape and extent of interconnecting openings. Permeability cannot exist without porosity but porous rocks may lack permeability.

Permeability is measured by the rate at which a fluid of given viscosity can move a given distance through a porous medium in a given interval of time. The unit of permeability is the darcy. One darcy is by definition the measure of permeability which will permit movement of one cubic centimeter of fluid of one centipoise viscosity during a time interval of one second under a pressure differential of one atmosphere through a porous medium having an area of cross-section of one square centimeter and a length of one centimeter. The unit commonly used to describe permeability in subsurface rocks is the millidarcy (one-thousandth of a darcy).

The permeability of a porous rock is dependent on a number of variables but largely upon the pore geometry of the system. In unbroken matrix rock the porous system is limited to the intergranular pore space or small solution cavities but, in considering the whole rock in place in the subsurface, fracture permeability may be a factor. As is the case with porosity, where matrix porosity and gross porosity may be different, permeability of a disposal formation can be largely matrix permeability or largely fracture permeability or some combination of the two types. Permeability measurements of rock samples usually determine matrix permeability, but rarely can laboratory tests measure the gross permeability of a disposal formation. These measurements are best made in the bore hole by production or injection tests.

Increases in the permeability of a formation near the bore hole are often possible by inducing fractures through hydraulic pressure or by leaching carbonate rocks with acid pumped into the formation. Decreases in permeability may also result if solid materials in the disposal stream or chemical precipitates plug openings in the formation.

Fluids in the Earth's Crust

Ground Water

Ground water is an excellent solvent, and when fortified by various naturally occurring acids (humic, carbonic, etc.), it has the ability to dissolve certain rocks, notably limestones. Solution cavities range in size from a micron or less to the size of natural caverns. Much of the porosity and permeability of limestone originated in the geologic past when the rocks were exposed to ground waters at or near the atmosphere/ground water interface.

Potable ground water normally is found near the present land surface, but fresh water occasionally occurs at considerable depths. Occasionally, surface waters are contaminated by natural salts but, as a general rule, brackish water or natural brines lie below a zone of fresh ground water.

Brines and Connate Water

Most sedimentary rocks are of marine origin; they originated as sediments in ancient seas. These seas contained salts in solution, as do modern oceans, and they were probably of similar composition. Some of the interstitial water in the sediments was immobilized as films on the individual grains and remains as true "connate" water in the formation. However, most of the water in the earth's crust is mobile. Compaction of the sediment with burial squeezes most of the original water out of the sediment to areas of lower hydrostatic pressure, usually upward or laterally.

The movement of water in the earth's crust is slow, and may for long periods remain static. Throughout geologic time crustal movements have imposed new hydraulic gradients on the fluids in the pores of the rocks and these fluids then seek to move to positions of lower potential energy. The chemical composition of a natural brine in the pores of rocks in the earth's crust depends not only on its source and nature before it enters the subsurface but also on the many events and the changing environment to which it is exposed through movement in the course of geologic time.

CHAPTER 27

ELECTROPHORETIC AND ELECTROCHEMICAL WATER PURIFICATION

Electrophoretic and Electrochemical Methods in Water Purification

Electrophoresis along with other electrokinetic and electrochemical phenomena appears to be a very promising method of water purification. Very simply, electrophoresis is defined as the migration of charged suspended particles in a d.c. electric field. This process of water purification is made possible by the fact that most naturally occurring substances carry a negative charge.

Clays derive their negative charge primarily by substitution of ions of lower valence within the crystal lattice. This involves primarily the substitution of Mg^{++} and Fe^{++} for Al^{3+} in the octahedral layer of the clay crystal and Al^{3+} for Si^{4+} in the tetrahedral layer. Bacteria and algae are normally negative because of dissociable polar groups on the cell surfaces which ionize in water. These groups are primarily carboxyl and amino groups and the degree of ionization depends largely on the pH of the system.

Only in the last decade has electrophoresis been experimentally applied to the purification of water, primarily for the removal of suspended particulate matter. It has also been found to decrease other colloidal and solute impurities (4, 9, 11, 15, 16).

Bier (4), in a laboratory study of forced-flow electrophoresis, completely eliminated suspended particulate matter of the types previously mentioned and radically decreased collodial impurities such as iron, humic acid, proteins, and detergents. In addition he reported substantial desalting of the purified effluent.

Significant color removal may also be achieved by electrophoresis. Most authors believe color in water is colloidal in nature and thus should possess a negative charge (7). Black and Christianson (6) have shown that the materials responsible for color in water exist primarily in colloidal suspension in water. They suggest that the color in water is due to light scattering and flourescence rather than molecular absorption of light energy.

Electrophoresis in combination with electrochemical methods also offers considerable potential in specialized water treatment problems where other methods are not applicable or effective. Such a situation was reported by Boyd (9) in the cleanup of heavy water at the Hanford Plutonium Recycle Test Reactor. Another indication of the tremendous potential of electrophoresis for water purification is the fact that as of 1960, the purest laboratory processed water reported to that date was obtained utilizing electrophoresis. Haller and Duecker (14) using a modified electrophoretic procedure for ion removal obtained an electrical conductivity of 0.039×10^{-6} mhos cm^{-1} at 18° Centigrade. This was three times lower than the lowest previously reported value which was obtained by 36 consecutive vacuum distillations. This ultra low-conductivity water was prepared by recirculating water through an electric field of 1,000 volts cm^{-1} maintained between two ion selective membranes.

Tertiary wastewater treatment is another possible use of this method. Faust and Manger (12) determined the electrophoretic mobility of particulate matter in domestic wastewater from 707 observations. They reported values ranging from 0.55 to 3.75 microns sec^{-1} $volts^{-1}cm$ with an average of 1.73. This mobility is on the same order of magnitude as most clays.

There exists a very definite potential for application of electrophoretic and electrochemical methods to the purification of waters. The design, development, and testing of economically feasible systems yet remain.

SYSTEM DESIGN PRINCIPLES

Three basic design theories or principles are advanced here and were incorporated into several electrophoretic and electrochemical clarification system models for evaluation. These design principles concerning essentially different electrode arrangements and flow regimes are obtained primarily from prior theoretical relationships and parameters which have been shown previously to influence electrophoretic or electrochemical removal of pollutants from water. Each design criterion was aimed at obtaining both good operational utility and acceptable economic efficiency.

Parallel Plate Design

Hiler (16), in evaluating electrophoretic clay removal between two parallel plates, was concerned primarily with the deposition of the colloids on the positive plate. This is likewise the concept inherent in Boyd's design (9). The design advanced herein differs somewhat from those which advocate sediment removal almost entirely by deposition on the anode. Instead this design was intended to obtain a downward movement of suspended material toward a turbid effluent outlet with a minimum of deposition on the electrode plates. A given portion of clear effluent was removed from the top portion of the unit. Minimizing deposition is highly desirable in order to extend the time period between cleaning cycles (polarity reversal and flushing), thereby increasing the efficiency of the system.

As clay moves to, and is deposited on, the anode by electrophoretic transport in a clay-water suspension it becomes stationary. However, it is not free from electrokinetic effects, the electric field being still present. The phenomenon is, however, changed from electrophoresis to electroosmosis. In the latter case, due to the electric field, water moves past the stationary charged clay particle rather than vice versa as occurs with electrophoresis. As this happens, clay layers deposited next to the anode slowly become dryer as the water moves away from the positive plate toward the liquid medium. This drying will continue progressively outward from the plate building up an ever thickening layer of drying deposited clay. As this happens the resistance to current flow is substantially increased and thus a lowering of the electric field strength occurs. When the field strength decreases significiantly it becomes necessary to reverse the voltage polarity to remove the clay layer and then flush the system. Thus it is easily seen why it is desirable to keep clay deposition to a minimum.

Fig.27-1 illustrates the principle upon which the parallel plate model was designed. A downward movement of colloids with minimum clay deposition was achieved by constructing banks of horizontal, parallel plates with a given vertical separation between oppositely charged plates in the same bank. A slight drop from one bank of electrodes to the adjacent bank of electrodes in the direction of flow gives a downward gradient to the plates toward the effluent end of the model.

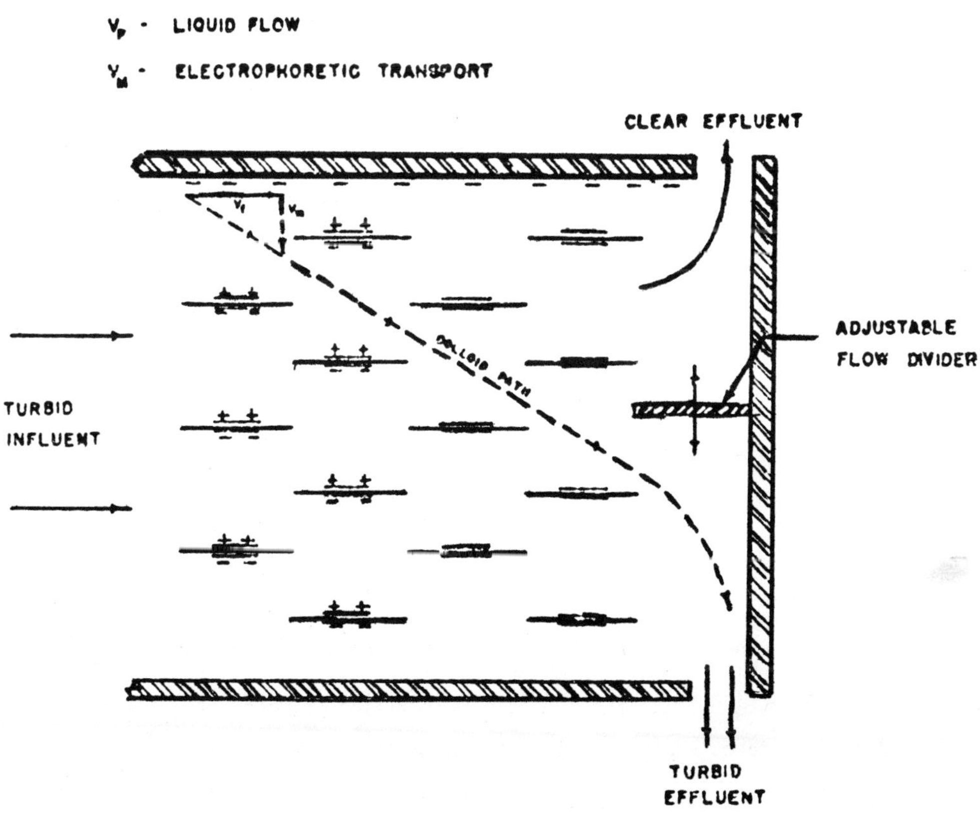

Fig.27-1 Schematic diagram of parallel plate design principle.

Electrophoretic transport velocity is directly proportional to the electric field strength between two electrodes as is shown in the Helmholtz-Smoluchowski equation for the electrophoresis of suspended particles:

$$V = \frac{D \chi \zeta}{4 \pi \eta},$$

where

V = transport velocity,

D = dielectric constant,

χ = electric field strength,

ζ = zeta potential, and

η = absolute viscosity.

Thus the hydraulic flow rate in the model and the electric field strength between the plates may be adjusted so as to give a downward gradient to the suspended material that is equal to the downward slope between the parallel electrodes, resulting in minimum clay deposition on the plates.

It is obvious that a prior knowledge of the zeta potential or electrophoretic mobility of the colloidal particles is necessary in order to determine the proper electrical field so that the resulting transport velocity when added vectorially to the flow velocity will give a resultant equal to the gradient of the parallel plates. The electrophoretic mobility may be defined as the transport velocity when the electrical field strength is unity and has the units microns sec^{-1} $volts^{-1}$ cm. These mobilities were determined by the microscope method of electrophoresis and are discussed later.

Porous Filter Design

A porous filter electrophoretic clarification model was designed much along the same line as the Bier unit but with some definite modifications. The Millipore and Gelman type filters were replaced by a porous sintered stainless steel filter which allowed the filter itself to become the negatively charged cathode. By making the filter the negative electrode, the passage of the electric field through the purified water is eliminated. This substantially lowers the electrical potential necessary for a required field strength especially if the conductivity of the treated water has been lowered due to the electrophoretic process. A cost reduction of at least one-half should thus be realized.

Fig.27-2 gives a schematic diagram of this electrophoretic system design. Turbid influent suspension enters the unit under a constant head and at a constant rate at the top of a model. When operating electrophoretically, at low conductivities, the negatively charged particles, as they pass between the electrodes in the inlet compartment, will migrate toward the positive electrode due to the electrophoretic transport. A portion of the water is drawn through the stainless steel filter at a constant rate and exits the model. If the velocity of particle migration is maintained equal to or greater than the hydraulic flow through the filter in the opposite direction, the colloids should not pass through the filter nor should the filter clog. In effect, the result is a continuously backwashing filter. Turbid effluent is withdrawn from the inlet compartment at the bottom.

When this model is operating electrochemically, at high electrical conductivities, the filter should likewise stop the flocculated particles, allowing them to be carried downward by gravity and swept out the bottom outlet. This model may necessitate periodic backwashing when operating in this manner. The gaseous products of electrolysis escape out the top of the unit.

Electrode Grid Design

Two parallel electrode grids constitute the third electrode arrangement. Water enters the unit from one to four directions between the grids where it divides going toward two effluent outlets. This is illustrated in Fig.27-3. This design facilitates very close electrode spacing without appreciable flow velocity and turbulence. The electrode grid rather than a porous plate also takes advantage of any coagulating electrochemical reactions without the possibility of a plugging problem.

A computer program utilizing a method of numerical relaxation for solving three dimensional flow nets was developed in order to obtain the optimum dimensions of this type of electrode grid. It was also used to determine the relationship between grid potential and electric field strength between the grids.

Consider that the steady state electrical flow field between the grid electrodes is composed of a very large but finite number of grids or series of lines running parallel to each of the three axes.

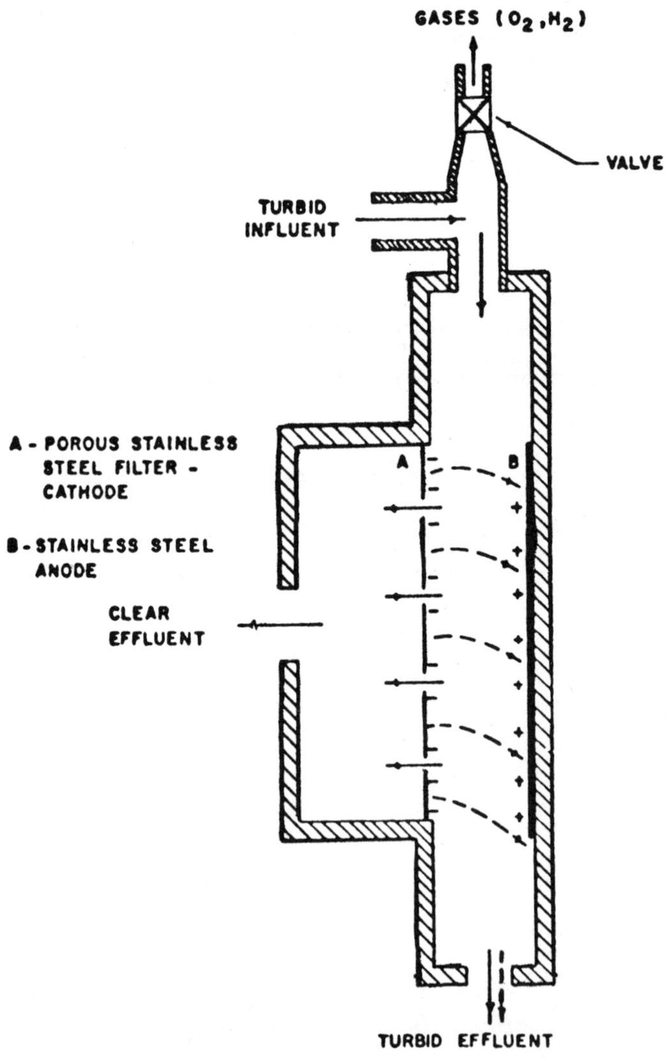

Fig.27-2 Schematic diagram of the porous filter system design. Solid arrows indicate water flow. Dotted arrows are colloidal transport.

Next consider an arbitrary point 0 in the flow field with a potential of ϕ_0 surrounded at distance "a" on each axis by points 1, 2, 3, 4, 5 and 6 with their respective potentials. This is illustrated in Fig.27-4. Next consider point A halfway between point 0 and point 1 and Point B halfway between point 3 and point 0. This is also shown in Fig.27-4.

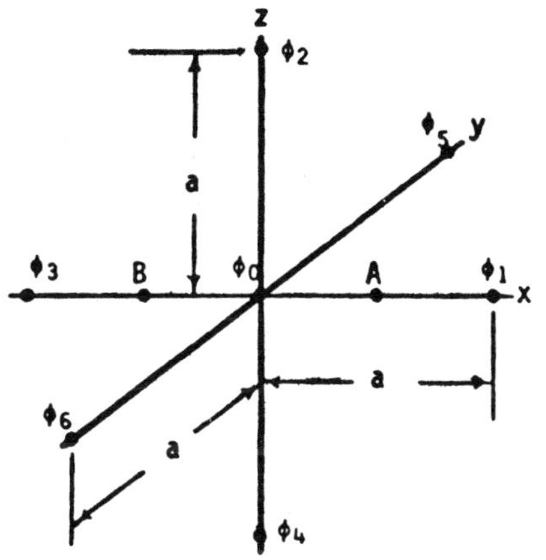

Fig.27-4 Arbitrary point 0 in hypothetical flow field for computer analysis.

The following partial derivatives may be represented in finite difference form as follows:

$$\left.\frac{\partial\phi}{\partial x}\right|_A = \frac{\phi_1-\phi_0}{a}, \quad \left.\frac{\partial\phi}{\partial x}\right|_B = \frac{\phi_0-\phi_3}{a},$$

therefore

$$\frac{\partial^2\phi}{\partial x^2} = \frac{\frac{\phi_1-\phi_0}{a} - \frac{\phi_0-\phi_3}{a}}{a} = \frac{\phi_1-2\phi_0+\phi_3}{a^2}$$

Likewise in the y and z directions

$$\frac{\partial^2\phi}{\partial y^2} = \frac{\phi_5-2\phi_0+\phi_6}{a^2} \qquad \text{and}$$

$$\frac{\partial^2\phi}{\partial z^2} = \frac{\phi_2-2\phi_0+\phi_4}{a^2} .$$

Laplace's equation describing three dimensional steady state potential flow is

$$\frac{\partial^2\phi}{\partial x^2} + \frac{\partial^2\phi}{\partial y^2} + \frac{\partial^2\phi}{\partial z^2} = 0.$$

Rewriting in finite difference form gives

$$\frac{\phi_1-2\phi_0+\phi_3}{a^2} + \frac{\phi_5-2\phi_0+\phi_6}{a^2} + \frac{\phi_2-2\phi_0+\phi_4}{a^2} = 0$$

and simplifying we have

$$\phi_0 = \frac{1}{6}(\phi_1+\phi_2+\phi_3+\phi_4+\phi_5+\phi_6).$$

The next step is to assume a network of values within the framework of the boundary conditions which exist. This was represented in the form of a three dimensional array in the IBM 360/65 digital computer. From the initially assumed network of values, through the elimination of residual errors by numerous iterations, the eventual satisfaction of both the differential equation and assigned boundary conditions was achieved.

The computer solution by the relaxation process is shown in Figs. 27-5 through 8. The maximum potential drop between the electrode and locations in a plane which coincides with the bottom of the electrode grid, will occur at a point in the center of each grid dimension.

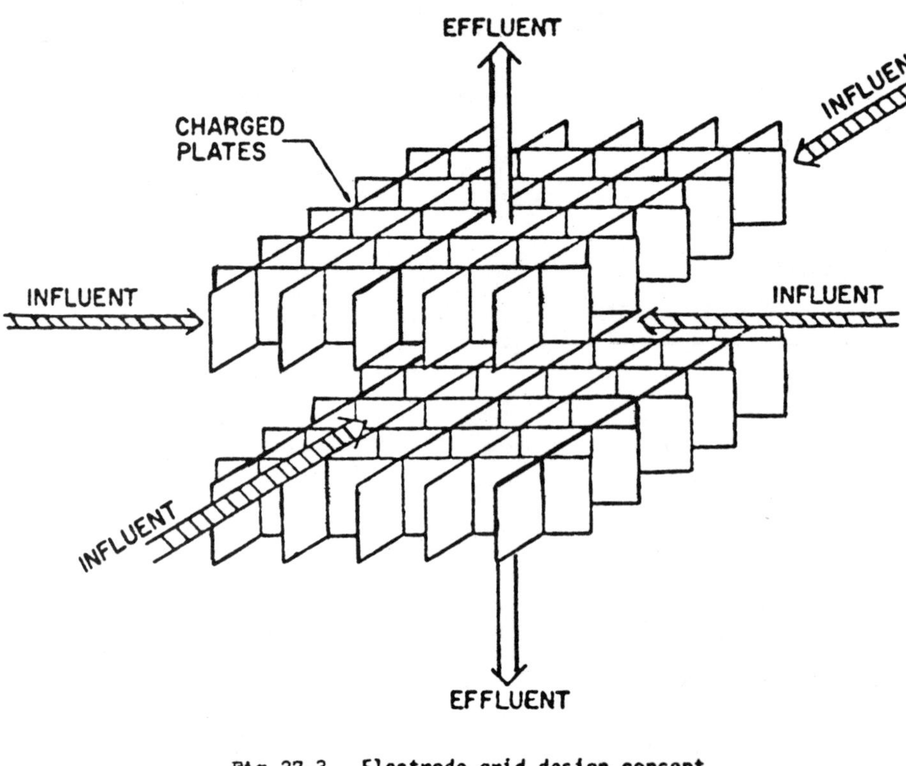

Fig.27-3 Electrode grid design concept.

The minimum field strength between the grids will occur between this point and an identical point in the oppositely charged grid. Figs. 27-5 and 27-6 express this as the minimum potential available in percent of the potential applied to the grid itself. This is plotted versus relative electrode spacing and relative vertical grid depth. The common denominator of these two dimensionless values is the horizontal dimension of each small grid section or is the side dimension of each square void space within the grid. The average potential available between the grids, again expressed as a percentage of grid potential, is presented in Figs. 27-7 and 8 and is plotted versus the same relative values.

The curves in Figs. 27-5 and 7 are of course asymptotic to 100 percent available potential. However, for practical purposes, any electrode spacing to grid width ratios of 2.0 or over will give efficient service. Likewise ratios of less than about 1.2 should be avoided. Figs. 27-6 and 8 indicate that a vertical grid depth to grid width ratio greater than approximately 1.6 is not necessary. In addition ratios of less than about 0.8 will substantially lower the efficiency of the system.

The models constructed for evaluation incorporated designs for pollutant removal entirely by electrophoresis on one hand to designs which could accommodate both electrophoretic and electrochemical processes on the other. One model was tested which was designed for electrophoretic pollutant removal only. This was the horizontal parallel plate model which was tested with very low conductivity water in order to limit the process occurring in the model primarily to electrophoresis.

A porous stainless steel filter model was designed which could function either electrophoretically or electrochemically. To limit the process to electrophoresis, again low electrical conductivity water was used. When operating electrochemically at higher conductivities, this model essentially combined the entire spectrum of conventional water treatment operations into one unit. A problem with clogging, however, indicated that filtration should be made a separate operation. As a consequence, an electrode grid model was also designed to operate either electrophoretically or electrochemically and combined the same operations as did the porous filter model with the exception of the filtration process. It was tested with two electrode materials, stainless steel and aluminum.

Clay suspensions using commerical kaolinite clay in various concentrations were mixed in order to evaluate suspended particulate matter removal by each of the models. These measurements were made by turbidimetric means. Sewage effluent was mixed with

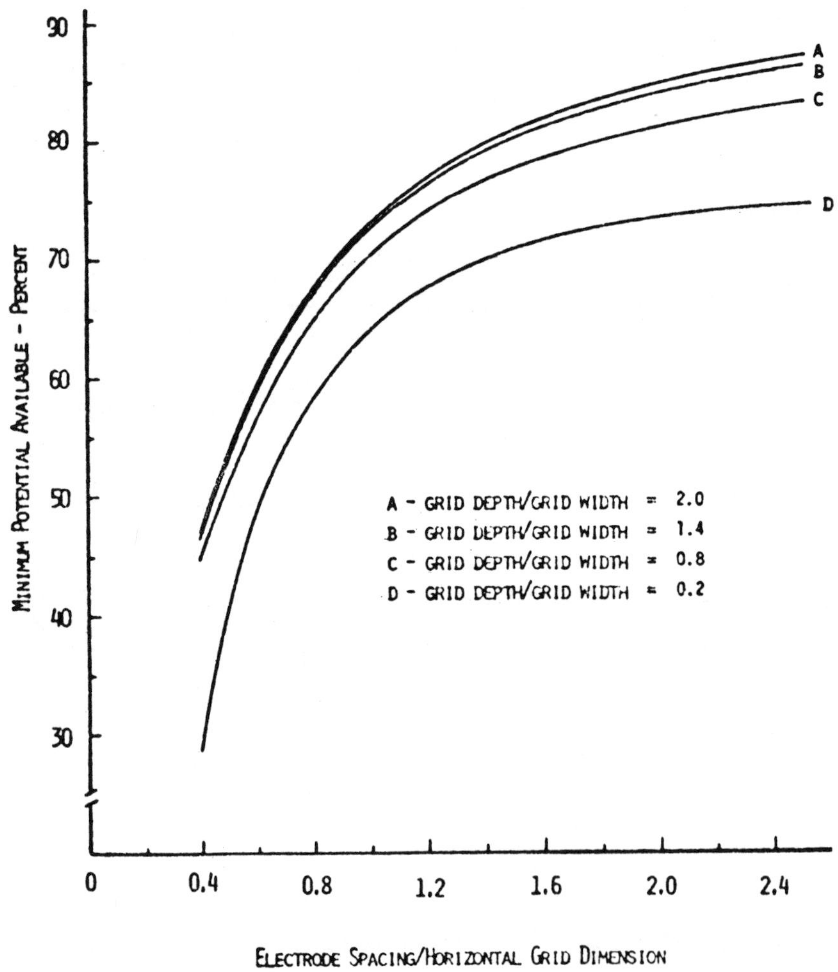

Fig. 27-5 Minimum potential available versus relative electrode spacing.

the suspension in order to evaluate bacteria removal or disinfection as measured by the most probable number (MPN) of coliform bacteria. In addition, electrophoretic mobility determinations were made on several Texas soils for the purpose of characterizing them for ease of electrophoretic removal. These mobilities were determined by the microscope method of electrophoresis.

As a result of exhaustive testing of the various models and the design concepts which they represented, the following con-

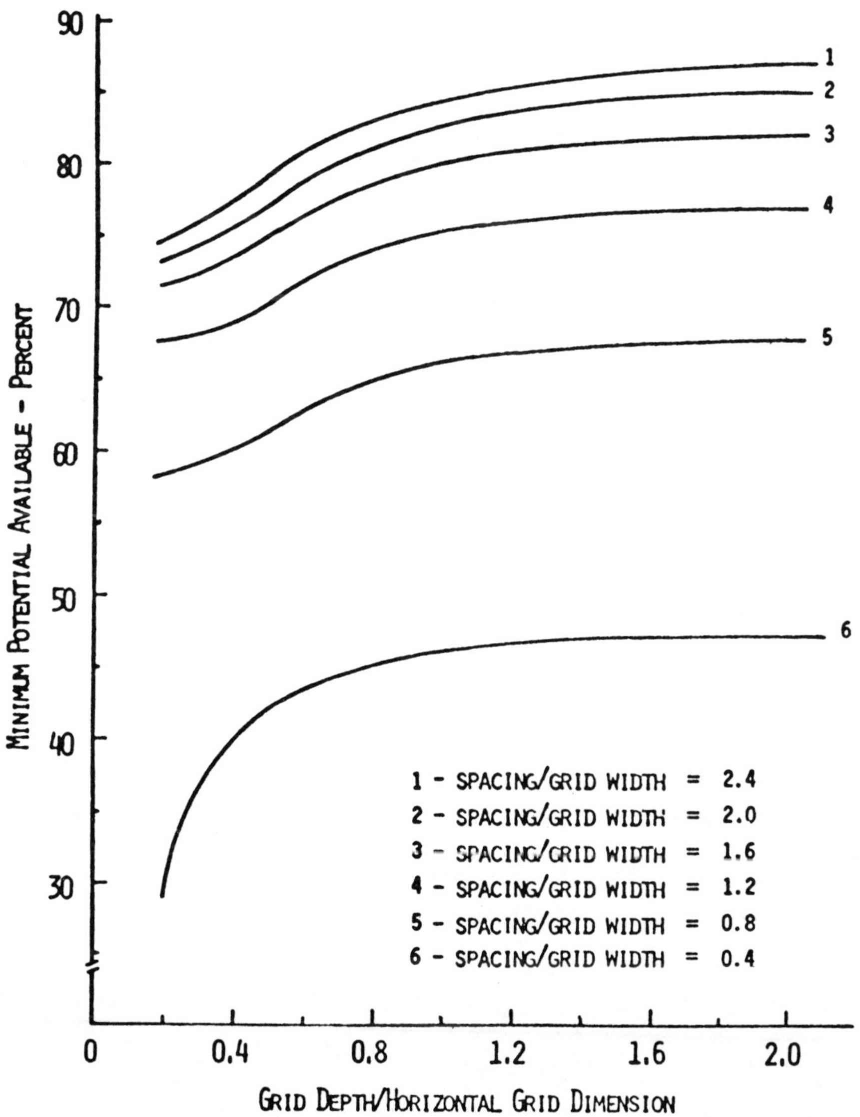

Fig. 27-6 Minimum potential available versus relative grid depth.

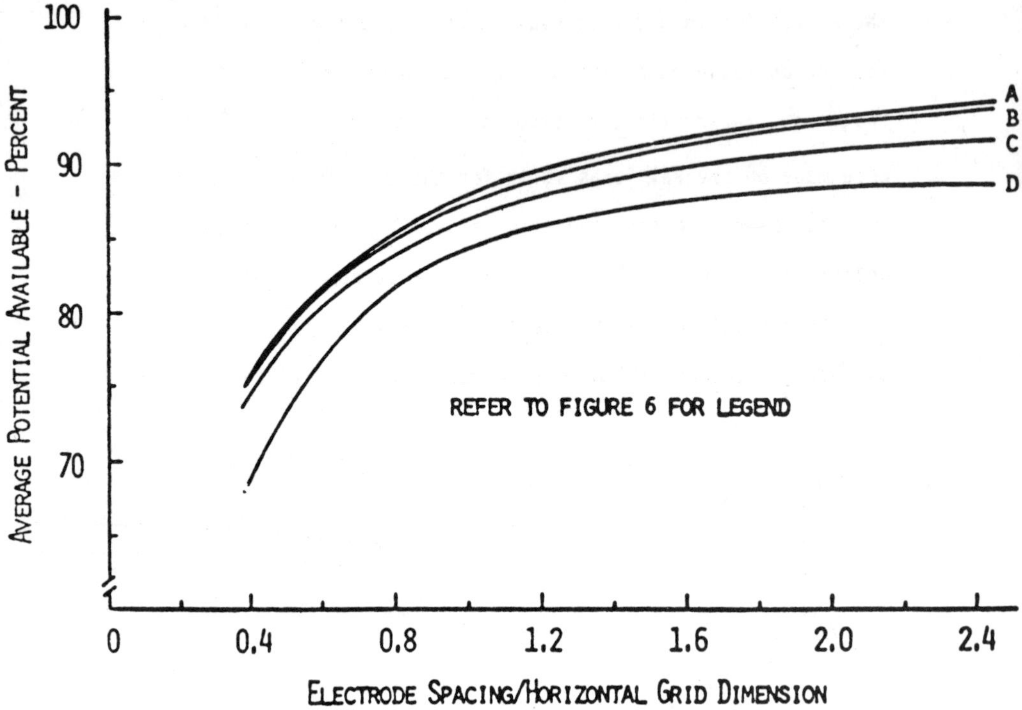

Fig. 27-7 Average potential available versus relative electrode spacing.

clusions are drawn:

1. Turbid waters can be successfully clarified by electrophoresis, but the process is limited to waters of very low electrical conductivity if the electrodes are exposed to the bulk suspension being clarified. At higher electrical conductivities, the electrolysis products are sufficient to disrupt entirely the electrophoretic transport. In addition, primary and secondary chemical reactions of the electrolysis products cause changes in the properties of the suspension. In some cases, they neutralize the negative charges of the colloidal particles rendering them unaffected by the applied electric field. Under this situation, the use of electrophoresis as a sole means of treatment for natural waters is virtually impossible.

2. In order to obtain clarification of natural waters entirely by electrophoresis, the electrodes must be separated from the bulk colloidal transport area. The method described by Bier

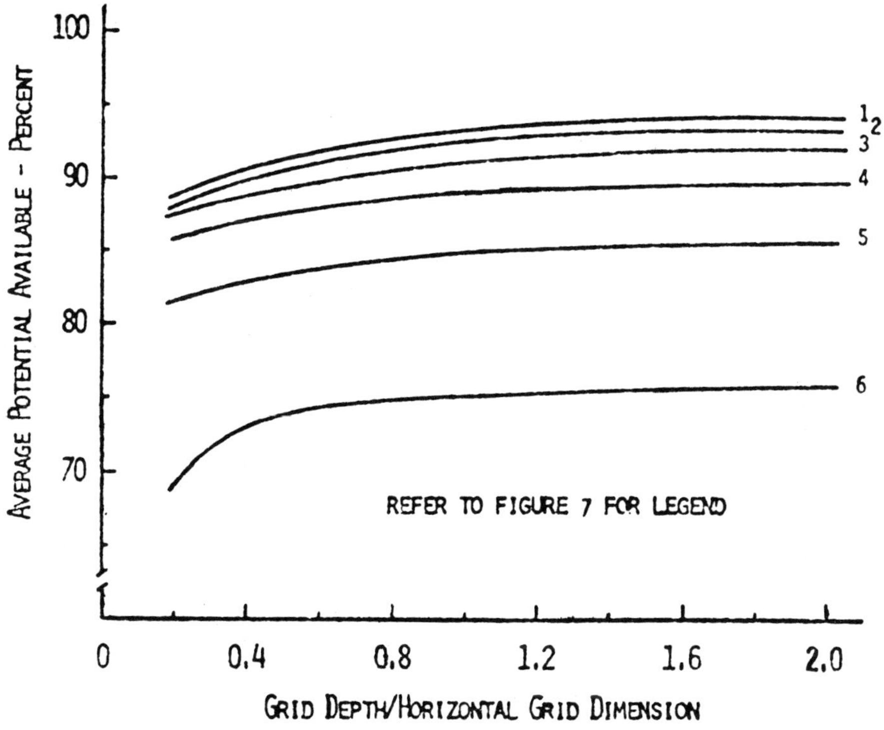

Fig. 27-8 Average potential available versus relative grid depth.

(3,4,11,19) appears to be a logical way to accomplish this.

3. The economic justification of treatment purely by electrophoresis is debatable if suspended colloidal particulate matter removal is the primary purpose of treatment. The operating costs are much too dependent upon the electrical conductivity of the water. This method would probably be limited to waters with electrical conductivities of 200 to 300 micromhos cm^{-1} and below. Even at these conductivities, the price paid for convenience and automation over conventional methods would be considerable. However, if ionic pollutant removal is also necessary, the utilization of electrophoresis along with dialysis might be a practical dual purpose use of the applied electric field.

4. Water clarification by electrochemical means was successful both operationally and economically. For optimum economic operation, a design should consider only electrochemical flocculation without regard to any additional electrophoretic clari-

fication capabilities.

5. Aluminum appears to be better than stainless steel as an electrode material for electrochemical flocculation.

6. When electrochemical flocculation is used for clarification, the costs are roughly proportional to the pollutant concentration and inversely proportional to the electrical conductivity.

7. Disinfection as well as flocculation is possible by electrochemical means. It is accomplished by chlorine gas evolution at the anode due to electrolysis and should be a separate unit operation for maximum efficiency and effectiveness.

8. A small practical water system appears entirely feasible which consists of electrochemical flocculation, settling, and electrochemical disinfection operations. The use of filtration would be optional but if desired, a replaceable cartridge type filter might be well adapted here. In this case the entire system could be automated with the exception of periodic replacement of flocculator electrodes and filter if used. The settling basin may even be designed so that it flushes itself automatically at given intervals. The physical addition of chemicals is entirely eliminated with this system.

REFERENCES

The publications referred to in Chapter 27 may be found in "Electrophoretic Clarification of Water", by E. A. Hiler, et al., Water Resources Institute, Texas A & M University, Aug. 1970, pp. 108-110

CHAPTER 28

RADIOACTIVE WASTE TREATMENT

Low-level radioactive wastes, consisting of water with low concentrations of salts and radionuclides, are produced at most nuclear installations. Conventional decontamination methods include scavenging precipitation, ion exchange, or distillation, followed by discharge of the purified water to the environment. The objectives of the work reported here are: (1) to determine the technical feasibility of recycling the purified water for reuse (e.g., to determine the effects of the concentrations of contaminants in recycled water; and (2) to develop improved decontamination methods. Recycle appears desirable since it would eliminate the cost of purifying the water that is taken from rivers for use in the nuclear station; also, it would allow the nuclear station to become self-contained. Improved decontamination would, of course, decrease the flow of radionuclides to the environment. The radioactive concentrate could be incorporated in asphalt or polyethylene prior to disposal in a government burial ground.

Some of the solids in the waste are present in the form of colloidal particles. Thus, efficient clarification is important as the first treatment step. Zeta-potential control was developed as a regulatory method since previous workers had reported that it showed considerable promise.

The second step is demineralization for efficient removal of soluble salts and radionuclides. This portion of the process differs from the previously developed process in which a high degree of decontamination, but only partial demineralization, was achieved. The third step is sorption on activated carbon. This sorption system was evaluated in an effort to increase the removal of radionuclides, such as ruthenium and cobalt, which had proved to be refractory in previous studies.

The water recycle process is compatible with present public opinion and with government policies that encourage the recirculation of waters and the minimization of discharge of both radioactive and nonradioactive salts to the environment.

This type of process should also be applicable to the recycle of nonradioactive industrial and urban waste waters that require clarification and demineralization.

PROCESS FLOWSHEET

The water recycle process, which was developed for the treatment of low-level radioactive waste having a low salt content (Fig. 28-1), consists of the following steps: coagulation and clarification, demineralization, and treatment with activated carbon. The decontaminated and demineralized water is returned to the nuclear installation for reuse, while the concentrated radioactive waste is denitrated, if necessary, and then incorporated into asphalt and buried.

The flowsheet(Fig. 28-2)includes:(1) coagulation by zeta-potential-controlled additions of the coagulant, alum, and the coagulant aids; activated silica; and a nonionic organic polyelectrolyte; (2) clarification by upflow through a fluidized bed of sludge (upflow clarifier), followed by filtration through a bed of anthracite coal and sand; (3) demineralization by cation-anion exchange; and (4) sorption of the remaining radioactive and nonradioactive contaminants on granular activated carbon. Continuous operation of the coagulation-clarification system under optimum conditions is important to the maintenance of high decontamination factors (DF's) for radionuclides such as cobalt and ruthenium, which, in neutral wastes, exist in sev-

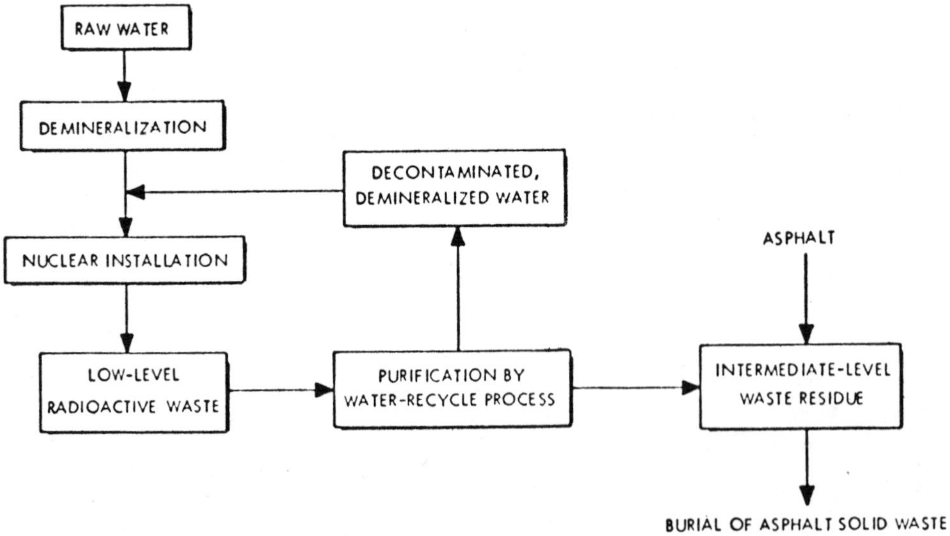

Fig. 28-1 Treatment of Low-Level Radioactive Waste and Disposal of Residues.

eral forms including ionic and nonionic colloids. Efficient removal of these colloids depends on the neutralization of their mutually repelling surface charges, that is, their zeta-potential (ZP). The successive addition of activated silica (0.5 ppm as SiO_2) and alum [2 to 3 ppm as $Al_2(SO_4)_3$] in the flash mixer increased the ZP of synthetic recycle water from -20 mv to +3 mv. Stirring at 1750 rpm effectively enmeshed the colloids within the freshly formed aluminum hydroxide floc. The addition of 75 ppb of a nonionic organic polyelectrolyte* at the flocculator enhanced the growth and settling properties of the floc and accomplished the desired filtering action in the upflow clarifier (the primary filtering medium) by maintaining a fluidized bed of sludge. The final ZP was -2 mv. After filtration through anthracite and sand, the water had a clarity comparable to that of distilled water, representing a greater than 100-fold decrease in turbidity.

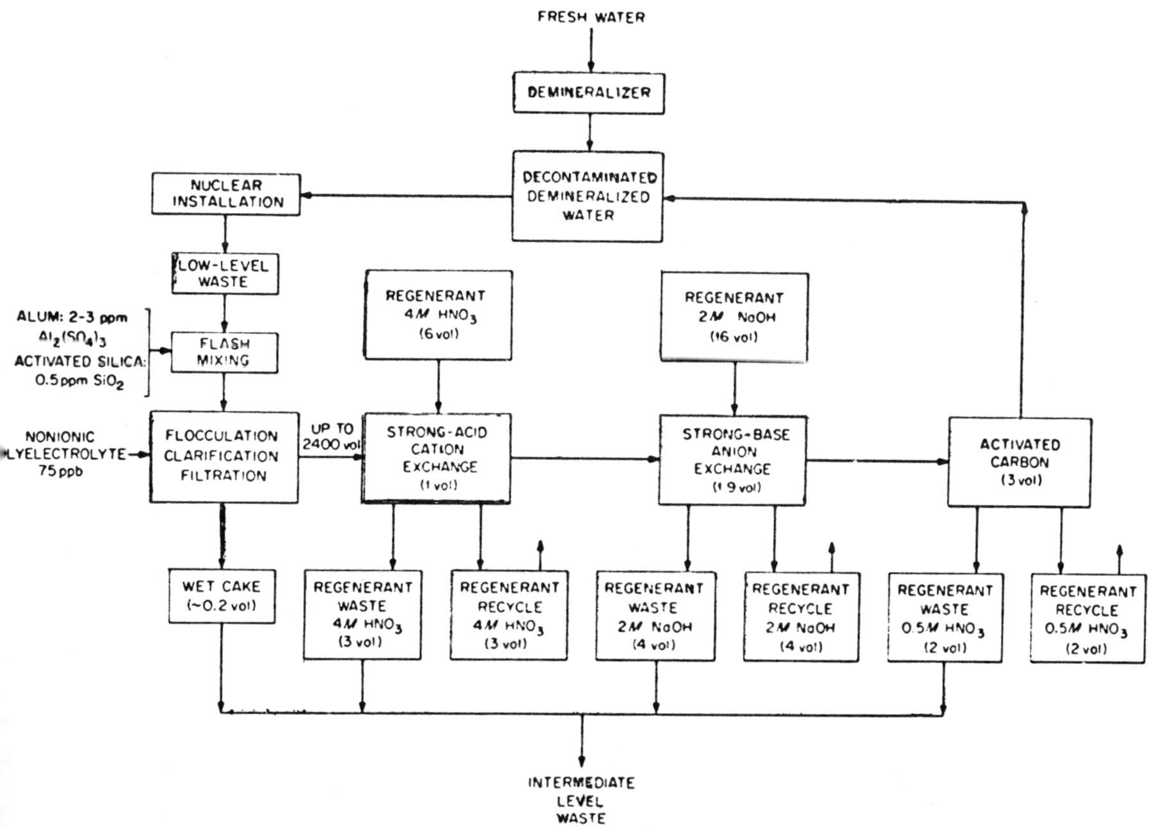

Fig. 28-2 Flowsheet for the Water Recycle Process.

*Purifloc N-12, product of the Dow Chemical Co., Midland, Mich.

CHAPTER 29

REDUCING POLLUTION FROM COAL MINES

Stream pollution by the acid and the dissolved and suspended solids in coal mine drainage is extensive in many areas of the Appalachian region. It severely limits the polluted streams for recreational, industrial, and municipal uses. Pollution from mine drainage is characterized by a low pH and a high concentration of metallic salts, chiefly iron. These are formed by the oxidation of exposed pyritic material in coal and rock after mining. Water entering an underground mine through the overburden leaches the salts and acid from the oxidized surfaces and transports them to the surface streams.

The purpose of mine air-sealing is to prevent or reduce oxidation of pyrite in the coal and adjacent strata and thereby to prevent the formation of sulphuric acid and iron sulfate. Therefore, the term "air-sealing" is defined as any means of excluding the entry of air into a mine, whereby the oxygen content in the mine atmosphere is sufficiently reduced so as to retard or prevent the oxidation of pyrite.

Conventional air-sealing entails the closing of all openings to the mine with impermeable material or structures, except that an air trap is placed in one or more of the lower seals. This prevents the inflow of air but permits the outflow of water. The methods used to air-seal will also serve to reduce in varying degrees water entrance into the mine and thereby reduce the effluent discharge rate and acid load.

In mines above drainage, with the coalbed dipping downward from the portal at the outcrop, and in shaft mines, air-sealing can generally be accomplished by complete inundation. However, if the coalbed rises from the main portal at the outcrop, inundation, whether partial or total, will depend on the extent of seepage through the outcrop. Mines above drainage and totally outcropping can be inundated after sealing generally only to the elevation of the lowest surface subsidence holes or fractures.

Effective mine sealing in which the oxygen content is reduced is sound in theory and supported by laboratory tests at the Bureau of Mines[3] and unpublished data, which determined that oxidation rates of pyrite are a function of oxygen concentration in the presence of water, with some catalytic effect from bacteria. Mine sealing has been practiced sporadically by industrial firms, research organizations, and State and Federal agencies since 1923 with varying degrees of success.[4] It remains a controversial water pollution abatement measure, however, because adequate means of evaluating the results were lacking.

In an effort to further evaluate the effectiveness of mine air-sealing, the Bureau of Mines sealed a small, abandoned, highly acid above-drainage (drift) mine in western Pennsylvania. This type of mine is representative of those which are responsible for approximately 35 percent[5] of the stream pollution in the coal regions of Appalachia.

Air-sealing requires a comprehensive program of engineering which may be difficult and expensive under adverse conditions, with no guarantee of total success. Mine sealing is difficult to accomplish where undiscernible joints and subsidence fractures exist in shallow overburden, resulting in the phenomenon of mine "breathing." However, modern sealing techniques are available, which can be effectively used if air and water channelways into the mine can be located. Even with improper sealing, some improvement in effluent quality should be achieved.

The first requisite for evaluation of air-sealing and water diversion at an abandoned mine is to secure accurate mine maps and locate all air and water entrances into the mine. The quality and quantity of all water issuing from the mine should be monitored continuously for at least 1 year to determine seasonal variations and obtain a base for comparison. In the present experiment, the mine effluent was monitored for 2 years 7 months prior to sealing and an equal time after air-sealing in May 1966. The mine atmosphere was sampled at the seals and at one core hole for oxygen depletion after sealing. Because air moved freely through the mine opening before sealing under natural ventilation, no oxygen determinations were made at the time.

The overburden was core drilled at four sites to determine the pyrite distribution, stratigraphy, rock condition, caving in the mine, and water table. The surface above the mine was photographed by an infrared scanning technique to locate zones of high permeability to air and water. Earth resistivity and soil infiltration tests were conducted for additional data, and differential air pressure across the seals was recorded for evaluation of a variety of environmental factors.

[3]Lorenz, W. C., and E. C. Tarpley. Oxidation of Coal Mine Pyrites. BuMines Rept. of Inv. 6247, 1963, 13 pp.

[4]Braley, S. A. Acid Mine-Drainage. VI. Control of Oxidation. Mech., v. 18, No. 6, 1954, pp. 105-107.

Chapman, C. L. Progress in Mine Sealing. Univ. West Virginia, Eng. Exp. Sta., Tech. Bull. 11, 1938, pp. 79-85.

Hodge, W. W. The Effect of Coal Mine Drainage on West Virginia Rivers and Water Supplies. Univ. West Virginia, Eng. Exp. Sta., Res. Bull. 18, 1938, 28 pp.

Leitch, R. D., W. P. Yant, and R. R. Sayers. Effect of Sealing on Acidity of Mine Drainage. BuMines Rept. of Inv. 2994, 1930, p. 2.

Moulton, E. Q., and others. The Acid Mine Drainage Problem in Ohio. Ohio State Univ., Eng. Exp. Sta., Bull. 166, 1957, 158 pp.

Shumate, K. S., and E. E. Smith. Development of a Natural Laboratory for the Study of Acid Mine Drainage Production. Proc. 2d Symp. on Coal Mine Drainage, Mellon Inst., Pittsburgh, Pa., May 1968, pp. 223-235.

U.S. Public Health Service. Supplement C, Acid Mine Drainage Studies, 1942, 68 pp.

[5]Hill, R. D. Mine Drainage Treatment. State of the Art and Research Needs. U.S. Dept. of the Interior, FWPCA, December 1968, p. 2.

CHAPTER 30

TREATMENT OF TANNERY WASTES

A waste treatment process was developed and tested, in pilot plant scale, for the treatment of the tannery wastes in combination with municipal sewage. The process consisted of the following steps in the order employed; equalizing and mixing of the alkaline and acid wastes; primary sedimentation; carbonation followed by upflow sedimentation; addition of screened municipal sewage; activated sludge treatment and secondary sedimentation of the mixed wastes; and chlorination. The sludges resulting from the treatment of the wastes and sewage were dewatered by centrifuge and were found to be suitable for burial. Design factors for the various steps of the process were developed and are presented in the attached report. Studies were made of the fundamental systems and reactions which form the bases for the processes employed in the pilot plant.

The results of the pilot plant investigation indicate that by use of the methods recommended, mixtures of chrome tannery wastes and municipal sewage can be treated successfully. It may be anticipated that by the employment of the methods recommended, that mixtures of tannery wastes and municipal sewage can be treated to remove more than 90 percent of the BOD and suspended solids together with about 65 percent of the total solids. Furthermore, the treatment will remove or convert 99 to 100 percent of the sulfides, remove about 97 percent of the chromium and about 65 percent of the calcium. Chlorination of the effluent will reduce the coliform bacteria concentration to less than 100 per 100 ml.

The processes recommended are conventional sewage treatment unit processes, with the exception of carbonation and solids contact sedimentation. Modification of conventional equipment will be necessary to accommodate the special characteristics of tannery wastes. Fine screening will be needed to remove hair. Calcium carbonate incrustation of equipment, especially screens and carbonation equipment, will have to be considered. Extra strength sludge collection and pumping equipment will be needed since the sludges are large in quantity and are denser than sewage sludges.

DESCRIPTION OF THE TANNING PROCESS

Tanning is the process whereby the normally putrescible proteins in animal skin are preserved. A brief outline of the histology of skin will give an insight into the sources of organic load in the wastewater in the tanning operation which derive from the skin itself.

The skin as it arrives at the tannery consists of two layers, the flesh and the derma. Strictly speaking the flesh is not a part of the skin proper. Varying amounts of flesh may be attached to the skin, depending on the processing employed at the meat packing plant, and the skill of the flayer who removes the skin from the carcass.

The flesh layer must be removed by the tannery in the early stages of processing. This flesh contains muscle tissue, fat cells, blood vessels and adipose tissue, some of which finds its way into the wastewaters from the tannery.

The derma layer is the portion of the hide or skin which ultimately becomes leather. The derma is a complex material, consisting of two broad subdivisions; the epidermis and the corium.

The epidermis is the top layer of the skin, and contains the hair, hair follicles, and oil glands which surround the hair roots. The surface of the ipidermis is covered with a very thin layer of non leather-making protein. Running through the epidermis is a connecting network of collagen, the leather-making protein. During the early stages of leather processing all of these undesirable materials are removed from the skin. They are keratin, elastin, mucoids, albumens, and other proteins and alkaloids; and a significant amount of fat.

The corium is the major part of the skin and is composed primarily of the protein collagen. This collagen for the most part is an interwoven network of fibers and fiber bundles. It is the objective of the tanner to cleanse the skin of as much of the non-leather making debris as possible, and to process or "tan" the remaining collagen fibers to render them strong, flexible and non-putrescible.

As can be seen from the brief description above, the chemistry of skin is essentially the chemistry of protein. During the early stages of the leather manufacturing process many of these proteins are hydrolyzed or otherwise broken down and removed from the skin. Much of the colloidal and dissolved protein breakdown products appear in the wastewater. Very little is known about the susceptibility of these waste products to bacterial attack in the biochemical oxygen demand test for pollution studies. It may be that of efficient biochemical oxidation of these proteins requires an acclimatized culture containing specialized enzymes.

The Tanning Process

In order to understand the problems involved in treating the effluent from tannery, it is necessary to include at this point a description of the tanning process.

In a chrome side upper leather tannery, cowhides and steerhides are converted into a leather by the process known in the trade as chrome tanning, as distinguished from other methods such as alum or vegetable tanning. The following is a general description of the various steps in the process. Fig. 30-1 is a flow diagram of the tanning operations indicating the sources of the various types of wastes.

The Beamhouse

The hides are received from the meat packer or hide broker in bundles and have been salted or brined to preserve them. They may also have been "prefleshed" at the packing plant. The freshly flayed hides, if they are to be prefleshed are processed through a machine which removes the greater portion of the flesh from the skin. Following this step the hides are treated with sodium chloride. The common method of doing this today is to immerse the skins in a circulating saturated brine, which also serves to cleanse the skin of much of the manure, dirt, blood, etc., which is associated with it. Treatment with sodium chloride dehydrates the protein and renders it less susceptible to spoilage. The hids are then packed in bundles for shipment to the tannery.

At the tannery the bundles are opened up and the hides are cut down the backbone into two halves or "sides". As a consequence of the preservation method the next step at the tannery consists of removing much of the salt from skins and restoring moisture to render them flexible and receptive to processing chemicals. This process is referred to as soaking. The sides from the hide storage area are placed in large pits equipped with paddle wheels which serve to circulate the contents. The pits contain water to which a small amount of a wetting agent may be added. From time to time, the sides are agitated by operating the paddle. Midway through the first soak small quantities of sodium sulfide and lime are added. After about 14 hours the paddle pits are drained and refilled with fresh water, and the stock is washed with running water for a short time.

Fleshing follows soaking. Fleshing consists of passing the sides through a machine with rotating helical blades which remove the excess flesh from the sides. Water runs on the blades to keep them free of debris. The solid fleshings are collected in a box equipped with drain holes, the liquid waste passing to the sewer.

The next operation is unhairing. This operation consists of loosening the hair by treating the sides with a chemical solution described below, followed by mechanical removal of the hair in a machine similar to a fleshing machine. The sides are placed in a "drum" or "wheel", a large cylindrical container which is rotated at a slow speed. Water, sodium sulfhydrate, lime, and small amounts of other chemicals are added to the drum. After a suitable length of time in this solution, the hair has been loosened from the skin. The liquid is drained and may be passed to a holding tank to be refortified and reused. The sides are washed with running water, dumped, and forwarded to the unhairing machines. The hair removed in the machines is retrieved from the water and goes to a hair reclamation area where it is washed, dried and baled. All washings, containing small particles of protein solids, grease, sulfide and other chemicals go into the effluent stream.

Liming follows the unhairing step. In this step the sides are immersed in a lime suspension in a paddle and occasionally agitated to insure a uniformity of treatment. Sodium sulfide in a small amount may be added to the lime pit also. This operation swells the skin and opens up the finely interwoven collagen fiber bundles. Non-leather making protein and other undesirable debris is loosened from the skin so that it may be removed. Following a suitable treatment time in this solution the sides are pulled from the paddle and are passed to a scudding machine. "Scudding" consists of putting the sides through a machine which scrapes the grain surface with a dull blade. This serves to remove remaining hair roots, epidermal keratinous tissue and other surface debris. This debris passes to the effluent stream.

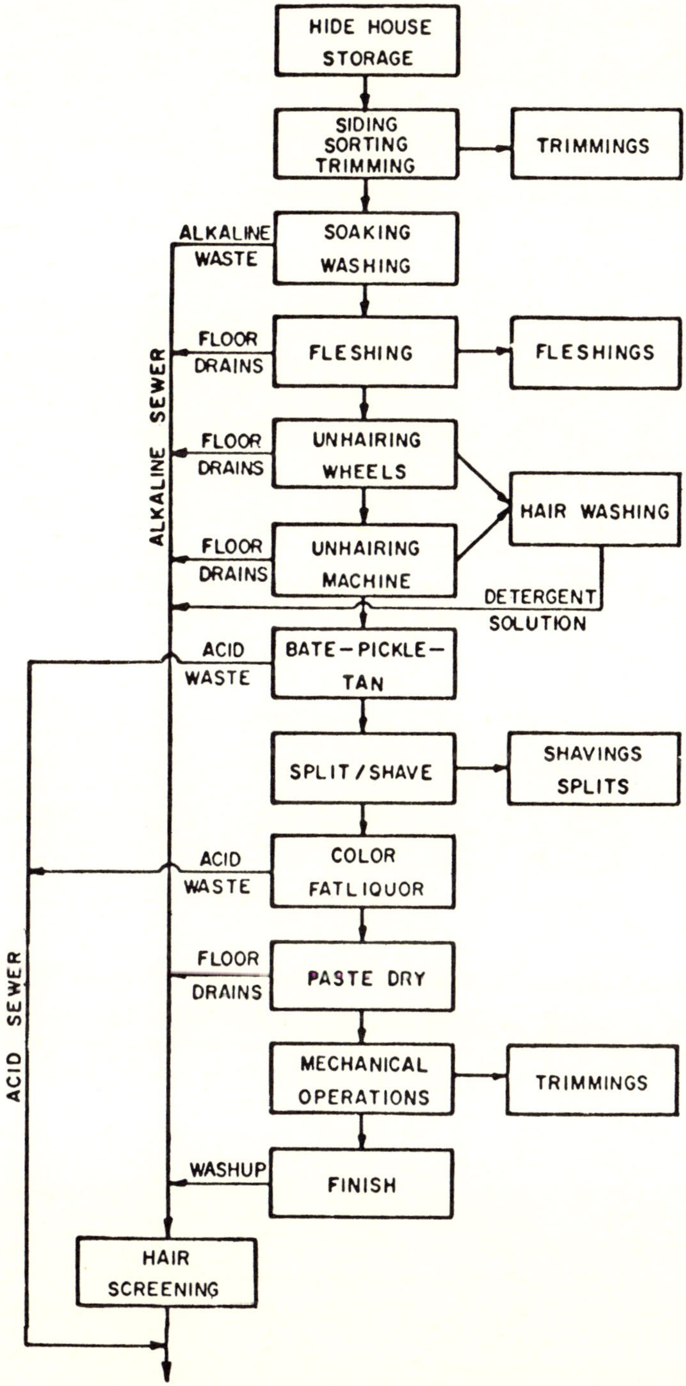

FIG. 30-1 TANNERY PROCESS FLOW SHEET

The above described operations constitute what is known as the beamhouse operation. The following part of the leather-making operation is known as tanning and coloring and is described below.

Tanning and Coloring

As a result of the several processes in the beamhouse, the sides are now in a highly swollen alkaline state. Before the collagen will accept and react with the tanning materials, it must be made acid and the swelling must be removed.

The first step in this process is bating and deliming. In this process the sides are first placed in a drum and washed thoroughly. This is followed by treatment with a proteolytic enzyme at a pH of about 9.0. This pH is attained by the addition of ammonium sulfate which acts as a deliming salt. A small amount of detergent may also be added. The enzyme acts on the noncollagenous proteins in the grain surface of the leather and contributes to the final soft pleasant grain character achieved in the finished leather. Following a suitable length of time in this solution, the stock is washed in running water to stop the bating action and remove the solubized lime as calcium sulfate.

Pickling follows and consists of lowering the pH of the stock to about 2.0 with sulfuric acid. This is done in a sodium chloride solution, which acts to repress the swelling and keep the stock in a firm, "fallen" state.

When the pickling has proceeded to the desired point, the tanning material is run into the rotating drum. This consists of a complex salt of chromium resulting from the reduction of sodium bichromate with sulfuric acid and an organic compound such as sucrose. This tanning solution is allowed to penetrate the skin following which the pH is slowly raised by the addition of sodium bicarbonate to a pH of about 3.5. This causes the chromium salt to combine chemically with the collagen, thus producing leather. The excess solution is drained to waste and the leather is washed in running water.

Following tanning the leather is subjected to several mechanical operations as follows. First, the leather is wrung to extract as much moisture as possible. The leather then goes to a splitting machine, which is essentially a horizontal band knife. This machine splits the leather into the grain, or top layer and the split, or bottom layer. This bottom layer referred to in the trade as a "blue split" or "chrome split" is processed into shoe linings, insoles, slipper soles, etc. The grain proceeds to a shaving machine, where the back is further leveled and smoothed by shaving off any unevenness which may be present. The chrome shavings thus produced are retrieved, baled and sold as a byproduct. A certain amount of the fine shavings adhere to the moist, fibrous back of the leather and are carried into the next operation which is coloring and oiling, and eventually find their way into the effluent.

Following shaving the sides are put into the coloring and oiling or "fatliquoring" wheels for further processing. It is here that the leather may be retanned with chromium salts, synthetic tanning materials (syntans) and the fibrous structure partially filled with clay or other fillers. These operations are all performed in a water medium and washing usually follows each processing step. Here also dye solutions are applied depending on the shade desired in the finished leather. The dyes are customarily fixed or made fast with formic acid. Finally oils are introduced into the drum, usually as oil-in-water emulsions. As these emulsions are broken at the leather surface the oil is worked into the leather. The oils commonly used are cod, sperm and other animal and fish oils, usually sulfated sufficiently to cause them to be self-emulsifiable. In some cases additional emulsifiers and stabilizers are added.

From the above description of the coloring and oiling procedure it is apparent that a great variety of both organic and inorganic pollutants find their way into the tannery effluent from this source. The next processes at the tannery are completed with the coloring and oiling procedure.

A relatively non-polluted effluent is discharged from the pasting and oiling. Pastings is the method employed to dry the leather. The sides are affixed to smooth porcelainized steel plates which have previously been sprayed with a water dispersible paste. The plates with the leather adhered to them are passed into a heated drier where the moisture is driven out of the leather. After the leather has been stripped from the plates, they are passed through a washer which removes the old paste, sending it to the sewer. A fresh paste is then applied to the surface.

After drying, the leather is subjected to a variety of dry mechanical operations and ultimately arrives in the finishing department where the final processing steps are performed.

Finishing

In finishing side leather for shoe uppers it is necessary to apply a pigment finish to the surface of the leather to make the surface uniform in appearance and of the shade desired. "Pigment finish" is simply pigment, organic and inorganic, dispersed in a suitable emulsified resin binder. It may be modified with plasticizers, waxes, etc., to give the handling characteristics and surface feel desired. A small amount of waste, principally machine clean-up water is discharged in this operation.

With the exception of plant clean-up water and discharges from compressor cooling and boiler room and the above constitute the sources of waste from the tannery.

RECOMMENDATIONS

Equalization Basin

Wide fluctuations in the rate of discharge and the quality of the wastes discharged by a tannery render the treatment of the waste on a continuous, as received basis, unsatisfactory. One of the first steps in the treatment process should be an equalization of the wastes discharges relative to both flow and quality. Fig.30-2 is a plot of the outflow concentrations of calcium from an equalization basin having detention times of 4 and 8 hours as compared with no equalization. From Fig.30-2, it may be seen that the maximum fluctuation in the calcium concentration from an equalizing basin, having approximately 4 hours detention time, will be from about about 400 mg/l to 1700 mg/l. From an 8-hour retention time in the equalization basin, the maximum variation in the effluent calcium concentration would be from about 700 mg/l to 1500 mg/l. On the basis of this study, it is recommended that the equalization basin have a capacity of approximately 4 hours retention time, equivalent to a volume of 210,000 gallons for a design flow from the tannery of 1.25 mgd. It is estimated that the maximum concentration of the various constituents in the effluent from the equalization tank will be:

Calcium	1,700 mg/l
Sulfide	190
Total Solids	19,500
Suspended Solids	4,320
Alkalinity	1,240
Chromium	460
5-day, 20°C BOD	2,100

Special designs may permit the incorporation of the equalizing function in combination with carbonation, primary sedimentation and other processes. The functions which may be incorporated in a single basin is a decision which will have to be made during final design when a selection is made of the specific equipment to be used. Suggested designs are shown later but these designs should be investigated in more detail by the final designer in order to develop a unified system.

The operating cycles of the tannery are such that for periods of time up to 10 hours, the discharge of the tannery is predominantly acidic. In order to minimize the size of the equalization basin needed to prevent the pH of the mixed wastewaters from falling below 9.5 it is more economical to provide a separate storage basin for the acid waste and to add them to the equalization basin at a variable rate such that the maximum rate of addition of acid waste will coincide with the maximum inflow of alkaline wastes. For this purpose, it is estimated that the size of the acid wastes holding tank should be about 100,000 gallons. Mixing of the wastes during equalization is considered to be essential. Mixing is particularly important when acid wastes are being added. No local areas of low pH should be permitted to develop within the equalization basin.

Primary Sedimentation

In the pilot plant, primary sedimentation was found to be an essential feature of the treatment process. In the pilot plant, the primary sedimentation basin served two principal purposes. The first purpose was to equalize the fluctuations in influent quality so that minute to minute changes were ironed out. The second purpose was to remove the bulk of the settleable solids to prevent overloading the sedimentation section of the carbonation-upflow sedimentation basin which followed. These purposes are sucessfully accomplished. Similar functions are needed in a full-scale plant.

It is suggested that the functions of equalization and primary sedimentation may be carried out in the same basin but in different compartments. Fig.30-3 is a sketch of one possible type of basin. Alternative designs are being investigated.

The sedimentation section of the basin should be designed with an overflow rate of about 700 gpd per ft.^{2} Design on this basis will result in detention times of about 2 hours.

Special design consideration should be given to the removal of sludge from the sedimentation basins in the primary systems.

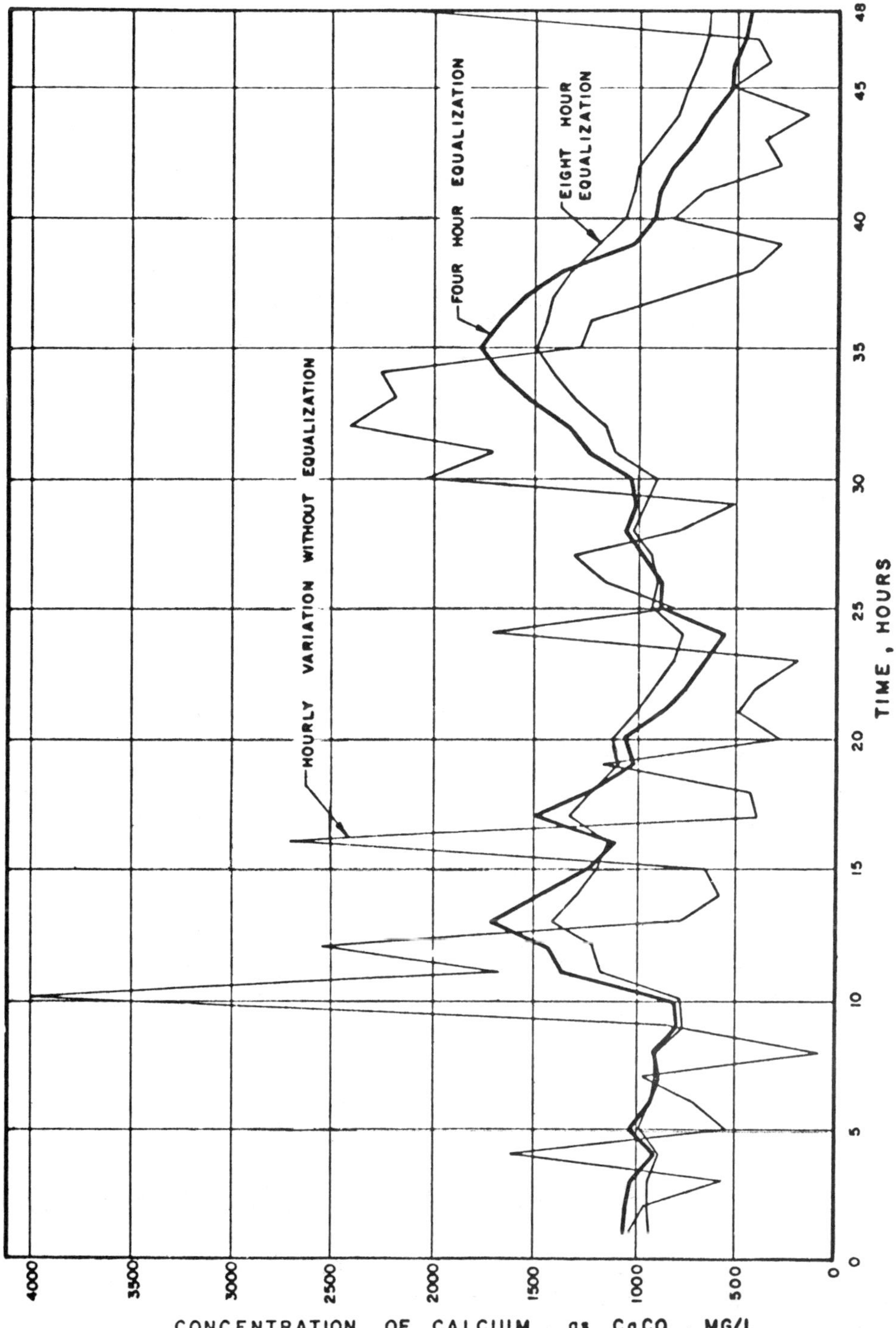

FIG. 30-2 EFFECT OF EQUALIZATION ON WASTEWATER QUALITY

The sludge may have a concentration in excess of 8 percent solids. The solids will be heavy and sticky and in many respects resemble very wet clay. Sludge removal mechanisms should be of extra strength design. Sludge draw-off piping would be equipped with water flushing connections which should be used to flush out the sludge draw-off lines after each use. If these draw-off lines are left full of sludge, the sludge is of

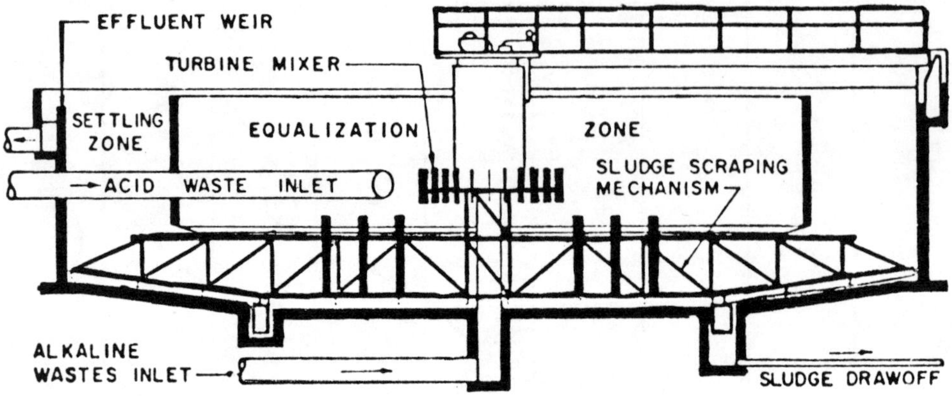

EQUALIZATION AND PRIMARY SEDIMENTATION BASIN

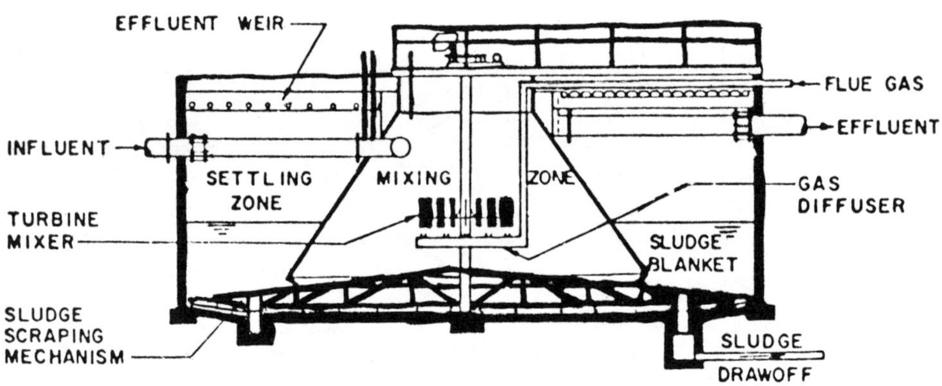

CARBONATION UPFLOW SEDIMENTATION BASIN

FIG. 30-3 -SUGGESTED TREATMENT PLANT EQUIPMENT

such a nature that it may set up and clog the pipes preventing any further draw-off of sludge until the pipes are cleaned.

Carbonation

The pilot plant tests show that the rate of absorption of carbon dioxide from the flue gas is very rapid. It is estimated, therefore, that a 20 minute contact period in a basin which was supplied with flue gas will provide an ample opportunity for full absorption of carbon dioxide by the tannery waste. Carbonation should be preceded by equalization because changes in the rate of flow and the quality of the various discharges from the tannery is so rapid that adequate carbonation is not possible on an as received basis. Carbonation might well be preceeded by primary sedimentation to remove undissolved lime which interferes with proper carbonation. Carbonation should be followed by sedimentation in order to remove the calcium carbonate which is precipitated by the flue gas. Inasmuch as carbonation will require only a few minutes retention time, this process might very well be carried out in a channel located between an equalization basin and sedimentation basin.

On the basis of the wastewater survey, it is estimated that a capacity for adding about 1,800 lbs of carbon dioxide per day should be provided. If it is assumed that the flue gas contains 11 percent carbon dioxide and the efficiency of using the carbon dioxide in the flue gas is 50 percent, the volume of flue gas required will be about 2,600,000 cu ft per day or 1,700 cu ft per minute. On the same basis, it may be computed that it will be necessary to burn about 1,150 lbs of oil per day in order to generate the required quantity of flue gas.

Aeration Basin Design

In the pilot plant, the aeration basin functioned in a manner normal to the activated sludge process. The only unusual reaction of the aeration basin was the formation of large quantities of a very stable foam. The foam was not a continuous problem but occurred from time to time. In the pilot plant the foam was controlled by the use of sprays and on occasion by anti-foaming agents.

It is recommended that the installation of the activated sludge process on a full-scale basis include foam control sprays and a chemical feed system for adding anti-foam agents to the spray system. In addition, surface aeration devices are recommended, because less foaming would be expected to occur than with the use of gas dispersion devices located beneath the liquid surface.

The aeration capacity to be provided should be sufficient to introduce about 10,000 lbs of oxygen per 24-hours into the waste. It is estimated that this will require aerators having a total power input of about 165 horsepower. The volume of the aeration basins should be about 1,000 cu ft for every 60 lbs of BOD treated. On this basis, the aeration basins would have a volume of 165,000 cu ft or about 1,250,000 gallons. Basins having a capacity of about 1,250,000 gallons would provide a 12-hour detention period for a waste and municipal sewage flow of about 2.5 mgd. It should be noted that this design provides for 1.25 mgd of tannery effluents and an equal amount of sanitary sewage.

Secondary Sedimentation

The activated sludge generated by the pilot plant was very light and difficult to settle. Based, however, upon the results of the operation of the secondary sedimentation basin in the pilot plant it is recommended that in the full-scale plant the secondary sedimentation basin be designed for the overflow rate of not more than 500 gpd per sq ft per day. Pumping capacity equal to the influent flow to the treatment plant, i.e., about 1,800 gpm should be provided to return the sludge which settles in the bottom of the secondary sedimentation basin back to the aeration basin. This entire capacity may seldom be used but should be available if needed.

Excess sludge from the secondary sedimentation basin should be returned to the primary sedimentation basin and resettled with the primary sludge. The reason for this recommendation is that the secondary sludge was very difficult to dewater alone. However, mixing the secondary sludge with the primary sludge resulted in a mixture which dewatered readily.

Sludge Dewatering

Although the sludge dewatered readily on sludge drying beds and there was no odor from the drying sludge, however for the following reasons it is recommended that sludge drying not be used at the proposed treatment plant;

1. A large volume of sludge will be generated by the treatment processes (about 38,000 gallons per day). Large quantities of sludge will accumulate in a short period of time if not continuously removed.

2. If unfavorable drying conditions, such as occur in the winter, persist for any extended period of time the volume of sludge which would accumulate could become overwhelming.

3. Sludge drying beds require a great deal of hand labor in clearing and hauling away the sludge. In view of the cost of labor, manual handling of the sludge is not economically attractive.

It is recommended that the sludge be dewatered by the means of a solid bowl centrifuge. Such a centrifuge will produce a cake containing about 20 to 30 percent solids and will be sufficiently dry to truck to a final disposal site. Dewatering by centrifuge can be arranged in such a manner that the sludge is never handled manually. Mechanical dewatering of the sludge permits dewatering without regard to weather conditions. The land space required for mechanical dewatering of sludge permits a minimum use of land for wastewater treatment purposes.

It is recommended that the sludge be disposed of by the sanitary landfill method. It is estimated that about 2 acres of land per year will be needed if the sludge is placed in layers 6 ft deep.

COST ESTIMATES

Cost estimates have been made and are based upon design criteria set forth in these recommendations with flow and effluent characteristics existing at the studied location. The total capital and operating cost of the full scale treatment of combined tannery and municipal effluent is estimated to be 1.6 cents per foot of leather based upon present costs.

CHAPTER 31

REVIEW OF INDUSTRY PRODUCT RECOVERY

Product recovery from waste waters is an accepted practice from an economical and pollution control point of view. In many cases, recovery of by-products is not practiced as it is cheaper to dump the wastes rather than to process them. But with the increasing environmental control legislation and enforcement, product recovery technique is gaining high importance.

The practice of product recovery results in a net cost to the industry, but it has been proven frequently to be cheaper than other methods of disposal. The choice of a method of treatment depends on the economic considerations and degree of treatment desired. Removal of pollutants to the desired level is generally achieved through conventional processes, but it should be realized that the same degree of treatment can be achieved either by product recovery alone or by a combination of product recovery and conventional treatment processes. Advantage of reducing pollutional load through product recovery consists in the overall reduction of the cost of treatment and conservation of natural resources.

Unfortunately, the entire thrust of taxation and regulation has been negative with regard to by-product recovery. Too many of the state regulations are promulgated and permits issued with a specific reduction in a pollutant level regardless of the amount of discharge. Therefore, a company may eliminate a substantial quantity of pollutant by by-product recovery and still find itself faced with the same degree of required removal although on a smaller quantity. This has the effect of eliminating part of the benefit of by-product recovery. Only when recovery reaches 100% (a very unusual case) can total recovery of waste costs be achieved.

This problem is complicated by the fact that all pollution control facility tax exemption laws enacted in this country specifically exclude credits for facilities which recover any product for reuse or sale.

Thus, while it can be argued that while the social good of the country is in the direction of by-product recovery, the existing regulations and taxation are such as to discourage by-product recovery by industry in place of conventional waste treatment techniques.

The concept of product recovery is gaining far more widespread utilization than in the past because of the press and expense of increasingly tight environmental control legislation and regulations. Such an approach is being encouraged by both industries and governmental agencies for the following reasons:

1. A general reduction of the level of industrial pollutant discharge.

2. The possibility of reducing overall environmental control expenditures.

3. Reduction in the rate of material utilization and wastage by the economy.

It is, however, interesting to note that, in the past, many recovery methods proposed have been rejected because it is cheaper to dump the wastes than to process them. With stricter anti-pollution regulations and recovery processes making very good economic sense, pollution abatement through product recovery, either wholly or partially, should be promising in order that the natural resources of the country may be handed on to posterity undamaged and undestroyed.

Valuable by-products may include:

1. Water and other solvents for recycle.

2. A major product that is collectable in a commercial form, e.g., iron ore dust from blast furnace gas effluents.

3. A major product collectable in a non-commercial form, e.g., in dilute solution.

4. A saleable or presently useable by-product.

5. A by-product not presently an article of commerce.

Recycled water may ultimately be the major valuable product because of increasing water supply costs, increasing water treatment costs and mounting charges for using municipal sewerage facilities. As considerable work has been done and reported in the literature[1-12] on renovation and reuse of water, this subject will not be discussed. The recovery of product fines, useable water and thermal energy are important methods of reducing overall waste disposal costs and should be seriously considered in every case. Recovery of products from solid wastes is not included as it falls outside the scope of this report. There are many products that are recovered now but there are many more that are not being recovered now that could be recovered in the future by the so-called second generation recovery systems.

ECONOMIC, TECHNICAL AND PHILOSOPHICAL FRAMEWORK

The product of waste must be considered as an integral part of the manufacturing process and the cost of treatment of industrial wastes must be charged against the product. The waste disposal operations result in a net cost to the industry producing the waste, but product recovery and utilization practices reduce the cost of treatment and frequently prove to be cheaper than other methods of disposal.

While considering treatment of a waste, the choice of a method of treatment or process is made that gives maximum benefits or returns. In many cases, the industrial wastes contain valuable products such as high value metals, acids and other products which could be used for producing by-products and these, when recovered, will give high economic returns in addition to removal or pollution load from the waste stream. Further, in some cases, water recovered from these wastes are of good quality and can be reused, resulting in the reduction of waste and cost of treatment. The renovation and reuse of waste water becomes a necessity when the natural water supply source is limited and water needs are increasing, due to growing population and expanding industries.

By-product recovery frequently accompanies water reuse and water conservation. Recycled water may ultimately be the major valuable product because of increasing water supply costs, increasing water treatment costs, and mounting charges for using municipal sewerage facilities.

Industrial wastes contain toxic chemicals such as cyanide, free acids and heavy metals. Disposal of these wastes into surface waters are objectionable because they cause fish kills, retard self purification of surface waters. These wastes cannot be discharged into the sewage treatment plants as they contain toxic materials. The reduction of waste load is of major importance in the waste treatment operations. Waste load reduction is achieved by removing the spent grains in a semi-dry state, by removing the yeast from fermenters for filtration and drying and by removing the sediment from the chill storage tanks as a slurry for separate disposal. In many cases, these operations have yielded a saleable by-product which has more than offset the increased cost of recovery.

Frequently, waste streams can be eliminated or reduced by process modifications or improvements. A notable example of this is the use of save-rinse and spray-rinse tanks in plating lines. This measure brings about a substantial reduction in waste volume and frequently a net reduction in metal dragout.

Segregation of waste streams is a necessity at times, not from the product recovery point of view but from the operational point of view. An example of this is segregation of acidic metal rinses from cyanide streams to avoid the production of toxic hydrogen cyanide (HCN) and thus eliminate potential safety hazards.

From the industries' philosophical point of view, the prime requirement of waste treatment, by-product recovery and water reuse is that the principal product or products of the plant is satisfactory to the consumer and the secondary requirement is that the operation of the plants be efficient and economical.

The urgent problems facing the industries are how to recover by-products from waste materials inherent in every industrial operation and what to do with the by-products. Confronted with the growing dangers of air and water pollution, the anticipation of government regulation, as well as loss of valuable materials through unprofitable waste disposal methods, industries are forced to develop sophisticated refining methods for processing chemical and industrial by-products and even develop markets for by-products.

The industries, in general, are becoming more and more aware of the necessity for pollution abatement and product recovery, not only because of its effect upon the general welfare of the public, but also because of its own dependence upon rivers and streams for suitable water for manufacturing processes. Industries are also increasingly aware of the fact that benefits accruing from pollution abatement through product recovery may be quite significant.

PULP AND PAPER MILLS

Recovery of chemicals from cooking liquor and process water reuse are the two important efforts in the conservation and pollution control in this industry and the other efforts are recovery of by-products from spent liquor and bark.

Cooking liquor recovery was developed in order to reduce the cost for cooking chemicals and reduce the waste water load being discharged to stream. In the Kraft process of pulping, economic considerations dictate recovery and reuse of chemicals and, therefore, stream pollution from Kraft operation is less prevalent when compared to other types of pulping.

In the modern technology, liquor recovery system is an essential part of the pulping process and the recovery technique is different with each type of pulping method.

1. Kraft Liquor

   A major part of the chemicals in the spent cooking liquor (black liquor) is usually recovered in a series of operations: evaporation, burning and causticizing. The dissolved organic residues from the wood resins can be used for heat and power generation. Some soda is lost in the washing operation and, therefore, salt cake is added to the recovered liquor.

2. Sulfite Liquor

   Sulfite liquor recovery techniques are more complex and difficult than those employed in the sulfate system. In fact, the feasibility of liquor recovery is dependent on the base involved in the sulfite liquor. It is generally conceded that recovery is not practical with the calcium base. When a magnesium base is used, the spent liquor is evaporated and burned and chemicals recovered for reuse. However, for the sodium base, it is possible to recover the inorganic chemicals with a fairly complex process. Spent chemicals from the ammonium base sulfite pulping process can only be partially recovered but the process does not appear economically promising.

   Numerous products including Torula yeast and alcohols are produced from waste sulfite liquor by fermentation. Phenol may be produced by pyrolysis, vanillin by lime addition, while acetic acid, formic acid, dimethyl sulfoxide (DMSO) and lignosulfonates are recoverable directly.

3. Semi-chemical Liquor

   Liquor recovery in the NSSC (Neutral Sulfite Semi Chemical) process is practiced in some independent NSSC pulp mills using a complicated carbonation method. In the integrated semi-chemical Kraft mills, the recovery system is also integrated. The "cross recovery" method is based on the sodium and sulfur values in the NSSC liquor that can be used as chemical make-up in place of salt cake. The NSSC spent liquor can be combined with Kraft liquor for evaporation and combustion but the NSSC cooking liquor has to be made from fresh chemicals.

The KSC spent liquor is recovered in the same process that is used in a regular Kraft pulping process.

Large quantities of water are essential in the pulp and paper manufacturing processes such as pulp processing, washing, dissolving or mixing the various loading, sizing, and color ingredients, and carrying the fibers through the screens and refiners to the paper-making machine. Water is also used to convey by-products and undesirable wastes and to generate power and steam. The maximum reuse of water becomes one of the basic approaches in the effective reduction of waste water quantities.

There are definite limits in the amount of mill water which can effectively be reclaimed and reused. Fresh water has to be added in certain operations to prevent the build-up of undesirable dissolved solids, temperatures, and slime from bacteria and fungi growth.

Each mill has its own approach on how best to reuse water in the various operations. However, the following process waters are most frequently collected and reused:

1. Water from the log flumes and barkers.

2. Evaporator condensate from the liquor recovery area.

3. Bleach plant washer filtrate.

4. White water from the paper machine.

5. Washer water from coarse screens.

An accepted practice is the reuse of waste water in wood handling and storage such as in the log flumes, hot pond, hydraulic or wet drum debarker, and for the showers prior to chipping. For these applications, heated effluents discharged from evaporators, bleach plant, or paper machine are preferred because the heat increases bark removal efficiency. Recycling barker effluent, which has been cleaned and screened, is a widely practiced water reuse procedure.

TABLE 31-1

BY-PRODUCT RECOVERY IN THE

PULP AND PAPER MILL INDUSTRIES

Process	By-Product Recovery			
	By-Product	Quantity	Value - Recovered Products	Diminution of Pollutional Load
Kraft	Turpentine	1.5-4.3 gal/ton	$0.20-0.40/ton	1%
	Tall oil	180-300 lbs/ton	$9-15/ton	2%
	Lime	2.0 tons/ton	$40/ton	95%
	Soda ash	1.5 tons/ton	$40/ton	95%
	Fiber recovery	Not relevant		20-70%
Sulfite	Torula yeast	500 lbs/ton	$25/ton	90%
	Various alcohols			
	Lignosulphonates	1500 lbs/ton	---	70%
	DMSO		$10/ton	1%
	Vanillin	500 lbs/ton	$40/ton	90%
	Phenol			
	Acetic acid	30 lbs/ton	$3/ton	4%
	Formic acid	8 lbs/ton	$0.80/ton	1%
	Fiber recovery	Not relevant		
NSSC	See Kraft and Sulfite Processing			
Debarking operations	Bark	100 lbs/ton	$100/ton	10%
Holo-pulping	Sodium hydroxide	Quantities not confirmed in full test.		

STEEL INDUSTRY

The disposal of waste pickle liquor is a major problem that has been plaguing the steel industry. The steel companies frequently find it costs almost as much to dispose of the waste as it does to buy the acid.

There are various methods for treatment, disposal and product recovery for pickle liquors such as lime neutralization, evaporation-crystallization, controlled neutralization, electrolysis and dialysis. Lime neutralization, though very commonly used, is an expensive method without any product recovery. When evaporation-crystallization is practiced, only free acid amounting to half the requirement for pickling is recovered. Further, even though the cost of evaporation-crystallization is not excessive, the disposal of ferrous sulfate ($FeSO_4$) poses further problems because of its inadequate market. Electrolytic schemes for complete acid recovery and plating metallic iron are being used with advantage.

Controlled neutralization with lime (duPont process) produces a smaller volume of sludge and easy to dewater solids. Magnetic iron oxide formed is easily recoverable and crystalline gypsum may be used for wallboard or cement manufacture. If local conditions do not justify recovery, solids may be piled as land fill.

Recovery of sulfuric acid and iron from waste pickle liquor in the steel industry has been technically proved feasible in the Ruthner process. This recovery system is a non-paying one with high investment. Roasting of ferrous sulfate monohydrate (crystallized from pickle liquor) with carbon and air to produce iron oxide and sulfur dioxide has been proposed. On a large scale, this system appeared more economical than the Ruthner process. The acid produced here would be shipped back to the steel mills or other customers. Iron oxide, to be of value, would have to be sintered and shipped back to the mills. Sulfuric acid recovered is the only significant credit here and recovery of iron in a ready-to-melt form, rather than iron oxide, is very desirable.

Evaporation-crystallization has been used in producing ferrous sulfate crystals. Waste pickle liquor is evaporated at 160° F. and ferrous sulfate is then crystallized in a vacuum crystallizer and cooled down to 30° F. The crystals are separated from the mother liquor by centrifugation. The mother liquor is diluted to 25% acid, thereby reducing the residual ferrous sulfate to 4.5% and this solution is returned to pickling.

Dialysis, a separation process utilizing diffusion as its driving force, may be used for the economical recovery of acids. In the viscose rayon industry, dialysis provided the means for recovering the bulk of the wasted caustic soda. About 0.85 pounds of caustic were recovered per pound of rayon produced, sufficient to have a real effect on the production cost. Recovery of sulfuric acid was done in two copper refineries on full scale plant units. The economic feasibility of dialysis for stainless steel pickle baths has also been proven.

Dialysis is a simple unit operation requiring a minimum of operator attention. The operating cost is almost nil as no power or chemical is used. However, it should be noted that dialysis alone will not resolve a waste treatment problem, but its role to remove raw materials and to reduce plant waste is significant. Some acid systems may allow recovery of as much as 70 or 75% of currently wasted acid, but even recovery of as little as 20% of the wasted acid may justify the process from the economical point of view.

The pickling of steel in the United States is predominantly done with sulfuric acid. The acid used is consumed and produces a waste pickle liquor which is both uneconomical and extremely difficult to regenerate. The waste pickle liquor is normally neutralized and/or dumped. This method of disposal is becoming more and more difficult to pursue. A practical answer to this problem is the use of hydrochloric acid instead of sulfuric in the pickling operation. Major steel strip manufacturers in the United States have already started the trend toward hydrochloric acid due to the following reasons:

1. Hydrochloric acid prices are dropping while sulfuric acid prices are rising.

2. Hydrochloric acid pickling is faster and produces a better quality product.

3. Most importantly, hydrochloric acid waste pickle liquor can be and is being regenerated economically.

Switching to hydrochloric acid is not practical in the open-batch pickling of wire, pipe and plate due to the production of corrosive vapor. In the continuous strip pickling operations where hoods and ventilation systems are provided, the switching is relatively easy. The vapor problem has been avoided in Europe by using enclosed tunnel and Ottocleve picklers.

TABLE 31-2

BY-PRODUCT RECOVERY IN THE

IRON AND STEEL INDUSTRIES

	By-Product Recovery			
Process	By-Product	Quantity	Value	Diminution of Pollutional Load
Pickle liquor	Iron oxide	25-35 lb/ton	$0.10-0.20/ton	100%
	Gypsum	40-50 lb/ton	$0.10-0.20/ton	100%
	Sulfuric acid	30-40 lb/ton	$0.60-0.80/ton	100%
	Hydrochloric acid	30-40 lb/ton	$0.60-0.80/ton	100%
Grey iron foundry	Flue dust Mill scale	---	---	60-80%
Blast furnace scrubber waters	Mill scale	25-50 lb/ton		60-80%

PLATING AND ALLIED INDUSTRIES

The rinse water from plating industry mostly contains some or many of the following contaminants usually in intolerable amounts: hexavalent chromium, sodium cyanide, complex cyanides of the heavy metals, such as cadmium, copper, zinc, sometimes silver and gold, soluble nickel salts, strong mineral acids and strong alkalis. Of these, the most toxic are hexavalent chromium and the cyanide ion. Aquatic life is also sensitive to salts of the heavy metals.

There are several basic methods of treating this waste for final disposal or recovery of products. The oldest is the batch treatment method. In this method, the rinse waters are collected for treatment and disposal. The destruction of toxic chemicals is accomplished by oxidation of the cyanides and reduction of chrome. Chlorine (Cl_2) is used to destroy cyanides (CN) and the chlorinated rinse water is pumped into a reduction tank containing sulfuric acid and a reductant-like ferrous sulfate ($FeSO_4$), sodium bisulfite ($NaHSO_3$) or sulfur dioxide (SO_2). Chromium is reduced from its hexavalent state (Cr^{+6}) to trivalent state (Cr^{+3}) and is precipitated along with metals as their hydroxides by the addition of lime. The clear water from the clarifier would be ready for discharge into a sewer and the sludge is disposed by any convenient method.

Another method of treatment of rinse water is the continuous flow method. To be effective and fool-proof, this requires excellent instrumentation to monitor the flow and deliver the proper amount of reactant. Also, such a system may not allow enough reaction time to allow slow reactions to go to completion.

Where the volume of chromic acid and sodium or potassium cyanide used in the industry is great enough, the system of recovery of the cyanides and chromic acid by ion-exchange and evaporation has obvious economic advantages. However, when the plating operation is split up into several small, different operations involving chrome plating, several different kinds of chromating baths and several types of cyanide baths, this method becomes practical.

Lancy's integrated system of treatment of cyanide and chromium wastes is found to be suitable for a plating operation split up into several operations. The system involves countercurrent rinsing and the basic advantage of this system is that the toxic contaminants are destroyed before they can enter into the rinse water, rather than having to be removed later.

When reclamation of chemicals is desired, ion exchange plays an important role. The rinse waters are kept separate in the following categories:

1. Hard chrome rinse waters
2. Cyanide rinse waters containing silver, gold or any other valuable metal.
3. Miscellaneous wastes such as alkaline cleaners and acids.

The hard chrome rinse waters are passed through a cation and a strong base anion exchanger. The contaminating metals such as copper, nickel and trivalent chrome are exchanged on the cation exchanger, while the hexavalent chrome is exchanged on the anion exchanger. Upon exhaustion of the system, the cation exchanger is regenerated with an acid and the anion exchanger regenerated with caustic soda.

The regenerant effluent from the cation exchanger can be discharged to a collection chamber for neutralization. The regenerant effluent from the anion exchanger will contain sodium chromate and some caustic soda. This material can then be converted to chromic acid by passing back through the cation exchanger so that the two exchangers may be alternated between service and conversion.

The ion exchange system for silver recovery would consist of a strongly acidic sulfonic cation exchanger and a strong base anion exchanger. The effluent from the system is de-ionized water and can be reused in the rinsing operation. Sodium hydroxide is used as a regenerant and the regenerant effluent contains sodium cyanide (NaCH) and sodium hydroxide (NaOH). This solution is passed downflow through the exhausted bed of the cation exchanger. The cation exchanger effluent contains sodium silver cyanide and NaOH. This regenerant solution may be used in the make-up of silver bath. Otherwise, addition of 15%

sodium hypochlorite (NaOCl) solution precipitates the silver as silver chloride (AgCl) and destroys the cyanides. The AgCl is recovered and the supernatant liquid dumped in the sewer, cyanide free. Sodium silver cyanide can be recovered by this method for a chemical cost of approximately 18¢/pound of silver recovered.

The rinse water from gold plating line contains enough gold to make recovery economical. The recovery of gold is accomplished by passing the rinse water through a strong base anion exchanger of the hydroxide form. After exhaustion of the resin, the gold is recovered by burning the resin. The residue remaining is metallic gold. One cubic foot of anion exchange resin, when exhausted with gold cyanide, contains approximately 125 ounces or $4,000 worth of gold.

The rinse water in the platinum plating industry contains platinum in the form of chlorophatinic acid. Strongly basic anion exchanger is used and when the resin is saturated with platinum, the recovery is achieved by burning the resin.

Nickel may be recovered from plating rinse waters by employing a cation exchange process. The effluent from the service cycle of the hydrogen exchanger can be passed through a weak base anion exchanger to obtain deionized water for plant use. Sulfuric acid is used for regeneration and the nickel is eluted from the resin as nickel sulfate. The capacity of the cation resin is 5.0 pounds of nickel per cubic foot as nickel sulfate and the chemical cost for the process would be 20¢/pound of nickel sulfate recovered.

Brightening of aluminum is done by the use of concentrated phosphoric acid (H_3PO_4) and a small amount of nitric acid (HNO_3). Rinse water from this operation contains aluminum and dilute H_3PO_4. It is estimated in an industry in Illinois that two to three million pounds of H_3PO_4 were lost each year. Cation exchanger was used to remove aluminum and the aluminum-free solution is evaporated in a multiple effect evaporator to obtain concentrated H_3PO_4 (85%) and good quality condensate. The recovery of H_3PO_4 in this industry provided a net savings of $8100/month above the operating expenses. Dialysis has also been studied in the recovery process.

In the film processing industry, tons of valuable metals go down the drain and recovery of silver alone could effect large savings. Some private laboratories, like those processing TV film, do recover silver but, in many cases, the quantities at any one location are very seldom large enough to warrant recovery measures.

MINING INDUSTRY

Acid mine water results from the exposure of sulfur and iron-bearing materials to processes of erosion and weathering. The drainage is generally acidic and contains sulfuric acid, ferrous, ferric, aluminum and manganese salts. In addition, the mine drainage may contain calcium, magnesium and sodium salts, carbon dioxide and silicic acid. The volumes of coal mine drainages encountered range from a few hundred thousand gallons/day to several tens of millions of gallons per day. An estimate of the magnitude of the problem in the United States is that four million tons of sulfuric acid equivalent per year is being discharged to inland water courses. The regulatory agencies, in an attempt to control mine drainage, usually stipulate that the discharge should be alkaline with a pH range of 6.0 to 9.0 and a maximum iron content of 7 mg/l.

Recovery of sulfuric acid has been proved uneconomical and, therefore, there have been many processes reported for the treatment and disposal of acid mine drainage. These can be considered as three classes for convenience:

TABLE 31-3

BY-PRODUCT RECOVERY IN PLATING AND ALLIED INDUSTRIES

	By-Product Recovery			
Process	By-Product	Concentration Range (mg/l)	Value ($/1000 gal.)	Diminution of Pollutional Load
Rinse waters	Nickel	150-900	0.80-7.00	80-100%
	Copper	2-20	0.01-0.10	80-100%
		100-500	0.60-3.00	80-100%
	Zinc	70-350	0.40-2.40	80-100%
	Silver	50-250	1.60-9.00	80-100%
	Chromium	400-2000	3.20-15.00	80-100%
	Cadmium	50-250	1.20-6.00	80-100%
	Brass	(Cu) 40-250	0.20-1.20	80-100%
		(Zn) 10-60	0.20-1.20	80-100%
	Tin	100-600	0.80-4.80	80-100%
	Gold			
	Platinum			
Aluminum bright dip	Phosphoric acid	10%	$4,000	80-100%
Film processing	Silver			80-100%

1. Lime neutralization process

2. Unorthodox or alternative processes

3. Limestone neutralization processes

The principle of these processes is that lime or other strong alkali is mixed with the acid mine drainage to neutralize the acids and to precipitate the contaminating metal salts which can be separated by sedimentation. However, the sludge formed has a high water content and presents a difficult disposal problem.

The acid mine drainage has been treated in many cases by lime to produce water for coal preparation plants and dust suppression purposes. The costs of treatment reported in one case was $1.09/1000 gallons, equivalent to a cost of 5.2¢/short ton of the coal produced. As seen from this cost figure, lime neutralization processes are not economical, even though in some specific cases they may prove economical. Attempts to prepare marketable by-products such as rouge from acid mine drainage have been made without any success. Use of organometallic compounds for the production of pigments from acid mine drainage have also been reported. The possibility of distillation and reverse osmosis processes have been discussed but the costs are prohibitive. In summary, it can be said that none of the alternative processes has yet been shown to be cheaper than the lime process for general application, but it is possible that processes which lead to the recovery of potable water might be commercially viable in districts which are particularly short of water.

PETROLEUM INDUSTRIES

Petrochemical Industries

Petrochemical industries have been faced with the dealing of large amounts of organic wastes in their early stages of development. The reduction of pollutant load has been achieved by using organic

TABLE 31-4

BY-PRODUCT RECOVERY IN COAL BY-PRODUCTS INDUSTRY

	By-Product Recovery			
Process	By-Product	Quantity	Value	Diminution of Pollutional Load
Coke plant ammonia liquor	Phenol	2 lbs/ton coke	$0.20/ton	40%
	Ammonium sulfate	24 lbs/ton coke	$1.00/ton	50%
	Ammonia	8 lbs/ton coke	$0.25/ton	50%
	Naphthalene, Benzene, Xylene, Cresols	20 lbs/ton	$2.00/ton	100%
Flue gas scrubbing	Sulfur, Sulfuric acid, Ammonium sulfate	60 lbs/ton of coal burnt	$1.20/ton	80-100%

wastes as fuel until recovery processes have been developed and a higher market for recovered chemicals developed. This approach has been followed in the production of almost every large volume petrochemical waste.

The more important recovery methods and products are presented below. Hydrogen is used in ammonia production, refinery operations, such as hydrotreating and hydrocracking, and many other purposes. Four recovery methods have so far been reported and the choice depends upon plant size, location, feed composition and hydrogen and use.

1. Wet scrubbing: is used to remove CO_2 and H_2 streams in conventional syntheses-gas plants. Scrubbing solutions include hot, carbonate-containing aqueous solutions, ethanolamines, and high pressure water.

2. Adsorption: is used on molecular sieves or activated carbon to separate hydrogen from hydrocarbons.

3. Cryogenic scrubbing: removes higher boiling impurities by condensation from low boiling hydrogen (-252.8° C.). This process is useful where a refrigerant, such as liquid air or nitrogen is readily available, as in partial oxidation synthesis gas plants.

4. Membrane permeation: is based on palladium's selectivity in passing hydrogen and is used to recover hydrogen from petroleum effluents.

Other petrochemical recovery processes:

1. Distillation: is used to recover chlorinated hydrocarbons following the chlorination of methane, olefins from the thermal production of ethylene, acetaldehyde following the production of vinyl acetate, and solvents from rubber latexes.

2. Crystallization: is used to separate xylene from ethyl benzene.

3. Adsorption: is useful for the recovery of carbon disulfide, carbon tetrachloride and other solvents.

4. Extraction: is useful for the recovery of isobutylene, naphthalenes, paraffins and phenol from a variety of wastes, and organic dyes are extracted following adsorption on fuller's earth.

5. Ion exchange: recovers basic amino acids, amines, alkaloids and organic acids.

6. Oxidation: regenerates spent caustic.

Petroleum Refineries

The recent trend in the newly constructed petroleum refineries is full integration of in-plant procedures and multi-stage waste water treatment combined with reuse and reclamation of purified water, resulting in considerable progress in pollution abatement and a drastic reduction of clean water intake requirements.

Use of hydrogenation processes, such as hydrosulfurization, yields products low in sulfur and, therefore, require minimum subsequent treatment. Application of chemical treatment processes, especially for gas scrubbing, allow complete regeneration and recovery of the solvent. The amount of spent chemicals requiring treatment and disposal is brought down to a low level by these processes.

Separate collection of waste water is employed in sewer systems according to their subsequent treatment. Efficient pretreatment such as evaporation, stripping of sour water condensates and equalization of flow and strength of waste waters are some of the processes used to improve product recovery and reduce waste load. Treated waste water is used as make-up water for recirculation cooling systems. In consequence of these measures, specific water consumption and waste water discharge per production unit has been in the remarkably low range of 9 to 22 gallons/barrel oil processed and 3 to 8 gallons/barrel of oil processed, respectively.

There are many by-products that could be recovered from petroleum refining wastes, even though all of them may not bring economic benefit. Therefore, the concept of by-product recovery is limited to those materials which, if recovered, would accrue some economic benefit, but not necessarily enough to cover the cost of recovery.

Based on this definition, the major by-product is sulfur, which is recovered from sour waters and from the hydro-treating process. The value realized through sulfur recovery is expected to increase greatly in the near future, due to the increased demand for low-sulfur fuels brought on by more stringent urban air pollution controls.

A number of refining process wastes have been recovered or reused and they are as follows:

1. Recovery of H_2SO_4 from sludges produced in the acid treatment of oils. Hydrolysis of the sludge produces a dilute (30-60%) black acid of rather limited utility.

2. Reuse of spent alkylation acid in the treatment of oils and waxes, with subsequent regeneration in captive or outside acid plants.

3. Sale of high-phenol waste caustics from treatment of catalytically cracked naphthas.

4. Use of sprung phenols as refinery fuels. These materials come from acid springing of spent caustics from cracked naphtha treatment.

5. Use of various waste acids in slop oil treatment.

6. Recovery of aluminum chloride from hydrocarbon sludges.

7. Recovery of acid oils by reaction of waste caustics with acids.

8. Use of boiler feed water treatment sludge in the neutralization of waste water.

9. Reuse of treated waste water to supplement normal refinery water supply.

10. Recovery of ammonia and hydrogen sulfide from sour water stripping for use as raw materials in the manufacture of fertilizer grade ammonium sulfate.

Production of single cell proteins from petroleum has been shown feasible and the proteins can be used as animal or human feed supplements. The price of a 50% protein single cell is estimated at 6 to 8¢/pound and hoped to be competitive as a potential animal feed supplement. For human consumption, the protein could command a price in the range of 30 to 40¢/pound. But it is feared that the cost of production of proteins is likely to mount up because of the uncertainties involved in additional processing. Another big hurdle will be probably in marketing or acceptance of such products due to factors like flavor, color, appearance, personal and cultural idiosyncrasies of consumers.

A $5.6 million plant is going into action for the commercial production of protein from petroleum. The concept consists basically of letting microorganisms grow big in bulk by nourishing themselves on hydrocarbons; the bulked product can be processed for foodstuff use. About 100 tons of gas oil will be needed to produce 100 tons of protein.

A single stage extraction vessel is used as a petroleum refiner's latest weapon for taking obnoxious phenols out of water from catalytic cracker distillate drums before it is let into the sewer. The extractant is light catalytic cracker cycle oil which is subsequently processed to finished heating oil. Using this principle, Humble Oil Company in Baton Rouge, Louisiana, is accomplishing a 75% reduction in phenol.

Aluminum chloride waste results from the petroleum and petrochemical industries, due to the use of aluminum chloride as a catalyst, and this cannot be discharged to stream or municipal sewer system, due to its low pH and high dissolved solids. Spent aluminum chloride is available in the form of about 27% aluminum chloride (32° Be') solution. The handling and disposal of such inorganic salts are troublesome and are not in the realm of current economic feasibility. However, such troublesome products can be used to control or eliminate other environmental problems. One such case[57] is the use of waste aluminum chloride solution in the field of waste water treatment for effective solids separation, phosphate removal and sludge conditioning. The present estimate is that some one hundred to two hundred million pounds/year will be available as 32° Be'solution and that more than 90% of it is recoverable. At the current cost of 32¢/pound of aluminum, the cost of 27% aluminum chloride solution is estimated at 4.4¢/pound and the total value of the waste aluminum chloride amounts to 6.6 million dollars.

Recovery of Oil From Waste Waters

Spills of oil constitute a major pollution threat to the water resources of the nation. Both water and land-based facilities are sources of this danger to our streams and rivers. Much damage has already been done from accidental or indiscriminate spillage of crude oil, petroleum and its by-products. Such spills have contaminated water supplies, killed fish and wild life, created fire hazards, and destroyed or reduced the usage of recreational areas.

The major sources of oil pollution include: gasoline service stations, oily waste industries, offshore oil and gas operations, oil tankers and industrial transfer and storage of oils. Cleaning up an oil-contaminated area is time consuming, difficult and costly. In addition, the losses due to destruction of fish and other wild life, damage to property, contamination of public water supplies and any number of other material and aesthetic losses are involved. These losses may be very great and extend for months or years, sometimes for decades, depending on the quantities and kinds of oil involved.

It is estimated that about 350 million gallons of used motor oil must be disposed of annually by the more than 210,000 gasoline filling stations. These stations are the key suppliers of used oil re-refiners. Re-refined oils are used in railroad journals, to freeze-proof coal, as dust control for rural roads and as motor oils and industrial lubricants. In the recent years, more than half of the re-refiners have gone out of business due to changes in labeling requirements and in the tax laws. Also re-refining has now become a marginal business. As the demand for used oil for re-refining diminishes, more of the waste oil will be disposed in other ways leading to pollution of natural water courses. Therefore, it is imperative, at this stage, that proper measures should be developed to provide incentives to collect and reuse these oils rather than disposing of them. The disposal of waste mineral oils and greases, such as those from garage and filling stations, to sewer systems are being prohibited by ordinance due to the treatment problems and their flammability and potential explosive hazard.

There are about 10,000 industrial plants in this country having significant quantities of oil in their wastes. Technology is now available to cope with wastes having either floating or emulsified oil. The more or less standard procedure of removing floating oil is with API separator. The performance of API separator is limited to the separation of solids and immiscible liquids which are susceptible to gravity separation. Stable emulsions and substances in solution cannot be separated by gravity differential principle. The use of H_2SO_4 for pH control and aluminum sulfate for floc formation is practiced for removal of soluble oil in metal-working plants. The skimmed oil is sold to an outside concern who reclaims it for various commercial uses. For emulsions, strong acid or salt may be used to de-emulsify and then the oil is separated by conventional method. Dissolved air flotation has also been used for removing the oil and improving the quality of the separator effluent with the reduction of oil and free solids content.

The Oil Pollution Act of 1924 and subsequent ammendment in 1966 require that willful or grossly negligent dischargers of oil from a vessel remove the oil from the navigable waters and adjoining shorelines immediately. To minimize pollution, such as the Torrey Canyon disaster, clean-up measures must be taken promptly and these are expensive. When the responsible parties cannot be identified immediately, or fail to act expeditiously, the initiation of effective clean up and recovery of clean-up cost poses a problem due to the present fiscal and legal difficulties.

A new process developed by Edgar Clarke, under a FWPCA grant, is said to recover high quality, fine particle carbon black (valued at

8 to 12¢/pound) that used to be a waste product. The process involves the following steps:

1. dehydrate, neutralize oil with caustic soda,

2. dilute with light naphtha and remove solids,

3. distill, and recycle naphtha,

4. vacuum distill off products,

5. heat with activated clay; use spent clay for land fill.

Berks Associates (Douglassville, Pennsylvania) using Clarke's process, report that their products obtained by re-refining used lube oil are fully comparable to those produced by major oil companies and the cost of producing them is less than the 20 to 22¢/gallon reported by the majors. There are recovery units in operation ranging in size from 20,000 to 100,000 gallons of lube oil per day. Schofield and deVries (Toronto) is marketing a process for recovering heating oils, and Comprino N.V. (Copenhagen) is planning to build 100,000 gallon/day units across Europe. A 150,000 gallon/day unit is under planning for the reprocessing of industrial wastes, such as cutting oils, soluble oils and certain rendering-house fats and oils. The end products from this new unit will be non-lubricating oils with a variety of industrial applications.

Engine plants of the Ford Motor Company at Cleveland recover and reuse oil from waste water. The disposal, efficient recovery and secondary utilization of waste products presented a unique and difficult challenge to the plants. The output of the 0.8 million gallon/day waste treatment plants consists of clarified water, a sludge slurry and skim oil. The sludge slurry from the clarifier is transferred to holding lagoons for disposal. Soluble oil is manufactured from the reclaimed oil following the process,

The performance of the reconstituted soluble oil demonstrated it cannot only match the performance of the proprietary soluble oils but also excel in some qualities such as rust protection, non-formation of "leopard spots", cleanliness and odorlessness. Thus, the recovery and reuse of what was once considered a practically worthless waste product was shown productive and can bring significant cash savings.

MISCELLANEOUS INDUSTRIES

Breweries and Distilleries

In the brewing industry, settled yeast from a previous fermentation is used in the fermenting of beer to initiate fermentation in the next batch. There is always an excess of yeast above the requirements of the brewery and only part of this yeast is satisfactory for further use in the brewery as the yeast, while settling, carries down some materials that are not desirable in beer. The undesirable part of yeast is generally discharged to the sewer. Though it is not harmful to plant operation, it has a very high BOD and becomes objectionable. This product could, however, be dried and used for cattle feed.

Distilleries discharge their wastes from molasses fermentation to the sewers or receiving waters. When this waste is evaporated and dried, it was found to be an excellent source of vitamin B complex and showed a substantial profit above the cost of evaporation and drying for use in producing cattle feeds.

SOME LEADING OIL-SPILL TREATMENTS

Developer	Product Name(s), Composition Action & Treatment Level	Unit Cost & Estimated Cost/Gal. of Oil Treatment	Where Used
Alken-Murray Corp. New York, N. Y.	Alken OSD Dispersant, non-ionic/anionic detergent mix, in solvent carrier, emulsifies oil. Treatment level: 7 to 10% by volume (higher for heavy oil in cold seas).	\$3.00/gal. (25¢)	Offshore drilling rigs; jet fuel spills; loading docks & paper mill spills.
American Machine & Foundry Co. Stamford, Conn.	Underwater injection of air repels oil, debris & trash.	---	A Connecticut beach.
American Oil Co. New York, N.Y.	Oil, absorbed in boat-mounted polyethane foam, is squeezed out.	---	Tests on inland waters.
Cabot Corp. Boston, Mass.	Cab-O-Sil, silane-treated silica (collodial) acts as a wick in thicker oil spills. Treatment level: 1 to 4% by weight.	\$2.00/lb. (25¢)	Tests on inland & sea waters.
Cyprus Mines Corp. United Sierra Div. Trenton, N. J.	Mistron Vapor, micronic talc powder for beach clean up via agglomeration. Mistron ZSC, zinc stearate coated grade designed for the open sea.	\$3.50 for a 50 lb. bag.	Ocean Eagle spill.
Drew Chemical Corp. Marine Div. New York, N. Y.	Ameroid Oil Spill Emulsifier #1, wetting and emulsifying agents in an oil-soluble liquid.vehicle. Treatment level: 3 to 5% by volume.	\$2.77-3.25/ ga . (10¢)	Torrey Canyon, General Colocotronis Ocean Eagle

Wastes from brewery and distillery fermentation have been converted to valuable by-products for many years. The two major by-products are high-priced animal feed and brewer's yeast. Spent grain mash used to be a serious water pollution problem, but the distilling industries have recovered more than 90% of their fermentation residues for sale as animal feed. The residue has a high concentration of vitamins and protein that, when mixed with other materials, makes a premium cattle feed.

Residual yeast from brewery and distillery fermentation is also valuable as an animal feed because it contains vitamin B complex, protein and essential minerals. Drying stabilizes the residual yeast by deactivating the enzymes in addition to improving the yeast's digestability. However, it is believed that a brewery must have at least about one million pounds of yeast solids per year to justify drying equipment.

The process of manufacturing tartaric acid and tartarate salts from winery wastes during World War II is considered uneconomical now. The pomace portion of the waste has been used as a vineyard soil conditioner or sometimes dehydrated and sold as a cattle feed

SOME LEADING OIL-SPILL TREATMENTS (Continued)

Developer	Product Name(s) Composition Action & Treatment Level	Unit Cost & Estimated Cost/Gal. of Oil Treatment	Where Used
Economics Laboratory Magnus Marine Div. New York, N.Y.	Mix of non-ionic solubilizers and dispersants in a penetrating carrier. Treatment level: As low as 1/2% by volume.	\$2.00-2.58/ gal. (5¢)	General Colocotronis
Enjay Chemical Corp. Linden, N.J.	Corexit 7664, mix of proprietary & purchased food-grade emulsifiers. Treatment level: As low as 1 to 2% by volume.	\$3.50-4.00/ gal. (10¢)	General Colocotronis, Esso Essen (Capetown, South Africa)
Guardian Chemical Corp. Long Island, N.Y.	Pyraxon, liquid & powder, mixed with oil, promotes burning by wicking and catalytic oil breakdown. Polycomplex A-11, used to disperse unburned oil. Treatment level: 13 to 25% by volume.	\$2.90/lb (10¢)	Laboratory & field tests.
Midland Silicones Ltd. (subsidiary of Albright & Wilson) Aberthaw, Wales	Silicone-treated flyash is dusted on the spill, sinking it. Material biochemically degrades on sea bottom. Treatment level: 250% by weight.	\$24-36/ton (40¢)	Lagoon near Aberthaw generating station.

supplement and plant mulch.

Food Industries

The food-processing industries have been faced with the realisation that both water supply and waste disposal are limiting factors in growth, operation, and product cost. Water-saving techniques have, in many cases, permitted increases in plant capacity without additional burden on fixed resources. In some cases, these have brought significant savings in total production expense.

Where water conservation is practiced, there are concomitant changes in waste quantity and character. With increased production, the total organic waste load has inexorably risen and the problems attendant to disposal have become increasingly difficult to solve. Process advances have produced some new waste constituents and water economics have resulted in concentration of waste loads in certain process streams.

Faced with a general expansion in the potential waste load which must be treated or transferred to receiving waters, regulatory authorities have been forced to tighten controls and stimulate industry's critical examination of its contribution to the pollution problem. Municipalities which receive the wastes of their allied industries assume responsibility for augmenting treatment capacity and impose the industries with service charges to transfer costs of expanded facilities and operation.

The seasonal character of wastes from the food-processing industries and their relatively high proportion might suggest that it might be more economical to treat the separate components rather than the combined flows. Industries, in seeking ways to reduce costs, are becoming increasingly aware of their responsibilities in pollution control and have begun to explore the economic feasibility of "in-plant" modifications to reduce waste and of "on-site" facilities to provide treatment.

The state of the art in food processing plants appears to involve continuous recirculation of waters with limited make-up and blow-down for a particular subprocess. For example, in the canning industry, wash waters for most produce can be continuously recycled within an optimum system. Final rinsing removes the residual dirty water and this serves as make-up for the initial scrub washing subprocesses. Similarly, flume water used for the carriage of fruits and vegetables is continuously recycled and builds up a high organic content. Low pH and appropriate bactericides can be used as a means to control bacterial growth.

Cooling tower recirculation is a common method of conserving water. The cooling tower blow-down can be used for supply water with the various subprocesses of the packing plant. Evaporator waters may be reused for processes, especially initial washing, etc. The increasing costs of water and sewer service charges based on flow volume and pollution content, are acting as incentives to optimize the conservation and reuse of water within the food-processing industry. Recent surveys of eleven California canneries indicate that water conservation and reuse can reduce water use per unit of product up to 40%. The sludge from the food industries is a major concern to the industries because of high moisture and fibre content plus a lack of stability which complicates storage of the seasonal production. However, the industries have shown keen interest in using waste products to increase profits and reduce air and water pollution.

The use of waste activated sludge from the citrus industry has been evaluated as a source of vitamin B and protein. Vitamins of the B group contained in this sludge include: thiamin, B-12, riboflavin and niacin. After the sludge is filtered, dried and pulverized, it is used as an animal feed supplement.

Citrus pulp has been extensively used by itself as a cattle feed. After dehydration, the pulp has been sold for $30 to $45 per ton. Without dehydration, citrus wastes have been sold as cattle feed for $4 per ton or converted to molasses and sold for $45 per ton. Pectin is also made from citrus wastes and used in jams and jellies.

The disposal of wet solids from the canning industry has been a major problem. Composting has been investigated and seems promising. A portion of the total canning waste has been dehydrated and used for animal feed. The total cost of producing this feed was reported to be $47 to $53 per ton of feed. Even though there is a net cost to the industry, the by-product utilization is cheaper than other methods of disposal.

The potato processing pulp is used as a cattle feed. Dewatering and drying of pulp is difficult because the solids are hydrophilic. In large plants, the 5 to 13% solids pulp is dewatered to 30 to 35% solids by pressure filtration after conditioning the pulp with lime. In an attempt to keep the peels away from water-borne wastes, dry-caustic peeling process is used. In this process, potatoes are sprayed for 50 to 100 seconds with 20% lye solution at 170° F. and drained. Then potatoes are subjected to infrared heat for 2 to 5 minutes following the holding at room temperature for 5 minutes. The peels are removed in the rotating rolls and during tumbling. The discarded peels are sold as cattle feed.

Pressure on food and pulp process has generated a parade of recent processes to upgrade liquid and solid wastes. The impetus behind this flurry of activity is the need to find dollar incentives for pollution prevention. Leading among the new processes are those that use yeast, bacteria and even laser beams to convert pollutant beans into edible products. Swedish Sugar Company has a process that will treat a potato-starch filled waste stream with a symbiotic combination of endomicopsis and candida yeasts. The end product suitable as animal feed is 50% protein, a mix of 95% candida yeast and 5% endomicopsis.

The wastes produced in this industry are largely organic and of a highly biodegradable nature. Therefore, product recovery is very much limited. However, solid wastes, in a few instances, have been utilized in many ways. Tomatoes have been pressed and dehydrated for use as hog or cattle feed. Pea vines, corn husks and corn cobs have also had limited sales. Citrus peel wastes may be pressed for molasses which can be processed, dried and sold as cattle feed. Installation of flash evaporators to produce the molasses has significantly reduced the waste load discharged. Also, certain types of pits and nut shells have been converted to charcoal.

Due to the low value of these by-products, the primary benefit to the food industries at this time is the avoidance of the cost of hauling wastes to a landfill. With the increase in cost for solid waste disposal, it is anticipated that new by-products will be developed. It should be noted here that there is a definite indication that many of the small firms are hard pressed to compete and the percentage of small firms will decrease in the future, as a result. The principal difference between old, prevalent and newer technology in these industries is not in the operational sequence, but rather in the speed and efficiency of the machinery performing the operations. New production machinery is expensive and can be justified by adequate production volume only.

Fertilizer Industry

The phosphoric fertilizer industry is faced with a difficult task of improving its gaseous effluents. Fluoride-containing gases are produced from the reactors and evaporators and collected by using spray boxes, wet impingement, wet-centrifugal spray and wet-impingement tray scrubbers. The fluoride content of these liquors, present as hydrofluosilicic acid (H_2SiF_6), represents a problem and an opportunity. The fluoride evolved during wet acid, phosphoric acid, superphosphate and triple superphosphate production in the United States is equivalent to the total United States production of hydrofluoric acid. Hydrogen fluoride is the source of fluorine compounds used in the production of aluminum, fluorocarbons and other chemicals. The availability of high grade fluorspar ore, the main source of hydrofluoric acid, is limited and these liquors represent an increasingly important source of fluorine. The hydrofluoric acid, 25% in concentration, can be used as fluorinating agent for public water supplies and may be converted into sodium silicofluoride for both this purpose and for ceramics and glass manufacture. Hydrofluosilicic acid has been suggested as a fluorinating agent and the use of strong dehydrating and decomposing agents for producing hydrofluoric acid from fluosilicic acid has also been studied.

In the recent years, attention has centered on conversion of hydrogen fluoride synthetic cryolite (Na_3AlF_6) or aluminum fluoride (AlF_3) for use in Hall cells. There are two commercial routes of conversion:

1. The acid is contacted with an alkali such as ammonia to produce the alkali fluoride and a filterable powdered silica that can be marketed for use as a thickening agent, for rubber and plastic additives and for other purposes. The alkali fluoride is reacted with an aluminum salt to produce the alkali cryolite which may be converted to sodium cryolite by fusion with a sodium salt.

2. The acid is reacted with aluminum hydroxide [$Al(OH)_3$] to produce aluminum fluoride (AlF_3) and silica (SiO_2)[84].

In the ammonium phosphate plant, the exhaust air stream from ammoniator contains high ammonia and, therefore, it is recovered by using acid stream scrubbers from both economical and a pollution control point of view. The initial cost and operating expenses for such installations are very high and the returns from the recovery of ammonia are just enough to compensate for operating cost. Therefore, fertilizer industries are generally reluctant to install such systems that cover only operating expenses. However, industries do expend large capital outlays in order to reduce complaints and improve public relations.

Most of the fluoride wastes effluents from the fertilizer industry are aqueous type and some are airborne. The airborne fluoride compounds become waterborne when they have been scrubbed with water or with aqueous alkaline reagents.

When product recovery is not practiced, due to the small quantity of waste or too low a concentration for recovery, the waste is treated and discharged to receiving water course or sewer. The treatment facilities essentially involve the following steps:

1. neutralization of waste water with hydrated lime at a pH value slightly above 11, which results in the formation of relatively insoluble calcium fluoride,

2. separation of the insoluble product of the reaction in a continuous clarifier,

3. vacuum filtration of the sludge, and

4. acidification of the clarifier effluent to lower the pH value to within the range of seven to nine.

Clarified effluents contain less than 20 mg/l of fluoride ion and by allowing it to stand for several hours, post precipitation of fluoride salt occurs resulting in the further decrease in the fluoride concentration. Where required, additional fluoride removal may be obtained by passage of the fluoride waste through a bed of activated alumina regenerated with sulfuric acid or through bone char regenerated with sodium hydroxide.

The other important by-product from sulfuric acid-based phosphoric acid plants is gypsum ($CaSO_4$) which has been used in wallboard and cement and for the production of sulfur via reduction with coal, followed by fixing of the lime--either with silica or by reaction with ammonia. Such practices are in use in sulfur-short locations and may become less attractive as the sulfur supply situation improves.

Gypsum, derived from the production of wet process phosphoric acid poses major problems because it does not have a good market. Storing of gypsum is practiced in some cases and this is becoming expensive because of the increasing land value.

A new process developed in West Germany involves recrystallization under controlled conditions by the addition of certain compounds. The product can be dried and sold to a plaster of paris distributor or directly cast as plaster board or construction blocks. The final product from the synthetic gypsum compares favorably with most products made from natural gypsum. In addition, the requirements for labor, space and utilities are very much less.

In order to conserve the sulfur or sulfuric acid, nitric acid is used in the digestion of phosphate rock. Phosphoric acid (H_3PO_4), calcium nitrate $[Ca(NO_3)_2]$ and hydrofluoric acid (HF) are formed in this process and the unwanted calcium nitrate is precipitated as gypsum by adding ammonium sulfate $[(NH_4)_2SO_4]$. The latter is regenerated and reused by treating the gypsum with carbon dioxide and ammonia and thus sulfate recycle is accomplished. Ammonia remains in solution as ammonium nitrate, a fertilizer ingredient.

Sulfur shortage has sparked renewed interest in nitric acid treatment of phosphate rock to produce fertilizers. There are a number of processes available. Nonsulfur-consuming nitrophosphate generally involves larger investments but smaller operating costs than the sulfur-consuming processes.

Use of gypsum for sulfur production to fill the sulfur gas is being studied by variations of standard ore-winning techniques or bacteriological digestion methods. These studies are directed to produce sulfuric acid or ammonium sulfate. In the wet process for phosphoric acid production, production of uniform gypsum crystals permits fast filtration, minimum sealing and modest requirements for site space, power and maintenance. Therefore, it was possible to cut down the capital and operating cost by 15%.

Nitrophosphate processes, which use nitric acid to dissolve phosphate rock, are becoming increasingly popular as a result of the current sulfur shortage. A Norwegian firm has made major improvements in this field and is producing a wide range of fertilizers with high water solubility.

Dorr-Oliver has come out with a process that uses sulfuric acid acidulation of phosphate rock without appreciable consumption of sulfur. Instead of letting sulfur out as gypsum, the new process converts gypsum into calcium carbonate ($CaCO_3$) and reacts co-product ammonium sulfate $[(NH_4)_2SO_4]$ with fluorosilicic acid vapor derived from the phosphate rock to generate sulfuric acid for recycle to the rock acidulation step.

Animal Products

Meat-packing industry is a major one in this category and includes the following:

1. Slaughtering plant.

2. Meat-packing plant.

3. Meat-processing plant.

A slaughtering plant is a killing and dressing plant which does almost no processing of by-products. A meat-packing plant is both a slaughtering house and a meat-processing plant. Packing houses involve cooking, curing, manufacture of sausage, rendering of inedible fats into greases and many others. A meat-processing plant does no slaughtering at all.

Blood is one of the major sources of BOD and, therefore, the recovery of blood is an important subprocess to the process of killing. Failure to recover blood increases the BOD by 72%. It is estimated that about 96% of the meat-packing industry is practicing blood recovery. Blood can be concentrated or dried and utilized in feeds and fertilizers.

Hairs of hogs are removed by mechanical scrapers and they are sold, hauled to offsite disposal or dissolved in caustic solution. The trend seems to be toward sale because it can be used in the manufacture of foam rubber and felt. In summer, when the hair and bristles are short, they are disposed of at the incinerator as they are of little value.

Viscera are removed and distributed to the proper channels depending upon whether they will become an edible or inedible product.

Rendering process is of two types: edible and inedible rendering. In the edible rendering, fats are converted into lard and edible tallow. Edible rendering is applied primarily to hog fats, but also to some beef fat. In the inedible rendering scraps, trimmings, and inedible organs are converted into inedible fats which are used in soap, in the manufacture of grease and in animal feeds. Inedible rendering also provides an outlet for those parts of an animal which have been condemned by federal meat inspectors. Through rendering, condemned material is sterilized and converted into recoverable products. In recent years, the disposal of paunch manure is a vexing problem. Paunch manure contains very little fat and a large quantity of partially digested material. The paunch content of cattle is estimated at 40 to 60 pounds per head and contains about 1/3 pound of BOD. In most cases, paunch solids are disposed of directly to farmers as fertilizer or as land fill.

Scraps generated in the manufacture of sausage are used in the rendering process. Grease and fats recovered in meat processing are valuable raw materials for soaps, gelatin and glue.

The wastes from industries producing fatty acids and glycerine contain a large amount of grease. Improper design of interceptors result in the discharge of a large fraction of grease. It was shown in many cases that by recovering the grease through the installation of proper interceptors that the yield can be increased by about 3% and the problems connected with sewer clogging, treatment process upset are solved satisfactorily.

In the wool pulling and wool scouring industry, valuable sheep skins and wool fibres have been recovered by the provision of proper screening arrangement.

The major concern in the poultry waste disposal is the high concentration of solids and fat. Centrifugal solid separators are used for the recovery of high solids and fat. An approximate materials balance of the system indicates that at a waste input of 3200 gph, about 400 pounds per hour of solids are removed and 125 pounds per hour of fat recovered. The insolubles content of the influent was reduced from 10,000 mg/l, or higher, to 1500 mg/l. The fat content was reduced from 0.9% to 0.04%. The fat, thus recovered, is used in the dog food manufacturing process and solid matter disposed of at local rendering concerns. The returns from the by-products are above the operating costs and the recovery system proved economical.

In areas where there are no rendering firms, the by-products are removed from the gross pollution load of the plant and are given free in exchange for hauling them away.

Item	Use
Trimmings, bellies and others	Used for edible purposes, oil production after rendering, protein feed after rendering and gelatin manufacture
Hair	Used in manufacturing, upholstering and rug backing
Fleshings	Glues
Degreasing exhaust Drum liquors	Reuse of solvent in tanning, soap
Spent lime liquors	After settling, the sludge can be mixed with other plant wastes and sold as fertilizer
Pickle solution wastes	In the past, this solution has been reused within the tannery for pickling a number of times. However, tendency of late has been to omit this reuse through advent of drum bate, pickling and chrome tanning
Chrome tan liquors	a) Holding and reusing in tannery b) Precipitate chromium hydroxide $[Cr(OH)_3]$ filter redissolve chromium with sulfuric acid. Considered economically impractical
Spent vegetable fans	Evaporated and sold as boiler compounds
Spent tan bark	Used as floor coverings for horse shows, circuses, playground. Sometimes used in paperboard manufacturing or in making white lead

In the leather tanning and finishing industry, very little water is reused. It appears that the practice of water reuse as a pollution reduction measure has been neglected in the average industry. The methods that could be used to reduce the quantity of process water are as follows:

1. Use of countercurrent washing techniques.
2. Clean-up and reuse of process waters.
3. Wash sprays instead of baths.
4. Automatic controls on process water.
5. Dry waste disposal instead of water carriage.

The following chart details normal by-product utilization in a tannery.

<u>Organic Chemicals Industry</u>

In the organic chemicals industries, the wastes produced are high in COD and their treatability varies from treatable to most difficult to treat. These wastes contain valuable by-products and these are recovered.

New technology and process changes are, in general, directed towards better yield and, thereby, reduction in waste load. In cases where the by-product is valuable, product recovery is always practiced.

The organic chemicals industry is a complex one due to the number of products and processes utilized in the industry. All of these plants practice product recovery and waste reduction in one form or the other. To include all of them will be too elaborate and, therefore, the major ones are included here.

1. Acetaldehyde

Acetaldehyde is produced by the Wacker process and waste discharge averages 1200 gallons per ton of product. The waste contains primarily chlorinated aldehydes and the COD is in the range of 10,000 mg/l. The waste is difficult to treat biologically unless it is diluted with other waters. Generally, several facilities handle these wastes by means of a deep well rather than by biological treatment.

The waste discharge is reduced to 150 to 200 gallons/ton of product by suitable changes in the still design for final disposal either through deep well injection or incineration. In the concentration process, there is no reduction in the organic load and, therefore, the only possible means of reducing the organic load is recovery and use of the chlorinated aldehydes, dechlorination of the aldehydes or improvements in yield. The present trend seems to be in the emphasis on the improvements in yield and recovery and it is expected that the average yield will reach 97% in five years thereby reducing organic loading by about 40%.

2. Acetic Acid and Anhydride

Light petroleum gas (LPG) oxidation results in the production of substantial amounts of other acids such as formic and propionic which must be disposed of by incineration, resale or biological oxidation. Unfortunately, resale of these materials does not appear to be practical because the markets for these chemicals are quite limited. Waste flow from this system is in the order of 1000 gallons/ton of product and the organic concentrations are in excess of 30,000 mg/l. In some operations, the stream is neutralized with caustic and the sodium salts of the acid are recovered.

Acetic acid and other higher acids are produced by oxidation of acetaldehyde. Water flows are of the same magnitude, but organic load is about 50% of the load generated by the oxidation of LPG. The Reppe process is another process for the production of propionic and other acids and the liquid waste amounts to about 50 gallons per ton of product including drains. The amounts and character of the waste would not appear to create serious waste treatment problems or expenditures. There will be improvements in the yield picture which will assist in reducing the waste management problems but it is not anticipated that any changes will radically change the waste situation.

3. Acetylene

The key factor in waste management in the production of acetylene from hydrocarbons centers around the control of the cracking furnaces to maximize the production of valuable products. The efficiencies are generally low--in the order of 40 to 50%--and efficiency improvements are therefore vital to the reduction of pollution load. Further, recoveries of waste products in burnable form is also vital to the effective operation of the unit. In some cases, a solvent is used to recover the acetylene. As solvent losses are usually significant, careful selection of the solvent and improved design of the vacuum stripper would assist in reducing the pollutional load.

4. Ammonia

The major streams arising from an ammonia plant are process condensate and site drains. The waste flow is 300 gallon/ton of product and ammonia lost in waste stream is 0.1 lb/ton of product. Removal of ammonia from waste streams is a necessity to eliminate eutrophication problems and recovery of ammonia has been proven practical. By air stripping of the waste after adjusting the pH with NaOH (about 0.1 lb/ton of product), it was possible to reduce the ammonia loss by about 60%. It is possible to remove greater amounts of NH_3 (90 to 95%) by increasing the air flow in the range of 500 to 1000 scf/gal., but such an approach is not economical.

5. Carbon Tetrachloride and Other Chlorinated Hydrocarbons

The major waste stream results from the production of hydrogen chloride as a by-product of the reaction. There is a great need for routes to the economic recovery of chlorine from hydrogen chloride or through the reuse of hydrogen chloride. A number of new oxychlorination processes have recently been developed which enable the user to utilize hydrogen chloride together with chlorine in a balanced facility. As an example, for the production of carbon tetrachloride (CCl_4):

$$CH_4 + 2Cl_2 \longrightarrow CCl_4 + 2H_2$$

$$CH_4 + 4HCl + O_2 \longrightarrow CCl_4 + 2H_2O$$

This approach eliminates the problem of hydrogen chloride (HCl), but does produce about 60 gallons of waste containing minor amounts of chlorinated methanes and HCl/ton of product. It is anticipated that, in the future, process changes will substantially eliminate the HCl being generated in these operations.

6. Cellulose Acetate

Major waste source arises from the stills used to recover acetic acid, acetone and other solvents. In addition, considerable amounts of degraded cellulose, phosphates (used as catalyst) and sulfuric acid (from spills) are generated. There are numerous opportunities for waste reduction in cellulose acetate production and the more important ones are the following:

a) Recovery and reuse of CA fines.

b) Improvements to achieve higher yields on cellulose.

c) Improved operation of acetone and acetic acid stills.

d) Use of additional stills to improve recovery.

7. Ethylene

In the production of ethylene, about 15 gallons of waste/ton of product are generated and this contains 3.5% sodium hydroxide ($NaOH$), 1% sodium sulfide (Na_2S) and 6.6% of phenols. Studies are underway to recover the alkali value from the spent caustic. There are no attempts at present for the recovery of sulfides but sulfide oxidation is generally practiced to reduce the immediate oxygen demand of this stream.

The process condensate stream may be reprocessed by being sent to a stripper and stripped by live steam. This will strip most of the non-phenol contaminants and about 20 to 25% of phenol. By contacting with fresh feed, it is possible to strip out most of the phenol which can be sent with the feed to the cracking furnace. The phenol-free waste water can be steam stripped to remove residual volatile hydrocarbons. The water stream, free of contaminants may be reused in steam generation in the plant.

The major products in the ethylene plants besides ethylene are propylene, low density polyethylene and alfa olefins. Spilled pellets and fluffs of polyethylene constitute a major pollution problem because of its unsightliness and reduction in the rate of oxygen transfer. A surface drainage separator with a basin and a baffle extending three feet below the water surface is used to trap the polyethylene. The trapped polyethylene is sent eventually to land fill for disposal.

Handling of quench water which constitutes 30% of the total plant effluent is difficult because it contains 1000 to 10,000 mg/l of oil in addition to finely divided coke and heavy tars. A part of the aromatic distillate co-produced in the plant as a solvent is used to extract the oil. The water is separated from the solvent-oil mixture, filtered, degassed to remove hydrocarbon gases and sent to a surge tank. This water is used for cooling water make-up, which accounts for 10 to 15% of the cooling water for the plant.

8. Phenol

Major quantities of aqueous wastes containing organics and inorganics create considerable difficulties relative to waste treatment. A typical 100,000,000 lb/year phenol plant based on cumene produces a stream of 200,000 gpd of waste water containing 13,200 mg/l COD and 180 mg/l of phenol. Inorganics are produced at the following rates:

Sodium carbonate	-	5,000 lb/day
Sodium formate	-	500 lb/day
Sodium bicarbonate	-	500 lb/day
Sodium sulfate	-	22,000 lb/day

The waste production from the direct oxidation facility might be considerably less than the older process.

The following are practiced to reduce the waste load:

a) Careful control and modification of the process involved.

b) Recovery of inorganics by crystallization.

c) Recovery of phenol by solvent extraction.

As the sulfonation and chlorination approaches are reduced

in importance, it is likely that the amount of waste generated per pound of phenol produced will be greatly reduced.

9. Urea

All of the major processes involve the high pressure reaction of CO_2 and ammonia. Typical wastes amount to about 120 gallons/ton of product with the following composition:

	lb/ton
Ammonia (NH_3)	3.6
Carbon dioxide (CO_2)	0.6
Urea (NH_2CONH_2)	0.8
Ammonium carbamate ($NH_4CO_2NH_2$)	0.1
Total	5.1

It is normal practice to remove about 50% of the ammonia by airstripping after adjusting pH with alkali. Recently, two new processes have been developed which strip the ammonia and carbamate by contacting with either NH_3 or CO_2 and sending the overload to the reactor. This approach reduces ammonia losses by 50% or more with a more drastic effect on carbonate.

10. Production of Chlorine from Hydrochloric Acid

In the conventional electrolysis process for producing Cl_2, 1.13 tons of NaOH are produced for every ton of chlorine produced. There is a greater demand for chlorine compared to NaOH and, therefore, there exists a chlorine-caustic imbalance. In order to avoid the imbalance, a new process called "KEL-CHLOR" has been developed by the Kellogg Company from the Deacon process and this process shows the possibility of making chlorine compounds. Hydrochloric acid produced as by-product in industries and not utilized may be used for the efficient and economic production of chlorine.

The following is a list of sources for HCl:

a) Production of HCl from inorganic sources.

b) Organic chlorination produces HCl as a by-product.

c) $MgCl_2$ can be hydrolyzed by hot steam to give MgO and HCl.

d) Ammonium chloride, a by-product of the Solvay process, can be dissociated to NH_3 and HCl and the two separated by a suitable method.

e) Hargreaves process, which uses SO_2 and air instead of H_2SO_4, could be used to produce HCl from NaCl.

f) When KCl is converted to potassium phosphate and nitrogen oxides in order to eliminate plasmolysis of the plants, HCl is produced as a by-product.

11. Anhydrous Hydrochloric Acid from Chlorinated Organic Waste

A variety of chlorinated organic residues are left behind as unwanted co-products after the manufacture of some big-volume chemical products. Examples: ethylene dichloride from vinyl chloride, benzene hexachloride from some insecticides, tetrachloro-ethylene from herbicides, and dichloropropylene from propylene glycol, chlorinated elastomers and nylon. The disposal of these wastes is a difficult problem and the methods already tried include: deep-well injection, ocean dumping, burning in open pits, lagooning, and burying. Disposal in deep wells is expensive and has hazards in some form or the other. Because of the toxic fumes emitted, open-pit burning of these wastes has been forbidden in several states. Thus, there is a pressing problem of disposing of the one billion pounds of waste generated each year in the United States in the production of chlorinated organics.

TABLE 31-5

BY-PRODUCT RECOVERY IN THE ORGANIC CHEMICALS INDUSTRY

Process	By-Product Recovery			
	By-Product	Quantity	Value	Diminution of Pollutional Load
Acetaldehyde	Acetaldehyde	5 lbs/100 lbs	$0.50	40%
Acetic acid	Formic & Propionic acid	200 lb/ton	$20.00	60%
Acetylene	Solvent	10lb/ton	$5.00-10.00	10-30%
Ammonia	Ammonia	0.1 lb/ton	Negligible	60%
Carbon tetrachloride	Chlorine	0.4 lb/lb	1.2¢/lb	50-100%
Cellulose acetate	Cellulose fines	0.1 lb/lb	$0.01/lb	30-40%
Ethylene	Caustic	2 lb/ton	$0.04/ton	30-50%
Phenol	Sodium carbonate	2 lb/ton	$0.04/ton	15%
	Sodium formate	0.2 lb/ton	$0.04/ton	1%
	Sodium bicarbonate	0.2 lb/ton	$0.04/ton	1%
	Sodium sulfate	8 lb/ton	$0.16/ton	50-80%
Urea	Ammonia	3.6 lb/ton	$0.10/ton	50%
	Urea	0.8 lb/ton	$0.02/ton	20%
	Ammonium carbamate	0.1 lb/ton	$0.01/ton	10%
Chlorinated organics	Hydrochloric acid	Varies	Varies	80-100%

Textile Industry

The textile industry, as a whole, is a major factor in the American economy. The finishing operation of the textile industry produces a large amount of liquid wastes containing pollutants. Due to the tremendous growth rate in this industry, it is anticipated that more efficient manufacturing processes will be utilized, a larger percentage of the waste will be treated due to the increased pressure by regulatory agencies at all levels.

Process water reuse and recovery of by-products are the two major attempts being made in this industry in addition to process modifications to reduce pollution load and increase the economy.

The textile industry is grouped in three major categories for convenience:

1. Wool weaving and finishing.

2. Cotton textile finishing.

3. Synthetic textile finishing.

In the wool weaving and finishing industry, approximately 5% of its process water is reused. It is anticipated that countercurrent scouring can reduce the amount of water required by as much as 6000 gal./1000 lb of wool and use of rinse waters in the wash after fulling can reduce water requirement by 4000 gal./1000 lb of wool. By-product recovery is limited because of the poor market for the products recovered. It is estimated that fifty to one hundred thousand tons of wool grease and 20,000 to 40,000 tons of suint could be recovered for production of lanolin and potash from wool and grease, respectively.

In the cotton textile industry, about 16% of the process water is reused. It appears that the process water reused will increase in the future because newer machinery is often of continuous or countercurrent design. Also, because process water is becoming more expensive in many areas, increased emphasis for reuse of process water is made. There is no significant by-product use of wastes in this industry. Many attempts have been made in the past to develop economically feasible methods for recovery of the expensive dyeing compounds but were unsuccessful. It is not hoped, in the foreseeable future, that any by-product be recovered for use on any significant scale.

About 10% of its process water is reused in the synthetic textile finishing industry. Even though there appears to be an adequate market for wastes reclaimed in this industry, there are no economically feasible methods available now. All liquid wastes contain chemicals used in the finishing itself and could be reused if reclaimed. All carriers are recovered and reused. Chemicals such as spent developed dye bath are not reclaimed as it is not economical. In the future, thermal wastes can be reused by heat transfer method.

Mercerizing of cotton yarn (treating with concentrated solutions of sodium hydroxide), results in a waste containing about 2% sodium hydroxide. This represents an important loss of this valuable chemical because of the relatively high chemical costs. Also, the alkalinity of the waste was very high requiring increased consumption of alum.

In a textile mill in Argentina, the waste stream containing sodium hydroxide was segregated and sodium hydroxide recovered by subjecting the waste to physical processes. Filtration of the waste prior to evaporation and two-stage evaporation were found to cause corrosion problems due to high temperatures. In the first stage of evaporation, the unit works at atmospheric pressure and, in the second stage at 25 in. Hg vacuum. The average concentration at the end of each stage is 5.2 and 25.0% by weight, respectively. The concentrated sodium hydroxide is filtered through a pressure filter and stored for reuse. The main goal of recovery of sodium hydroxide was the economics due to the high replacement cost of this chemical. Total amount of sodium hydroxide recovered amounted to 140 tons/year and the net profit obtained through recovery amounted to $22,000/year. An additional benefit resulting out of sodium hydroxide recovery is the decrease in consumption of alum at the treatment plant.

Nuclear Plants

The recovery of uranium and other fissionable materials from depleted nuclear-power fuels is necessary since only a fraction of the fissionable material is consumed prior to removal from the reactor.

The fuel is dissolved in nitric acid or any other acid before recovery. Sometimes fuels are mechanically decladded or oxidized prior to acid attack. The various products such as uranium, plutonium and other products, are separately recovered by solvent extraction employing an intricate set of pH adjustments. Efforts to recover fission products are now commercial. Most radioisotopes are not prepared in this manner but could be recovered from wastes by chemical action or ion exchange.

The Canadian nuclear power industry uses heavy water reactors fueled with natural uranium. The spent fuel contained plutonium, a potential fuel, but the cost of recovering it was such that it was not competitive with natural uranium which is in abundance in Canada. The spent fuel is stored in zirconium cladding, under water at the reactor site for final disposal. If the price of uranium rises sufficiently, it will become profitable to recover the plutonium.

A fluoride volatility process has been developed in which the fuel is treated with oxygen, hydrochloric acid or chlorine. This removes the cladding material as the volatile chloride. Fluorine is then added and uranium and plutonium leave as the hexafluorides which may be separated due to differences in their relative volatilities.

A reclamation process in use, extracts and converts the plutonium content of metal scraps and sundry wastes from nuclear plants into a high purity plutonium nitrate solution. The scrap is initially dissolved in hydrofluoric acid, nitric acid and aluminum nitrate monohydrate. The dissolved plutonium is extracted with organic solvents and the plutonium-rich organic phase is stripped with nitric acid and concentrated to a final product having a plutonium of 200 to 300 gm/l.

Milk Industry

In the milk industry, approximately fifteen billion pounds of fluid whey results per year. Because of its high COD and production of undesirable conditions, the whey has to be evaporated, processed and some returned to the farmer as slops.

Cheese whey has an average composition of about 4.8% lactose and 0.9% protein and, therefore, many useful by-products could be obtained. However, the demand for these by-products is not large enough to utilize the surplus whey. Conversion of whey to yeast to be used as a source of protein in animal nutrition is very attractive, but the cost of production is expensive. However, the economic factor becomes less important from the point of view of disposal of a potent waste.

CHAPTER 32

OVERVIEW OF SEWER OVERFLOW

1. It is now well established that overflows from combined sewers constitute a significant part of the nation's total water pollution problem. The separation of sanitary and storm sewers has been considered to be the ultimate solution for the elimination of such overflows. However, the mere fact that the systems are separated does not necessarily reduce the number or pollutional effect of overflows. Separate sanitary sewers can act as combined sewers due to the excessive infiltration of surface runoff and produce overflows of untreated sewage.

2. A study was conducted on 25 percent of Roanoke, Virginia's separate sanitary sewerage system which is probably representative of most existing separate sanitary sewerage systems installed prior to 1950. The study revealed that overflows from this separate system amount to one to two percent of the annual average untreated sewage. Overflows are even more significant on an individual storm basis, amounting to as much as 14 percent of the daily untreated sewage.

3. Overflows from sanitary sewers are not controlled; therefore, they occur indiscriminately throughout the watershed causing additional health and safety problems. Overflowing manholes in streets and residential property and flooded basements are esthetically objectionable as well as potential health hazards and should be eliminated.

4. All sanitary sewer overflows are not directly into a stream or water course. Only 25 percent of the pollution from such overflows could be actually traced to entering a stream. The remainder either re-entered the sewer after the storm, ponded, or entered the ground water table. However, the possibility of some of the pollutants reaching the stream through undetected leaks or through ground water cannot be ruled out.

5. The greatest cause of the sewer overflows is excessive infiltration. During storms sanitary sewers become, in effect, storm sewers due to the numerous entry points for surface runoff. Contrary to original beliefs, relatively few downspouts and storm drains were directly connected to the sanitary sewers.

6. Storm water entry points ranged from perforated manhole covers

to broken pipe sections exposed to surface stream flow. Crushed pipe resulting from other types of construction operations, broken service laterals, and service laterals poorly connected to the sewer main were other major sources of infiltration.

7. No logical relationship could be found between the topographical characteristics of the drainage area and the rates of infiltration or number and type of points of infiltration.

8. Flow hydrographs indicated that peaks in the sanitary sewers occurred simultaneously with those in the storm sewers. A rainfall intensity of 0.04 inch in 1 hour was required to cause measurable infiltration.

9. Throughout the range of storms monitored, the average rate of infiltration was proportional to the average rainfall intensity, as shown on Figure 19; thus, the higher the rainfall intensity the higher the rate of infiltration. Rates of infiltration as high as 24,000 gallons per inch of pipe diameter per mile per day were recorded, which is 50 to 100 times the normally specified allowable amount.

10. Total volumes of overflow and infiltration for any particular rainfall event are a function of storm duration and average rainfall intensity. Average rainfall intensities greater than 0.4 inch per hour generally have durations of less than two hours.

11. Many municipalities rate the capacity of the interceptor sewer as a multiple of the average dry weather flow. Capacity allowances on such a basis have no relation to the actual problem other than merely providing some excess. Rather, capacity allowances for infiltration in the design of interceptor or trunk sewers should be based on the total length of upstream sewer.

12. In addition to localized overflows from the sewerage system, overflows occur at the Water Pollution Control Plant on the average of 10 times per year. This overflow contributes as much as 27 percent of the total pounds of untreated BOD as a result of a single storm event. When added to the overflow from the sanitary sewer, the total overflows contribute 43 percent of the daily untreated BOD.

13. About 77 percent of the total annual overflows occur between May and October, and constitute about 75 percent of the total overflow volume. This corresponds to the period of lowest stream flows, and the peak recreational water use season when water pollution would be the most critical.

14. Higher than normal river flows cannot be relied upon for dilution of the overflows. Localized thunderstorms produce sewer overflows without significantly affecting the flow of the river.

15. While no in-depth study was made on the pollutional potential of storm surface runoff, the associated data indicated that its quality was similar to that of secondary treated effluent. This storm runoff increased the BOD concentrations in the streams three to five times over that of normal dry weather flow. Considering the volumes of storm runoff and its quality, the amount of BOD contributed to the Roanoke River from storm runoff alone could easily be 200 percent of that of the sewer overflows. Therefore, it appears that pollutional effects of storm runoff are appreciable and clearly point to the need for better housekeeping practices in our urban watersheds. A more detailed analysis may indicate a need for treatment of urban runoff. At any rate our small urban streams can no longer be given the traditional "backdoor" approach and treated as conveyors for our refuse.

16. Due to the losses of untreated sewage through overflows, the effectiveness of pollution control cannot be gauged solely by the efficiency of a treatment process. The installation of tertiary treatment facilities to remove additional fractions of pollutants appears unwarranted when large volumes of the sewage are lost through overflows and never reach the treatment process. Therefore, the entire "system" including lines and treatment plant, should be evaluated as a whole.

17. It is concluded that a reduction of pollution from sewer overflows in Roanoke can best be obtained by a combination of remedial measures. These measures include more complete sewer separation, replacement of critical lines to increase capacity, detention of peak flows within the drainage basin, controlled release based on the system's downstream capacity, and treatment of all flows at the Water Pollution Control Plant. The key is the reduction of infiltration by separation or elimination of storm surface entry points, utilizing the computer to achieve optimum designs of the system.

18. The possibility of removing all storm water from existing sanitary sewerage systems, especially older ones, is remote. However, it is believed that as much as 80 percent could be eliminated through a program of inspection and repair. The total estimated cost of such a program is $61 per capita; far below the cost of complete line replacement.

19. The use of the computer permits evaluation and optimum selection of detention basin size and location and pipe capacities under a variety of trial conditions and storm events. Unfortunately, a considerable amount of data is required on the existing sewerage system, much of which must be determined in the field as it is often not available from office records.

20. The detention basins selected will detain the overflow for treatment during off peak periods from at least 90 percent of the storm events causing overflows in a five-year period; this corresponds to 90 percent of the overflow volume during the same period.

RECOMMENDATIONS

1. Further sewer separation is essential and it is recommended that an intensive inspection and repair program be undertaken on a selected drainage area basis.

2. Sewer joint sealing procedures are not new, but there are many sealing materials which are unproven. It is recommended that the repair program for the initial drainage basin selected, be developed as a demonstration project to evaluate the materials, methods, and effectiveness of the overall program. This demonstration area will lay the groundwork and supply much needed information for future improvements to the remainder of the City's sanitary sewer system.

3. A significant amount of infiltration originates from storm water inlet points on private property, through either illegal connections or breaks. The City's program should provide for a cooperative arrangement between the private owner and the City for these repairs and separation of storm water. This could be established through uniform charges for such work or by licensed contractors with established rates.

4. A sewer leak detection program is of little value if results are not properly recorded and acted upon. The results of a leak survey should be promptly evaluated and action taken. Private owners of property with violations or deficiencies should be promptly notified and periodic follow ups made to insure corrections have been made.

5. Pollution from overflows at the pollution control plant can be reduced by in-house changes in process piping. It is recommended that digester supernatant and sludge thickner overflow piping be revised so that they return to a point other than the overflow manhole; thus, these extra pollutants are not flushed into the river during storms.

6. It is recommended that more study be directed to the impact of the pollutional effects of urban storm surface runoff. Such studies would relate detrimental effects of storm runoff to other known sources of potential pollution and develop any recommendations for reducing pollution from this source.

7. Studies involving stream and sewage flow measurements, rainfall measurements, and sampling require equipment that often is not readily available. It is, therefore, recommended that the allowance for initial start-up time for such studies be a period of three to six months. Further, such studies are dependent upon weather for the gathering of data; therefore, the time frame should be flexible to insure a representative season has been observed.

PART III

SOLID WASTE DISPOSAL

CHAPTER 33

ANALYSIS AND COMPOSITION OF SOLID WASTE GENERATED

The United States presently produces about 180 million tons of solid wastes per year exclusive of those solid wastes which are normally handled by sewerage systems and which are not the subject of this section. If present trends continue by 1980, this figure will be 250 million tons, and by the end of the century it will be 475 million tons.

The present generation rate is equivalent to 6.9 lbs/capita/day. Table 33-1 below shows the sources of this waste, on a lbs/capita/day basis.

Table 33-1

Solid Wastes Generated in the United States

	Lbs/Capita/Day
Residential	2.4
Commercial	1.0
Bulky Waste	.3
Sub-total	3.7
Industrial	3.2
Total	6.9

Of this, approximately 5.1 lb/capita/day is disposed of in private (non-industrial) and municipal sites. Urban dwellers produce more waste than their rural counterparts.

This waste stream can be broken up into its material components as shown in Table 33-2.

Table 33-2

Composition and Analysis of an Average Municipal Refuse from Studies Made by Purdue University

Component	Percent of all Refuse by Weight	
Rubbish:		
Paper	42.0	
Wood	2.4	
Grass	4.0	
Brush	1.5	
Greens	1.5	
Leaves	5.0	
Leather	0.3	
Rubber	0.6	
Plastics	0.7	
Oils, paints	0.8	
Linoleum	0.1	
Rags	0.6	
Street Sweepings	3.0	
Dirt	1.0	
Unclassified	0.5	
TOTAL		64
Food Wastes:		
Garbage	10.0	
Fats	2.0	
TOTAL		12
Noncombustibles:		
Metals	8.0	
Glass and Ceramics	6.0	
Ashes	10.0	
Composite Refuse, as received		
TOTAL		24
All Refuse		100

The bulk of this refuse has been handled by landfill sites within the community generating the waste, and by incineration. In the past,most landfill sites were little more than open dumps, and incinerators haphazardly dispensed ash and noxious gases into the atmosphere. The general public has become aware of the dangers posed to the environment and themselves by uncontrolled disposal. This has resulted in air pollution control requirements for incinerators and the replacement of some open dumps by sanitary landfills.

This upgrading of solid waste disposal practices has resulted in higher disposal costs. Last year over three billion dollars were spent in this country for the collection, transportation and disposal of solid wastes. These costs are steadily increasing. Many cities are running out of nearby landfill sites, and are transporting their refuse, farther and farther away. Many municipal incinerators are overage and inefficient and will soon have to be replaced. Other incinerators will have to have better air pollution control equipment installed, which will increase the cost of incineration by 25%.

These factors are becoming especially acute in the coastal metropolitan belts where approximately 40% of the American population resides. These communities are looking to the sea as a disposal site for their refuse.

For example, present landfill sites in New York are expected to last at most seven years and according to some authorities only four. New York has traditionally relied on coastal landfill but finds it politically infeasible to obtain additional shoreline locations for landfills. Boston faces a similar situation. At the same time, development of shredding, compacting and bailing machines has advanced to the point where general garbage can be compacted to densities higher than sea water. This opens up the possibility of dumping at sea with little or no probability of any of the garbage returning to land.

At present, the ocean is being used as a disposal site for construction and dredging wastes, as well as certain types of industrial waste.

Thus, advances in technology combined with the increasing costs associated with the possible alternatives have greatly increased the pressure to dispose of solid wastes at sea.

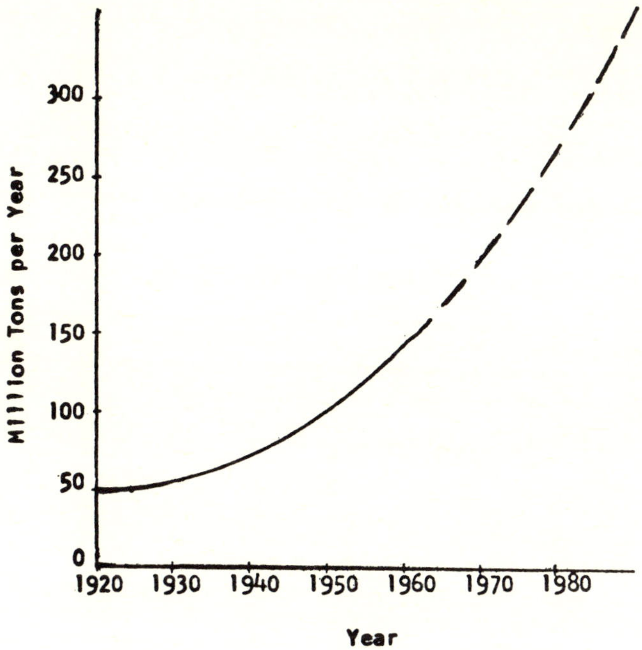

Fig. 33-1 TOTAL REFUSE PRODUCTION IN THE U.S.

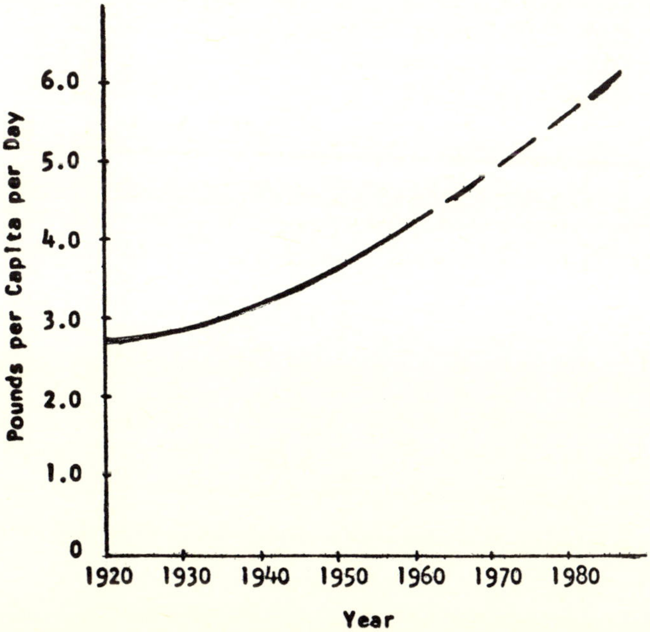

Fig. 33-2 PER CAPITA REFUSE PRODUCTION

TABLE 33-3 COMPOSITION OF HOUSEHOLD GARBAGE

	Percentage by Weight					
	Germany	England	Holland	Switzerland	California	Montreal
Rags, leather, rubber	1.3	1.8	1.9	2.2	1.6	5.7
Paper, cardboard	4.5	12.6	15.5	22.6		50.1
Bones	0.7	0.5	1.0		68.4 (Paper, cardboard through Other organic wastes)	
Organic kitchen wastes	18.0	12.5	18.5	50.0 (Bones through Other organic wastes)		10.2
Other organic wastes	10.0	2.2	9.7			8.6
Iron and metals	2.5	3.7	5.3	3.5	10.7	8.6
Glass & glass fragments	3.0	2.8	3.4	3.8	11.7	3.9
Ashes, china, clinker unfit for composting	60.0	63.9	44.7	17.9	7.6	12.9

CHAPTER 34

LAND-BASED DISPOSAL METHODS

Incineration

Incineration is a refuse reduction, rather than a refuse disposal, process. The solid waste residue must still be ultimately disposed of after the incineration process.

Incinerator capacity has steadily increased since 1950, as has the percentage of solid wastes that are incinerated before disposal. However, despite predictions of further growth of incinerator capacity, incineration as a method of solid waste reduction has many problems. Many of the incinerators in large cities will either have to be replaced or closed down because they are overage (1). The capital costs for a large municipal incinerator have been steadily increasing because of higher land values, construction costs,and the added expense of efficient air pollution control equipment.

The analysis of incineration is presented under the subheadings Volume Reduction, Air Pollution from Incinerators, Incineration with Heat Recovery, and High Temperature Incineration.

Volume Reduction

The average ton of refuse has a volume of 13.3 cubic yards (150 lb/cu. yd.) (28) at the generating source. This volume is usually reduced 80 to 90 percent by incineration (4). In this process, usually 98 to 99 percent, by weight, of the combustible materials can be converted to water vapor and carbon dioxide (28). The total weight reduction is commonly between 75 and 80 percent (28). Incineration before landfill will greatly prolong the life of the landfill site and provide a more stable and compact fill material.

In some instances ferrous metal salvaged from the incinerator residue is sold to the copper smelting industry (2). Glass and non-ferrous metals may also be recycled if a market exists. At the Stanford Research Institute, a method of grinding incinerator residue and using it as a "cement" of sorts is being investigated The Bureau of Mines is actively investigating a system of mechanical, and magnetic separation of incinerator residue. This is discussed in more detail in the section on recycling.

Air Pollution from Incinerators

The objective of incineration is to convert refuse moisture and organics to components of the atmosphere by controlled and enclosed combustion. The chimney gases, which are the primary products of incineration, consist of carbon dioxide, water vapor, nitrogen, a solid residue of ceramics, glass, metals and various other ashes, sulphur oxides, and other inorganic gases.(26) The carbon dioxide and water vapor are easily absorbed into the atmosphere, but the large particles, the dust and the noxious or corrosive gases constitute pollutants. The general public is increasingly demanding that air pollution be controlled.

A wide variety of air pollution control equipment has been developed. These include settling chambers, mechanical cyclones, wet scrubbers, electrostatic precipitators and baghouse filters.

Incineration with Heat Recovery

Refuse is not an ideal fuel. The composition of refuse varies over a wide range, as does the heating potential it could produce. In some European incinerators with heat recovery, coal is added to the refuse to produce a more consistent and higher quality

* Some scientists have expressed concern over the rising level of carbon dioxide in the atmosphere pointing to possible effects on the earth's energy balance. They are described in Part I dealing with air pollution problems.

fuel. The steam would have to be sold to a nearby user, who would require steam on a 24-hour basis. The most obvious users of this steam would be electricity generating plants. Other uses could be community steam heating or air conditioning.

Some cooperation between the users of the steam and the designers of the incinerator would be necessary. In many cases, the extra expense of boilers and a steam distribution system will outweigh the return from the sale of steam.

The following table compares a 4,000 ton per week refractory incinerator with and without heat recovery. The costs include only variables that are affected by the decision to have or not to have heat recovery.

	Refractory	Refractory with Boiler
Initial Cost (26)	3,583,000	5,737,000
Annual Cost (26)	656,000	1,007,000

Assuming a 20-year life and a 5% interest rate, this leads to a unit cost of $4.55 per ton for the plant without heat recovery. In order for the plant with heat recovery to match this figure, this steam must be worth $1.04 per thousand pounds. The best kind of steam that can be expected from an incinerator boiler is 200 p.s.i.g. saturated. All modern electrical generation stations are based on superheated steam so that, unless and older turbine was available nearby, generation of electricity from this steam would require a separate electrical plant which would be less efficient than its competitors both because of the thermodynamics and the economics of scale encountered in power generation.

High Temperature Incineration

Recently an extremely high temperature process, the Melt-Zit incinerator, has been studied. This system uses temperatures of 2600 to 3200 °F to reduce all non-combustibles to an inert molten slag. A Bureau of Solid Waste Management study of a pilot plant operation indicates that the system is not yet sufficiently well developed for full scale implementation. Problems with high coke and limestone consumption and deterioration of the lining were encountered in the test runs. Initial economic analyses indicated that if the system could be made to operate up to specifications, the incineration costs would be about $1.00 a ton more than conventional high temperature incineration due mainly to increased fuel costs (25). This difference would have to be made up in savings in input segregation costs, savings in disposal costs due to the inertness and high density of the residue.

Land Disposal

Land disposal is by far the most widely used method of solid waste disposal in the United States today. Available records indicate that there are 90,000 recognized land-disposal sites. Of this number, about 19,000 were planned, and some 12,000 are subject to a degree of local control that would identify them as "sanitary" or "modified-sanitary" landfill sites (3). The vast majority of land-disposal sites constitute a public nuisance and a national disgrace.

Land-disposal sites, when properly managed, can form an attractive disposal system. These sites take unsorted refuse of varying composition and provide a final deposit for the refuse. This section will provide a discussion of the hazards of uncontrolled land-disposal and the costs and benefits of sanitary landfill.

Open dumps are the most prevalent type of disposal site used in the country. This type of operation is usually accompanied by continuous or periodic burning. Open dumps require little capital or operating costs where land is available. On the other hand, they are health and fire hazards, unsightly and malodorous, require substantial amounts of land which is hard to utilize after the dump is closed, adversely effect neighboring property values and are a cause of air pollution.

The disadvantages and hazards outweigh the advantages of the open dump in even a moderately urbanized area. For the coastal municipalities we have in mind we do not regard open dumping as worth costing. Most of these communities have stopped uncontrolled dumping some time ago.

The sanitary landfill is described by the American Society of Civil Engineers as: "A method of disposing of refuse on land without creating nuisances or hazards to public health or safety, by utilizing the principles of engineering to confine the refuse to the smallest practical area, to reduce it to the smallest practical volume and to cover it with a layer of earth at the conclusion of each day's operation, or at such more frequent intervals as may be necessary."

Landfill disposal often offers the most economic method of disposal, as well as being a system that can accept almost all refuse and dispose it in a safe and beneficial manner. Landfill can often be integrated into planned land reclamation, whereby lands which formally had little or no economic value can be made to yield recreational and economic benefits and lands which have been despoiled, quarries, sand pits, etc.,can be restored to use.

A sanitary landfill operation requires a good deal of planning. The landfill must be well integrated into the total refuse collection and disposal system. Operating procedures should be carefully outlined, and the economic and technological facts about the operation should be studied. Preliminary planning should include an active public information program to explain to the public what makes a sanitary landfill work well and what benefits can be expected. In most cases, public acceptance or rejection of a disposal system is the most important factor in the choice and planning of that system.

Site selection is an important engineering step to establishing a sanitary landfill. The land area - or more important the volume of space - required is primarily dependent upon the character and quantity of wastes to be disposed. This varies of course with each project, but as a rule of thumb, 7 acre-feet (11,293 yd^3) per 10,000 population per year is frequently used or a little over a cubic yard per person per year.

Zoning restrictions and accessibility are also factors in choosing a landfill site. These are often conflicting constraints, and the distance to a landfill site is usually considerable. The transportation element in any disposal system is an important part of the total cost, so that it is important to investigate transfer stations and alternative systems.

The availability of cover material should also be investigated, because the additional cost of hauling cover material to the site over long distances may be restrictive. Sandy loam is considered to be an excellent cover material since it contains 50 to 60 percent sand and the remainder is clay and silt in equal amounts with good workability and compaction qualities. A soil containing too much clay presents operational problems during wet weather and can crack in dry, exposing the garbage to vermin, surface run off and releasing odors. Too granular a soil will not prevent passage of flies and may create a water pollution or erosion problem by permitting surface run off.

Over the years three general methods of landfilling have been developed: the area method, the trench method, and the ramp method.

Abandoned quarries, strip mines, gravel pits, borrow pits, gullies or rolling land are suited to the area method. In the area landfill a day's refuse is dumped in one spot. A bulldozer then spreads and compacts the wastes. A six inch layer of cover material (ideally taken from the adjoining slope or working face) is placed over the fill to form a "cell". Successive cells are built next to or atop each other until the landfill is completed. A final two to three foot seal of cover material is spread and compacted over the entire area to finish off the fill.

In a trench landfill a progressive trench is cut into the ground and solid wastes are dumped into this trench. The solid wastes are spread thin, compacted and covered with earth excavated from the trench. Level or gently sloping land is best suited for the trench method if the water table is not near the ground surface. The material excavated from the trench can be used for the next day's covering operation, which is an advantage over hauling the cover material to the site. A disadvantage is that more equipment may be necessary for a trench-type landfill than for the area type.

In the ramp or slope method, the refuse is dumped on an existing slope. After spreading and compacting the material on the slope it is covered. The process is repeated until the landfill is exhausted.

During the operation of any type of landfill attention must be paid to proper compaction, the size of the working face, the depth of the cells and the cover.

Certain materials may have to be excluded from sites. These may include explosives, dangerous chemicals, demolition wastes, dead animals, and so on. However, almost all ordinary household refuse in acceptable.

The most common piece of equipment used at disposal sites is a track-type tractor with a bulldozer blade. The various manufacturers of landmoving equipment generally can provide a wide variety of equipment to suite various conditions. Some road maintenance and fire control machinery may also prove to be necessary.

REFERENCES

The publications referred to in Chapter 34 may be found in "Economic Aspects of Ocean Activities" Vol. III, by MIT for National Council on Marine Resources & Engineering Development, Sept. 1970, pp. VII-1 to VII-8

CHAPTER 35

SANITARY LANDFILL

1. AREA LANDFILL

a. PROGRESSIVE EXCAVATION

The simple continuity of the progressive excavation method is its most distinguishing feature. Cover material is excavated directly in front of the working face and is placed directly on the previously compacted fill. The cover is excavated only as needed to properly cover the fill material. This type of landfill is usually serviced by bulldozers or clam-shell machines and utilizes a ramp type working face. Draglines are sometimes used to operate a trench-type progressive excavation.

In a ramp type project, it is easier to see the work and control the spreading of the refuse if it is discharged at the base of the working face and spread from the bottom up. This method of operation also has the tendency to screen the operation from the public view and minimize the nuisance of blowing paper.

b. CUT AND COVER

When the material excavated from a trench is stockpiled adjacent to the site and later used for cover over the compacted refuse, the operation is called a "cut and cover" type of area fill. In some cases, the excavated material may exceed the requirements for the cover needed and the extra material is sold. The rate of excavation bears no relation to the rate of refuse disposal and in many cases long parallel trenches are opened considerably in advance of the need for refuse disposal. This is sometimes a big advantage where there is excessive rainfall or the ground may become frozen in the winter.

The cut and cover method for operating an area fill is well suited to sites where excavation may be made below the water table. A dragline is essential in an operation of this type and the refuse is discharged at the top of the working face. The dragline may be used to spread and compact the refuse.

c. IMPORTED COVER

The imported cover type of landfill is used when depressed areas of land are available but sufficient cover cannot be secured at the site. Fills in rock quarry pits are a good example. The refuse is placed and compacted by a bulldozer or dragline as in other types of landfill. Cover material is usually stockpiled or delivered as needed to the site. Waste sand from nearby gravel mining operations and earth from building site excavations or from highway excavations are a common source of cover material.

2. DEPRESSION LANDFILLS

In depression fills the total depth of refuse generally exceeds the depth for a single layer of lift operation. Each stratum or lift is constructed by the placing and compacting of the refuse so that cells are constructed, with fill material on all sides to prevent travel of fire through the mass and for the control of rodents, flies and odors.

Pit and quarry sites are normally used for depression landfill operations. The distinguishing characteristic of all pit and quarry sites is that they are lower at all points than the surrounding terrain. The pits are usually of such a depth that several lifts or strata are necessary to bring them to grade. If there is available cover material and no drainage problems are created, there is no reason why the area may not be filled to a level much higher than the original ground, particularly if the area was originally characterized by rough terrain.

Refuse is transported to the working face by way of access roads. If compaction is done by bulldozing from the bottom up in sloping layers, a high density can be achieved. Maximum capacity can also be secured because the weight of several lifts will help create a greater density than would normally be achieved in a cut and cover area landfill. It may be advisable to operate the first lift by the progressive excavation method of landfill. The fill should be above the ground water level unless it can be shown that any pollution of the ground water will not adversely affect adjacent areas.

It is important that the pit and quarry sites include enough suitable excavatable cover material around their perimeters or that there is enough overburden and non-marketable materials available on the site to provide a volume of cover equaling at least 25 percent of the refuse.

Sanitary landfills must be first class operations. Facts and plans will do most to insure acceptance -- a carefully thought-out master plan which will illustrate the potential benefit to the community is needed. These plans must be flexible to accommodate changes in real estate development. Cooperation of cities and counties as well as their respective departments is a must. Municipalities or contractors should select and buy land in advance for future disposal needs. It is important that the land deed state that the area will be used as a refuse disposal site for a specified number of years. Thus forewarned, housing developers and citizens can plan accordingly.

The American Public Works Association Research Foundation has prepared a method for determining "landfill area required". The formula used is as follows:

$$V = \frac{FR}{D} \left(1 - \frac{P}{100}\right)$$

where:

V = Landfill volume in cubic yards required for refuse disposal per capita per year.

F = A factor which incorporates the cover material, averaging 17 percent for deep fills and 33 percent for shallow fills with corresponding F values of 1.17 and 1.33.

R = Amount of refuse contributed in pounds per capita per year.

D = Average density of refuse in pounds per cubic yard delivered at the landfill (about 325 for collection by compactor trucks.)

P = Percent reduction of refuse volume in the landfill, varying from zero to 70 percent.

D. POTENTIAL PROBLEMS

There are several problems that can arise at sanitary landfill sites. The most significant include the following:

1. WATER POLLUTION

A principal hazard of sanitary landfill disposal is the possibility of fluids leaching from the fill and polluting the streams. Such problems can be minimized, or prevented, by constructing the fill in such a manner that it does not become saturated. This can be accomplished by filling all space below maximum ground water level with inert material, providing an impervious dike around the fill to exclude flood waters or surface drainage from adjacent higher ground, and covering and grading the top of the fill to drain off much of the precipitation which falls on its surface. Studies have also been conducted to determine the risk to ground water from refuse tipped into dry and wet pits.

Besides the problem of ground water pollution, the filling of swamps of flood plain lands can have an adverse effect upon flood conditions. Flood control is essentially a space allocation problem, and under natural conditions flood plains and swamps provide natural channel storage areas for surplus water. If these areas are filled, they are no longer available for flood water storage, and space needed to accommodate flood waters can be obtained only by raising flood stages. In addition, enroachment in the channel cross-section reduces hydraulic efficiency of the waterway and may cause it to back up behind the constriction. Although the use of sanitary landfills is sometimes extolled as a desirable way to "reclaim flood plain areas", care must be exercised to see that the operations are properly designed and located so that they will not cause adverse effects on flood stages.

2. GASES

The amounts of gas produced seem to be directly proportional to moisture content of the fill. Temperatures and climatic conditions are also factors. No way has yet been found to prevent methane gas production in sanitary landfills. Odorous gases, particularly hydrogen sulfide gas, have been noticed in sites where contractors have disposed of gypsum board and similar building materials having a sulfate content. This is a result of the formation of hydrogen sulfide by the action of sulfate-reducing bacteria on these matérials. An incident in a Chicago landfill which was completed twenty-five to thirty years ago illustrates the extent of gas production and danger. The sanitary landfill site was covered with two feet of snow, sealing the natural openings through which the gas normally escapes. The result was to force the gas into sewers serving homes in the area, causing an explosion.

3. FLYS AND RODENTS

Fly and rodent control is a problem in any landfill operation. Fly control is best accomplished by using a cover material with a binder which is well compacted. A well compacted cover of 2 5/8 inch prevents fly emergence. With an uncompacted earth fill of 5 foot thickness, 90 percent fly emergence has been noted. Covering the refuse every night eliminates the attraction of rodents.

4. FIRES

Landfill fires also present significant control problems. Fires can be controlled by proper supervision of personnel and dumping procedures. Minimum amounts of area should be open at one time to prevent wide spread of fires. Watering down of refuse tends to reduce possibility of fire while at the same time aiding compacting.

5. DECOMPOSITION

Studies on factors which affect the rate of decomposition of organic matter in sanitary landfills have been made in several areas. These factors include moisture, soil mixture, depth of fill, type of soil, aeration and temperature. In Los Angeles County, such factors prevent decomposition of refuse in landfills. However, in Seattle, Washington where there is heavy rainfall, decomposition does take place. More investigation is needed to determine the effects of moisture and seasonal changes on decomposition.

In a landfill site near Chicago, after seven years, a sanitary landfill site was excavated and almost complete deterioration of material was noted. This material made excellent cover and supported the growth of grass. It was an excellent soil additive. Even material buried only four years showed considerable decomposition and a low moisture content. The sites were on an old lake bed, but the fill itself was considerably above ground level. Whether decomposition is desirable or not must to some extent depend on what future use is planned for the site.

E. COSTS

Costs of sanitary landfill method for refuse disposal cover a wide range. The total cost includes the land cost plus site development plus operating and equipment costs. Since the land should increase in value, even in a remote area, the cost of the land from a long range point of view is sometimes neglected.

Data obtained in 1954 showed that the total cost of plant and equipment required for sanitary landfill for a community of 10,000 persons was approximately \$8,000, for a community of 50,000 persons \$20,000, and for 90,000 persons \$25,000. More recently, there is almost unanimous agreement that costs will usually range from \$.80 to \$1.50 per ton. A cursory evaluation of landfill costs yields the following equation (not including transportation costs).

$$\$/\text{ton} = .50 + \frac{6{,}000}{\text{tons per year}}$$

Transportation costs need to be added to operating and equipment costs[12] to obtain total cost per ton. There are two methods of figuring transportation costs. The better method is based on hauling time; the second method is based on mileage.

The first method of figuring costs is the "ton minute". The unit cost of "ton minute" is arrived at by dividing the dollar cost per minute by the net tons on the load. It does not matter whether the vehicle travels 15 miles per hour on surface streets or 60 miles per hour on the freeway. What does matter is the length of time the trip requires, since the vehicle is being paid for on a time basis rather than mileage base. The cost of any direct haul vehicle can be represented on a graph. A line drawn through the origin with a slope equal to the cost per ton per minute represents the cost for hauling a ton for any length of time.

The hauling time method is also helpful in determining when a transfer station type of operation is economic. In any transfer operation, the cost of owning and operating the station itself is not productive of moving refuse to the place of final disposal. This cost must be "earned back" by the greater efficiency of the haul vehicle being used. There also might be other unproductive expense at the disposal site, since the transfer vehicle will quite likely be less maneuverable, require the removal of covering tarpaulins, and may require a longer time to unload than a collection vehicle.

The second and older method is based on cost/ton/mile, i.e., 15¢/ton/mile, 30¢/ton/mile. These costs are based on average speeds of 15 miles per hour on city streets and 45 miles per hour on freeways, full operating, maintenance, and depreciation cost of equipment, labor costs including fringe benefits and overheads. Adding to these figures a cost of $1 per ton for dumping, and assuming a ten mile haul distance from collection route to dump, the total cost per ton would come to: twenty miles at $.15 = $3 plus $1 or $4 per ton total cost.

F. ADVANTAGES AND DISADVANTAGES

The advantages of sanitary landfill over other solid waste disposal methods are:

1. Less capital investment.

2. Can accommodate peak refuse quantities readily.

3. Combined refuse collection, including garbage, ashes, combustible rubbish and non-combustible rubbish is possible thus reducing collection costs.

4. Operations can easily be terminated without a great loss in equipment or land. The equipment is of a type which can readily be used for other municipal functions and the land at any stage is no worse, and usually is better than it was before the operation began.

5. Sanitary landfill requires less land than open dumping because the refuse is compacted to between 40 percent and 50 percent of its original volume and can be deposited to a greater depth by digging ditches. Approximately one acre per year is required for 10,000 persons (seven acre feet).

6. Unusual materials and bulky articles do not usually cause difficulties of operations.

7. Sanitary landfill can be established immediately upon the purchase or rental of standard digging and compacting equipment and authorization to use the land. No plant has to be built before operations can begin as is true of other solid waste reduction methods.

The disadvantages of sanitary landfill are:

1. Large amounts of land are required.

2. Sites located outside of a city are usually under some other Governmental jurisdiction.

3. Winter operations present difficulties.

4. Prevention of ground water pollution may be costly.

5. If the distance to the sanitary landfill site is very great, the cost of transfer operations may be high.

CHAPTER 36

OPEN TRENCH BURNING

The burning of refuse at dumps is commonly considered to be an undesirable practice. Smoke and other air contaminants normally emitted cause nuisances in nearby developed areas and contribute to the community-wide air pollution problem. As a result, dump burning is being prohibited by air pollution control legislation in populated sections of many states. However, controlled burning methods are being used in some locations which avoid some of the undesirable features commonly associated with dump burning. Such methods are not a "cure all" that would make it possible to establish burning sites in or very near populated areas. These practices, however, can reduce the amount of smoke produced with little additional cost. Also, they may furnish an interim solution to the refuse disposal problem even in areas where burning may ultimately be prohibited for air pollution control reasons.

The three types of controlled burning dumps used reflect the influence of the topography of the site selected. They are the single lift fill, the trench fill, and the canyon fill.

A. SINGLE LIFT FILL

The single lift fill is used where level low ground can be improved by filling. Roadways and a bank with a safety berm are constructed so that the completed fill will fit the contours of the surrounding land.

B TRENCH FILL

The trench fill is constructed by excavating a trench in level ground. It is operated as if it were two single lift fills facing each other. The excavated material can be used to cover the completed fill or to provide fill material for other work. The cost of excavating the trench makes this type of operation slightly more expensive than the other two.

C. CANYON FILL

In a canyon site, conditions similar to the single lift fill can be created by cutting two shelves along the side of the canyon. The upper shelf is maintained for burning by occasionally bulldozing the ashes down to the lower shelf where they create a solid and permanent fill. The main purpose of the lower shelf is to limit the dumping area and facilitate salvage. Burning is confined to the upper shelf.

D. LAND REQUIREMENTS

Conveniently located marginal land can often be found that is suitable for the establishment of one of these types of controlled burning dumps. Regardless of the terrain, there are five requirements to consider.

1. Sufficient isolation so that surrounding residents will not be affected.

2. Caretaker (full time preferably) to supervise the dumping and police the operation.

3. Clean level roadways and a safety berm for trucks and cars to back up against. Together with good housekeeping, these encourage the cooperation of the public in dumping over the bank.

4. A bank approximately 15 feet high with a 45 degree slope to cause the load dumped for a collection truck to properly loosen and scatter down the bank. This procedure allows enough air to get to the refuse for efficient combustion.

5. Sufficient length of dump face for proper segregation of materials.

Trench fills or single lift fills on low ground should be so located that no portion of the fill intercepts ground water. Problems of water pollution, stuck equipment, and difficulties in removing salvage are common in marshy areas. Provisions must be made for handling drainage water from the fill and roads, as well as any surface water that normally flows through the site. A slope of one to two percent is sufficient to prevent difficulties on the roadways and in the trenches. Access roads should be constructed so that they are passable throughout the year. Fire breaks must be carefully constructed and maintained to prevent the spread of fire to surrounding property.

E. SEGREGATION OF REFUSE

Separation reduces objectionable smoke and odors, and permits maximum salvage. By providing separate areas for dumping, it is usually possible to accomplish a great deal of segregation as dumping proceeds. When the people using the site have become accustomed to the system, signs alone may be sufficient. The degree to which segregation is carried must be fitted to the area in which the dump is operated and the degree of perfection that is required. The following separation into five classes of materials has been successfully used in a number of operations.

1. Household rubbish, mixed refuse, paper, cardboard cartons, cans, bottles, toys, and similar materials. After the non-ferrous materials are salvaged, the remaining refuse can be burned daily with almost no smoke. The tin cans and other ferrous metals can be periodically salvaged by using a mobile crane and electromagnet. With salvage of cans, the useful life of the trenches can be extended significantly over their life without salvage.

2. Stoves, refrigerators, washers, tanks, drums, beds, and other large items. These items are almost totally salvable and should be kept separate to conserve space.

3. Tires, ground rubber, roofing paper, linoleum, other heavy smoke producing materials, such as concrete and bricks. Salvageable items can be removed and the remaining material buried without burning. If these heavy smoke producing materials must be burned, particular attention should be paid to selecting a time when atmospheric conditions are the most favorable to disperse the smoke.

4. Lawn clippings, brush and tree trimmings. These green materials need to be thoroughly dried before burning or they will only

smolder. In order to avoid fly production in cases where large quantities of green lawn clippings are received, they should be buried immediately.

5. Dirt and ashes. By providing a separate storage area, dirt and ashes can be saved for cover material.

F. BURNING PROCEDURE

When refuse is dumped in a pile and set afire, excessive amounts of smoke are commonly produced. However, with little additional expense, proper conditions for controlled burning can be established. Gravity is utilized to do most of the work of loosening and scattering the refuse on a properly constructed bank.

Usually the caretaker uses a long handled hook fork to finish breaking open and spreading the dumped refuse. After the refuse is properly spread, the fire is started on the downwind side. This keeps the fire from smothering itself, and so results in a cleaner burn. At the tail end of the burn, any matted material such as newspapers or grass clippings, will have a tendency to smolder and produce smoke. These matted materials are broken open at this time with a long handled hook fork so that the air can get to them and complete the burning quickly. When the burn is conducted in this manner, the refuse spread out on the bank can get sufficient air for efficient combustion and yet it is still concentrated enough to generate the necessary temperature to sustain good combustion. The bed of cans and bottles acts as a grate, allowing air to get underneath as well as around the refuse. The time required for combustion of one truckload of refuse is reduced from an hour or more to approximately 15 minutes. Also, the amount of smoke produced is markedly reduced. Best results are obtained when a maximum of four or five truckloads of refuse are burned at one time.

G. ADVANTAGES

Controlled burning dumping offers the following advantages:

1. A minimum amount of land and equipment are required, thus minimizing the capital investment.

2. Dump sites last longer.

3. Proper bank slope and depth promote complete, clean, fast burning with a minimum of smoke.

4. Maximum salvage of metals helps maintain the operations, conserves dump space, reduces costs and conserves natural resources.

5. The final residue is held to an absolute minimum and the land can be returned to immediate use.

6. The safety berm prevents accidents and encourages dumping over the bank.

H. DISADVANTAGES

The disadvantages of controlled burning dumping are:

1. Some smoke and air contaminants are produced, so that an isolated site is required.

2. Burning is a fire hazard to surrounding property.

3. Fly control is not as effective as at properly operated sanitary landfills.

4. Public cooperation is necessary in separation.

5. Dead animals, swill, cannery wastes, wet manure, and other wet wastes must be specially handled.

CHAPTER 37

COMPACTING

The compacting of refuse can occur at any or all three of the areas involved in refuse disposal. Compacting units have been built for on-site use. That is, at apartment buildings, schools, super markets, institutions, etc. where large amounts of refuse are generated. The collection of refuse today is largely made by collection vehicles with compactor units. The final area of compacting is at the disposal site where heavy equipment is run over the refuse or, as in the case of North Tonawanda, New York, a special vehicle is used to provide compacting.

Compacting of refuse is easily accomplished by the machinery available today. Hydraulic or pneumatic cylinders can exert forces as high as 90,000 pounds, reducing the original volume of the refuse by 60 to 80 percent. However, refuse is not like scrap metal which retains its compacted shape. Refuse must be restrained in its compacted shape by keeping it under pressure, either by bagging, baling, or some other means.

A. ON-SITE COMPACTORS

There are several basic systems of on-site compaction. One is a proprietary system which utilizes a pneumatic ram to compact the refuse into paper sacks or plastic containers. The principle of operation is relatively simple. The machine compressor and paper bag holder are mounted below a refuse chute. As the refuse enters the bag, it triggers a mechanism that shunts the bag under the pneumatic ram, places a clean bag under the chute, compresses the collected refuse and the partially filled bag stands ready to be shunted back under the chute until full.

A four-bag machine, operating under high compaction pressures (3,000 psig), would cost approximately $3,000, including the air compressor unit.[18] Each bag will hold approximately 3.5 cubic feet of refuse having a total weight of 75 pounds. The bags will cost approximately twelve cents each. Therefore, for a typical 100 unit apartment housing 250 people and assuming a waste generation rate of 2.25 pounds of refuse per capita per day, 560 pounds of refuse would be generated each day. This amount could be handled by eight bags at 8 x 12 or 96¢ per day for bags.

The paper bag compactor has a number of significant advantages in that it:

1. Provides a sanitary method of collection.

2. Reduces the volume to one-third of its original volume.

3. Produces refuse packages that are relatively easy to handle.

4. Reduces the amount of on-site storage required.

5. Reduces the need for expensive compaction equipment on the municipal collection vehicles.

The disadvantages are:

1. The requirement for a high-pressure (300 psig) storage vessel on the premises.

2. The critical dependence on electrical energy.

3. The possibility of poorly manufactured paper bags of low strength. As these systems are installed, it is possible that the relatively high bag costs (12¢) will stimulate the manufacture of cut-rate bags.

4. Unknown system reliability and maintenance costs.

5. No reduction in weight of refuse.

The other system of compaction in general used today consists of a horizontal ram and compression area connected to a wheeled detachable container reportedly reduces the volume to about one-quarter of its original volume.

The system is reported to cost approximately $4,500. When the detachable container is fully loaded, a signal is energized and it will not accept any additional refuse. The standard detachable container will hold approximately 1,000 pounds of compacted refuse which then can be wheeled to the area for municipal or private collection. It has all of the advantages and disadvantages of the bag compactor except for the size of the refuse container and several additional disadvantages:

1. The location of the collection chutes must be controlled to make sure that the loaded refuse container can be moved to outside collection.

2. It requires modification of, or special types of municipal collection vehicles.

3. Because of the size and weight of the detachable container, it probably will require the services of at least two custodial personnel if it must be removed to another location by the collection truck.

There are also on the market a number of systems in which a heavy duty crusher or disintegrator crushes or chews up the refuse. The crushed or disintegrated refuse is then delivered to a baling machine where it is wrapped and sealed for later pick-up. This type of system might lend itself to a central station concept wherein refuse from a number of chutes is conveyed to a central station for compaction and baling. In general, it has little application to a single multi-family unit.

B. COLLECTION OF REFUSE

Compactor type collection vehicles are used extensively throughout the United States today. There are approximately thirty manufacturers of these vehicles. At present, about three-quarters of all refuse trucks in operation are of the compactor type, and virtually all vehicles sold by 1980 will compact waste. Needs exist for a higher ratio of payload to dead weight in such trucks -- today's models are so big and heavy and compact materials so effectively, that they often exceed legal weight limits when full. Another improvement needed is some device or system that retains the compaction achieved in the truck. At present the compressed material regains much of its bulk when dumped, and must be recompressed at the landfill site. Trucks in operation today can exert up to 90,000 pounds of force on the refuse and compact 60 cubic yards or 10,000 pounds of refuse into a 20 cubic yard truck.

C. DISPOSAL SITE COMPACTORS

Almost all compacting done at landfill sites is accomplished by dragging a crane bucket (dragline method) over the refuse or running over the refuse with a bulldozer. The amount of compaction obtained would depend on the make-up of the refuse (moisture content), amount of previous compaction either on-site or in collection, the number of times the bulldozer runs over the refuse and the spring-back of the refuse before it is finally covered with earth.

Another method has been developed for compacting and is currently being used in North Tonawanda, New York. A large automated vehicle, which has a compactor built into it, moves along the landfill site digging a trench about 4 feet wide by 8 feet deep. It accepts truck-loads of refuse, compacts the refuse and extrudes it into the trench. While still under pressure, earth is filled in over the refuse as the vehicle moves along so that the refuse has no opportunity to spring-back. The manufacturers of this equipment claim that one machine of this nature could service a town of 80,000 to 100,000 people.

The economics of refuse disposal are reasons enough for compacting. However, with stricter air pollution laws which are in some cities calling for the shut-down of apartment building incinerators, on-site compacting is becoming a necessity. Compacting is a space and time saver. Space, whether on-site, in collection trucks or at landfill sites, is at a premium. In addition, the compacting in collection trucks enables collection crews to collect three to four times as much refuse as without compactors, allowing a substantial saving in labor costs.

CHAPTER 38

RAILHAUL DISPOSAL OF SOLID WASTE

The concept of railhaul disposal of solid wastes is gaining support in both the government and private sectors, and is one of the few concepts on which thorough, comprehensive studies are becoming available. Most of the data and projections presented in this section were provided by the American Public Works Association (46) and the New York State Pure Waters Authority.

These studies list the main advantages which can be gained through railhaul solid waste disposal.

1) A strong possibility for a low-cost total disposal system. Railhaul permits strategic location of transfer stations, and the economies of scale available in a regional (as opposed to local) system.

2) A reduction of air and water pollution and other environmental health hazards.

3) A solid waste disposal system with high degrees of both reliability and flexibility.

4) A system which will have widespread application throughout the U.S.

The Railhaul concept allows a great deal of flexibility in designing the overall system structure. The following systems have been proposed: System R1 - Railhaul Sanitary Landfill, System R2 - Railhaul Incineration, and System R3 - Railhaul Ocean Dumping; these are discussed in the following sections.

System R1 - Railhaul Sanitary Landfill

This system consists of the following three steps:

a) Trash is delivered by primary collection vehicles to centrally located transfer stations and is compacted into bales.

b) Bales are transported by rail to sanitary landfill sites.

c) Bales are placed in sanitary landfill.

The first step begins with packer or compaction type trucks entering an enclosed dumping area and discharging their contents into a storage bin. An overhead crane transfers the refuse from the storage bin to a conveyor belt which feeds the shredding units. (Shredding proved to be neccessary to assure high-density compaction.) From the shredding unit the material is transferred by conveyor to the baling unit, where it is compressed into high-density (approximately 65 lbs/ft^3) bales. These baling units include integral automatic strapping equipment. These bales are then placed in the rail car by a traveling crane.

The second step begins when refuse bales are transported in specially designed side-loading enclosed rail cars. Rail car manufacturers have indicated that the design will present no particular problems.

The third step consists of unloading of the bales at the disposal site. They are then carried to the open face of the landfill operation, and disposed of.

System R2 - Railhaul Incineration

This system consists of the following four steps:

a) Trash is delivered by primary collection

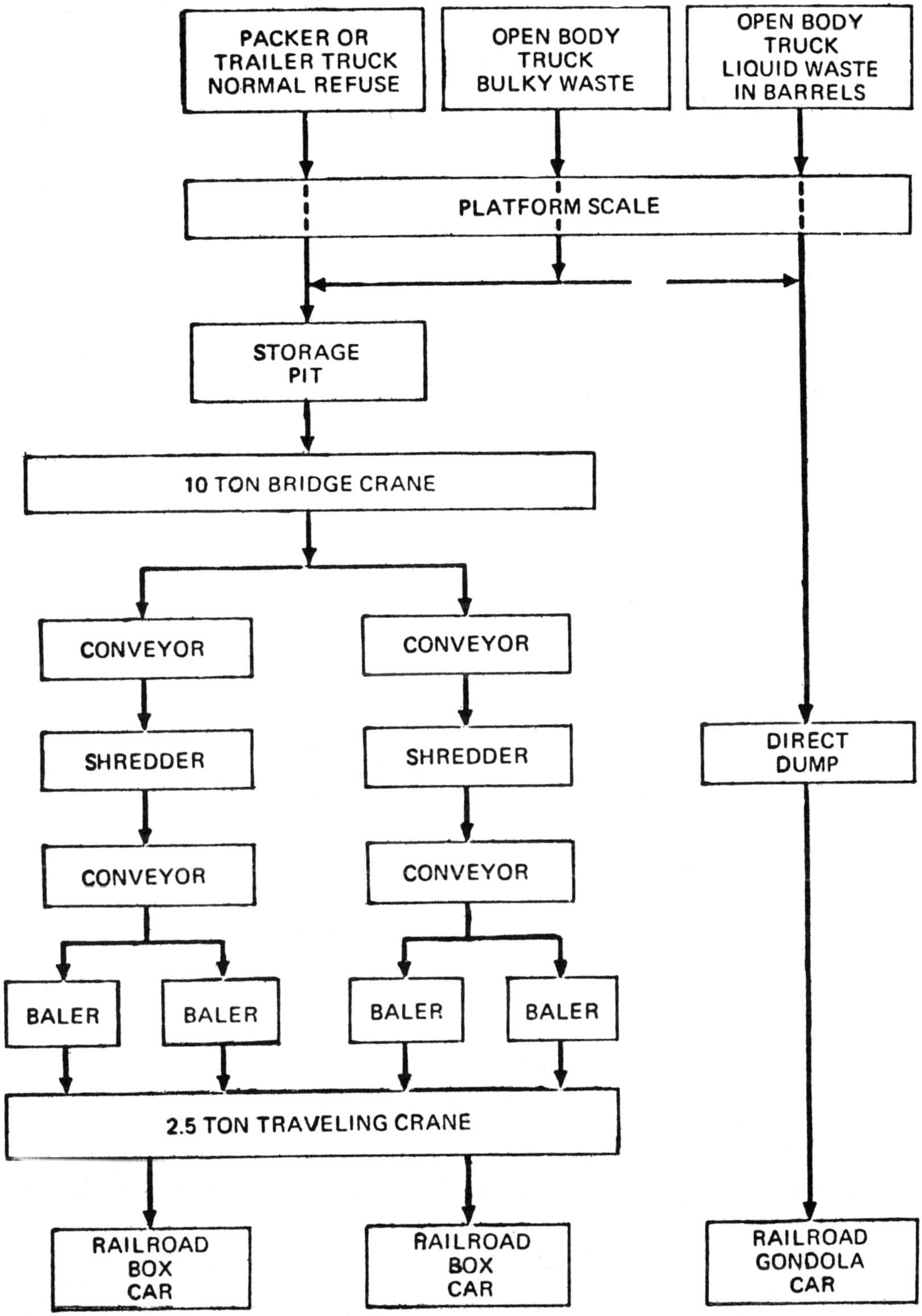

Fig. 38-1 RAILHAUL SANITARY LANDFILL CONCEPT

vehicles to centrally located transfer stations and is compacted into bales.

b) Bales are transported by rail to regional incinerator.

c) Bales are unstrapped and incinerated.

d) Incinerator residue is disposed of by sanitary landfill or sea dumping.

This system differs from system R1 only in the final disposal operation, incineration rather than landfill.

System R3 - Railhaul Ocean Dumping

This system consists of the following three steps:

a) Trash is delivered by primary collection vehicles to centrally located transfer stations and is compacted into bales.

b) Bales are transported by rail to dock-side facilities and loaded onto barges.

c) Barges are transported to disposal sites and bales are deposited on sea bed.

REFERENCES

The publications referred to in Chapter 38 may be found in "Economic Aspects of Ocean Activities" Vol. III, by MIT for National Council on Marine Resources & Engineering Development, Sept. 1970, pp. VII-1 to VII-8

CHAPTER 39

RECYCLING AND COMPOSTING

We have already mentioned the inherent attractiveness of recycling in the face of the ecological effects of conventional disposal methods and in view of the non-renewable nature of much of the material in refuse. The more obvious candidates for reclamation are paper and paper products, ferrous metals, aluminum and other non-ferrous metals, glass, and rubber. In addition, organic refuse which composes 20 to 25% of the total in typical municipal waste streams (3), can be composted and, at least potentially, sold as a plant nutrient. We will consider each of these in turn.

Paper and Paper Products

Approximately 50 million tons of paper and paper products have been used by Americans annually. Roughly 80% (6,20) of this was a one-time use after which the products have been discarded. Paper products constitute between 40% and 50% of all solid wastes (3). Of the 40 million tons of paper waste, 25% or 10 million tons were recycled. For the last few years, paper manufacturers have been meeting 20% of their raw material needs from recycled paper (the rest of the raw material needs were met primarily by wood pulp). (6,20) This rate has been slowly declining recently. (6,20) At one time 50% of the paper industries input needs were furnished by salvaged paper. If the price were competitive and the paper were in useful form, the paper industry could meet up to 80% of their raw material needs from recycled paper products, putting considerably less pressure on our forest resources (6).

Of the 10 million tons of paper reused, a substantial portion was industrial wastes from printing activities, manufacturing scrap, and large commercial, industrial, and govermment establishments. A much smaller percentage is from domestic, small commercial schools and office sources. Collection from these sources cannot provide a reliable input for a large paper mill.

There are technological problems in the reuse of magazines and most intermediate grades because of the filler materials. These problems are not overwhelming and indeed could be overcome if an incentive were provided. There are almost no technological barriers to the reuse of paperboard and newsprint.

Separation at the generating source and efficient collection would result in more paper being recycled. Development of machinery to separate paper from mixed refuse would similarly increase the use of salvaged paper. The paper industry will grow 100% in the next 16 years. If 50% of its new demand could be met by recycled paper, it would release 91.5 million acres of forest land for other uses (6). For the U.S., this could result in considerable savings in foreign exchange. Paper products are among the easiest to recycle and a major effort should be made to take advantage of this.

Ferrous Metals

In the past, attempts to recycle ferrous metals have focused on the recovery of tin cans. Before and during World War II, many municipal incineration installations were able to profitably sell their tin cans. In the early 1950's Los Angeles was able to finance the collection of the residue of backyard incinerators by selling the rights to the metals (principally tin cans) in this residue. The principal market for tin cans is the copper smelting industry, in which application the existence of tin in the metal is not a disadvantage. Recently, there has been a trend away from tin can recovery.

The short run trend has been that the costs of separating and processing the scrap have risen faster than the value of the scrap itself.

Aluminum

Four aluminum cans have a scrap value of one cent Production of aluminum from raw bauxite is an expensive process requiring large amounts of power. The recycling of aluminum scrap may be economically feasible in the near future given development of inexpensive means of segregating aluminum from other non-ferrous metals. Some efforts are presently being made to recycle aluminum, but these are on a very small scale. There are almost no technological barriers to reuse of aluminum.

Other Non-Ferrous Metals

Lead, copper, zinc, and tin have high salvage values. However, these metals are often found in alloys or other forms which make reuse difficult. The quantities involved are also small. In general the recycling of these metals will require advances in technology. The Bureau of Mines is presently conducting research on the separation of valuable non-ferrous metals from incinerator residues and fly ash (56, 57). However, this work is still at the laboratory stage.

Glass

Fourteen million tons of glass is present in the waste stream (3,4). Glass has a low scrap value, and manufacturers have been reluctant to accept salvaged glass, unless it is sorted by color and grade, and contains no metal. The main barriers to the recycling of glass seem to be economic, not technological in nature. The basic raw material, sand, is not in short supply in many parts of the country.

Rubber

Rubber enters the waste stream mostly as discarded tires. These would seem to be easily salvageable. However, since there are so many grades of rubber, no large scale recycling is taking place today. Rubber in other forms is difficult to segregate and salvage.

Plastics

Plastics form a small, but increasing, part of the waste stream. At present most plastics are non-degradeable. Increased attention should be given to the development of bio-degradable plastics. Reuse of plastics apart from recovery of their heat value, would be very difficult.

Automobiles

Salvage of automobiles is profitable in most instances and private industry, with government incentive should be eventually able to solve this aspect of solid wastes disposal

More study in this field is needed, however, as well as continued government incentives.

The Bureau of Mines is presently conducting research into more economic methods of segregating automobiles into their component raw materials and possible uses of the results. Among the latter is the use of car bodies as a reductant in the processing of nonmagnetic taconite, a presently unusable resource(56). Large scale recycling of car bodies is a fact and we expect the trend to continue. Of course, the municipal refuse collection stream presents a considerably more difficult segregation problem than that associated with car bodies.

It is symptomatic of the present state of the art in refuse segregation that the bulk of the segregation at composting plants is by hand picking. This is clearly infeasible on the scale required by our large cities. Efforts to develop mechanical separators are underway. At MIT, a laboratory system is being studied which uses a combination of electronic and electromagnetic garbage sensors to segregate the garbage stream according to its electrical properties.

The most promising attack on the segregation problem appears to be that undertaken by the Bureau of Mines (58). This effort is aimed at incinerator residues and at present is limited to and takes advantage of the volume reduction inherent in incineration, which is a substantial limitation given the costs of incineration.

The problem of segregation could be made considerably easier by packaging and manufacturing design which accounts for the disposal problem. Here we have in mind not only the avoidance of non-biodegradable products, of composite products which cannot be separated into their material components, but also the physical or electrical tagging of objects to facilitate segregation. For example, it might be possible to identify hard-to-segregate materials by radioactive tags which could easily be sensed by segregation equipment.

PROBLEMS CAUSED BY PACKAGING MATERIALS IN WASTE DISPOSAL

If one were asked to sum up in a single word the basic problem which packaging materials represent for waste collection and disposal facility operators, that word would be "cost." Packaging materials are a relatively recent large volume waste product, whose handling can increase solid waste collection and disposal expenditures. Increased costs may show up as a requirement for more collection equipment, the need for a larger labor force, as a requirement for new landfill sites, as higher maintenance and labor costs in incineration, and as a requirement for the addition of air pollution control equipment.

If packaging materials represented a health or safety hazard, the justification for government action aimed at packaging would exist without question. However, with some very minor exceptions—polyvinyl chloride materials which decompose into chlorine compounds when burned—packaging materials do not represent a health hazard. They may give rise to air pollution if they are burned in incinerators which are improperly operated. But pollution is created by the incinerator operator, not packages. Glass or metal containers thrown away carelessly may cut tires or people's feet. But here again, the fault lies with those who litter, not with container manufacturers. Finally, food wastes or chemical residues left in containers can create ground water pollution or serve as nutrient to insects and rodents; in such instances, also, the containers do not cause the health hazard.

As used in industry, the term refers to quality rather than quantity; in industrial jargon, a product is overpackaged when a more costly material is used than is necessary.

Materials Research

Least fruitful, it appears, would be effort expended on changing the characteristics of packaging materials. The primary reason for this is that exactly those characteristics which make a package difficult to handle in disposal are those which make it desirable as a package. This is a way of saying that any container which is easily disposed of is a poor container; and while such a generalization could not be applied to all packages, it is applicable to those packaging categories which create difficulty in disposal.

Changes in materials and containers which would be desirable from a disposal point of view are these:

(1) *Plastics*—should have better burning rates so that they will not create trouble in incineration; should be degradable in soil; should be made easier to separate by some mechanical means.

(2) *Steel*—if tin and lead were eliminated from steel cans the steel would be more acceptable for salvage; nonferrous metals (aluminum ends on steel cans) make the steel less acceptable in salvage; containers should be made in such a manner that they can be flattened, crushed, or collapsed with modest pressure by the housewife without use of tools or machinery.

(3) *Aluminum*—currently two or more types of aluminum alloys are used to make a can; a single-alloy can would be more desirable from the salvage point of view.

(4) *Paper*—synthetic coatings (clay, plastics) and photogravure inks create problems in repulping or deinking; coatings and inks should be created which provide presently obtainable characteristics but do not cause problems in reuse of the paper.

(5) *Glass*—containers should be easier to break without losing their strength in use; when broken, glass containers should fall apart into uniform pellets, not slivers and shards that create safety hazards.

As can be seen from the above, some of the desirable characteristics are not obtainable without significantly changing the molecular structure of the materials, thereby changing their vital characteristics (more flammable plastics and new paper coatings). To change other characteristics would call for the creation of new container types which would cost considerably more than currently available packages (glass and collapsible steel). Yet other changes are much more practical—tinless steel and single alloy aluminum, which are already beginning to be used commercially in packaging.

Most of the difficulties created by packaging are due to inadequate technology or the absence of technology in waste disposal. So long as wastes are landfilled without prior shredding and grinding, glass, plastics, and aluminum containers will be deposited in unaltered form and will retain their form. Ground up thoroughly, small bits of glass, metal, and plastics would be much more acceptable as fill material, even if they did not degrade or decompose. In incineration, more sophisticated combustion techniques could eliminate problems caused by plastics. Or separation of plastics by some automatic means followed by separate burning in specially designed incinerators to alleviate the problems of grate fouling which are encountered when shock loads of plastics are burned could be used as a solution In the same vein, development of effective, low cost, automatic materials separation technology would be a more practical aid to composting than development of degradable plastics—and would probably be less costly and more likely to succeed as well.

Salvage Technology Research

The research goal most worthy of pursuit would be the development of automatic materials separation techniques. At present, only steel cans can be separated from waste effectively, using magnets. Air float systems can be employed to sort out light materials—but plastic films and paper are indiscriminately mixed in such systems. Inertial and ballistic systems can sort out heavier nonferrous materials, but these too produce mixtures of glass aluminum, plastics, wood, etc. To achieve effective separation, invariably involves hand sorting of wastes, which is a costly activity and unappealing to the workers.

Absence of systems which could selectively and automatically separate wastes is one of the real bottlenecks in waste handling and reuse. Development of such a technology would seem feasible and would involve the mating of technologies developed in several industrial activities: (1) sensing techniques used in the process industries, aerospace, and medical electronics; (2) materials handling technology developed for agriculture (automatic harvesters); for food handling (e.g., American Machine & Foundry's automated short-order kitchen); for packaging; for metals fabrication, etc.; (3) computer and/or numerical control technology used in metal working, electronic data processing, and process control; and (4) materials tagging, marking, and tracing techniques.

Research efforts in this area would involve identification of sensing techniques which could recognize and classify waste materials with little or no modification; marking of those materials which cannot be sensed, for example, by infrared sensitive materials; and combination of sensing techniques with materials handling equipment in an operating system.

Table 39-1 —*Package costs of selected products*

As a percentage of product sales price, f.o.b. factory

Product and packages	Percent •
Paint in an aerosol can	16
Paint in a conventional metal can	5
Toy in a film-overwrapped carton	14
Toy in a blister pack	8
Motor oil in a metal can	26
Motor oil in a fiber can	10
Small appliance in a corrugated carton	6
T.V. set in a corrugated carton	1
Beer in a tinplate can	43
Beer in a one-way glass bottle	36
Frozen food in a boil-in-bag and carton	10
Frozen fish in a carton	5
Moist pet food in a metal can	17
Dry pet food in a carton	9
Cereal in a folding carton	15
Cornmeal in a paper bag	5
Analgesic in a plastic bottle	10
Antibiotic in a plastic bottle	1
Baby food in a glass jar	36
Baby juice in a metal can	33

COMPOSTING

Composting has not been an effective means for municipal solid waste reduction and disposal in this country. Despite its very early use as a soil conditioner and fertilizer in Europe, composting has made little progress toward an industry of any significance in the United States. Several reasons for its limited progress are: (1) the low cost and wide use of chemical fertilizers, (2) the high operating and distribution costs of compost, (3) the limited number of successful composting operations in the United States and (4) the reluctance of municipalities to enter commercial ventures.

At the present time, there is an increasing interest in composting stimulated by proponents of a natural organic fertilizer method of crop production, soil biologists, and horticulturalists and equipment manufacturers or process designers-licensors. This current interest is supported by a forecast of increasing amounts of municipal waste, the decrease of economic landfill sites and the belief that composting operations can be operated profitably.

Compost results from aerobic decomposition of municipal refuse. When organic material is decomposed in the presence of oxygen, the process is called "aerobic". Under proper conditions, municipal refuse will yield a compost in the form of a granulated material resembling coarse coffee grounds. Certain materials such as tin cans, glass, and plastic materials will not convert to compost and must be either salvaged or disposed in a landfill site.

Composting is a logical consideration for solid waste reduction because it converts municipal refuse into a useful soil conditioner and because it can be used to treat not only solid refuse but sewage sludge as well.

Between 80% and 85% of all refuse is compostable (52,3). There are a number of processes to produce compost, but all composting operations can be broken up into three basic steps: refuse preparation, stabilization, and product upgrading.

Refuse preparation includes the receipt of material, sorting, and salvage. Sorting is required in most composting plants to remove non-compostables, bulky items, and material with salvage value. In conjunction with a complete recycling scheme, only non-salvageables would be composted. Some systems include inertial or magnetic separation of refuse, but most rely on hand-picking, an expensive, inefficient, and time-consuming method.

Grinding is required for efficient composting. There are various grinding systems used with power requirements from 3 to about 30 h.p. per ton-hour grinder capacity (3). Two stages of grinding are normally required.

Stabilization, or aerobic digestion, is usually done by either a mechanical process or by windrows in the open. Five to six days is the average decomposition time for ground refuse in U.S. mechanical plants. Windrow systems require from two weeks to three months for adequate stabilization. Since mixed refuse has a very high paper content, the carbon-to-nitrogen ratio of the ground product usually exceeds 70%. This should be adjusted to approximately 40% for rapid stabilization. This is usually done by adding sewage solids or nitrogen solutions. The moisture content should be approximately 55% and the refuse ground to a particle size less than one inch. Finally, the temperature achieved during composting should exceed 140°F for at least four days (52).

Product upgrading operations which follow digestion consist of some or all of the following: curing, grinding, screening, pelletizing or granulating, drying, magnetic separation, and bagging. Storage of the compost will result in slow decomposition without upgrading of some sort.

The windrow operation requires a good deal of land and is best suited to small cities where a local market for compost exists. Mechanical plants are better suited to large cities. They should be tied into a transportation system that can economically deliver refuse and distribute the compost.

Mechanical plants are fairly expensive. The cost of building the 150-ton/day facility at Gainesville, Florida, was $1,100,000; and the 360-ton/day Houston plant costs $2,000,000.

The major problems in composting seem to be:

1) segregation of non-degradables

2) the size of the market for compost.

If composting were to be adopted on a large scale by a coastal metropolis the output would certainly saturate the market. At that point, the only benefit of the composting would be that it makes a more desirable fill than raw or compacted garbage.

The private market for composting is very small. Compost is not well suited as a fertilizer for heavy agriculture. It is well suited for home gardening, but this market would be quickly saturated. It is also suited as a fertilizer for landscaping and parks. In Europe it has been used for years in vineyards and "garden agriculture" (small vegetable plots). Again, these markets are usually saturated very quickly.

Compost manufacturers are beginning to think of their product as a filler for heavy agricultural fertilizers. This requires close cooperation with fertilizer manufacturers to create a suitable compost. It also requires a steady and very large supply at somewhat lower prices.

Compost has been used by governmental agencies for large-scale landscaping and filling. The use of compost for landfilling would prolong the life of a landfill. The completed landfill would be a more useful product. Cover would not be necessary during the landfill operation and fire and health hazards would be decreased. In summary, until a much larger market is visable, it appears that the processing costs of composting which have been estimated at about $6.00 per ton would, on a large scale, do nothing more than convert garbage to a more desirable fill material. It would be an unusual situation where a community would be willing to pay this price for this conversion.

In summary, recycling with or without composting is not a feasible alternative for large scale disposal of mixed municipal refuse streams. The rapidity with which this inherently attractive alternative becomes economic will depend as much on the development of markets for the output of the processing as it will on the development of the processing technology itself. Given the increasing scarcity of raw materials and an aggressive research and development effort, we can be sure that both these events will occur in the not too distant future.

This relatively pessimistic view of recycling is, of course, not an argument against research and development of recycling. Quite the contrary, it points to renewed and strengthened efforts in reclamation technology.

COMPOSTING PROCESSES

Compost is produced as a result of natural fermentation that occurs in moist cellulosic materials, and is available as a granulated dark brown material resembling coarse coffee grounds. The finished product, after drying, contains organically bound nitrogen, a large portion of carbonaceous matter (humus), micro-organisms, and important micro-nutrients. The compost is usually dried to 10 percent to 15 percent or less moisture.

The major raw materials today employed in domestic commercial natural organic fertilizers are cow manure, bedding straw and other animal excrement. Municipal dried activated sewage sludge is also marketed and can be blended with compost raw material to enhance the fertilizer value of the product.

Municipal refuse, containing much cellulosic material is successfully composted today. This new material, however, contains a high ratio of cellulosics to nitrogenous compounds, and more digestive bacterial cycles converting carbon to carbon dioxide (gas) are required to reduce carbon content to the accepted ratio range of 20 to 30: 1 parts, C/N. Blending with higher nitrogen sewage sludge hastens the process by improving the ratio and also upgrades the final product.

The relative importance of this category of natural organics (and its compost) can be judged by Department of Agricultural data, shown in Table 39-2.

There are numerous processes in operation both here and abroad that can produce compost. They vary in scope, complexity, and first cost from the most primitive form of windrow (long mounds) to those employing enclosed digestors, quantitatively designed, and on which processes and equipment patents are held.

Prior to describing the major processes in use, it will be well to review the agricultural, horticultural and sylvicultural value of compost.

There are six properties of compost that are of interest and concern: (1) organic nitrogen content, (2) humus content, (3) micro-organisms present,

(4) micro-nutrients contents, (5) presence of incorporated solid extraneous matter, and (6) the possible variation of the chemical and physical proportions of the various ingredients (attributable to its raw material).

The organic nitrogen is a water insoluble form of nitrogen. Thus rain, irrigation or watering will not leach away this portion of the total nitrogen not immediately taken up by the roots. It will be slowly converted to the active state by the action of the micro-organisms present. Thus, the 5 percent to 7 percent nitrogen in an enriched compost is longer lasting and is made available over a longer period of time.

The humus content improves the soil structure enhancing growth and increasing water absorption and retention, especially important in clay soils.

Table 39-2

CONSUMPTION OF COMMERCIAL FERTILIZERS IN THE UNITED STATES[2]

(tons)

Blood, dried	2,762
Compost	36,134
Castor pumice	3,498
Cotton seed meal	6,864
Manure, dried	357,009
Sewage sludge, activated	91,996
Sewage sludge, other	41,221
Tankage	10,497
Other	12,495
Natural Organic Material	562,476
Mixtures, N-P-K, N-P, N-K, P-K	19,658,957
Nitrogen Materials	8,779,205
Phosphate Materials	2,781,565
Potash Materials	1,288,624
Secondary & Micro-Nutrient Materials	1,461,388
Total	34,532,215

The micro-organisms present act as soil revivifiers. They convert chemically prepared nitrogen found in synthetic fertilizers into an available form for the plant and in this fashion the nitrogen participates in the nitrogen cycle.

The micro-nutrients, also called trace elements, serve as the necessary catalysts (enzyme activators), insuring plant health and growth. The exact mechanism is not completely understood, but seven chemical elements have been identified in this category.

These four properties of compost comprise the advantages of compost to the soil scientists and users. These properties are common to manure and dried, activated sewage sludges. The detractors of compost, and natural organics, generally, however, point to two additional properties associated with compost that is prepared from refuse material.

Since the raw material for this discussion is municipal refuse, it will contain all manner of debris (i.e. rubber tires, tin cans, other ferrous and non-ferrous materials, glassware and a host of plastic containers and wrappers). These materials, either all or part, can and do find their way into the final product, despite varying amounts of care directed to their removal. The risks engendered by their presence are the jamming or breaking of spreaders and tillers, or the ingestion by grazing animals of the broken glass.

Since the refuse raw material is of varying composition, the percentage of the basic nutrient elements varies, which, in the case of two of the seven micro-nutrients, is critical -- boron and manganese. Variation in primary nutrient content, nitrogen, phosphorous or potassium, is not critical to the plants, but raises legal problems concerning minimum guarantees of chemical composition of the compost, which are required by state registration laws governing the sale of fertilizers.

Thus, we have compost that can be produced from municipal solid wastes (whose value can be enhanced with nutrient additions when blended with sewage solids during its processing), capable of making positive and lasting contributions to plant soil environment. Why is it not a more important contributor to the fertilizer statistics?

A. GENERAL TECHNICAL DISCUSSION

Composting in the United States generally employs processes which involve aerobic decomposition. When organic material is decomposed in the presence of oxygen, the process is called "aerobic". In aerobic stabilization, living organisms, which utilize oxygen, feed upon the organic matter and develop cell protoplasm from the nitrogen, phosphorus, some of the carbon, and other required nutrients. Much of the carbon serves as a source of energy for the organisms and is burned up and respired as carbon dioxide (CO_2). Since carbon serves both as a source of energy and as an element in the cell protoplasm, much more carbon than nitrogen is needed. Generally, about two-thirds of the carbon is respired as CO_2, while the other third is combined with nitrogen in the living cells. If the excess of carbon over nitrogen in organic materials being decomposed is too great, biological activity diminishes and several cycles of organisms may be required to burn up most of the carbon. When some of the organisms die, their stored nitrogen and carbon become available to other organisms. The utilization of the nitrogen from the dead cells by other organisms to form new cell material once more requires the burning of excess carbon to CO_2. Thus, the amount of carbon is reduced and the limited amount of nitrogen is recycled. Finally, when the ratio of available carbon to available nitrogen is sufficiently low, nitrogen is released as ammonia. Under favorable conditions, some ammonia may be oxidized to nitrate. Phosphorus, potash, and various micro-nutrients are also essential for biological growth. These are normally present in more than adequate amounts in compostable materials and present no problem hence a discussion of their metabolism by the biological cells will not be included.

The natural cycle of nitrogen and carbon in aerobic decomposition is the one which takes place on ground surfaces such as the forest floor, where droppings from trees and animals are converted into a relatively stable humus or soil manure. There is no accompanying nuisance when there is adequate oxygen present for the bacteria. The energy released in the form of heat in the oxidation of the carbon to CO_2; a gram-molecule of glucose dissimilated under aerobic conditions, 484-674 kilogram calories (kcal) of heat may be released. When the organic material is in a pile or is otherwise arranged to provide some insulation, the temperature of the material during fermentation can rise to over 70°C. (158°F.). If the temperature exceeds 65° to 70°C, however, the bacterial activity is decreased and stabilization is slowed down. When the temperature exceeds about 45°C. (113°F.), thermophilic organisms, which grow and thrive in this range, develop and replace the mesophilic bacteria in fermenting the material. Only a few groups of thermophiles carry on any activity above 65°C. (150°F.). Oxidation at thermophilic temperatures takes place more rapidly than at mesophilic temperatures and, hence, a shorter time is required for stabilization. The high temperatures destroy pathogenic bacteria and protozoa, hookworm eggs, and weed seeds in the material that are detrimental to public health and agriculture.

Complete or perfect aerobic oxidation of organic matter produces no objectionable odor. When odors are produced, the process is not entirely aerobic, and the carbon is converted to methane and any sulfur present to malodorous chemicals. Aerobic decomposition can be accomplished in silo digesters, pits, bins, stacks, or piles, if adequate oxygen can be provided. Turning the material at intervals, or other techniques for adding oxygen are necessary to maintain aerobic conditions.

To determine how this material can and is being promoted to be made commercially available, we shall review the current domestic available processes. Basically, they fall into two major categories -- open windrow fermentation, with various modifications and enclosed mechanical digestor fermentation. Both methods provide and claim, with varying degrees of success, processes that are aerobic, provide optimum bio-digestion conditions that maximize yields, with minimum time and plant area requirements, and to have eliminated burdensome sanitation problems (i.e. pathogen destruction, odor-free, vermin-free and insect-free processing).

B. OPEN WINDROW, PILES OR VENTILATED CELLS METHODS

1. WINDROW

The windrow process is conducted in open air and relies on natural ventilation with periodic turning to insure aerobic conditions.
One plant of this type operating in the United States, is located at Wilmington, Ohio, and was built in 1963. It processes twenty tons of refuse per day. "A minimum of sorting is provided and two hammermill grinders are used in series. The ground refuse is composted in windrows where it is turned weekly by means of a front-end loader until the material is converted to compost. Difficulties have been experienced in selling the finished compost."

2. VENTILATED CELL

Ventilated cell composting employs a multi-story building with a vertical arrangement of progressive cells. Mixing and aeration occur when the material drops from cell to cell. Several different methods have been tried based on the ventilated cell process. In the United States the Naturizer Process, the Riker Process, the Frazer-Ericson Process, and the Fairfield-Hardy Process have all been tried with little success.

a. THE NATURIZER PROCESS

The first plant using the Naturizer Process was built in Norman, Oklahoma in 1959. It was closed in 1964 except for experimental purposes. "Two unique features of the plant were a specially designed swing hammermill and the Naturizer Composter. The digester system provided six day retention in two, three story buildings with a movable floor in each story. Considerable time and effort were expended to build up a market for the product, with sales activity extending as far away as Dallas, Texas." The second plant using this process was located in San Fernando, California and began continuous operation in July of 1963. It was shut down in October of 1964. "The plant had a capacity of 70 tons of refuse per day, but operated at only about 40 tons per day. It was an improved version of the process developed at the Norman, Oklahoma plant. Appreciable salvage operations were performed. It was necessary to provide an afterburner on top of the composting unit to comply with Los Angeles County Air Pollution Control District requirements to insure that no odors would be discharged. Compost was sold in bags and in bulk. A unique development for the use of compost was called 'Sta-Soil' which was a water suspension of compost, chemical fertilizer, grass seed and shrubs. It was sprayed over denuded slopes, or cuts and fills, to provide a blanket against soil erosion and a 'soil' in which grass and shrubs would quickly take root." The latest plant using this process is located in St. Petersburg, Florida. The International Disposal Corporation operators of the plant have a twenty year contract with the city of St. Petersburg. This contract call for the disposal of 100 tons of refuse per day, six days a week, at a cost of $3.24 per ton to the city.

The digester is a five story building with conveyors running the length of each floor. In operation the conveyors will travel their length (150 to 165 feet) in twelve hours and remain stationary for twelve hours. After passing through the digester and final screening, the material is dumped in the finishing yard where it is left for two weeks. The material is marketed under the name of "Cura" and finds favorable acceptance for use in citrus groves, golf courses, commercial nurseries, and with industrial landscape architects. The St. Petersburg plant cost $1.5 million.

b. THE RIKER PROCESS

A plant using this process was built in Williamston, Michigan in 1955. It was closed in 1962 when the only customer it had stopped purchasing the compost. "This four tons of refuse per day plant treated garbage, vacuum-filtered raw sewage sludge and corn cobs. Ground garbage and corn cobs, mixed with sludge cake were composted for twenty-one days in two four compartment vertical composters."

c. THE FRAZER-ERICSON PROCESS

A plant operating under this process was built in Springfield, Massachusetts in 1954. It was closed in 1962 when the city failed to renew the contract. The plant had a capacity of twenty tons of garbage a day. The garbage, being too wet to compost aerobically, was held for about two days to dewater. This frequently created fly, rat and odor problems.

d. THE FAIRFIELD-HARDY PROCESS

A plant located in Altoona, Pennsylvania began operations in 1950 using the windrow method, but was converted to the Fairfield-Hardy Process type operation in 1963. "Refuse is ground in a wet pulper, followed by dewatering presses (Fig.39-1 before it is fed into the Fairfield-Hardy Digester for a five day cycle." "Stirring is provided by augers suspended from a rotating bridge in a circular tank. Air is provided by means of a blower and air pipes embedded in the floor of the tank. Normal operating temperatures are about 140 to 160°F. The plant has a capacity of 45 tons per day but normally operates at about 18 to 23 tons per day. Successful experiments have been run in which digested sewage sludge was added to the ground refuse." The city of Altoona paid the operators of the plant $4.63 per ton for processing their garbage. From the 7,000 tons of garbage processed a year, 3,150 tons of saleable compost are obtained. A total of 1,300 tons of coarse compost sells for $5 per ton and 1,850 tons of the dried and pelletized compost sells for $20 per ton if it has a 5 percent to 17 percent moisture content. A plant of this type which could process 100 tons per day would cost approximately $900,000, including land. A prototype plant built in Largo, Florida originally utilized the Fairfield-Hardy Process, but was rebuilt in 1963 by the National Composters Company who installed a digester of their own design. "This plant has a capacity of 50 tons per day. Operating five days a week, it treates 40 to 50 tons per day of mixed refuse to which is added 1,000 to 1,500 gallons per day of essentially raw sewage sludge." The essential difference in the digester is that the material is turned, not by the augers, but by a rail-mounted bucket elevator moving in either direction. The product sells for $16 per ton. Another plant has been built in Houston, Texas, by the builders of the Largo, Florida plant. This $1.75 million plant has been designed to process 300 tons of garbage per day, six days per week. The city of Houston pays $4.51 a ton for this service. The end product of the operation is called "Metroganic 100". It is being sold to rice farmers, citrus growers and others engaged in agriculture, particularly those in the Rio Grande Valley.

C. CONTINUOUS MECHANICAL COMPOSTING

This process involves continuous mixing with gradual particle size reduction and positive aeration. There are several systems that use continuous mechanical composting -- the Dano Method, however, is the only one used in the United States.

1. THE DANO BIOSTABILIZER SYSTEM

The Biostabilizer is a large drum, 9 to 12 feet in diameter and 60 to 100 feet long. It is designed to effect continual mixing, aeration, grinding and decomposition in one unit. Refuse material first passes through sorting and magnetic separation before being charged into one end of the slowly revolving unit (.25 to .8 rpm). water and sludge are also added at the inlet. Slow grinding is accomplished by the tumbling shearing action of the Biostabilizer which, with decomposition, processes the refuse so it will pass out through coarse perforations (4 inch diameter) at the outlet end. This operation is followed by a second magnetic separation, vibrating screen and a ballistic or gravity separator before the partially decomposed refuse is arranged in windrows for curing. It is said that five to seven days' composting in the Biostabilizer unit is equivalent to three to four weeks of windrow composting with several turnings, although material coming out of the Biostabilizer is not sufficiently decomposed to be stable and will reheat again when moist. For this reason, windrow composting is used in the finishing stage. The first plant to use the Dano System in the United States was built in Sacramento County in 1957. It had a capacity of about 40 tons per day. It was closed in 1963 for lack of a market for its product. The second plant using this system was built in Phoenix, Arizona in 1962, originally scheduled to process 300 tons per day with two Dano Biostabilizers using a three day cycle. However, two additional units were added and actual capacity was still only 175 tons per day. The plant was closed in 1965 because it could not compete with the cost of landfill operations.

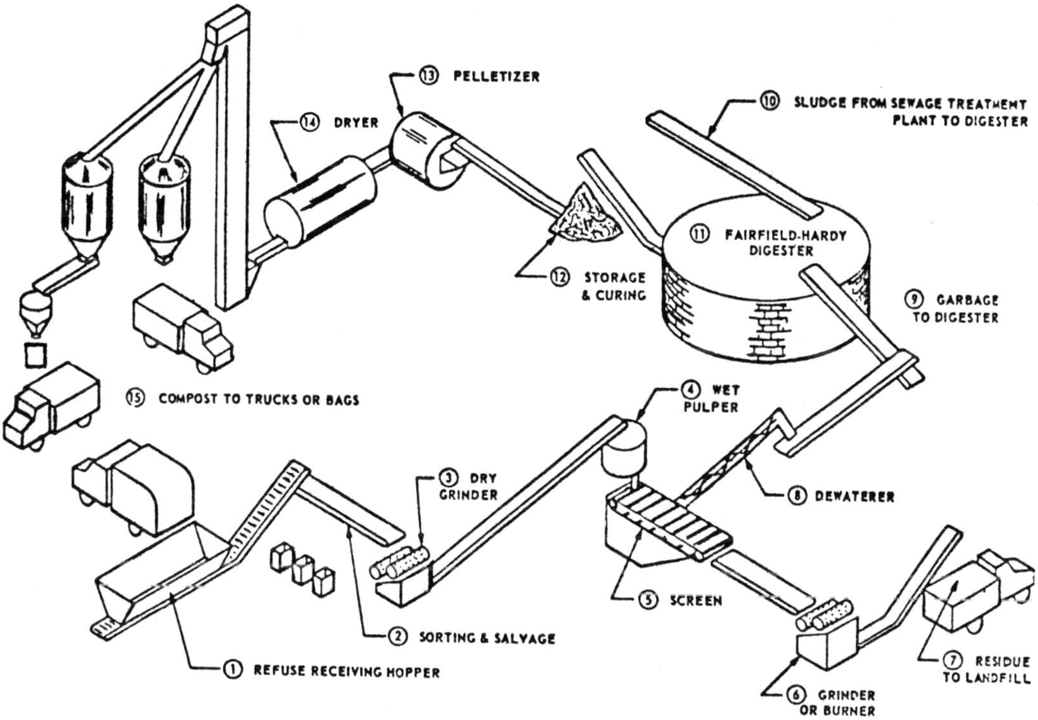

Fig. 39-1

SCHEMATIC OF COMPOSTING PLANT UTILIZING FAIRFIELD-HARDY DIGESTOR.

REFERENCES

The publications referred to in Chapter 39 may be found in "Economic Aspects of Ocean Activities" Vol. III, by MIT for National Council on Marine Resources & Engineering Development, Sept. 1970, pp. VII-1 to VII-8

CHAPTER 40

WASTE DISPOSAL FOR APARTMENT HOUSES

WASTE DISPOSAL METHODS FOR APARTMENT HOUSES

A. INCINERATION

This method will reduce the volume of the residue to 10 percent or less of the refuse charged, and the weight to 25 percent or less on the same basis. There is therefore a major reduction in the handling and storage requirements at the building, as well as in the pickup and disposal facilities to the dump area. The residue is sterile and odorless compared with the refuse, so that tenant complaints, and health hazards from vermin are minimized. Fire danger is also eliminated.

Incineration at the building will reduce the number of cans of refuse to be stored from ten or fifteen to one required for the residue.

B. COMPACTION

The compaction volume may be reduced to one-half or one-third the volume of the original refuse, so that storage room requirements are decreased with no change in weight. There are several systems for this purpose. In the first, refuse is forced into metal containers by mechanical pushers, stored until dumped into the pickup trucks, during which operation it may regain much of its original volume.

Another compaction system forces the refuse into heavy paper bags, which are then sealed and deposited in the truck, so that the initial reduction in bulk is retained. This is satisfactory for landfill disposal, but experience has shown that the bags must be broken apart if delivered to a municipal incinerator, to avoid delayed combustion in the furnace and high unburned combustible loss. In a third system, the refuse is compressed in portable metal containers, which are picked up and carried to the disposal facility for unloading, and returned to the building, at added hauling cost.

In the above compaction systems, there is no reduction in the weight of the refuse, and a temporary decrease in bulk. The refuse remains in a putrescible state and is inflammable, which may require fireproofing the storage room to avoid fire, and cooling to avoid odor and vermin problems.

C. SHREDDERS, GRINDERS, AND PULPING

By addition of water after maceration, these methods reduce the volume, but add materially to the preparation cost. The ultimate disposal problem may be complicated because the finely ground material has been found to pack, and burn too slowly in an incinerator, especially when water has been added.

D. GARBAGE GRINDERS

This method requires separation of the food products in the refuse from most of the other material. As the former comprises only ten to fifteen percent of the total refuse, and is the least bulky for storage, the saving in disposal costs is relatively small. Such equipment is prohibited in New York City, because of the added load on the sewage disposal system, into which the ground material is discharged.

E. BALING NEWSPAPERS

When there is a market for scrap paper, which may comprise up to 20 percent by weight of the total refuse, the cost of separation and handling can be justified. Loose collection in cans or bags requires the greatest storage room space, and handling labor, as well as being most subject to fires, odor and vermin problems.

F. BULK DENSITIES

Bulk densities of the several materials involved have been measured as follows, subject to variations in moisture and other factors. Weight of the refuse as deposited in the incinerator through the usual charging flue is 4.1 pounds per cubic foot or 111 pounds per cubic yard. Residue as removed from the incinerator, including some moisture used for quenching is 15.4 pounds per cubic foot or 416 pounds per cubic yard. This includes 65 percent to 70 percent metal, mostly cans, and glass such as bottles and jars.

INCINERATION REQUIREMENTS

There are two general types of incinerators available; the major difference being the method of feeding the refuse from the various floors into the furnace.

A. FLUE-FED INCINERATORS

Flue-fed incinerators are those in which the refuse is charged through hopper doors on each floor into a refractory flue, the bottom of which opens directly into the top of the furnace or combustion chamber. In the single flue design shown in Fig.40-1, the combustion products flow upward through the charging flue and are discharged above the roof of the building, whereas in the double flue type the refuse is charged down one flue, and the combustion products normally flow upwards in the parallel flue, both flues being lined with refractory material.

At suitable intervals the hot gas from the auxiliary burner may be discharged upward through the charging flue to purge or sterilize it of any odors or vermin that may have accumulated on the walls. For this reason, both flues must be open at the top for free exit of whatever combustion products are produced. As an alternate to this hot gas purging the charging flue may be sterilized by use of suitable spray nozzles and detergents.

B. CHUTE-FED DESIGNS

Chute-fed designs are those in which the refuse is charged through hopper doors on each floor as above, into a metal chute, collecting in a basement hopper, from which it is transferred either manually or mechanically to the incinerator furnace (see Fig. 40-2 .) The combustion products pass up and out through a refractory flue, above the roof.

C. UPGRADING EXISTING INCINERATORS

Many of the existing incinerators can be brought up to satisfactory performance by incorporating the design factors that have been found beneficial in the new units, without major changes or costs. Because of the intermittent nature of refuse charging, the incinerator normally operates for only three to four hours per day. By spreading this burning period, the load for each burn is reduced, so that even an undersize furnace can be made to function satisfactorily.

Many cities have permitted burning only during daylight hours, or when the day porter was on duty. In New York City for example, the burning time was 7 A.M. to 5 P.M. One result was that between 5 P.M. and 7 A.M. the refuse would pile up in the charging flue to the second or third floor, which causes a serious pollution problem when the porter ignites the refuse the next morning.
The introduction of automatic controls for igniting the refuse in the furnace and controlling the combustion air, has permitted extending the burning time, in some cases to 18 hours or more so that the overnight pile-up is eliminated and four or more burns are accomplished, with major improvement in the performance. The principal elements included are listed below.

1. Auxiliary burner in the furnace or primary chamber to ignite the refuse and maintain desired temperature in conjunction with the heat from the refuse.

2. Overfire air fan with manifold and nozzles to assure adequate turbulence and complete burnout of the volatile combustibles.

3. Programming electric clock with 24 hour dial and adjustable contact pins to permit starting and stopping of the above items at preset intervals. All controls to be enclosed in tight steel box.

4. Fly ash removal equipment adequate to meet the local ordinances.

The reduction in emissions by the methods described above is approximately as follows:

Equipment	Reduction of Emissions in Percent
1. Overfire air jet system	40
2. Overfire air system plus auxiliary burner	62
3. Items 1 and 2 plus wet or dry cyclone collector	85 - 90
4. Above items 1 and 2 plus wet scrubber, impact or plate type	94
5. Items 1 and 2 plus electrostatic precipitator, bag filter or venturi scrubber	99

The installed cost of adding items 1 and 2 ranges between $1,500 and $2,000. Items 3 and 4 would cost between $5,500 and $8,500. Item 5 ranges from $12,000 to $15,000 based on what little data is available. Conversion of single flue incinerators to the double flue type should be done whenever possible, as the separation of refuse charging and gas emission flues has been found to be very beneficial.

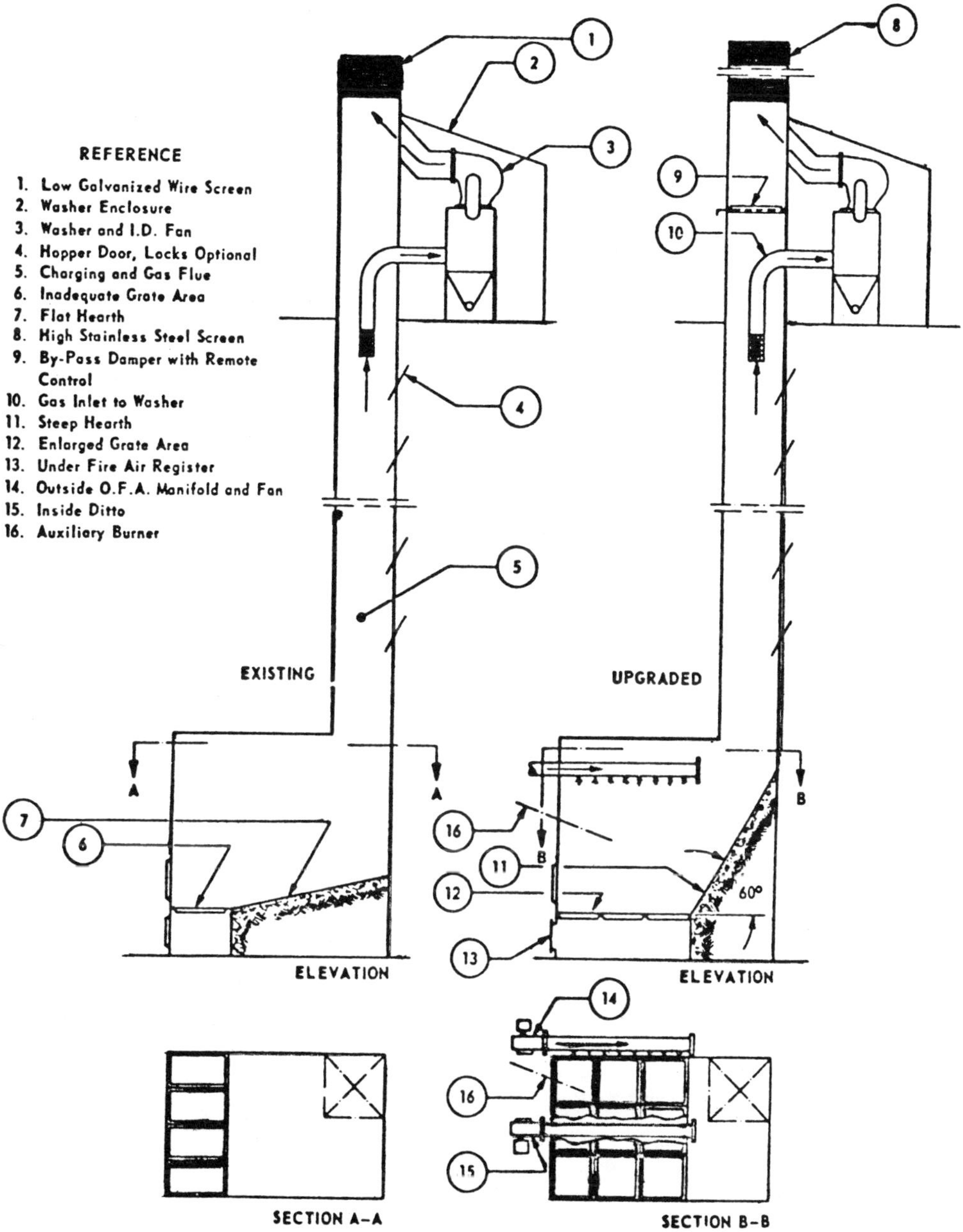

Fig. 40-1 Single Flue with Washer or Precipitator on Roof

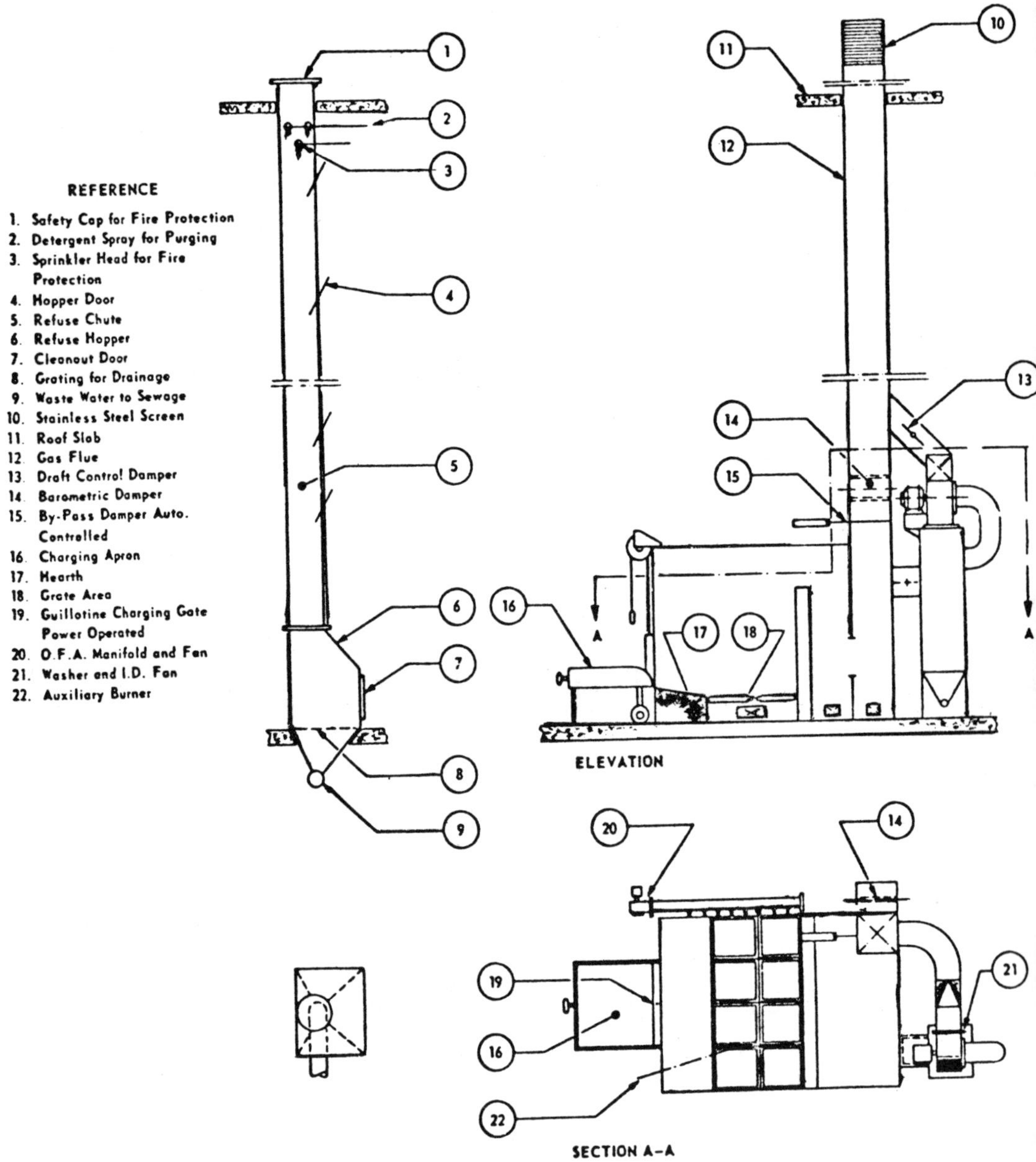

Fig. 40-2 Direct Fed Incinerator and Charging Chute

PERFORMANCE STANDARDS

Satisfactory performance of new incinerator installations may be accomplished by the use of either of several types of filing requirements, such as performance standards or construction specifications, or possibly a combination of the two. In the former, minimum requirements are prescribed, such as fly ash, smoke and other emissions, as well as unburned combustible in the residue, while in the latter, specifications covering furnace dimensions and construction, flue sizes, controls and air pollution removal equipment, are established. In either case, complete filing of plans and details is necessary.

Many localities permit installation in accordance with the filer's plans and specifications, subject to final approval upon completion of satisfactory performance tests, or based on previous tests of prototypes. In some cases the manufacturer's guarantee may be accepted, tests to be required only if violations are experienced. Usually the cost of such tests must be paid by the installer rather than by the control agency.

Incinerator emission tests are relatively expensive and require specialized equipment and personnel. The cost will range from $500 to several thousand dollars and take from one to several days for preparation, setting up equipment, and actual testing which should be done as closely as possible under normal operating conditions and at the design capacity.

Development of continuous monitoring devices which is now underway to a limited extent, will aid both the operator and the control agency. The former now has no way of telling whether or not the incinerator emissions are excessive, while the latter can vouch for performance only at the time of his visit. The value of indicating and recording instruments would be to show whether or not a new installation was ready for approval and to determine the validity of complaints which at present is often difficult or impossible to resolve.

CONSTRUCTION SPECIFICATIONS

These specifications are similar to the specifications issued by architects or consulting engineers and are in accord with building and equipment codes. Construction specifications have been developed through the years to set up minimum standards that will assure satisfactory performance, regardless of variations in the details of construction of the several bidders.

Rather than being restrictive, a well developed and flexible construction criteria benefits the manufacturer of adequate, well constructed and possibly more expensive designs by setting up the minimum requirements that will be accepted. Such criteria are also helpful in examining and approving applications, reducing the possibility of one examiner in the control agency turning down a design which another examiner might have recently approved, based on inadequate knowledge or experience.

It should be noted that approval of an application under the above system does not relieve the filer or user of the equipment of responsibility for satisfactory performance as the best of equipment must be properly operated to avoid violations.

An example of the implementation of the above is shown in the "Criteria for Apartment House Incinerators" currently in use in New York City. Similar criteria are issued by a number of air pollution control agencies in cities such as Los Angeles, Chicago and Milwaukee.

CHAPTER 41

THE STATE OF ART OF MUNICIPAL INCINERATORS

Municipal refuse incineration equipment and systems have been developed and demonstrated to perform the necessary thermal reduction function efficiently, reliably and economically. Further fundamental technological "breakthroughs" are not required to provide satisfactory incineration service to municipalities. Modern system engineering can do it if the following principles are observed.

A. Standards of performance, expressed in clear, quantitative engineering terms must be created and acknowledged. The fear that municipal mixed refuse is so variable that standards are impossible has been allayed by several studies which have shown that most refuse can be defined within reasonable ranges of chemical and physical terms. Incinerator performance can be based on refuse in these ranges with the occasional excursion outside these ranges handled by special control equipment.

B. Specifications for procurement of incinerators should be expressed in terms of engineering performance and not in terms of ambiguous perfection ("shall be odorless, smokeless and perform to the satisfaction of city officials") nor in terms of equipment dimensions and materials ("furnace shall be xxx inches wide and refractories shall be of xxx material").

C. Methods of testing performance to specification must be established, acknowledged and implemented. Statistically reliable tests of refuse, flue gas and residue composition, tests of system capacity and of power and water used are expensive and are therefore often curtailed to the point that true system performance is not known. There is need for development of simplified test instrumentation and techniques, and perhaps for Government financial support of incinerator acceptance test programs.

D. Economic consideration should be given to total cost evaluation, including the sum of capital investment, operating costs and maintenance costs to meet specified performance for the life of the investment, rather than to simple consideration of lowest first cost for each separate item of equipment.

E. Unification of responsibility for incineration system design, construction and performance is necessary with the assurance of standby financial resources to "debug" and develop new incinerator system installations until they are performing to specification.

ELEMENTS OF AN INCINERATION SYSTEM

Many systems of incineration have evolved over the years as technology improved, as living habits changed, and as solid waste collection practices developed. Although considerably different in construction and operation, it is clear that all successful systems of central municipal mixed refuse incineration include provisions for carrying out the following essential functions:

1. Receiving loads of mixed refuse at varying rates of supply.
2. Measuring the quantity of refuse received.
3. Storing a "buffer" amount of refuse in a sanitary, accessible, yet nuisance-free manner.
4. Sorting, mixing or otherwise preparing refuse for incineration.
5. Feeding the refuse to a furnace at a controlled rate.
6. Drying the refuse sufficiently to permit ignition of combustibles.
7. Burning the refuse to produce essentially inert solid residue and tolerable gases.
8. Dissipating the heat of combustion.
9. Collecting, cooling and removal of non-combustible residual solids.
10. Cleaning and discharging effluent gases and liquids in an acceptable form.
11. Controlling the process for safety, efficiency, economy and community acceptance.
12. Protecting the personnel and equipment from the elements, from dangerous refuse and from careless operation.

Before proceeding to the description of the principles of the equipment used to perform each of the above twelve functions, it is necessary to clarify the popular method of "rating" modern municipal incinerators, since often there are different kinds of equipment used for different size-rated incineration systems. It has become common practice to rate municipal incinerators in tons of burning capacity per 24 hour day. This is an unfortunate and often ambiguous scale of measurement, because a ton (2,000 pounds) of mixed refuse may easily have as much as 50 percent variation in its heating value (Btu/pound) due to different amounts of paper, plastics, wood, vegetable matter, moisture, cans and bottles, etc. (The effect of variation in heating value on the effective capacity of an incinerator is discussed in another section of this report.) Also, in the absence of uniform standards for measuring the degree of completeness of combustion, the "capacity" in tons per day burned in a given incinerator may be considerably increased if less complete combustion is accepted. Furthermore, relatively few incinerators are operated for full 24 hour days, and the starting up and shutting down operations, at partial load, lead to confusing concepts of the amount burned "per day". Nevertheless, it does not appear likely that the tons per 24 hour day (TPD) method of rating will be easily changed, so it will be used as a classifying parameter in this section.

Municipal incinerators have been built or offered in size ranges of 30 to

1,200 tons per day (TPD). In all but the smallest installations, it is considered desirable to provide at least some measure of duplication of important pieces of equipment, so that a single "outage", whether due to equipment breakdown or normal maintenance, will not prevent the plant from operating at partial capacity. Generally, the larger incinerator plants consist of groups of two, three or more 150 TPD to 500 TPD furnaces fed by a common loading system, with either common or separate gas and solid residue discharge systems.

A. REFUSE RECEIVING FACILITIES

In the United States, practically all refuse is delivered to central incinerators by motor vehicles, usually in "packer" trucks of 16 to 26 cubic yard capacity which results in loads of three to six tons of moderately compacted refuse. The more modern packers have means for mechanically ejecting these loads onto a level floor or into a pit, but many of the older types dump their loads by tilting the truck body so the refuse slides out the back. These are suitable for dumping (tipping) their loads into a pit, but may not eject their entire load onto a level floor unless the truck is driven forward with the body tilted. Many smaller communities, and larger communities in emergency situations, still use standard three to five cubic yard open dump trucks for refuse delivery, and in the aggregate, there are many deliveries made by all sorts of private vehicles ranging from the family sedan with a pail of refuse in the trunk, to light trucks with crates of refuse or old furniture, to large vans loaded with special industrial wastes.

The receiving facilities of modern incinerators are designed primarily to accommodate packer trucks, but they must also be able to handle the other types of vehicles. There are two principal types of receiving stations:

1. The floor dump, which is an open paved floor area on which the trucks deposit their loads. A tractor then pushes or lifts the refuse into conveyors or feed hoppers or piles the refuse in storage heaps for later disposition.

2. The bin dump which is a concrete lined pit with a curbing at one edge to which trucks can back their rear wheels and discharge the load to a level below grade, from which it is later lifted by a crane.

Although there is much to be said in favor of enclosed tipping areas -- protection from the elements, reduction of windblown refuse, containment of noise and odor and night storage of vehicles -- a significant number of incinerators, in the interests of low first cost, make do with only a canopy over the tipping bays, or nothing at all.

Traffic control and personnel safety are important considerations with heavy trucks backing into close quarters to dump their loads, and many receiving areas attempt to provide multiple dumping areas (three to six bays usually) to handle peak delivery loading, and separate exits so the "empties" don't have to thread their way back through incoming traffic. Sometimes a special bay is reserved for cars and light trucks to keep them out of the way of the big packers.

B. REFUSE MEASURING EQUIPMENT

Refuse received at a municipal incinerator is usually measured by weight, by volume, or by number of delivery vehicles, in order to:

1. Compare performance against system rating.

2. Establish patterns of refuse receipt for planning purposes.

3. Measure the quantities and rates of refuse collection from a controlled area.

4. Establish a basis for service charges to various sources of refuse.

5. Detect unusual or undesirable loads.

The most common method of measurement is by weighing on a truck scale which is usually located so that all incoming vehicles must cross it to reach the tipping floor or bins. The weighing equipment varies from direct reading mechanical scales to the more sophisticated scales including load cells, remote indicators and automatic print-out devices. Usually, the scale is attended by a weighmaster who checks the credentials of the reufse truck driver. Many incinerators restrict their service to specific towns, licensed private collectors, homeowners who have obtained a permit, etc. He checks the empty weight of the vehicle, and inspects the load for obviously undesirable refuse like demolition masonry, logs, large metal home appliances or explosives which may damage the incinerator.

The newest weighing systems can be made fully automatic so that the driver of the incoming vehicles inserts a coded license card or key in a box to open barrier gates, and the load is weighed, recorded and admitted to the dumping floor, and finally the empty truck is weighed and released through another barrier gate, all without need for a scale attendant.
In some cases, the scales are used to record weights of loads of brush, furniture, home appliances and other refuse which is taken directly to a land fill.

Another method of weighing is by use of a load cell on the bucket suspension of the travelling crane. This can accurately measure the weight of refuse actually fed into the incinerator furnaces. However, it affords no control over the amount of refuse received from each incoming truck.

Residue leaving the incinerator is usually measured by estimating volume, if it is measured at all. In most cases, the residue is wet from the quenching process, and the amount of water is quite variable, so weight measurements are not considered reliable indications of the residue weight. The residue is frequently carried off to a land fill in open body dump trucks of three to five cubic yard capacity, and it is therefore measured by the number of truckloads. If a sample is dried and weighed, the volume estimate can be converted to weight in order to calculate the average percent reduction of incoming refuse weight.

C. STORAGE OF REFUSE

It is accepted practice to collect refuse during the daylight hours, and since collection trucks start their rounds about the same hour, and are filled about the same hour, refuse is likely to arrive at the incinerator in cycles, with two or three peak periods during the day. Where the incinerator operates more than one shift, or more days than the collection service, the refuse collected must be stockpiled for burning over an extended period. Also, in case of incinerator shut-down for maintenance or repairs, or in the event of holidays, disasters or other causes of generation of large quantities of waste material, there is need for the incineration plant to accept and hold quantities of refuse until it can be handled by the incinerator working at its steady rating.

The most common storage device is the concrete bin, set below ground level so that trucks can dump directly into it. Bins are usually constructed of reinforced concrete and often have steel rails or facings set in the sides to resist damage from the crane buckets. Unfortunately, incinerators are often erected on otherwise marginal land which may be poorly drained or may be the site of an old refuse dump. This can lead to expensive construction of the bins as well as expensive foundations for the rest of the structure. The bins must be water proof, yet must be provided with sewers for draining of accumulated liquids or flushing water.

Bins tend to be large. For example, the bin for a 500 TPD plant with provision for one full day's storage when filled up to ground level, and one and one-half day's storage with refuse piled by the crane, might be about 75 feet long, 25 feet wide and 50 feet deep. For a 1,000 TPD plant, the change would probably be mostly in increased length. Bins are usually carefully shaped to permit full access by overhead cranes, and to facilitate cleaning. Since dust, odors and fires have been recognized as problems, some bins are equipped with sprinklers, ventilation suction systems and doors to close them off from the drive-in and tipping areas.

The other storage method, the floor dump, is not nearly as common as the bin, and appears to be used only in smaller incinerators. It consists of a concrete or bituminous surfaced floor sheltered or enclosed by a simple structure. A bulldozer equipped with a lift bucket is used to move and pile the refuse on the floor, sometimes ten feet high, until it is fed into the incinerators. The sides of the enclosure are constructed to withstand the side thrusts of the refuse piles, and there are sewers for drainage and often sprinklers and ventilation systems to cope with dust, fire, and odor.

In case of temporary emergencies, excess refuse is sometimes piled outside the tipping floor, or is taken directly to a land fill, until floor or bin space becomes available.

D. SORTING, MIXING AND PREPARING THE REFUSE

Some communities attempt to segregate refuse during collections, while others take anything that will go into a packer truck. Still others bring furniture and large metal objects like refrigerators, stoves, bedsprings and bicycles to the incinerator. Sometimes there are commercial wastes like large packing crates, or industrial wastes like large wooden pallets, or rubber tires, or spoiled batches of foodstuffs. Occasionally, non-combustibles like concrete slabs, or china sinks or rolls of fence wire appear. Obviously, certain of these items should be kept out of an incinerator, while others must be treated in some special way to get them into and through the incinerator without causing damage.

Gross separation of objects unsuited for incineration is done by watchful weighmasters and furnace loaders. Breaking or crushing of bulky combustible items like furniture and crates is often done by the travelling crane or the bulldozer, whichever is used in the particular plant. A few incinerators are equipped with chippers, wood hogs or hammermills for disintegrating pallets, demolition lumber and logs. The resulting "chips" are then mixed in with the rest of the refuse going into the furnace. Those disintegrators that are acceptable for this severe service are very heavy duty machines provided with large throats to receive large items. They generally require more than 50 horsepower and are quite noisy. The unavoidable inclusion of metals and stone in the refuse takes a heavy toll of the knives or hammers of the machines. Nevertheless, they are very effective toward making bulky items suitable for incineration and thus saving land fill.

The problem of unusually "wet" refuse like lawn clippings, or spoiled loads of fruit; or of high heating value refuse like plastic scrap or rubber tires is handled by judicious mixing on the tipping floor or in the storage bin. The crane operator or bulldozer operator tries to "average out" the unusual loads, into a reasonably combustible mix.

E. FEEDING REFUSE TO THE FURNACE

Moving the refuse from the place it is stored to the inside of the furnace is an unusual and exacting materials handling task, for which several types of specialized equipment have been developed. The task is difficult because the material to be handled is composed of so many different sizes, shapes, weights, textures, hardnesses, slipperiness and resiliencies. There is paper in all forms, from telephone books to carton boxes to greasy garbage wrappings. There is cloth, wood, pieces of metal machinery, all shapes of cans and bottles, grass and brush clippings, earth, dust and occasionally significant amounts of waste foodstuffs. In good weather, the conglomerate waste (mostly paper) may be quite dry and fluffy, but during prolonged rainy spells, or in snow storms, the refuse received at the incinerator may be soggy wet. Also, the service imposed on the material handling equipment can only be classified as "severe".

The refuse, while generally lightweight when compared to earth or minerals, nevertheless is handled in such large volumes and by such large equipment that high tonnages must be grasped, pushed, lifted, and carried. For example, a typical crane bucket, large enough to "grab" 1,000 pounds of mixed refuse at a "bite", weighs about 5,000 pounds in itself, in order to withstand the jarring, abrasive service conditions. Hence, the crane and its drums, bearings, cables, motors, gears, brakes, etc. must all be designed to lift 6,000 pounds each time, and to do it continuously and reliably. Similarly, a tractor used to push and stack refuse on a charging floor might handle over 1,000 tons per day, in the course of feeding 500 tons per day into the furnaces.

Another requirement of the feeding system is to supply a controlled flow of refuse to the furnaces with minimum interference with air supply for combustion, and with maximum protection against flashbacks of fire or gases through the charging opening.

The most popular feeding system, by far, is comprised of a below-grade storage bin and a travelling crane with a grab bucket which lifts and carries the refuse high above the furnace, and releases it into a funnel shaped hopper which leads to a chute that allows the refuse to slide into the furnace under the action of gravity. A few smaller incineration plants use monorail hoists in which a single fixed overhead "rail" supports a "trolley" which can travel the length of the rail on wheels. The trolley contains electric motors and brakes, and drums of steel cable which suspend and raise and lower a "clamshell" type bucket or grapple, usually two to four cubic yard capacity, over the refuse storage bin. (The halves of a bucket may be visualized as two cupped hands with the fingers of each tight together, while a grapple would look like two hands with the fingers spread apart in a claw configuration.) With monorail cranes, the bin is narrow with steep sloping sides to make the refuse fall under the line of travel accessible to the bucket, and the furnace feed hoppers must also be in this line of travel, to accept the refuse from the crane.

The larger incinerators have bridge cranes in which two parallel overhead rails mutually support a cross structure, or bridge, on wheels, so the bridge can travel the length of the rails. The bridge, in turn, supports a "trolley" which suspends and operates the bucket or grapple as described above. The bucket of the bridge crane can reach any point in the area between the support rails and can therefore handle refuse in wide bins and can reach furnace charging hoppers in more locations. Usually, the crane is operated by a man in a cab mounted right on the bridge or trolley. The cabs are well ventilated, often air conditioned, and are frequently provided with communications systems, since the operator must work in an environment of dust, odor, heat and noise, yet his judgment and performance in sorting and mixing refuse, and rate of feeding the furnaces are extremely important to the successful operation of the incinerator. A few cranes are operated from a fixed "pulpit" on ground level or above, and some monorail types are controlled by a man walking on the ground alongside the trolley and using electrical switches on the end of a cable, or mechanical controls activated by a rope.

Good cranes are costly equipment because of the sophisticated controls, the severe duty and the need for reliability, since a crane stoppage shuts down the entire plant. They also require special provisions in the buildings which house them, such as strong, true mountings for the rails, headroom and side clearance for the trolleys and bridges, heavy duty, well protected electrical power source to the trolley (either the "third rail" type or festooned retractable cables), and sometimes storage space for standby bridges, trolleys and buckets.

Tractors with bulldozing blades and lifting buckets are simpler and less costly than travelling cranes, but they can only be used to feed a furnace hopper where the refuse does not have to be taken out of a below-grade bin, and lifted above the furnace. If a tipping floor is at an elevated level with respect to the furnace, as is possible with a hillside location or with a manmade ramp, the tractor operator can mix, sort and feed refuse to a bank of incinerators just as efficiently as a crane operator, and, in case of a breakdown, the machine can be quickly replaced with another tractor, or even, temporarily, with a snowplow on a truck.

Continuous chain, bucket or belt type conveyors appear to be rarely used in feeding incinerator furnaces. This is not because of inability to transport the waste material, but is perhaps due to the dismal prospect of a disabled conveyor buried 50 feet deep under 500 tons of mixed refuse, or to the inability of a single conveyor to stack and mix refuse like a crane or tractor can do. In at least one small incinerator, a tractor is used to push refuse from a tipping and sorting floor into a shallow floor trench onto an apron type metal conveyor which carries the refuse up an incline into the furnace.
In some older types of furnaces, the refuse is charged directly through an opening in the top of the furnace and falls on the firebed. Such openings are usually fitted with refractory lined horizontally sliding charging gates, often hydraulically activated, located just below the charging hopper. A load of refuse from crane, dozer, or directly from a truck is dumped into the hopper and covers the charging gate. When the gate is withdrawn horizontally, the load of refuse forms an air

seal until it falls into the furnace, and then the charging gate is slid back to its closed position as quickly as possible. This type of batch feeding permits a quantity of cold air to enter the furnace with each charge and can lead to erratic combustion and smoking. A true air lock can be achieved with two sliding charging gates, one above the other, but this appears to be rarely used.

The most frequently used charging method consists of a smooth metal lined chute extending several feet down from the "throat" of the hopper into the furnace, and terminating above one end of the stoker or hearth. The resultant column of refuse forms an air seal, and the lower end of the column of refuse is exposed to the heat of the furnace for drying and ignition. The stoking action starts the ignited refuse on its way through the furnace, and new refuse from the column replaces it. The "buffer" quantity of refuse in the chute and hopper permits the actual feed rate into the furnace to be controlled by the stoker action and allows the crane to be used for stacking refuse or feeding other furnaces for reasonable intervals of time.

The end of the chute in the furnace is water cooled or refractory coated. To cope with occasional backfires of the refuse in the chute, the hoppers for these chutes are either equipped with sliding charging doors at their throats or with metal covers which can be quickly applied to seal them off.

An innovation to the gravity fed chute has been the addition of an hydraulically activated horizontal ram at the bottom of the chute to push "slugs" of mixed refuse into the furnace. This affords positive control of the feed rate and serves as the chute seal in place of the charging gate. The ram is arranged so as to push a load of refuse onto an exposed drying and ignition hearth, and in the next stroke, the new load tumbles the dried refuse over a parapet onto the actual stoker.

F. DRYING AND IGNITION OF REFUSE

Most materials that truly burn, i.e., combine with oxygen, must first be converted to their gaseous form by heating. Much of the material in mixed refuse has moisture on it (surface moisture) or absorbed in it (inherent moisture), and any heat that is applied first turns this liquid water to steam but leaves the burnable material too cool to volatilize and ignite until most of the water has been driven off. To perform the drying function and to prevent smothering a going fire with undried and non-combustible material, most furnaces have some provision for exposing newly charged material to radiant heat energy and hot gases to drive off and absorb the moisture. As previously mentioned, these provisions take the form of exposure at the bottom of a feed chute, or on a drying stoker or hearth, or brief suspension in hot gas as the refuse falls onto the stoker through the opening in the roof of an incinerator furnace.

After the moisture has been driven off, the heat radiated to the refuse by the hot gases and hot surfaces, and conveyed to the refuse by the motion of hot gases, increases the temperature of the refuse until the hydrocarbon compounds vaporize and decompose and begin to combine with oxygen. This is the ignition process which starts the burning. In most furnaces, the original ignition is done by a match or a pilot oil or gas burner, and thereafter the burning refuse ignites the incoming refuse. Subtile design features like positioning of grates with respect to heat reflecting walls, or guiding of flaming gases over the incoming refuse, are employed to ensure autoignition. In some furnaces, to ensure ignition when there are unusually wet loads, auxiliary gas or oil burners are positioned to play their flames on the incoming refuse.

During the drying and ignition phase of combustion, there are large quantities of steam and gases liberated and the fire heats and expands these to many times their original volume. Therefore, furnaces usually provide for unrestricted flow of these gases away from the ignition zone, so that fresh air can get in to supply oxygen and prevent smothering of the flame.

G. BURNING THE REFUSE

The burning process consists of the continued heating and vaporizing of the elements that will combine with oxygen until only inert minerals and metals remain as ash. Actually, there are many additional complex reactions during burning, and most of the metals actually oxidize to some extent, as do other elements like sulfur and even nitrogen in the intense heat of the furnace, but these are minor effects and would only confuse the general picture if discussed here.

The heart of the incineration system is the furnace, which consists of a chamber to contain the gaseous reaction, a stoker to transport the refuse through the furnace and agitate the refuse to expose new surface to oxygen and heat, an air supply to furnish oxygen for combustion, and a pressure differential (draft) to cause the gaseous products of combustion to flow out of the chamber.

Furnaces come in many shapes and sizes, but the three most common configurations in modern municipal incinerators are the upright cylindrical (like an oil drum), the rectangular (like a shoe box) and the multi-chamber rectangular (like two shoe boxes joined, with one lying flat and the other on end). They are generally constructed on concrete foundations with either a structural steel framework supporting inner walls and roof arch of refractory material, (supported wall and suspended arch construction) or typical masonry with bricks laid one atop another (gravity walls) and self supporting arched roofs made of keystone shaped bricks (sprung arches). The supported wall and suspended arch are almost universally used in modern incinerators. Metal or refractory hooks secure the refractory to the structural steel, and a layer of insulation and an outer sheet metal casing usually complete the wall structure.

Three different forms of refractory are used; fired refractory bricks, which are laid up with a very thin layer of refractory cement between bricks, plastic refractory which is supplied in a damp clay-like consistency and is spread and pounded into place against lath or hooks, and castable refractory which is poured into temporary molds, like concrete. The latter two forms are usually dried out at low furnace heat and then assume their final vitreous strength when the furnace is brought up to its normal operating temperature. There are many compositions of refractory material, basically clays of silica or alumina. These are called fire clays and are used in various grades for various conditions of heat, erosion, and chemical resistance. Special high performance refractories like silicon carbide or aluminum oxide are used for extremely severe duty such as furnace linings at the edges of the stoker bed where there is intense heat and severe abrasion.

In practically all municipal incinerators, the refuse rests on grates while burning. These grates are, in general, metal surfaces with holes or slots through which combustion air enters. Usually the grates are movable by mechanical means so they can move the refuse through the furnace, agitate the refuse to promote combustion, and remove the ash from the furnace. These mechanical grate systems are called stokers, since they perform the function which used to be done by men who tended the fire using long metal hoes and stoking bars. Most furnaces are equipped with doors in the sides through which manual stoking can be done when necessary.

Upright cylindrical furnaces have a floor of rocking grates above the ash pit. These rocking grates are pivoted on axles so that when they are rocked large spaces open up for the ash to fall through. Above the floor of grates, mounted on a vertical axis, is a star shaped rabble arm which rotates slowly and spreads and tumbles the refuse on the grate during combustion. In these furnaces, new batches of refuse are dumped in from the top at intervals, and they are, therefore, classed as batch type furnaces.

The flow-through furnaces are equipped with stokers which receive refuse at one end, either continuously or in batches, and continuously move it horizontally and downward through the furnace, finally depositing the ash in a receiver at the opposite end of the furnace. During the journey through the furnace, the refuse is supplied with "underfire air" which comes up through the grates from a "windbox" under the stoker. Fine particles of ash (siftings) often fall through the holes in the stoker surface and must be caught and removed to avoid eventually clogging the windbox and the grate openings.

There are four principal types of flat-bed stokers:

1. The travelling grate stoker which is essentially a moving chain belt carried on sprockets and covered with separated small metal pieces called keys, so that the entire top surface can act as a grate while moving through the furnace, yet can flex over the sprocket wheels at the end of the furnace, return under the furnace, and re-enter the furnace over a sprocket wheel at the front. The sprockets drive this chain conveyor, and are in turn driven by electric motors at slow speed.

2. The reciprocating grate stoker, which is a bed of bars or plates arranged so that alternate pieces, or rows of pieces, reciprocate slowly in a horixontal sliding mode and act to push the refuse along the stoker surface. These are driven through links by electric motors or hydraulic cylinders.

3. The rocking grate stoker which is a bed of bars or plates on axles so that by rocking the axles in a coordinated manner, the refuse is lifted and advanced along the surface of the grate. This type of stoker is actuated by linkage driven by hydraulic or electric motors.

4. The inertial grate stoker, which is a fixed bed of plates with the entire bed carried on rollers and activated by an electrically driven mechanical drive which draws the bed slowly back against a spring and then releases it so that the entire bed moves forward until stopped abruptly by another spring. The inertia of the refuse carries it a small distance forward along the stoker surface and then the cycle is repeated.

Each of these types of stoker may be flat or inclined down in the direction of flow, and may be used as a single stoker or as a series of two or three units arranged in stair-step array so the refuse is tumbled and agitated as it moves from one section to the next. Often the individual sections can be operated so as to advance the refuse at different speeds to control drying and burn-out of the refuse.

The grate surfaces are made of sturdy iron or steel castings, alloyed and designed to resist distortion, growth, cracking and oxidation. However, in well designed furnaces, the grate surfaces do not characteristically operate at temperatures even near that of the fire because they are protected by unignited refuse, by ash, and by the cooling underfire air passing through them. In the drying and ignition zones, the volatile combustible gases, water vapor and smoke are driven off and flow into the secondary combustion chamber where they are mixed with air and retained long enough to complete combustion. After the ignition zone of the stoker, the residual refuse burns off its fixed carbon with a clean hot flame which radiates heat energy to facilitate the proper burning of the still-combustible gases and airborne particulate matter.

The rotary kiln is really a combination furnace and stoker and is very effective in gently tumbling the burning refuse until complete combustion is achieved. The kiln is a large metal cylinder with its axis horizontal or slightly inclined. It is lined with firebrick and mounted on rollers so that electric motors can slowly rotate it about its horizontal axis. As used in municipal incinerators, the refuse is always first passed over flat bed type drying and ignition stokers in a furnace, and then, when most moisture and volatile constituents have been driven off, the burning residue is fed into the kiln for final burn-out. In such an arrangement, the volatiles driven off in the ignition chamber are led through a passage above the rotating kiln and join the hot gas effluent from the kiln in a secondary combustion chamber where combustion of the gases and airborne particulate matter is carried out.

Another kind of furnace which is acceptable to some municipalities is the tepee burner. As the name implies, this is a large sheet steel walled structure that looks like an Indian tepee with an opening at the top to let the smoke out. Instead of stokers, these have either earth or concrete floors, or sometimes fixed gratings with air supplies under them. Refuse is fed in either by a conveyor which enters the tepee half way up its side wall, or by a bulldozer which pushes the refuse in through a ground level door, to make a heap in the center of the floor. Residue is removed after the fire has burned out, by pushing it out with the bulldozer.

A few cities have recently built a specialized type of incinerator for burning logs, heavy brush and bulky objects like discarded furniture. This has taken the form of a large box-like furnace with typical furnace wall refractory and insulation construction, and large doors at one end. The floor is a simple firebrick hearth and one end of the furnace is constructed to collect the flue gases (the products of combustion) and treat them in a cleaning process before release to the atmosphere. The bulky refuse is pushed into the furnace by a bulldozer, ignited, and allowed to burn down to residual ash which is cleaned out of the incinerator by the bulldozer when the furnace has cooled.

Waterwall furnaces are one of the most recent innovations in municipal incinerator design. It was recognized that incinerator furnaces differ from other kinds of furnaces, in that ordinary industrial furnaces are built to conserve all possible heat for useful purposes. Incinerator furnaces burning American mixed refuse usually have excess heat which causes slag (melted and recrystallized ash) to accumulate on the refractory walls until it sometimes tears the walls down with its weight. To guard against slag, or other heat damage to the furnace, extra cooling air is introduced into the furnace, but this increases the total gas volume that passes out of the furnace for cleaning, and results in larger and more costly dust collectors, fans and stacks.

Waterwall construction consists of steel tubes laid side by side and welded together to form panels which are in turn formed into the walls of a furnace. Water is circulated through the tubes and absorbs heat from the incineration process. The water may turn into steam or into hot water, depending on the design, and this steam or water may be put to a useful purpose, or simply used to carry the heat away to the outside environment. This construction reduces the necessity for excess cooling air in the furnace, and as a result, the flue gas exhausting and cleaning equipment can be smaller and less costly. A layer of protective insulation and light metal sheathing is used outside the waterwalls to conserve the heat in the tubes and to protect personnel and adjacent equipment.

In contrast to the "underfire" air which enters the furnace through and around the grates, air which enters the furnace above the grates through the sides or roof of the furnace is called "overfire air". Overfire air is used to mix and burn with the combustible gases driven off from the refuse, and to cool the burning gases to temperatures which will permit reasonable furnace life. Overfire air is usually introduced in high velocity jets at specific points which vary with furnace design, and is directed so as to provide turbulence and thorough mixing of the gases for optimum combustion.

The flow of air and other gases through a furnace is caused by blowing the air in (forced draft), sucking the gases out so that atmospheric air flows in through the openings provided (induced draft), or heating the gases so they become lighter and rise (natural draft). Forced draft is generally used for underfire air and overfire air jets in modern furnaces, and is supplied by heavy duty industrial type fans driven by electric motors and controlled by dampers which are simply gates or flaps in the air ducts.

Furnace pressures are usually held slightly below atmospheric pressure so that if doors are opened, or if there are any leaks in the walls, atmospheric air will flow into the furnace, rather than permitting the hot, odorous and sometimes dangerous gases to flow out of the furnace to the surrounding workspace. This "negative draft" is maintained by sucking out slightly more gas than the air and gases that are blown in or generated in the furnace. The traditional way to do this is by natural draft, utilizing a tall stack. The heated gases in the stack are lighter than the cooler atmospheric air outside the stack, and so a positive upward gas flow is established by the same principle as sucking up liquid through a straw. Stacks are popular because they are simple and can be constructed to handle large volumes of gas and to withstand hot and corrosive gases. Also, they discharge the airborne products of combustion high in the atmosphere where the winds can quickly disperse them. After the first cost of erection, they require no further expense for power and usually need only nominal maintenance. However, there are practical limitations on the heights and costs of stacks, and these, in turn, limit the amount of suction available to draw the flue gases through efficient dust collection devices. For these reasons, the newer incinerators include induced draft fans which can provide almost any degree of suction required for good gas cleaning.

Induced draft fans, while usually less costly than a natural draft stack, are limited by their tolerance to hot and corrosive gases so that normally the gases are water cooled before the fan. The fans are usually quite large and require large electric drive motors which use considerable power. They can exhaust into relatively short lightly constructed stacks which serve only to carry the residual gases and particulates to a reasonable dispersal height. Sometimes provision is made for varying the fan speed to control the gas flow through the furnace, but more often dampers, which are less expensive, are used for this control.

The temperatures in a modern municipal incinerator furnace are controlled between reasonably close limits. Average temperatures in the burning fuel bed on the grates, and just above, may reach 2,100°F to 2,500°F. The temperature of the flaming gases falls from this level to about 1,200°F to 1,600°F in the time the gases flow through the primary and secondary combustion chambers. Generally, the average gas temperature of the furnace is kept above 1,400°F to ensure oxidation of all malodorous compounds, and below 1,800°F to prolong the furnace life.

The secondary combustion chambers in incinerators are seldom sharply defined chambers connected by passages to the primary combustion chamber. Rather they tend to be extensions or enlargements of the primary combustion chambers sometimes set off by half-walls or baffles that cause the gases to flow in turbulent eddys for a time long enough to complete the combustion process.

H. DISSIPATING THE HEAT OF COMBUSTION

Early incinerators which had to burn refuse of low heating value were designed primarily to conserve and reflect the heat of combustion so as to dry and ignite the refuse and heat the resultant gases above the deodorizing temperature with minimum use of supplementary fuel. The acceptable limit on furnace temperature was then imposed by the durability of the refractory lining. Hot flue gases were discharged directly through masonry stacks with refractory linings, with a "settling chamber" at the base and a "spark screen" at the top as the air pollution control. Early attempts to use waste heat to generate steam were unsatisfactory because there was relatively little "waste heat".

Modern municipal incinerators, burning refuse of much higher heating value, and emitting their flue gases through sophisticated air pollution control devices, are designed to dissipate the heat of combustion so that after achieving the 1,400°F to 1,800°F necessary for complete combustion in the furnace, the flue gases are cooled to 500°F to 700°F before they enter the air pollution control devices.

The use of air cooled and water cooled furnace walls has been mentioned previously. A full water wall furnace, or a waste heat boiler consisting of an array of water tubes in the path of the hot gases, can bring the flue gases down to 600°F and even lower, thus reducing their volume as well as their temperature so that economically sized exhaust fans and electrostatic precipitators, cyclones or bag filters may be considered. The heat of combustion is transferred to the steam and the energy is made available to do work.

Air cooled refractory furnaces have thus far been used mainly to prolong refractory life and reduce slagging rather than to reduce flue gas temperatures.

The two most common methods of reducing flue gas temperatures are by dilution with ambient air and by evaporation of water directly into the gas stream. In addition, there is always some direct conduction and rediation of heat through the walls of the furnace and ducting. Direct admission of air is a simple and easily controllable operation. A damper in the ducting allows outside air to be sucked into the flue gases because the draft of the chimney or of the induced draft fan causes a lower pressure inside the gas ducts than outside. However, even though cooling the flue gases shrinks their volume, the amount of fresh air added increases the net volume so that larger and thus more costly ducts, fans, air pollution control equipment and stacks are required.

If water is mixed with, or exposed to the hot flue gas stream, the water tends to evaporate into steam, and in so doing withdraws heat from the gas. In round numbers, the evaporation of one pound of water can lower the temperature of 40 pounds of hot gas by 100°F. The total volume of the cooled gas and the steam is less than the volume of the original hot gas, so smaller fans, ducts and pollution control equipment can be utilized. However, the water usually does not entirely evaporate and the residual liquid absorbs sulfur and chlorine from the flue gases and becomes acidic and attacks the metal mechanism and structure of the gas cleaning equipment unless they are designed to resist corrosion. Often, water quenching is combined with a wet scrubber for gas cleaning, since the water supply, distribution and containment systems are common. The quenching and scrubbing water may be introduced as sprays, on wet baffles which are obstructions placed in the gas duct with water flowing in thin films over the structures, or as wet bottoms, which are simply shallow water tanks which form the bottom of sections of the flue gas ducting. The effectiveness of evaporation cooling is usually dependent upon good mixing of the gas and liquid.

I. COLLECTING, COOLING AND REMOVAL OF RESIDUAL SOLIDS

Solid residue is generated at three places in a municipal incinerator: the incombustibles that come off or through the stoker, the siftings that fall through the grate openings, and the fly ash collected from the flue gas. Each of these solids occurs in different quantities and different form, and since, on the average 20 percent to 25 percent of the weight of municipal mixed refuse is glass, rock, cans and other metals and minerals, the collection, cooling and removal of the incombustible solid residue is a significant materials handling task. Fortunately, this solid residue comprises only 5 percent to 15 percent of the average volume fed into the incinerator, the rest having been converted to gas.

The principal residue is that discharged from the stoker. This consists of broken glass and ceramics, tin cans of all shapes and sizes up to oil drums, stones, earth, assorted hardware, some fused ash (clinkers) from melted glass, metals and minerals, all surrounded by flakes and dust of the light fluffy ash that typically results from burning paper or wood. Although some incinerators do a remarkably thorough job of burning the combustibles, practical limitations of the time that the average load can be left in the furnace permit some incompletely burned refuse to appear in the stoker residue. Telephone books, catalogs, and heavy bundles of newspaper, particularly if wet, are not readily penetrated by air for direct combustion, or by radiant or convective heat for drying and volatilization, so it is not uncommon to find bits of unburned paper in the ash. Heavy timbers or green wood may pass through the average 40 to 60 minute burning cycle with a core of unburned wood still present. Certain foodstuffs containing high water content and occurring in thick sections, like watermelon rinds, carrots, apples or waste meats may char on the outside and seal in liquids, thus permitting some putrescible material to appear in the ash. All this may be aggravated by an operator pushing wet refuse through an incinerator at maximum feed rate to handle a peak load, instead of slowing the stoker to permit better drying and longer combustion time for difficult wastes.

Nevertheless, a typical "good" residue contains less than 5 percent by weight of unburned carbon and less than 1 percent of putrescible organic material.

In cylindrical batch feed type furnaces, the stoker residue falls into refractory lined metal hoppers where it is quenched, then through a gate in the bottom of each hopper, into a truck for removal to a dumping site. The residue is usually steaming and sometimes still burning locally when it is removed, because of the difficulty of reaching all the internal surfaces with the water sprays.

The larger, newer incinerators with flow-through furnaces generally use conveyor systems to remove the stoker ash. The residue is usually discharged from the stoker grates into a water filled trough where it is thoroughly quenched and cooled. A metal drag link conveyor, consisting of a pair of endless metal chains with metal bars attached between the chains, like a rope ladder, is driven by an electric motor through sprocket wheels to drag along the bottom of the ash trough and capture the residue, pull it out of the trough, up a chute and into a waiting bin or truck. These drag link conveyors are heavily constructed of heat resistant cast steel drag link parts, with corrosion resistant metal or concrete tanks and chutes for containing and guiding the wet, abrasive residue. They travel slowly, thus allowing water to drain back into the tank from residue moving up an inclined chute. The unused portion of the drag link chain which is returning from the point of discharge of the residue back to the entrance to the quenching tank may require as much supporting, guiding and protective structure as the working section of the chain.

It is common practice to position drag link conveyor paths transversely across the discharge ends of two or more rectangular furnaces so that the residue of both can be carried to a common discharge point for economy of overall structure and material of the conveyor. Because the residue removal conveyors are such vital elements of the incineration system, and because they are big, heavy, and somewhat difficult to service, they are often built in parallel pairs with a diverting chute so that each of several furnaces can direct its residue to each of two residue conveyors, to ensure reliable service.

Another type of conveyor system is a metal mesh travelling belt on which the residue falls from the stoker, and is spray quenched. Other systems, using metal apron type conveyors, vibratory conveyors and rubber belts for cooled residue are in use, but the metal drag link type is the most popular. Manual handling of residue has practically disappeared, except for a few small older incinerators.

A few incinerators have equipment for separation of residue and salvage of metal, mostly tin cans. The separation is usually done in a metal drum or barrel with holes in its sides and mounted to rotate about its horizontal axis. This tumbles the residue and the ash falls through into a hopper, while the cans finally pass out the end of the drum to another hopper. Sometimes magnetic separators are used to collect the steel from the rest of the residue. The cans are usually pressed flat or shredded to make them into salable scrap. The non-metallic ash is a dense, inert material, almost like damp earth, and is sometimes sold for road fill.

The siftings are dust-like bits of ash, often still glowing, that drop through the grate openings and cover whatever is below. The amount varies with the type of stoker and the material being burned. Certain types of plastics that melt, or greases that can run through the grate openings may accumulate in the underfire air chambers and ignite and burn unless precautions are taken. Some stokers rely on manual cleanout through doors provided for the purpose. Others provide hoppers under the grates and means ranging from mechanical and pneumatic conveyors to water sluicing to move the accumulated siftings either to an outside collection point or into the quenching tank with the other solid residue.

An innovation to combine siftings removal with stoker residue removal consists of a "wet bottom" furnace in which the entire furnace foundation is made as a concrete basin which is filled with water. A drag link conveyor as wide as the furnace runs the entire length of the furnace and drags out the grate siftings which have fallen into the water, as well as the stoker residue which falls into the quench water at one end of the foundation.

Fly ash, in a municipal incinerator, is the particulate matter which is light enough to be carried out of the furnace by the existing gas stream. Dust, ash, fine burning particles, and even sizable pieces of burnt or burning paper are released from the grate by underfire air and fuel bed agitation, so that about 2 percent of the weight of mixed refuse charged into the furnace leaves the furnace as airborne fly ash. The various air pollution control devices, to be described in another section, collect from 30 percent to 95 percent of this fly ash, either by dry collection in hoppers or by capturing it in water. As in the case of siftings, the dry fly ash is conveyed by mechanical or pneumatic conveyors, or by sluicing with intermittent floods of water through pipes, either to a storage hopper for truck loading, or to the residue quenching tank for removal with the other solid residue.

Fly ash entrained in water has been most successfully collected by the use of settling tanks or lagoons. In either case, a large quiescent body of water permits practically all of the particulate matter to settle to the bottom in time, and the relatively clear water is drawn off the top for re-use, while the particulates are discharged or dredged from the bottom. There have been several attempts to separate the particulates from scrubbing water with centrifuges, mechanical screens and filters, but there has been little sustained success.

J. CLEANING AND DISCHARGING EFFLUENT GASES AND LIQUIDS

The gaseous products of combustion, in modern municipal incinerators, usually contain fly ash which is gas borne particulate matter, mostly incombustible, as well as air, carbon dioxide, water vapor, small amounts of sulfur oxides, and traces of other gases. The water vapor and carbon dioxide are natural and acceptable constituents of the atmosphere, but the other materials, the large particles, the dust and any noxious, odorous or corrosive gases tend to pollute the atmosphere and must be controlled within acceptable limits, if they cannot be eliminated.

The smoke, generally carbon particles less than one micron (.00004 inches) in size, and the odors, generally chemical compounds formed during the process of combustion, should not exist if there are the time, temperature and mixture with oxygen for complete combustion. Unfortunately, practically complete combustion is seldom achieved in the older incinerator designs, and even in the newer ones there are transient periods of start-up and shut-down, or unusually wet loads or unusual materials like tires or roofing that can cause smoke.

The particulate matter is collected in three classes of equipment in existing municipal incinerators: subsidence chambers, mechanical cyclones, and wet scrubbers, and these are listed in the order of increasing effectiveness. The "efficiency" of dust collectors is usually simply defined as the percentage (by weight) of the particulate matter entering that is caught. For example, if one pound of dust goes in, and 0.1 pounds comes out in the "clean" gas, then 0.9 pounds was caught and the efficiency is 0.9/1.0 or 90 percent. This is an unfortunate method of rating because it makes no allowance for the variability of size, shape and weight of the gas borne particulate matter. To illustrate, if the incoming gas stream contained a brick, several scraps of paper, and a thousand grains of sand, and if the collector removed only the brick, it would be rated as 99+ percent efficient by this method. Therefore, the effectiveness (efficiency) of particulate collectors is meaningless unless all collectors are evaluated on "standard" fly ash of known density and size analysis. Such tests are tedious and expensive, and until recently, were seldom run on incinerators. Instead, the acceptability of incinerator chimney effluents was often specified and is still often judged on the basis of a Ringlemann test. This is a visual comparison of the color and opacity of the "smoke" from a stack, with various samples of grey color on a printed card, and does not differentiate between carbon fumes, fine dust, large flakes of paper ash, or visible steam from condensed water vapor, nor does it allow for different colored sky backgrounds, atmospheric conditions, or the observer's color perception.

Subsidence chambers, which are also called expansion chambers, baffle chambers or settling chambers are in principle large volumes where the gases are slowed down and retained for a while to allow some of the particulate matter to settle out to the floor of the chamber by gravity. Large heavy particles drop out of the gas stream quite readily, particularly where there are changes in gas flow direction caused by baffles. Smaller particles settle out to a degree, if the gas is not very turbulent and ample time is allowed. However, for typical incinerator fly ash, with a high percentage of particles below 50 microns (1/500 of an inch) in size, the turbulence of the flowing gas stream holds most of them in suspension until they pass out the stack. Since more than half of the total particulate matter in the typical municipal incinerator fly ash as it comes from the furnace is under 30 micron size, it is understandable that even a large subsidence chamber, which may be larger than the incinerator furnace, cannot offer very high collection efficiency. Nevertheless, this is the type of dust collection system in use in the majority of present municipal incinerators.

An improvement in collection efficiency of subsidence chambers is obtained by the use of wet bottoms or wet baffles, which, as previously described, serve to cool the hot flue gases by evaporation, while trapping and retaining the dust particles that impinge on the wet surfaces. An obvious extension of the water cooling and impingement principles is the introduction of water in sprays or cascading curtains in the ducting and subsidence chambers. However, this rudimentary wet scrubbing does not statistically bring most of the dust particles into contact with water droplets or water surface and there is only moderate improvement in collection efficiency. About 25 percent of the municipal incinerators have water augmented subsidence chambers.

Subsidence chambers, with or without water augmentation, are constructed to have low pressure drops for the large volumes of gas flowing through them, and consequently they are suitable for natural draft installation where the motive force for gas flow is provided by the rising hot gases in a tall chimney, instead of by use of a fan.

Chimneys (stacks) for municipal incinerators are usually constructed of masonry, though steel or reinforced concrete are becoming more popular. Tall stacks, often 100 to 250 feet high, are the usual "trademark" of municipal incinerators, towering over everything but power plant stacks, and often discharging clearly visible plumes of smoke and steam which have given incinerators a poor reputation as air polluters.

Chimneys are highly specialized and costly structures, since they must be designed to withstand adverse terrestrial and atmospheric conditions, to permit access for inspection and maintenance and must be lined with heat, moisture and acid resistant materials to withstand the attack of flue gases. Tall stacks allow the discharged particulates and gases to be caught in upper air currents and dispersed, instead of settling in the immediate vicinity of the incinerator. Though more expensive in original cost than an induced draft fan installation, they do not require continuous use of power. These two features, dispersion and low operating cost, have probably led to the popularity of natural draft installations and thus have limited air pollution control devices to those that could operate with approximately one inch of water pressure drop.

Chimneys also are natural settling chambers. The flue gas usually enters through a horizontal duct (breeching) several feet above the base, and the expansion and turning of the gas as it starts to rise allows particulates to fall to the base of the stack where they can be removed through clean-out doors. Sometimes a wet base is provided in the chimney to trap particulates in water and flush them out hydraulically.

Mechanical cyclones direct the particle-carrying gas at high velocity, into cylindrical chambers which cause the gas stream to whirl around in a circular path. The heavier particles are thrown to the outside of the circle by centrifugal force and are collected in hoppers at the base of the cylinders while the cleaned gas spins out through the center of the chambers. The cyclones are usually made of steel and cast iron selected for corrosion and abrasion resistance. Occasionally, they are lined with wear-resistant refractory materials in critical sections. The smaller the diameter of the individual cyclone cylinders, the higher the centrifugal force that can be developed for a given pressure drop, but as the cylinders get much below 10 or 12 inches in diameter, the number required, the supporting and gas-guiding structure, and the number of small openings prone to plugging, all increase and so do the cost and maintenance problems. Cyclones are used with 3 to 6 inches of water pressure drop.

Well designed and properly installed cyclone systems can provide significant efficiency improvements over subsidence chambers, but it is only in the past ten years that cyclones have begun to find use in municipal incinerators. At present, fewer than ten percent of incinerators have them.

Wet scrubbers are the natural outgrowth of wet bottoms and wet baffles, in the evolution of gas cleaning systems for municipal incinerators. The principle of wet scrubbing, which is well known in the chemical processing field, is to bring all the gas into intimate contact with a liquid which can seize and hold the particulates and gases to be removed from the incoming gas stream. The liquid must then be separated from the clean gas. It should be recognized that this results in a stream of polluted liquid, but there are acceptable methods for processing liquids to remove the impurities and allow re-use of the liquids.

Almost any degree of gas particulate cleaning desired can be obtained in a properly designed wet scrubber, providing sufficient mixing energy is supplied to cause the particles to contact liquid. The energy is usually applied to breaking the liquid into droplets in various kinds of spray nozzles, or by moving the gas at high velocity through water, and in both cases turbulating the gas - water mixture long enough to statistically assure contact. Removal of the water droplets takes additional power and is done by centrifugal means (like a mechanical cyclone) or by flowing the gas through a zig - zag path between baffles which trap the droplets. For municipal incinerators, five to twelve inches of water pressure drop is usually used in the gas stream. Wet scrubbers provide favorable conditions for dissolving gases containing sulfur, chlorine and other trace chemicals out of the flue gas stream. However, these tend to form active acids, when in water solution, and the scrubbers, ducting, fans and stacks must be constructed of materials that can withstand acid corrosion.

It is only in the past few years that true wet scrubbing systems, as differentiated from crude spray chambers or wet baffles, have been applied to municipal incinerators, and fewer than ten percent of the incinerators have them. They are usually rectangular chambers, constructed of stainless steel or lined with rubber, or cement. Various configurations of fine sprays, or bubbling the gas upward through layers of water held on perforated trays or on a bed of spheres are the methods commonly used to obtain intimate gas - liquid contact.

There have been a few attempts to combine a cyclone and a scrubber into a wet wall or wet bottomed cyclone. The results have not been satisfactory for incinerator fly ash because dampened fly ash tends to cake into a mud and bridges over openings and blocks them. It has been necessary to keep scrubber surfaces flooded with liquid to prevent "mud" build-up.

The spent liquid coming from wet bottoms, wet baffles, residue quench tanks, pit drainage, spray chambers and scrubbers in many cases presents a disposal problem because it is chemically and physically contaminated, and a water salvage problem because of the cost of water. Where water supply is no problem, and where sewers or streams are unrestricted, the older and smaller incinerators sometimes dispose of the waste water directly without any treatment.

The most common method of water treatment, though used in less than 50 percent of municipal incinerators, is the use of settling tanks or lagoons. Particulate matter can be collected from the bottom of the tank or lagoon, and relatively clear water can be drawn off through a weir at the top. A few of the more recently built plants have installed mechanical screen strainers, filters, or centrifuges to clarify the water for recirculation, but the results have been less than satisfactory.

The waste water which has been in direct contact with flue gases tends to become acidic, while the water which has leached through fly ash or siftings may be quite basic. Some of the larger metropolitan incinerators treat the waste water with neutralizing chemicals before clarification treatment.

In general, reclamation and reuse of waste water, though desirable, is troublesome in today's municipal incinerators. Corrosion, abrasion and plugging of plumbing, pumps, valves and spray nozzles are common problems unless careful attention has been paid to designing for reuse of contaminated water.

K. CONTROLLING THE PROCESS FOR SAFETY, EFFICIENCY, ECONOMY AND COMMUNITY ACCEPTANCE

A modern municipal refuse incinerator operating at rated conditions releases an awesome amount of energy under conditions that could be destructive to the plant itself and dangerous or offensive to the community if not properly controlled. Moreover, like any other chemical process, the quantities and mixture of the reactants (refuse, air and water) and the conditions of the reaction (time, temperature, turbulence) must be controlled for satisfactory results. Also, the presence of potentially hazardous elements like high temperatures, deep pits, moving conveyors, heavy trucks and cranes and lethal fumes make it imperative that there be controls to safeguard the operating personnel.

The words "instrumentation" and "control" are often linked together as a common term, but they are really quite different and the difference is important. Instrumentation is the equipment used to indicate and/or record physical conditions like weight, temperature, position, flow, time, speed, voltage, etc. Instruments give signals, but in themselves they do not change the conditions of operation. Controls are mechanisms which change conditions of operation, like a valve which can change water flow, a switch which can turn on a fan motor, or a speed-up in the rate of refuse feed which can increase furnace temperatures.

A control "system" must have four basic elements: a standard of desired performance, a sensor (instrument) to determine actual performance, an intelligence to compare actual versus desired performance (error) and to make a correction, and a control device to cause a corrective change to occur. If these elements are integrated into an automatic system that controls the process to a set standard, like a household thermostat that holds 70°F temperature, it is a closed loop or automatic feedback control system. If, on the other hand, control is effected by making a change and then observing the result (like steering a car), it is an open loop system.

Present municipal incinerators generally utilize their instrumentation in conjunction with open loop control systems, although there are increasing instances of fully automatic closed loop control systems for controlling furnace temperatures and furnace draft.

There are many types of instrumentation used in incinerators, but they can be classed as follows:

1. Temperatures

   - Optical pyrometers for flame temperatures in the range of 2,200°F to 2,500°F.

   - Thermocouples (Chromel-Alumel) for furnace temperatures in the 1,400 to 1,800°F range and iron-constantan in duct temperatures down to 100°F.

   - Gas or liquid filled bulb thermometers for duct temperatures, below 1,000°F. and for ambient temperatures and water temperatures.

2. Draft Pressures

   - Usually only a few inches of water column.

   - Manometers and inclined water gauges for accurate readout close to the point of measurement.

   - Diaphragm actuated sensors where remote readouts are desired.

3. Gas or Liquid Pressures from 1 to 100 psi

   - Bourdon tube pressure gauges for direct readout.

   - Diaphragm actuated sensors for remote readout.

4. Gas Flows

   - Orifices or venturis with differential pressures measured by draft gauges.

   - Pitot tubes and draft gauges.

5. Liquid Flows

   - Orifices with differential pressure measurement.

   - Propeller type dynamic flowmeters.

6. Electrical Characteristics

   - Voltmeters, ammeters and wattmeters.

7. Smoke Density

   - Photo-electric pickup of a light beam across the gas duct.

8. Motion

   - Tachometers for speeds of fan, stoker or conveyor drives.

   - Counters for reciprocating stokers and conveyors.

9. Visual Observation

   - Vidicon cameras for closed circuit television for viewing furnace interiors, furnace loading operations or stack effluents.

   - Peep holes in furnace doors.

   - Mirror systems.

The general functions of control systems in present day municipal incinerators are:

1. To control underfire airflow despite varying furnace draft and varying restrictions to airflow through the grate openings.

2. To control overfire air as a means of controlling furnace temperature as the quantity and heating value of the charged refuse vary.

3. To control furnace pressure by varying the draft induced by the chimney or the I.D. fan.

4. To control gas temperature at the dust collectors by varying the amount of quenching water or diluting air introduced.

5. To control refuse drying and burning time by varying the conveying speed of the stoker mechanism.

Usually, the indicating and recording instrument readouts are grouped on a control panel centrally mounted on the operating floor, with a system of warning lights or bells to summon the operator when corrective action must be taken. Often, a duplicate panel of indicating instruments is placed in the plant superintendent's office for additional surveillance.

It is quite common to have recording type instruments which indicate on a removable paper chart, the conditions which exist over a 24 hour period. These records can be very useful in making statistical summaries of operating conditions, and of reviewing operating conditions which may correlate with a malfunction.

L. PROTECTING THE PERSONNEL AND EQUIPMENT

A modern municipal incinerator plant represents an investment of two million dollars or more in equipment and structure, and is expected to serve the community for about 25 years with reasonable maintenance. However, there are forces continuously at work tending to deteriorate the incinerator, namely, the environment, dangerous refuse and careless operation. The experience of engineers and operators has resulted in design features and operation procedures which serve to protect incinerators.

At the receiving bays and storage pits, clearly marked traffic lanes, high intensity lighting and heavy duty curbings and rails reduce the incidence of collisions between refuse trucks and the columns and structures of the receiving area.

Lining the concrete storage pits with steel bumping rails and attaching heavy wear-shoes to the crane bucket can greatly reduce mutual deterioration of pits and crane buckets of grapples. Enclosing the crane machinery and subduing the dust of the storage bins with ventilation and water sprays is a means to reduce crane "down time" and improve crane life. Adequate clearance room and automatic stops save damage from crane overtravels.

Storage pits are susceptible to fires from spontaneous combustion or from smoldering refuse inadvertently dumped in. Sprinkler systems or at least ready water hoses are employed in most incinerators.

Backfires in the loading hoppers of continuous feed incinerators may occur, particularly if the furnace pressure rises above atmospheric. Many incinerators provide metal covers or cut-off dampers on the loading hoppers or chutes so that backfires can be quickly contained and smothered.

Undergrate or windbox fires can be a source of damage if a slow burning meltable or powdered fuel, usually special industrial waste, is fed in large quantities. The combustible refuse may fall through large grate openings and the resultant fire from below may burn or warp stoker parts. Good grate designs and reasonable surveillance of incoming refuse are the measures usually employed to prevent this.

Overheating of the furnace chambers or hot gas ducting can cause serious damage in a short time. Multiple temperature sensors, closed loop control systems, and audible and visual alarm systems are employed to avoid this danger. Increasing overfire air, reducing underfire air and reducing refuse feed rate are control methods used to reduce furnace temperature.

Failure of the quench water supply, whether from clogged nozzles, pump failure, electrical power failure or lack of water, can be quickly disastrous to the flue gas cleaning system and fans. Auxiliary water towers and automatic bypass ducting that can dump the furnace gas directly to the stack are the usual provisions for such an emergency.

Loss of electrical power can be anything from an annoyance to a disaster, depending on the particular system involved. Several incinerators have standby auxiliary power supplies with gasoline, diesel or turbine engines.

In cold climates, freezing water can do damage to piping and to wet conveyor systems. The common practice is to enclose such sensitive systems in heated buildings, or to selectively heat sensitive lines.

Corrosion is the perpetual enemy of the metal structures and equipment of incinerators, particularly when even traces of sulfur, nitrogen and chlorine in flue gases can acidify moisture. Enclosing mechanisms in weather-proof shelters, good painting practice, use of corrosion resistant metals and stainless steel, and use of durable non-metallics where warranted, are the protective methods employed.

Much concern has been expressed over the danger of municipal incinerator explosions, but there appear to be no records of serious damage resulting from explosions of pressure cans, paints and solvents, chemicals, and even ammunition and dynamite. One reason for this seems to be that the furnace volumes are large and unconfined, so that even a sizable explosion cannot make much of a pressure wave. Furnaces are usually lined with thick layers of brick and are steel cased, and there is usually a "cushion" of several inches of porous slag on the inside surfaces. Also, many of the potentially explosive containers soften or burn before an explosive release of energy can occur.

Personnel safety and reasonably tolerable, if not attractive, working conditions are necessary to recruit and retain people competent to operate a modern incinerator efficiently. Guard rails around pits, hoppers and conveyors, good lighting, good ventilation, protective clothing and eye shields for working at furnace openings, and constructive safety training programs are to be found in the majority of installations. Also, well managed municipal refuse disposal plants are essentially odor free clean and dry, with well appointed locker rooms, lunch rooms, reception lobbys and administrative offices. The buildings generally have pleasing and functional architectural treatment with fences and landscraping designed to make them good neighbors.

A. RECEIVING, MEASURING AND STORING

It is expected that the principal changes in these areas will be those reflecting the ever larger quantities of refuse to be handled. Following the European developments, it is reasonable to suppose that refuse will be brought to central incinerators by larger trucks, by rail and by barge and there will be need for better methods of rapid unloading and nuisance-free storage. The bridge crane with faster travels and simplified controls will probably be supplemented by various types of conveyors, including pneumatic transportation systems. The storage pits will continue in vogue for the larger plants, but the smaller towns may go increasingly toward open floor dumping because of lower first cost. In both cases, there is likely to be better enclosure of the stored refuse, sprinklers for dust and fire control and mechanical ventilation for discharging the odorous exhaust air into the incinerator.

B. SORTING, MIXING, FEEDING

Considerable effort is being expended to improve these functions, and it is therefore likely that future municipal incinerators will include new equipment for preparing the refuse for efficient incineration. Breakers, crushers and shredders will be employed to reduce awkward size combustibles like wooden pallets, demolition lumber and furniture to smaller pieces, and obviously incombustible items like bathtubs, bicycle frames and kitchen stoves will be separated during the collection process. There are signs of growing interest in special bulky refuse incinerators for logs, brush and furniture.

It is not expected that there will be a significant trend toward pre-incineration salvage. Fairly sophisticated machinery has already been developed for this purpose for composting operations, but the salvaged materials, even when classified, are troublesome to market. "Tin cans" contain steel, tin, solder, plastics, aluminum, paper and residual food, and are not desirable as scrap. Clothes are often unsalvagable mixtures of synthetics and natural fibres, and the market for salvaged glass and even clean paper is undependable.

There is likely to be increasing interest in various types of ram and conveyor feeders as supplements to the gravity fed hopper and chute. The use of wheeled vehicles to load furnaces from floor dump operations is also expected to increase. As shredding and grinding of refuse become more common, more mechanical or pneumatic conveying of the prepared fuel into the furnace will probably be employed.

C. FURNACES

It appears that developments in municipal incinerator furnaces will proceed in several directions; refractory lined furnaces and water wall furnaces, both with conventional flow-thru stoker systems, and a number of advanced concepts. Some of these may prove to be impractical, but others may, in time, offer significant advantages.

There is a trend toward air cooled walls in refractory lined furnaces, and the use of silicon carbide or high alumina facings in the slagging and abrasion zones of the furnaces. Air cooling is accomplished by natural or forced circulation of air over the outside of the refractory lining, and often the air which has picked up heat from the lining, is discharged into the furnace as overfire air. The cooled walls seem to resist penetration by heavy glassy slags, so that the slag falls off or is easily removed, without causing spalling (breaking off layers) of the refractory surface. There is increasing recognition that with the high heating value of American refuse, it is not as necessary to heavily insulate the furnace walls to conserve all the heat of combustion. With thinner, more conductive air cooled walls, some excess heat can be directly dissipated through the furnace walls to reduce furnace temperatures and improve refractory life.

There is also likely to be increasing use of plastic and castable refractories for walls and arches (ceilings) to reduce the labor involved in the traditional methods of laying up firebrick.

Waterwall furnaces are expected to find increasing use, both as cooling surfaces to shrink gas volume so the flue gases can be more easily cleaned, and as heat absorption surface in steam generation systems. Following European incinerator practice, there may be refractory coated tubing used for corrosion resistance in critical areas of the furnace.

The stoking systems and grates used in these furnaces will probably undergo detail improvements to increase ease of servicing and durability. The new designs will tend toward controlled agitation of the residue to strike the best balance between full combustion and least generation of fly ash.

Some of the more novel approaches to incineration of municipal refuse under experimentation are suspension burning, melting with auxiliary fuel, pyrolysis, fluid bed combustion and pressurized burning. Suspension burning, which is the process widely used in power boilers, consists of blowing the finely divided fuel into a vortex pattern in a furnace chamber so that it burns while suspended in the turbulent air stream. It is efficient and can provide high heat release in a relatively small volume, without the necessity for supporting a burning fuel bed or a grate or hearth. Well controlled preparation of the refuse is required.

Melting of the incombustible residue can be accomplished if the heat release of the burning refuse is augmented by burning it with a high quality fuel like coke in a properly designed refractory chamber. The melted residue, including metals and minerals, can be run into a water bath where it solidifies and fractures into coarse crystals which are probably the ultimate in cleanliness, compactness and desirability as a residue.

Pyrolysis consists of decomposing the refuse by the application of high heat, without supplying oxygen for combustion. Theoretically, the refuse can be converted to combustible gases, fixed carbon (like charcoal) liquids and tars containing useful organic chemicals and inert ash, metals and minerals.

Fluid bed combustion is carried out in a bed of inert granular mineral (like sand) which is heated in a refractory vessel on a perforated plate. Air is blown upward through the holes in the plate at a controlled rate which churns the sand into a turbulent mass like quicksand. After preheating with a gas or oil burner, the refuse is introduced and burns while circulating in the hot sand. Theoretically, there is excellent control and complete combustion. Separation of the residue from the bed material presents a problem.

Pressurized burning may be performed by any of the above methods, in a smaller furnace than normal, by introducing the combustion air under high pressure. The same weight of air is thus contained in a smaller space, so combustion can proceed normally. The additional power required to compress the air is a drawback, but the existence of hot pressurized flue gas offers the thermodynamic possibility of directly operating a gas turbine engine to generate useful power.

D. DISSIPATION OF THE HEAT OF COMBUSTION

A fair amount of interest is being shown in the use of gas-to-air heat exchangers of tubular construction to cool and shrink the flue gases without the corrosion, mud-forming and waste water disposal problems of wet quenching. The heated ambient air can be directly released to the atmosphere without pollution problems, but the tubes must be designed so that they are not blocked or corroded by the raw flue gas emerging from the furnace.

If the heat absorbed in water walls is merely to be dissipated, it is likely that air-cooled condensers will be preferred, because they do not produce a cloud of visible steam in cold weather like cooling towers do, and they do not require large supplies of cooling water as water cooled condensers do.

Perhaps the most sensible prospect, and therefore the concept that may ultimately prevail, is to convert the incinerator furnace waste heat into steam which can be used for one or more of the following purposes.

1. For heat and power required in the operation of the incinerator plant itself.

2. For heat, power and process steam exported to nearby industry, institutions or municipal installations.

3. For power to be fed into commercial electric utility networks.

E. HANDLING OF RESIDUAL SOLIDS

As the burning process developed, there has been improvement in the "quality" of the residual materials, but the ever-increasing quantities of residue and the form of it -- dust, sludge and miscellaneous incombustibles -- make it likely that continued attention will be paid to application of modern materials handling techniques.

Hand cleanout of siftings will almost certainly disappear in favor of automatic siftings collection by mechanical, pneumatic, or sluice-type conveyors. Hot collection of stoker residue with quenching by water spray in a hopper will probably be replaced entirely by direct quenching of the residue in water filled sumps. The all metal drag-link conveyor, while presently the most popular residue handling system, is likely to see competition from the less costly European type systems of swinging metal pushers to lift the cooled residue from the quench tank, followed by rubber belts or metal plate type belts to transport it to the disposal point.

It is also probable that there will be continued efforts to separate metal cans and bulky items from the dust, sludge and minerals of the residue, not only for the possible salvage of the metal (burned out tin cans have certain value as scrap and for metallurgical processing), but to make the non-metallic residue more valuable as high quality fill. As an alternative, the cans and clinkers may be crushed or ground to reduce the volume and improve the homogeneity of the total residue.

F. AIR POLLUTION CONTROL

It is the consensus of the industry that improvements in air pollution control systems and equipment will be the most immediate significant change in the next generation of municipal refuse incinerators. Although municipal incinerators contribute only a few percent to the total air pollution of a metropolis, the clearly visible tall chimneys with their plumes of smoke, or even white steam, the lingering mental association between incinerators and smoky, odorous burning open dumps, the occasional fallout of dust and flakes of charred paper from old systems with little air pollution control equipment have made the public conscious of municipal incinerators as pollution makers rather than pollution controllers. The methods and equipment for air pollution as control to almost any degree desired are commercially available. The problem has been a lack of enforced air pollution control standards, and the desire to "economize" in the purchase and construction of central incinerators.

Engineering studies have indicated that in order to comply with the more stringent air pollution codes now in existence or being prepared, a modern municipal incinerator might require 94 percent overall collection efficiency with a fractional efficiency of about 75 percent by weight collected of all particles sized five microns and below.

These efficiencies can be obtained with high quality mechanical cyclones, wet scrubbers and filter bag collectors with a fan-powered induced draft system. With natural draft alone, they can only be achieved with electrostatic precipitators. Odors and true smokes of submicronic particles can be virtually eliminated by proper combustion control in the furnace, although bag filters and electrostatic precipitators can collect smoke particles, and wet scrubbers can dissolve out certain odorous or noxious gases.

It is expected that there will be a strong movement toward electrostatic precipitators and high performance wet scrubbers. The electrostatic precipitators operate on the principle of electrostatically charging the particles in a gas stream by passing them through the corona of a high voltage (upward of 40,000 volts) conductor. The charged particles are carried at very low velocity between oppositely charged electrical plates where they are attracted out of the gas stream and cling to the plates. At intervals, the plates are shaken and the accumulated dust falls into hoppers. Electrostatic precipitators are large and relatively expensive because they must handle large volumes of gas at low velocity. They can have difficulties if the particulates will not accept and hold the electrical charge because of high temperature or chemical composition. If the fats and greases of refuse have not been properly burned in the furnace, they can condense in the precipitator and cause malfunction and fire. Also, large flakes of burned paper can sail right through them if not otherwise trapped and collected. Nevertheless, they have been used quite successfully on European incinerators and will be used on American incinerators, both for upgrading existing installations and for new installations. Waterwall furnaces or waste heat boilers, which cool and reduce the volume of the flue gases will permit use of smaller and less costly electrostatic precipitators.

Wet scrubbers and also water quenchers, tend to saturate the hot flue gases with water vapor. As the flue gases emerge from the stack and encounter cold air, some of the water vapor condenses into visible steam. The public tends to confuse this harmless steam with smoke, so it is expected that future scrubbers will be equipped with heat exchangers to reheat the scrubbed flue gases using furnace heat, to reduce the incidence of visible steam plumes. If the scrubbing water is kept cold, the flue gases will be cooler and less moisture will be picked up in the scrubber. For this, settling ponds with spray coolers and ordinary cooling towers may be called upon to cool the recirculating water.

Mechanical cyclones should also appear in upgraded existing incinerators and in some new ones. Unfortunately, there has been some poor experience with improperly designed equipment, inadequate draft, and carried-over quench water and it will be difficult to convince the buying influences that these problems can be overcome through competent engineering.

Bag filters suck the flue gases through the woven fabric walls of special bags, like vacuum cleaner bags. A pre-coat of dust forms and thereafter catches just about all the solid matter entrained in the gas stream. The multiplicity of bags are contained in a large structure (baghouse) and equipped with automatic devices to divert the gas stream and shake the collected dust off the bags and into hoppers at appropriate intervals. Bag filter installations can do a near perfect job of dust collection, but they are expensive, and require considerable power and maintenance. It is believed that other methods of air pollution control will be accepted more readily on future incinerators.

It is also likely that there will be increased use of induced draft fans specially designed for municipal incinerator use. These will be of heavy duty, abrasion resistant and corrosion resistant construction for reliability; and will include flow control by variable speed or by efficient damper systems. Because the induced draft fans on large incinerators may require 100 or more horsepower, more attention will be given to fan efficiency and durability, rather than to lowest first cost.

G. CONTROLS AND INSTRUMENTATION

Industrial process instrumentation and controls are a highly developed art and future incinerators may be instrumented and atuomated to almost any desired degree. However, it is expected that municipal incinerators will not go very far toward fully automated or computer-controlled operation in the next few years because municipalities will continue to look toward their refuse disposal operations as sources of employment for many unskilled and semi-skilled people. Also, the fact that these people will be engaged in the operation of incineration systems will lead to the adoption of automatic controls for certain critical parameters like furnace temperature and percent of carbon dioxide in the flue gas; and very simple, perhaps color coded, "on - off" type signals and push buttons for the less critical functions. Read-out instruments giving the precise numerical value of each temperature and pressure in the system may not be considered worth their cost.

H. PROTECTION OF EQUIPMENT AND PERSONNEL

As the sophistication and cost of municipal refuse incineration system equipment increases, it is logical to suppose that continued attention will be paid to protection of the capital investment. For reasons of economy and efficiency, however, it may be that municipal incinerator structures will have less future emphasis on traditional architecture (like a town hall or a school) and tend more toward weather resistant unenclosed furnaces and ducting, and minimum functional enclosure of offices, workshops and critical equipment.

In keeping with the generally increasing social consciousness, it is expected that there will be additional safety provisions, training, comfortable work stations and uniformed technicians and skilled managers whose service to the community can be rendered with dignity and professional satisfaction.

INCINERATOR CONTROL EQUIPMENT

SYSTEM EQUIPMENT CONSIDERATIONS

In an incinerator, continuous, rather than batch, feeding of the refuse can reduce air pollution. Continuous feeding permits greater control over furnace temperature and therefore influences the completeness of combustion. Other design and operating factors which insure low fly ash emission from an incinerator furnace are:

1. Excess air should be maintained between 50 and 150 percent to insure proper burning and control of furnace conditions.

2. Furnace exit temperature should be held between 1,400°F and 1,800°F to minimize slagging and to eliminate stack odors.

3. Furnace design must allow sufficient residence time for combustion to be completed.

4. Overfire air should account for from 20 to 40 percent of the total air and should be introduced into the furnace in such a manner that it insures sufficient turbulence for complete combustion.

5. Gas quenching or heat removal should be arranged so that it does not affect combustion. If water quenching is used, it should be arranged to remove some of the larger fly ash particles.

With good combustion, the limited amount of fly ash produced is very fine and is difficult to capture in the collector. It is, however, well within the capability of present technology to capture most of the particulate matter.

The advent of polyvinylchloride (PVC) and other plastics that contain a high percentage of chlorine and/or fluorine (Teflon) may subject incinerator metals to severe corrosion attack. The problem is particularly acute during periods of start-up and shutdown when the chlorine and other halogens as well as sulfur combine with the plentiful moisture in the flue gas to form highly corrosive acids. During normal operation, the higher temperature diminishes the danger of formation of the corrosive elements. If water quenching or gas scrubbing is incorporated into the incineration system, special metals must be used to handle the corrosive acids formed and the water must be treated to neutralize the acid attack. Gas quenching is necessary in U. S. incineration practice, since water walls or other heat exchangers are not normally used, and since the furnace exit temperature is usually in excess of 1,500°F, the cost of fans and fly ash collectors which could withstand these temperatures is prohibitive.

Much peripheral equipment has a direct bearing on the performance of the fly ash collection equipment. Improper gas ducts leading to the collectors can upset gas and dust distribution, thereby affecting collector performance. The collection equipment designer must analyze the flow and incorporate proper ducting.

Dust hoppers are secondary collectors or settling chambers as well as dust storage space. The hoppers must be liberally sized and not allowed to fill completely if they are to accomplish the final state of gas-dust separation as intended. Ash removal systems must be properly engineered and automatic continuous hopper emptying is preferable. Continuous hopper discharge insures maximum hopper space for dust separation.

CHAPTER 42

CONVERSION OF REFUSE INTO USEFUL MATERIALS BY PYROLYSIS

Pyrolysis (destructive distillation) techniques developed at the Bureau of Mines to determine the yield and quality of the products from coal were used to convert municipal (household) and industrial refuse into potentially useful solids, liquids, and gases without adding to environmental pollution. The tests demonstrated that a ton of municipal refuse can be converted into 154 to 424 pounds of solid residue, 0.5 to 6 gallons of tar, 1 to 4 gallons of light oil, 97 to 133 gallons of liquor, 16 to 32 pounds of ammonium sulfate, and 7,380 to 18,058 cubic feet of gas. A ton of industrial refuse, consisting of paper, rags, and cardboard, can be converted to 618 to 838 pounds of residue, 1.5 to 3 gallons of light oil, 68 to 75 gallons of liquor, 12 to 23 pounds of ammonium sulfate, and 9,270 to 14,065 cubic feet of gas. The energy from the gas is more than sufficient to provide the heat for the pyrolysis.

PYROLYSIS PILOT PLANT AND TEST PROCEDURES

The experiments were conducted in the pilot plant shown schematically in figure 42-1. The plant consists essentially of an electric furnace, cylindrical steel retort, condensing and scrubbing train for product recovery, and gas-metering and sampling devices.

The electric furnace (2) is 26 inches inside diameter and 48 inches deep and is heated by nickel-chromium resistors spaced evenly in the furnace wall. The retort (3) is 18 inches in diameter and 26 inches deep and is made of 16-gage steel in the wall and 10-gage steel in top and bottom. Gases and vapors exit from the retort through a 2-inch-diameter offtake pipe and enter an air-cooled trap (4), where tar and heavy oils are collected. The gases and vapors are cooled to room temperature in two water-cooled condensers (5) connected in series, where additional heavy oil and liquor are collected. Final traces of heavy-oil mist are removed by one of the alternate electrostatic precipitators (6). The gas then passes successively through packed scrubbers, where ammonia is removed with sulfuric acid (7), and carbon dioxide and hydrogen sulfide are removed with caustic soda solution (9). The scrubbed gases pass to the large (11) and small (14) meters, which are geared together so that 99 percent of the gas passes through the large meter and is flared. The suction side of the small meter is cross-connected to the inlet side of the large meter so that 1 percent of the gas passes through a drying tube (12) and a condenser (13) immersed in acetone and solid carbon dioxide, where light oil is removed. Light oil recovered from the gas that passes through the small meter is calculated to the total gas yield. The gas from the condenser passes through the small meter (14) to the gas holder (15), and representative samples are taken from the holder for analyses. Steam is used to purge the condensers and piping at the conclusion of the test.

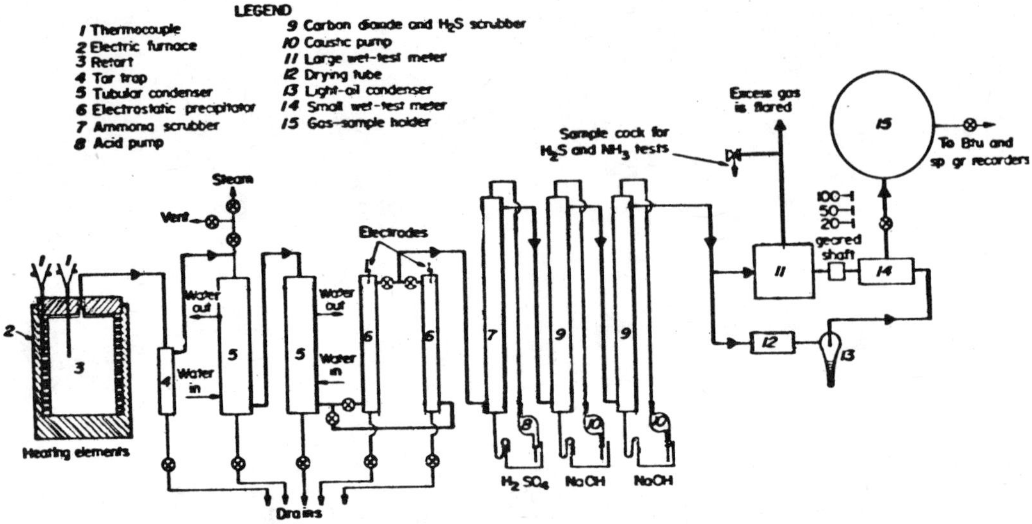

Fig. 42-1 **Flow Diagram of Pilot Plant Used To Pyrolyze Municipal and Industrial Refuse.**

YIELDS OF PRODUCTS FROM PYROLYSIS OF REFUSE

Tar, heavy oil, and liquor were collected from the air-cooled condenser, water-cooled condensers, and the electrostatic precipitator. Light oil was collected from the carbon dioxide-acetone condenser after the heavy oils had been removed and the gas had passed through the acid and caustic scrubbers. Residue remained in the retort after the test. Total gas, the sum of readings from both meters, was corrected to 60° F and 30 inches of mercury pressure.

A ton of municipal refuse pyrolyzed at 900° C yielded 154 pounds (7.7 percent of refuse charge) of solid residue, 0.5 gallon of tar, 114 gallons of liquor, 25 pounds of ammonium sulfate, and 17,741 cubic feet of gas. Pyrolysis at 750° C increased tar production to 2.6 gallons per ton and light oil to 2.5 gallons per ton, but showed a marked decrease in gas production to 9,628 cubic feet per ton, which was slightly more than half that obtained at 900° C. Pyrolysis starting at 500° C and increasing to 900° C yielded the largest amount of tar (4.8 gallons per ton), but the gas yield was 11,509 cubic feet, which was intermediate to that obtained at the other two temperatures.

Pyrolysis at 900° C of a ton of municipal refuse containing mainly plastic film yielded 382 pounds (19.0 percent of refuse charge) of solid residue, 1.4 gallons of tar, 0.6 gallon of light oil, 97.4 gallons of liquor, 31.5 pounds of ammonium sulfate, and 18,058 cubic feet of gas. At 750° C, gas production decreased to 7,380 cubic feet, less than half that obtained at 900° C, and tar and light-oil production increased. Pyrolysis through a temperature range of 500° to 900° C yielded the largest quantity of tar, 5.6 gallons per ton, and 11,545 cubic feet of gas.

A ton of industrial refuse pyrolyzed at 900° C and containing mainly paper, rags, cardboard, and small amounts of metals yielded 618 pounds (30.9 percent of refuse charge) of solid residue, small amounts of tar, 1.4 gallons of light oil, 68.5 gallons of liquor, 22.9 pounds of ammonium sulfate, and 14,065 cubic feet of gas. Pyrolysis of industrial refuse at lower temperatures generally increased tar and light-oil yields but did not result in the marked decreases in gas production that were obtained when municipal refuse was pyrolyzed at the lower temperatures.

The solid residue, a lightweight, flaky, carbonaceous material, represents a 90-percent weight reduction for the municipal refuse and a 65-percent reduction for industrial refuse. The residue from municipal refuse has the highest fuel value and can range from 10 to 17 million Btu per ton.

As mucn as 53 percent of the raw product gas is carbon monoxide, methane, and ethylene, and the volume of gas produced is ample to provide the energy for pyrolysis of the refuse. The high hydrogen content of the gas may also have promise as a source of hydrogen or methane.

The tar from the pyrolysis of refuse will provide an additional 736,000 Btu per ton of refuse, and the light oil produced is a potential source of benzene and toluene.

Summary of Comparison of Land-Based Systems

A survey of the market costs of land-based systems has indicated:

a) Recycling and reclamation cannot be expected to handle a major portion of municipal solid wastes for some time to come due to both processing problems and lack of markets for the output.

b) The final disposal costs associated with incineration meeting present federal pollution standards appear to be some 30% higher than disposal costs associated with railhaul and sanitary landfill for the large coastal city.

c) Thus, railhaul and sanitary landfills appear to be the land-based alternative of choice in most situations. However, the large coastal city faces very severe political problems in obtaining upland sites for sanitary landfill. In the past five years, many upland communities and regions have enacted broad-based regulations forbidding the importation of solid wastes.

Costs for Solid Waste Disposal

	Capital Cost (land excluded) Dollars per ton per day	Operating Cost Dollars per ton disposed
Sanitary Landfill	1000-2000	$1.25-2.25
Central Incineration	3500-7000	3.50-5.00
Composting	1500-10,000	2.00-7.00

Maintenance and operation of open dumps is estimated at 5 to 25 cents per ton.

CHAPTER 43

OCEAN DUMPING OF REFUSE

In the face of economic and political pressures outlined in the last chapters, the concept of dumping solid wastes at sea in an essentially raw state (allowing for some mechanical treatment but no change in chemical content or composition) has received increasing attention. This section will attempt to analyze the logistics of alternative methods of dumping at sea. However, before doing so, it is necessary, at least, to comment on the non-market effects of disposal at sea.

These effects can usefully be grouped under two headings: ecological and sociological, where ecological refers to the effect on fauna and flora that would be measured by the biologist and sociological refers to effects on the ocean which will have a direct impact on humans. The need for this distinction arises from the fact that the effects which, by any biological measure, are beneficial may be valued negatively by humans. For example, distribution of chopped paper in nutrient-poor surface waters may have a very favorable effect on the ambient fauna but be regarded as quite offensive by the human users of the area.

In the context of this section, consideration of the ecological effects implies that one must be able to guarantee that the value of any deleterious effects on plant and animal life due to dumping at sea will be smaller than the savings achieved by ocean dumping (the differential in costs between ocean dumping and the cheapest feasible alternative). This requires keeping the following two facts in mind. The ecology of the oceans is in a very dynamic and extremely delicate state of equilibrium. This system may be very insensitive to certain perturbations but extremely unstable with respect to others. Second, one must consider the long-term effects of the disposal of solid wastes at sea. Even a disturbance which appears to have a beneficial short-term effect may result in harmful changes in the long-term patterns of growth and distribution.

Consideration of sociological effects implies that the value of the detrimental effects of ocean dumping on recreational opportunities and enjoyment, on navigation and industry must also be weighed in any complete analysis of disposal at sea. In this context, it is important to remember that we are thinking in terms of rather large volumes of waste. New York City alone generates some 20,000 tons of municipal refuse per day.

The ecological and sociological effects of ocean disposal can be controlled by one of two rather differing philosophies.

1) An area of small ecological-sociological value is selected for a disposal site. Disposal and containment are handled in such a way that biological effects are as small as possible and are guaranteed to be restricted to the boundaries of the site. In this way, a site of low value is essentially "written off" and hopefully the ecology of the surrounding areas is not affected. In general, this philosophy points to disposal in deep waters well offshore, careful packaging of the wastes, and careful control of the dumping operation to insure dumping at the prescribed site. This is the philosophy which has governed most dumping to date.

2) The entire open ocean within range of the refuse generation area is divided into classified sectors, with a classification based on sociological ecological value. (e.g. prime fishing area, recreational area, spawning area, harbor area, ecologically rich area, ecologically barren area, area already used for dumping, etc.). For each area classification, an acceptable concentration of floating, suspended, and sinking solid waste would be computed. Dumping would then occur in each sector throughout the entire open ocean until the limiting concentration in a sector had been reached. Dumping operations in such a sector would then cease until the ecology had ab-

sorbed the waste and thereby lowered the concentration again.

This philosophy points toward dispersed dumping in shallow or surface waters of loosely packaged or ground refuse. It almost certainly implies some segregation of the waste stream to avoid particularly toxic or non-degradable materials. It attempts to take advantage of the ocean's biological activity to return the refuse to the life cycle quickly. In so doing, it will often find ecological and sociological values in conflict.

We will now review the logistics of several alternative systems which employ dumping at sea.

System A1 - Dumping of Loose Refuse

This system consists of the following:

a) Trash is delivered by primary collection vehicles to dockside transfer station;

b) Loose Refuse is loaded into barges;

c) Barges are transported to sectors designated for that day's dumping and refuse is discharged into water.

Such a scheme could be conducted only under the "concentration" philosophy. This means that a particular area of the ocean is classified according to its sociological and ecological value, and a correspondingly acceptable concentration of refuse is computed. Refuse is then dumped in the area until the limiting concentration is reached.

However, a preliminary estimate of the composition of the incoming refuse reveals that the refuse is approximately 50% (by weight) paper. This fact brings two opposite situations into play.

a) Since almost all of the paper products will either float or remain suspended near the surface, a volumetric concentration level becomes meaningless. Only the volume very near the surface will be utilized.

If one conceded that only the upper 100 feet of the ocean shall be considered for the volumetric concentration, the area required to dispose of metropolitan solid waste becomes enormous.

b) On the other hand, paper products are generally quite easily decomposed by the physical and biological ocean environment, and rather high temporary concentrations may be permissable.

System A2 - Dumping of Bales Compacted Dockside

This system consists of the following:

a) Trash is delivered by primary collection vehicles to dockside transfer stations and compacted into bales.

The primary collection trucks enter an enclosed dumping area and discharge into a storage bin, which should have capacity for one full operating day's (16 hours) accumulation of refuse. An overhead crane picks up the refuse and places it on a conveyor, which feeds the shredding facilities. The refuse passes through the shredding operation onto a second conveyor which feeds the baling press. The trash is compressed by the one-stroke baler into a bale, and strapped in the same operation.

b) Bales are loaded onto barges. The compacted bales are dumped by monorail into a barge. The barge loading facility is entirely enclosed within the station.

c) Barges are transported to disposal sites and bales are deposited on the sea bed.

System A3 - Dumping of Bales Compacted Inland

This system is almost identical to system A1. The only difference lies in the location of the transfer station. System A1 located the transfer station at the barge loading facility, and this seems reasonable for coastal municipalities, where the centroid of the solid waste generation area is near the coast.

However, if the system of compaction and dumping at sea becomes economically and/or socially superior to rail haul and sanitary landfill for inland municipalities, this system would provide a feasible solution.

This system consists of the following steps:

a) Trash is delivered to inland transfer stations and is compacted into bales. Packer or compaction type trucks enter an enclosed dumping area and discharge into a storage bin. An overhead crane transfers the refuse from the storage bin to a conveyor belt which feeds the shredding units. From the shredding unit the material is transferred by conveyor to the baling unit where it is compressed into high density bales. These baling units include integral automatic strapping equipment. These bales are then placed in the rail car by a traveling crane.

b) Bales are transported by rail to dockside facilities and loaded onto barges. The refuse bales are then transported in specially designed side-loading rail cars to the barge loading site. Refuse would be removed from the rail cars at the unloading area by a high capacity fork lift truck, and placed in the barges by means of a monorail loader.

c) Barges are transported to disposal site and bales are deposited on sea bed.

Barge Transportation Costs

In all of the ocean dumping systems, the cost of the barge-haul and dump procedure constitutes about 25% of the total disposal costs. The available proposals of solid waste disposal systems utilizing barge transportation have all recommended placing responsibility for this phase of the operation in the hands of a private contractor for a fixed fee. If considerations of the economies of scale of barge transportation were made part of the initial design parameters, instead of considering the barge haul operation as an external "black box", the scale of proposed sea-based solid waste disposal systems might be significantly increased.

An analysis of a barge system capable of handling 1,500,000 tons of refuse a year leads to the following results:

Summary of Barge Transportation Costs

Towing Distance	Towing Speed	Cost/Ton
20 miles	5 knots	$ .57/ton
50 miles	5 knots	1.30/ton
100 miles	7 knots	2.25/ton

Dispersal Effects of Dumping of Compacted Bales

Both Systems A2 and A3 employ the same ultimate disposal technique, the placing of compacted bales in a selected disposal site. This is an example of the "containment" philosophy, where an area of minimal ecological-sociological value, and hopefully minimal circulation is restricted as a dumping area. This concept places five constraints on the operation of the system.

1. The system must guarantee that each and every bale sinks to the bottom, without reaching a condition of neutral buoyancy. This means the compaction system must produce bales of a density greater than 64 lbs/cu. ft. High density compaction can be attained by the use of converted scrap-metal presses or extrusion-type bales similar to those used in the baling of paper and cotton.

2) The system must guarantee that the bale maintains its integrity during descent and upon impact on the ocean floor.

Indications are that the quality of the bales presently being produced may not be high enough to prevent extensive disintegration of the bale both in transit to the site and during the dumping operation. Inadequate preshredding allows many bales to contain whole bottles, cans, and other objects which would easily slough off under the influence of currents or wave action, and during handling procedures.

While the presently available shredding, baling, and strapping mechanisms appear to have the capability of producing bales of adequate integrity, further design improvement and intensive quality control procedures are clearly necessary.

3) The system must guarantee that the bales are deposited within the prescribed site limits. This will require suitable navigation equipment and aids to assure positive location of the barges above the dumping site. It also requires some knowledge of the descent trajectories of the bales under various sea-state conditions. This information is not yet available, but some sort of fathometric trace may provide adequate instrumentation to roughly determine descent trajectories.

4) The system must guarantee that the bales do not disintegrate in the ocean environment. While it appears that the bale will remain intact in its descent and impact phases, quick disintegration due to biological and chemical decomposition may still be a problem. At present no data on this problem is available.

5) The system must guarantee that the bales themselves will not be transported beyond the dumping site limits by current and tidal action. This problem must be solved by a suitable site selection criteria. Experience with incinerator residue indicates that a depth of 200 feet will ensure negligible ocean transport while fifty feet will not.

Incineration at Sea

The concept of incineration and/or the disposal of incinerator residue at sea has only recently been proposed as a sea-based alternative to existing solid waste disposal methods. As sites for the land-filling of incinerator residue become more costly and scarce, and as the need for the construction of new efficient incinerators appears to become more acute, a system for burning and disposal at sea may gather a great deal of backing in both the public and private sector.

Three main system alternatives present themselves for disposal of incinerator residue:

S1) a. Incineration at existing inland incinerators and proposed regional incinerators.

b. Transport of incinerator residue to dock facilities and loading onto sea-transport mechanism.

c. Transport of incinerator residue from dockside to dumping site and placement on sea floor.

S2) a. Transport of raw garbage in primary collection vehicles to dock-side transfer station.

b. Incineration at new incinerator facilities constructed at waterfront sites.

c. Transport of incinerator residue from dock-side incinerator to dumping site and placement on sea floor.

S3) a. Transport of raw garbage in primary collection vehicles to dock-side transfer station.

b. Loading of raw garbage onto a water-borne incinerator, and transport of raw garbage to dumping site.

S3) c. Incineration while en route and/or at dumping site and placement of incinerator residue on sea floor at dumping site.

These systems present alternatives regarding

a. amount of material transported and distance transported;

b. use of already existing facilities;

c. type and cost of initial capital equipment expenditures;

d. system reliability.

Before these alternatives are considered, it is necessary to review the biological effects and the question of dispersion of the residue of incineration.

Biological Effects

Any system which includes the ultimate deposition of incinerator residue on the sea floor, whether in a state of static or dynamic equilibrium, must guarantee a controlled minimal effect on the marine life there, both plant and animal, free-swimming and bottom dwelling. It must be realized that: 1) Many perturbations will appear to have little or no short-term deleterious effects but may have very serious and irreversible long-term (5-25 years) effects on the ecology of the sea bed; 2) Since an extremely delicate equilibrium is in operation in the food chain of the sea floor, the very slightest perturbation, even one which seems to have beneficial short-term effects may cause violent, irreparable damage; 3) Since the residue of the present state-of-the-art incineration is essentially non-biodegradable, the self-cleansing and renewing mechanisms of the sea ecology will have a long and difficult task in assimilating this residue.

Experiments to determine the chemical composition of a representative sample of incinerator residue established the percentages of nitrates, ammonia, sulphates and other components of importance in the nutrition of marine plankton. Special attention was also placed on the heavy metal content of incinerator residues, because of their toxicity to marine organisms, and potential concentration in the marine food chain. Bioassays were conducted on a variety of marine species under exposure periods ranging from one day to three years. Although survival rates constituted the main toxicity measurement, sub-lethal effects on hatching and larval processes and long-term growth rates were monitored.

The main thrust of the program was the determination of the level of concentration at which harmful effects were introduced, under the assumption that a residue concentration of 1% would require at least 25 years of very intensive dumping to accumulate. Results indicated that --

a) Quahog, winter flounder, shrimp, adult lobster, and millet were relatively immune to incinerator residue;

b) Menhaden, lobster larvae, and sea scallops showed some significant mortality rate increases due to incinerator residue;

c) Hard shell clams showed lower mortality

rates and higher growth rates after exposure to the residue;

d) Results regarding concentration of heavy metals were observed for all species tested, and it is suggested that this factor is unlikely to cause toxicity to exposed marine species; nor is there likely to be significant concentrations of these elements in food fish important to man.

The overall results of this testing program seem to indicate that the short-term effects on marine life outside the prescribed dumping area will be minimal, and no significant long-term effects are apparent.

Dispersion of Incinerator Residue

The second constraint common to all systems depositing incinerator residue on the ocean floor is the need for suitable site selection. Once you feel that you can guarantee minimal deleterious effects on marine life outside the immediate dumping area, you must be able to control the dispersion of the residue and gradual growth of the disposal site.

As part of the Harvard School of Public Study, several tons of actual incinerator residue were deposited at two locations off the southern shore of Rhode Island. A shallow water site, approximately 50 feet deep and 2 miles from shore, was selected as being representative of the minimum feasible depth for disposal. The area was visually monitored by SCUBA divers, and weather, current and wave height measurements were recorded. It was observed that the attenuation of violent free-surface wave effects was insufficient to prevent the motion of some lighter residue, such as tin cans. Heavy storm seas were capable of large mass transport of the residue, and the 50-ft. depth was classified as totally insufficient for the system.

A second deep water site, approximately 200 feet deep and 18 miles from shore, was monitored for a period of several months through the use of a small research submersible and instrumented buoy arrays. The maximum distance a half-gallon can was observed to have moved as a result of severe storm seas was approximately 50 feet. Since the greatest portion of the residue had remained essentially motionless for the entire summer, the study group stated:

"On the basis of these and other observations, it has been concluded that incinerator residues should be deposited in waters at least 200 feet deep to avoid significant migration from the assigned site under the combined effects of wave action plus tidal and bottom currents and that depressions in the ocean floor, where bottom currents are minimal, would be ideal."

Logistic considerations which may be added to the above conclusion, are:

1) As one goes to deeper disposal sites (and therefore sites further from shore), the problem of locating the site under poor operating conditions (storm seas, darkness, fog, etc.) becomes extremely difficult.

2) Once one has located the dumping site, the problem of positioning the unloading facility to assure that the trajectories of the falling residue carries it near to the center of a one or two square mile site also becomes exceedingly difficult. There appears to have been no study of the descent rate of the residue.

3) In trying to select a site on the basis of existing current data, one must realize that some migration rates are allowable. Most of the more easily transported residue components, such as tin cans, are also those that are most susceptible to disintegration from the abrasive action of the transport process. Observations of incinerated cans suspended in sea water indicate that within one year the metal disintegrates into a mass of small flat crumbly scale particles.

4) Assuming little or no significant migration occurs, use of existing sites presently designated for the disposal of explosives, dangerous chemicals, and mud, when these areas are within reasonable distance from collection sites, would represent a solution with smaller net ecological effects.

Coastal Landfill

Almost all the large coastal cities with which we are concerned in this report have in the past placed heavy reliance on landfills, (some sanitary, most not) along their shoreline. In some cases, such as

New Orleans, use of this alternative continues unabated. However, most such cities have found it increasingly difficult to locate politically feasible shoreline sites for landfill. The San Francisco Bay area communities are now prohibited by law from developing new disposal sites in the bay. New York City regards itself as unable to locate any new coastal landfill sites. Boston not only cannot locate a coastal site for landfill, but is even unable to find a politically feasible site for an incinerator along its long shoreline. This rapidly increasing opposition to use of the shoreline for disposal is a product of both the realization of the ecological importance of such traditional sites as marshes and flats and, more significantly, of the rapidly rising value of present shoreline real estate which would be removed from the shore by landfill. Nonetheless it is not clear that complete prohibition of coastal landfills is consistent with the values of the community involved.

The problem of how much of the coastal zone should be devoted to sanitary land fills is a special problem in the economic allocation of the shoreline which problem is more generally treated in considerable detail in Volume II of this series. Therefore, we will not consider the problem further in this volume other than to note that whether or not the large coastal cities have run out of coastal fill sites from the point of view of the economists, it appears that most of them believe they have run out of politically feasible sites. From a pragmatic standpoint then, it appears that coastal landfill will not play a major role in the future in the disposal of municipal solid waste.

A potential varient of shoreline landfill which avoids some of its political problems involves the construction of artificial islands using, in part, municipal wastes. We have not been able to give this alternative any analysis during this study. However, especially in areas which are considering the construction of such islands for such reasons as to accommodate the rapidly increasing drafts of bulk carriers (e.g. the Gulf Coast ports), it appears to deserve careful attention.

Conclusions with Respect to Disposal at Sea

There are two rather distinct philosophies which can be followed for dumping refuse at sea. One approach is to accept the ecological destruction of a designated area and attempt to confine all the effects of the refuse to that area. One, then, chooses areas which are judged to be ecologically unimportant. This has been the prevailing philosophy among marine biologists who argue for dumping off the continental slope at depths greater than 1,000 fathoms with the waste suitably contained to keep it from spreading. The other philosophy is to view the ocean as a link in the natural process of returning the wastes to the life cycle. Holders of this point of view point out that the ocean has considerable regenerative powers, that in proper concentrations much waste material can be regarded as useful nutrients for the lower levels of marine life. They point to even such unattractive materials as the acid mine wastes dumped in outer New York Harbor, noting that in the center of this dump the water is biologically dead while around the edges marine life flourished at levels higher than those in the undisturbed waters.

Obviously, there are very substantive differences in these two philosophies. The first calls for dumping in very concentrated areas, the second calls for a much more even distribution of the wastes. Holders of the first view point out that almost no bacterial action takes place at such depths. The apple on the ALVIN after being submerged at 4,500 feet for close to a year was perfectly edible when the boat was raised, as was the meat in the bologna sandwich. Thus, holders of the second view are led to recommending near-surface dumping in shallow water or at least in the photic zone. These are just the areas most valuable to man, and this places a much heavier burden on holders of the second philosophy since they must be much more concerned with the sociological effects of the garbage than those who call for confinement in deep waters. Their information requirements are much higher and presently this information does not exist.

There are a variety of ways of dumping solid wastes at sea:

1) Surface dump of the raw garbage

2) Surface dump of ground garbage

3) Compaction and dump of compacted bales with and without packaging

4) Pump raw garbage to depths where natural compression renders all the garbage denser than the sea water.

The first alternative was at one time practiced by New York City among others. However, New Jersey was able to obtain an injunction on any dumping which might result in the garbage returning to shore and New York found it uneconomical to take the garbage far enough to prevent its return. The economics have changed in favor of longer trips. However, this alternative is dominated by compaction because the shorter trip which compaction allows more than compensates for the costs of compaction. Thus, dumping raw garbage appears neither feasible nor economic.

The second alternative is the one most consistent with the philosophy of using the oceans biological powers to return wastes to the life cycle. It will run about $1.00 more per ton than the first alternative and probably require as long a trip. Therefore, at present it also appears to be dominated by alternative (3) with respect to costs of transportation. However, if the validity of the dispersal philosophy can be substantiated, then it will deserve serious attention.

Compaction of garbage to densities higher than that of sea water is presently at the edge of the state-of-the-art. Bales have been produced from municipal garbage with densities in the 70-75 pound per cubic feet range. However, in our view, no one has demonstrated the ability to consistently produce heavier than water bales over a range of garbages.It is felt that this ability can be achieved at little more than present baling costs as soon as the need is demonstrated.

Baling is the technological advance that has made dumping at sea feasible. No experimentation on the ecological effects of the dumped bales other than very short-term observations of a few bales on the bottom in shallow water have been undertaken. Nor has the differential in these effects associated with different packaging materials been studied. Covering the bales with polyethelene would cost about$1.00 per ton (40). Given a deep water dump, this would have little effect other than to prevent surface spauling during decent.

In summary, dumping of compacted bales at sea appears to be the most attractive of the marine alternatives with respect to solid waste disposal from the point of view of market costs. However, before it or any other sea based disposal system is placed in operation, we should know much more than we presently do about the ecological effects associated with this activity.

INDEX